# BEAM HALO DYNAMICS, DIAGNOSTICS, AND COLLIMATION

# Related Titles from AIP Conference Proceedings

**BIW**  Beam Instrumentation Workshops 1989-2000
Edited by Thomas Russo and Gary A. Smith, October 2003, CD-ROM: 0-7354-0144-6

**674**  Accelerator Instrumentation: First ICFA Instrumentation School at the ICFA Instrumentation Center in Morelia
Edited by Luis Villaseñor and Victor Villanueva, July 2003, 0-7354-0141-1

**650**  Beams 2002: 14th International Conference on High-Power Particle Beams
Edited by Thomas A. Mehlhorn and Mary Ann Sweeney, December 2002, 0-7354-0107-1

**648**  Beam Instrumentation Workshop 2002: Tenth Workshop
Edited by Gary Smith and Thomas Russo, December 2002, CD-ROM included, 0-7354-0103-9

**647**  Advanced Accelerator Concepts: Tenth Workshop
Edited by Christopher E. Clayton and Patrick Muggli, December 2002, CD-ROM included, 0-7354-0102-0

**642**  High Intensity and High Brightness Hadron Beams: 20th ICFA Advanced Beam Dynamics Workshop on High Intensity and High Brightness Hadron Beams; ICFA-HB2002
Edited by Weiren Chou, Yoshiharu Mori, David Neuffer, and Jean-François Ostiguy, November 2002, CD-ROM included, 0-7354-0097-0

**592**  High Quality Beams: Joint US-CERN-JAPAN-RUSSIA Accelerator School
Edited by S. I. Kurokawa, S. Y. Lee, J. Miles, E. A. Perevedentsev, November 2001, 0-7354-0034-2

**581**  Physics of, and Science with, the X-Ray Free-Electron Laser: 19th Advanced ICFA Beam Dynamics Workshop
Edited by C. Pellegini, S. Chattopadhyay, M. Cornacchia, and I. Lindau, August 2001, 0-7354-0022-9

**578**  Physics and Experiments with Future Linear $e^+e^-$ Colliders: LCWS 2000
Edited by Adam Para and H. Eugene Fisk, July 2001, 0-7354-0017-2

**530**  Colliders and Collider Physics at the Highest Energies: Muon Colliders at 10 TeV to 100 TeV, HEMC'99 Workshop
Edited by Bruce J. King, August 2000, 1-56396-953-X

To learn more about these titles, or the AIP Conference Proceedings Series, please visit the webpage http://proceedings.aip.org/proceedings

# BEAM HALO DYNAMICS, DIAGNOSTICS, AND COLLIMATION

29th ICFA Advanced Beam Dynamics Workshop on
Beam Halo Dynamics, Diagnostics, and Collimation
HALO'03

Workshop on Beam-Beam Interactions
Beam-Beam'03

Montauk, New York    19-23 May 2003

Sponsored by:

 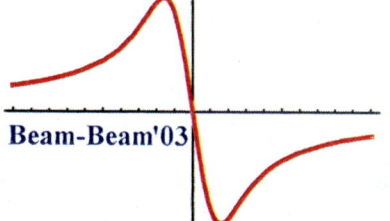

EDITORS
Jie Wei
Wolfram Fischer
Pamela Manning
*Brookhaven National Laboratory*
*Upton, New York*

CD-ROM INCLUDED

Melville, New York, 2003
AIP CONFERENCE PROCEEDINGS ■ VOLUME 693

**Editors:**

Jie Wei
Wolfram Fischer
Pamela Manning

Brookhaven National Laboratory
P. O. Box 5000
Upton, NY 11973
USA

E-mail: jwei@bnl.gov
wfischer@bnl.gov
pmanning@bnl.gov

The article on pp. 256-260 was authored by U.S. Government employees and is not covered by the below mentioned copyright.

Authorization to photocopy items for internal or personal use, beyond the free copying permitted under the 1978 U.S. Copyright Law (see statement below), is granted by the American Institute of Physics for users registered with the Copyright Clearance Center (CCC) Transactional Reporting Service, provided that the base fee of $20.00 per copy is paid directly to CCC, 222 Rosewood Drive, Danvers, MA 01923. For those organizations that have been granted a photocopy license by CCC, a separate system of payment has been arranged. The fee code for users of the Transactional Reporting Service is: 0-7354-0166-7/03/$20.00.

© 2003 American Institute of Physics

Individual readers of this volume and nonprofit libraries, acting for them, are permitted to make fair use of the material in it, such as copying an article for use in teaching or research. Permission is granted to quote from this volume in scientific work with the customary acknowledgment of the source. To reprint a figure, table, or other excerpt requires the consent of one of the original authors and notification to AIP. Republication or systematic or multiple reproduction of any material in this volume is permitted only under license from AIP. Address inquiries to Office of Rights and Permissions, Suite 1NO1, 2 Huntington Quadrangle, Melville, NY 11747-4502; phone: 516-576-2268; fax: 516-576-2450; e-mail: rights@aip.org.

L.C. Catalog Card No. 2003114530
ISBN 0-7354-0166-7
ISSN 0094-243X

Printed in the United States of America

# CONTENTS

Preface .................................................................................................... ix
Committees ............................................................................................... xi
Group Photo .............................................................................................. xiii

## PLENARY SESSION

**Mechanisms of Halo Formation** .................................................................... 3
    A. V. Fedotov

**Halo Diagnostics Overview** .......................................................................... 9
    T. Shea and P. Cameron

**Beam Collimation at Hadron Colliders** ........................................................... 14
    N. V. Mokhov

**Halo Formation due to Beam-Beam Interaction** .............................................. 20
    F. Zimmermann

**Dynamic Aperture for Single-Particle Motion: Overview of Theoretical Background, Numerical Predictions and Experimental Results** .............................................. 26
    M. Giovannozzi

**Space Charge Simulation** ........................................................................... 32
    C. R. Prior

**Beam Cleaning in High Power Proton Accelerators** ......................................... 38
    J. Wei

## WORKING GROUP I: HALO DYNAMICS

**Summary of the Beam Dynamics Working Group** ........................................... 47
    I. Hofmann and A. V. Fedotov

**CEA Studies on Halo Formation** .................................................................. 49
    N. Pichoff, P.-Y. Beauvais, R. Duperrier, G. Haouat, J.-M. Lagniel, and D. Uriot

**Sources of Beam Halo Formation in Heavy-Ion Superconducting Linac and Development of Halo Cleaning Methods** ................................................................................ 53
    P. N. Ostroumov

**Halo Formation and its Mitigation in the SNS Linac** ........................................ 57
    D.-O. Jeon, representing the SNS Project

**Halo Studies for the ESS and Linac4 Front-Ends** ............................................ 61
    F. Gerigk

**Self-Consistency and Coherent Effects in Nonlinear Resonances** ...................... 65
    I. Hofmann, G. Franchetti, J. Qiang, and R. D. Ryne

**Longitudinal Mismatch in SCL as a Source of Beam Halo** ................................ 69
    A. G. Ruggiero

**Observations and Simulation of a Fourth Order Resonance with Space Charge** ..... 73
    G. Franchetti, I. Hofmann, M. Giovannozzi, M. Martini, and E. Metral

**Halo and RMS Beam Growth due to Transverse Impedance** ............................. 77
    V. V. Danilov and J. A. Holmes

**Collective Effects and Collisions in Halo Genesis and Growth** ........................... 81
    G. Turchetti, C. Benedetti, A. Bazzani, and S. Rambaldi

**Effects of Halo on the AGS Injection from 1.2 GeV Linac** ................................. 85
    W. T. Weng, J. Beebe-Wang, D. Raparia, A. G. Ruggiero, and N. Tsoupas

**Beam Halo from Quadrupole Rotation Errors** ................................................. 89
    R. A. Kishek, S. Bernal, I. Haber, H. Li, P. G. O'Shea, B. Quinn, M. Reiser, and M. Walter

**Simulation of Halo Particles with Simpsons** ................................................... 93
    S. Machida

**Comparison of Particle Simulation with J-PARC Linac MEBT Beam Test Results** ..... 96
    M. Ikegami, T. Kato, Z. Igarashi, A. Ueno, Y. Kondo, R. Ryne, and J. Qiang

# WORKING GROUP II: HALO DIAGNOSTICS

**Halo Diagnostics Summary** .................................................................................................... 103
    P. Cameron and K. Wittenburg

**Physics Results from the Los Alamos Beam-Halo Experiment** ..................................... 108
    T. P. Wangler

**Observation of Emittance Growth at KEK PS** .................................................................. 114
    S. Igarashi, T. Miura, E. Nakamura, Y. Shimosaki, M. Shirakata, K. Takayama, and T. Toyama

**Tune-Based Halo Diagnostics** ............................................................................................. 118
    P. Cameron

**Wide Dynamic-Range Beam-Profile Instrumentation for a Beam-Halo Measurement: Description and Operation** ........................................................................ 122
    J. D. Gilpatrick

**Scintillator Telescope in the AGS Extracted Beamline** ................................................... 126
    D. Gassner, K. A. Brown, and I. H. Chiang

**Beam Tail Measurements using Wire Scanners at DESY** ................................................ 129
    S. Arutunian, M. Werner, and K. Wittenburg

**Collimation Experience at RHIC** ........................................................................................ 133
    K. A. Drees, R. Fliller, D. Trbojevic, and V. Kain

**SNS Longitudinal and Transverse Halo Measurement** .................................................... 137
    D. Gassner, P. Cameron, and R. Witkover

**The IPM as a Halo Measurement and Prevention Diagnostic** ......................................... 140
    R. Connolly, M. Grau, R. Michnoff, and S. Tepikian

**HERA Beam Tail Shaping by Tune Modulation** ............................................................... 144
    C. Montag

# WORKING GROUP III: HALO COLLIMATION

**Collimation Working Group Summary Report** ................................................................. 151
    A. Drees and N. V. Mokhov

**Beam Loss Control on the ISIS Synchrotron: Simulations, Measurements, Upgrades** ..... 154
    C. M. Warsop

**MARS14 Collimation and Shielding Studies for the 3 GeV Ring of J-PARC Project** ...... 158
    N. Nakao, N. Mokhov, K. Yamamoto, Y. Irie, and A. Drozhdin

**SNS Collimating System Design — Performance and Integration** ................................. 162
    N. Simos, H. Ludewig, D. Raparia, N. Catalan-Lasheras, J. Brodowski, and G. Murdoch

**Estimate of Dose and Residual Activity in the SNS Ring Collimation Straight** ........... 167
    H. Ludewig, N. Simos, D. Davino, S. Cousineau, N. Catalan-Lasheras, J. Brodowski, J. Tuozzolo,
    C. Longo, B. Mullany, and D. Raparia

**Handling High Activity Components on the SNS (Collimators and Linac Passive Dump Window)** ....................................................................................................... 172
    G. Murdoch, A. Decarlo, K. Potter, T. Roseberry, J. Schubert, J. Brodowski, H. Ludewig, J. Tuozzolo,
    N. Simos, and J. Hirst

**The Tevatron Collider Run II Halo Removal System** ....................................................... 176
    D. Still, J. Annala, M. Church, B. Hendricks, B. Kramper, and A. Legan

**Beam Loss and Collimation at LHC** .................................................................................... 180
    J. B. Jeanneret, O. Aberle, I. L. Ajguirei, R. Assmann, I. Baishev, J.-P. Bojon, L. Bruno, E. Carlier,
    E. Chapochnikova, E. Chiaveri, B. Dehning, S. S. Fartoukh, A. Ferrari, B. Goddard, J. M. Jimenez,
    D. Kaltchev, V. Kain, I. Kourotchkine, H. Preis, F. Ruggiero, R. Schmidt, P. Sievers, J. Uythoven,
    V. Vlachoudis, L. Vos, and E. Vossenberg

**Beam Loss Scenarios and Strategies for Machine Protection at the LHC** .................... 184
    R. Schmidt, R. Assmann, H. Burkhardt, E. Carlier, B. Dehning, B. Goddard, J. B. Jeanneret, V. Kain,
    B. Puccio, and J. Wenninger

**Collider and Detector Protection at Beam Accidents** ..................................................... 188
    I. L. Rakhno, N. V. Mokhov, and A. I. Drozhdin

**Crystal Collimation at RHIC** .................................................................... 192
    R. P. Fliller, III, A. Drees, D. Gassner, L. Hammons, G. McIntyre, S. Peggs, D. Trbojevic, V. Biryukov,
    Y. Chesnokov, and V. Terekhov

**Simulation Aspects of Beam Collimation and Their Remedies in the MARS14 Code** ................. 196
    M. A. Kostin, N. V. Mokhov, S. I. Striganov, and I. S. Tropin

**Comparison of the TESLA, NLC and CLIC Beam-Collimation System Performance** ................. 200
    A. Drozhdin, G. Blair, L. Keller, W. Kozanecki, T. Markiewicz, T. Maruyama, N. Mokhov, O. Napoly,
    T. Raubenheimer, D. Schulte, A. Seryi, P. Tenenbaum, N. Walker, M. Woodley, and F. Zimmermann

**Collimation for CLIC** ........................................................................... 205
    R. Assmann, H. Burkhardt, S. Fartoukh, J. B. Jeanneret, J. Pancin, S. Redaelli, T. Risselada, D. Schulte,
    F. Zimmermann, A. Faus-Golfe, H.-J. Schreiber and G. A. Blair

**Design and Performance of the Tesla Test Facility (TTF) Collimation System** ..................... 209
    H. Schlarb

## BEAM-BEAM'03

**Beam-Beam'03 Summary** ........................................................................ 215
    W. Fischer and T. Sen

### *Beam-Beam Limits*

**Coherent Beam-Beam Effects, Theory and Observations** ........................................ 221
    Y. Alexahin

**Beam-Beam Instability in Case of Strong-Weak Beam-Beam Interactions** ........................ 227
    L. Jin and J. Shi

**Summary of Beam-Beam Observations during Stores in RHIC** .................................. 231
    W. Fischer

**Beam-Beam Performance of the SLAC B-Factory** .............................................. 235
    W. Kozanecki, Y. Cai, F-J. Decker, R. Holtzapple, J. Seeman, M. Sullivan, and U. Wienands

**Collision with Finite Crossing Angle at KEKB** .................................................. 240
    K. Ohmi, M. Tawada, and Y. Funakoshi

**Luminosity Increase at the Incoherent Beam-Beam Limit with Six Superbunches in RHIC** ....... 244
    W. Fischer and M. Blaskiewicz

**Theory and Observations of Beam-Beam Effects at the Tevatron** ................................ 248
    T. Sen

**Tune Modulation from Beam-Beam Interaction and Unequal Radio Frequencies in RHIC** ....... 252
    W. Fischer, P. Cameron, S. Peggs, and T. Satogata

### *Beam-Beam Compensation*

**Progress Report on Beam-Beam Compensation with Electron Lenses in Tevatron** ............... 256
    V. Shiltsev, Y. Alexahin, K. Bishofberger, G. Kuznetsov, N. Solyak, M. Tiunov, and X. Zhang

**Wire Map and Applications to Long-Range Beam-Beam Compensation** ......................... 261
    B. Erdelyi and T. Sen

**Multipole Compensation of Long-Range Beam-Beam Interactions** ............................... 265
    J. Shi, L. Jin, and O. Kheawpum

**Beam-Beam Simulations in Four-Beam Scheme for High Luminosity $e^+e^-$ Colliders** ............ 269
    Y. Ohnishi and K. Ohmi

### *Beam-Beam Study Tools*

**Beam-Beam Simulations for Lepton Machines** ................................................... 273
    J. T. Rogers

**Parallel Strong-Strong/Strong-Weak Simulations of Beam-Beam Interaction in Hadron Accelerators** .................................................................. 278
      J. Qiang, M. Furman, R. D. Ryne, W. Fischer, T. Sen, and M. Xiao
**Weak-Strong Beam-Beam: Averaging and Tune Diagrams** ............................................. 282
      J. A. Ellison, H. S. Dumas, M. Salas, T. Sen, A. Sobol, and M. Vogt
**Numerical Calculation of the Phase Space Density for the Strong-Strong Beam-Beam Interaction.** .......................................................... 287
      A. Sobol and J. A. Ellison

## COMMENTS AND DISCUSSIONS

**Halo Generation and Beam Cleaning by Resonance Trapping** ......................................... 291
      A. Chao

**List of Participants** ........................................................................... 293
**Workshop Program** ............................................................................. 297
**Workshop Poster** .............................................................................. 299
**Photos** ....................................................................................... 301
**Author Index** ................................................................................. 315

# Preface

The 29th ICFA Advanced Beam Dynamics Workshop on Beam Halo Dynamics, Diagnostics, and Collimation, in conjunction with the Beam-Beam'03 Workshop, was held during the week of May 19 – 23, 2003, at Gurney's Inn at the eastern end of Long Island, New York. The Brookhaven National Laboratory, the Brookhaven Science Associates, the Spallation Neutron Source Project, and the US Department of Energy sponsored the workshop.

Beam halo limits the performance of modern particle accelerators. Impact of beam halo challenges the design and operations of high-intensity, high-brightness, and high-energy accelerators: ISIS operating at the Rutherford-Appleton Laboratory, the Spallation Neutron Source (SNS) and the Japan Proton Accelerator Research Complex (J-PARK) presently under construction, the Relativistic Heavy Ion Collider (RHIC), the Fermilab Tevatron, the Large Hadron Collider (LHC), and the proposed lepton linear colliders and proton drivers. Although progress has been made in recent years, physical understanding of halo dynamics is still far from comprehensive, and experimental benchmarking is just beginning. State-of-the-art techniques are required for the detection and diagnosis of the formation and development of beam halo, and technically demanding design and material selection are needed for the scraping and collimation systems. It has become urgently important to bring together theoretical and experimental physicists and engineers, with expertise on beam dynamics, diagnostics, and collimation design working on both linear and circular accelerators, for focused discussions and investigation of the subject. The HALO'03 workshop was intended to provide such a platform for experts from the fields of accelerator physics, diagnostics, engineering and material science.

The goal of the workshop was achieved by the effective planning and leadership of the working-group chairs, a close interaction between various working groups, and an active participation of our colleagues. There were 94 participants from laboratories and universities in Asia, Europe, and America, and 88 invited and contributed presentations. Each working day starts with a plenary session consisting of plenary talks and working-group progress reports followed by parallel sessions consisting of parallel talks and organized discussions. Joint sessions were arranged among each of the three HALO'03 working groups and the Beam-Beam'03 working group.

I would like to thank the local organizing committee, the program committee, the international advisory committee, John Jowett and the ICFA committee for their planning and organization of the workshop. In particular, I would like to thank our working-group chairs for their devotion and leadership, to thank Pam Manning for her coordination for more than a year before, during, and after the workshop, and to thank the participants for meeting the challenge of an unconventional workshop that bring together scientists of typically non-overlapping fields into close interaction. The working-group chairs were:

HALO dynamics:       Ingo Hofmann (GSI) and Alexei Fedotov (BNL)
HALO diagnostics:    Pete Cameron (BNL) and Kay Wittenburg (DESY)
HALO collimation:    Nikolai Mokhov (FNAL) and Angelika Drees (BNL)
Beam-Beam:           Wolfram Fischer (BNL) and Tanagi Sen (FNAL)

Viewgraphs of all presentations are available on the web site:
www.c-ad.bnl.gov/halo03/misc/title.htm, then click "View Talk" at the right of the page.

Jie Wei

## International Advisory Committee:

| | |
|---|---|
| R. Baartman | TRIUMF |
| W. Barletta | LBNL |
| A. Chao | SLAC |
| I. Gardner | RAL |
| H. Haseroth | CERN |
| S. Holmes | ORNL |
| N. Holtkamp | ORNL |
| R. Macek | LANL |
| R. Maier | FZJ |
| H. Okamoto | U. Hiroshima |
| C. Pagani | INFN |
| G. Rees | RAL |
| T. Shea | ORNL |
| R.H. Siemann | SLAC |
| A.N. Skrinsky | BINP |
| W.T. Weng | BNL |
| H. Wiedemann | SLAC |
| F. Willeke | DESY |
| J.-W. Xia | IMP |
| ICFA Beam Dynamics Panel | |

## Program Committee

| | |
|---|---|
| J.M. Brennan | BNL |
| W. Chou | FNAL |
| Y. Fedotov | IHEP |
| R. Garoby | CERN |
| S. Henderson | ORNL |
| I. Hofmann | GSI |
| J.M. Lagniel | CEA |
| N. Mokhov | FNAL |
| Y. Mori | KEK |
| A. Mosnier | CEA |
| C. Prior | RAL |
| T. Raubenheimer | SLAC |
| T. Roser | BNL |
| F. Ruggiero | CERN |
| H. Schmickler | CERN |
| K. Takayama | KEK |
| H. Thiessen | LANL |
| R. Wanzenberg | DESY |
| R. Webber | FNAL |
| J. Wei | BNL |

## Local Organizing Committee:

P. Cameron
M. Campbell
A. Drees
A. Fedotov
N. Franco
J. Hauser
L. Hoff
P. Manning
S. LaMontagne
W. McGahern
D. Raparia
N. Simos
J. Wei (Chair)
D. Zadow

## Beam-Beam Committee:

| | |
|---|---|
| Y. Cai | SLAC |
| W. Fischer | BNL |
| M. Furman | LBNL |
| W. Herr | CERN |
| M. Minty | DESY |
| F. Pilat | BNL |
| T. Sen | FNAL |
| R. Talman | Cornell |
| K. Yokoya | KEK |

# Plenary Session

# Mechanisms of halo formation

Alexei V. Fedotov

*Brookhaven National Laboratory, Upton, NY 11973*

**Abstract.** Uncontrolled beam loss leads to excessive radioactivation in high-intensity machines. At the same time, it strongly affects performance of high-energy accelerators and colliders. For the well controlled beam, the loss is typically associated with the low density halo surrounding beam core. There are many mechanisms which contribute to halo formation. Some of them are more important for linear accelerators while other are more relevant to circular machines. In order to minimize uncontrolled beam loss or improve performance of an accelerator, it is very important to understand what are the sources of halo formation, as well as which of them can have significant contribution for a specific machine of interest. In this paper, we overview various mechanisms of halo formation. We then specifically discuss which effects are expected to be dominant in linear accelerator and which effects dominate in rings, concentrating on high-intensity machines.

## BEAM LOSS AND BEAM HALO

The latest designs of high-current accelerators set strict requirements on beam-induced radioactivation of the vacuum chamber. Guided by such restrictions, significant efforts were made to understand various mechanisms which can contribute to beam loss. Because losses at a very low level became of concern, more detailed understanding of beam tails (low density region outside the dense beam core) was necessary. Naturally, such a low density region surrounding the dense core was referred to as beam "halo". At the same time, similar effects were studied using standard concepts of emittance growth and beam tails.

On the academic side, there is a tendency to come up with a precise definition of beam halo. The attempts were made to distinguish "halo" from "tails" of beam distribution. Such definitions become problematic as soon as one wants to describe various mechanisms of halo formation which may contribute to a beam loss. In fact, as far as the radioactivation is concerned, it does not matter whether the loss comes from beam tails or halo. From practical point of view, beam halo can be regarded based on its application:

1) If the concern is beam loss, then beam halo is just some number of particles of any origin which lie in the low-density region of the beam distribution far away from the dense core. Of course, the behavior of such particles will be very different depending on their origin.

2) The next step is to understand the mechanisms of halo formation. The structure of beam halo and its characteristics are different depending on the mechanism. The parameters of the driving mechanism may be enhanced to provide a clear signatures of halo formation. Although, in such cases with enhanced parameters, "halo" may be at a relatively high level of beam distribution, it is an essential stage towards understanding of halo dynamics. At this point, the attention is devoted to studies of halo formation before the beam loss actually occurs, typically by an observation of emittance growth or beam profiles.

Not surprisingly, there are many mechanisms which contribute to halo formation:

**in high-current linear accelerators:** 1) various sources of machine nonlinearities and misalignments 2) rms mismatch 3) space-charge coupling resonances 4) space-charge induced structure resonances 5) Coulomb scattering within the beam and on the residual gas 6) collective instabilities, etc.

**in high-current circular accelerators:** 7) injection, foil stripping and extraction 8) rf noise 9) machine nonlinearities 10) rms mismatch 11) space-charge coupling resonances 12) space-charge induced structure resonances 13) imperfection lattice resonances 14) gas scattering 15) collective instabilities 16) e-cloud effects 17) numerous "project-specific" effects ranging from different painting schemes for multi-turn injection to a closed orbit oscillation at extraction, etc.

**including short bunches:** 18) transverse-longitudinal coupling 19) effects from synchrotron motion, etc.

**including high-energy accelerators:** 20) intrabeam scattering 21) instabilities relevant for high-energies, etc.

---

[1] Work supported by the SNS through UT-Battelle, LLC, under contract DE-AC05-00OR22725 for the U.S. Department of Energy.

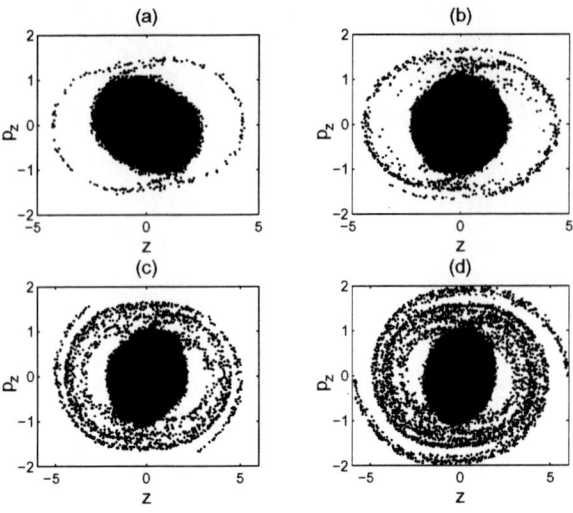

**FIGURE 1.** Mismatch dependence of the longitudinal parametric halo for a weak tune depression ($\eta_x = 0.93$, $\eta_z = 0.87$): a) 20% mismatch b) 30% c) 40% d) 60%.

**including colliders:** 22) beam-beam driven halo 23) etc.,etc.,etc.

And, yes, one needs to consider both the transverse and longitudinal halo, especially for short bunches typical for high-intensity linacs.

Many mechanisms listed under "linear accelerators" were specifically repeated for rings to emphasize that some effects in high-current linear accelerators are also relevant for high-current rings. However, due to a significant difference in the regimes of the tune depression, the time scale (growth rates) of the space-charge driven effects becomes very different. In this paper, we limit our discussion to generic mechanisms relevant for both linacs and rings. In addition, we briefly address some ring-specific and project-specific mechanisms, using an example of the SNS ring.

## BASIC MECHANISMS COMMON FOR LINACS AND RINGS

### Intrinsic incoherent resonances

The oscillating space-charge force can lead to a class of resonances where individual particles inside the beam can get into a resonance with an oscillating beam mode. These resonances are referred to as intrinsic or incoherent resonances of the individual particles. Such a parametric resonance mechanism was suggested as one of the dominant effects for halo generation in linacs [1]. Some literature on this subject can be found, for example, in Ref. [2]. In recent years, existence of self-consistent

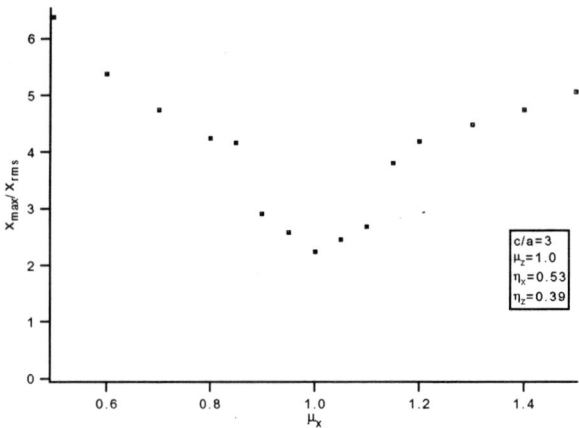

**FIGURE 2.** Extent of transverse halo for the 6D stationary distribution (tune depressions $\eta_x = 0.53$, $\eta_z = 0.39$).

three-dimensional computer codes allowed systematic study of this mechanism for realistic beam distributions [3],[4],[5]. Such resonances between the motion of the

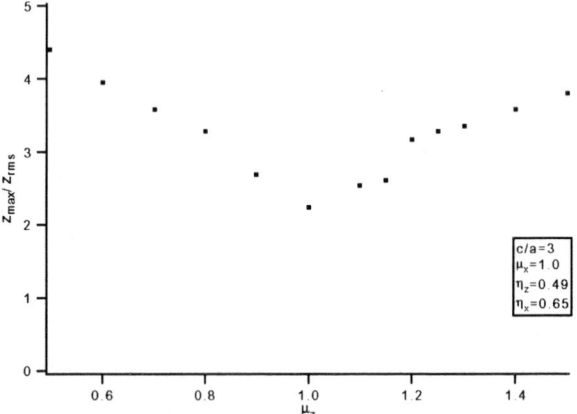

**FIGURE 3.** Extent of longitudinal halo for the 6D stationary distribution (tune depressions $\eta_x = 0.65$, $\eta_z = 0.49$)

individual particles and collective beam oscillations are governed by the rms beam mismatch. For example, in the case of a uniform density beam envelope, which is oscillating with the frequency $\Omega$ and the mismatch parameter $\mu$, the equation of motion for the individual particle has the form

$$x'' + q^2 x = \mu \frac{2\kappa}{a_0^2} x \cos\Omega\theta, \qquad (1)$$

where $\kappa$ is the space-charge parameter, and $q^2 = k^2 - \kappa/a_0^2$ is the depressed incoherent frequency without the approximation of small space charge. This equation has a primary parametric resonance at $q = \Omega/2$. When higher order terms are included, one also gets the non-linear parametric resonances [6]. The halo extent associated with the 1:2 parametric resonance (limited due to non-linearity omitted from Eq. [1]) was extensively stud-

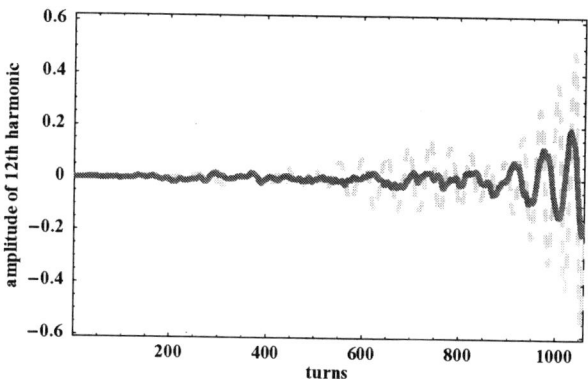

**FIGURE 4.** Time evolution of unstable harmonic: 1) pink color (solid line) corresponds to the chromaticity of $\xi = -7$ 2) green color (dash line) corresponds to $\xi = 0$.

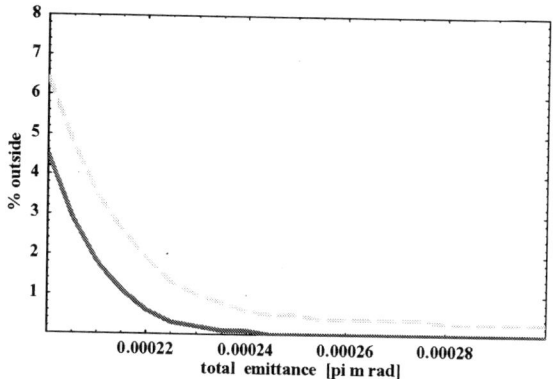

**FIGURE 5.** Halo due to the unstable harmonic of Fig. 4.

ied for the envelope modes in high-current linacs. The halo extent associated with this resonance can be large not only for a strong tune depression of the order of $\eta \sim 0.5$ (typical for linacs) but also for the tune depression of only a few percent $\eta \sim 0.97$ (typical for rings). Examples of halo extent for the 6D stationary distributions as a function of the mismatch parameter are shown in Figs. 2 and 3 for the transverse and longitudinal halo, respectively. For realistic distributions, the halo extent is slightly higher due to an additional redistribution [3]. Such a finite extent is a feature of bounded motion due to the nonlinearity for a single isolated resonance (simplified assumption). In real accelerator, with magnet misalignments, rf noise, interplay of various resonances, etc., one can expect to have some sort of diffusion process which may have impact on the halo extent if the rate is sufficiently high. Also, the extent of parametric halo can be enhanced due to a combination with other effects. The realistic prediction thus rely on computer simulations which can predict emittance growth/halo in a complex environment of a real accelerator. In the limit of zero space charge, the motion near the core is very regular, and the rate at which particles are driven into the 1:2 resonance becomes very small. As a result, for space charge with $\eta \sim 0.98$, it takes significantly more time for particles to be trapped into the 1:2 resonance [7]. The rate of halo development (which is a function of both the mismatch parameter and tune depression) becomes the most important question when one tries to estimate an effect of such a parametric resonance on halo formation in rings. In addition, when applied to accumulator rings, one should take into account many other effects such as the fact that the beam intensity is not a constant during accumulation, the phase-mixing effect due to multi-turn injection, etc. [7],[8].

The incoherent resonances, which were discussed above, and, which are also known as the mechanism for the "parametric halo", did not require any resonances with the lattice since the collective oscillation was induced by a mismatch. In rings, however, such collective beam modes may have a fast excitation as a result of both the space-charge and machine resonances. The resonances of the individual particles with such driven collective beam modes can be called "driven incoherent resonances", since the collective modes are first resonantly driven and then incoherent particles are trapped into the resonance with the corresponding collective mode. In such a situation, the incoherent resonances may play an important role in halo formation even in the limit of a weak tune depression [9].

**linear accelerators:** Space-charge tune depression is typically strong. As a result, the rate of such space-charge driven halo is very fast. This is an important mechanism - both the transverse and longitudinal mismatches should be minimized. Since bunches are typically rather short, it is necessary to consider the effect of the longitudinal-transverse coupling [3].

**circular accelerators:** Tune depressions are relatively weak (apart from "cooler rings" or specific small scale rings for space-charge studies). In addition, there are such effects as multi-turn injection (phase-mixing), redistribution, etc. - in most practical situation such a halo has little chance to develop. However, the "driven halo" is possible [9].

## Space-charge coupling resonances

Another type of a mechanism common for both high-intensity linacs and rings is the space-charge coupling resonances, which are driven by the space-charge potential itself rather than the field potential of magnets [9].

**linear accelerators:** Both the symmetric and antisymmetric resonances can be excited due to a possibility of a very different focusing constants - this results in the "equipartitioning charts", which suggest to avoid opera-

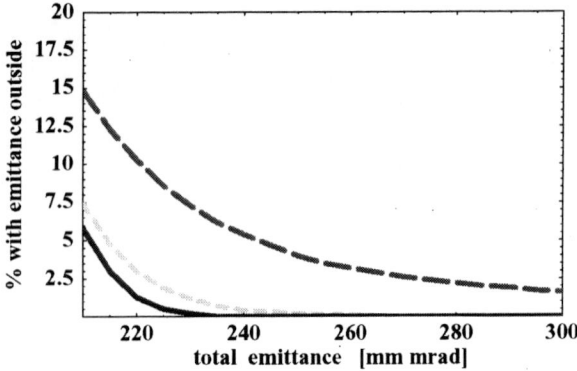

**FIGURE 6.** The SNS ring w.p. (6.36,6.22), $N = 2 \cdot 10^{14}$: no errors-blue (solid line); octupole and sextupole errors, no correction-red (long-dash line); correction of $3Q_x = 19$, $2Q_y - Q_x = 6$, and $2Q_x + 2Q_y = 25$ resonances-green (short-dash).

tion near such resonances [5].

**circular accelerators:** Resonances with the zeroth harmonic become important for beams with unequal emittances. In addition, there is now a possibility of asymmetric resonances with the error harmonic.

### Space-charge structure resonances

Collective modes of beam oscillation resonating with the lattice structure generate substantial emittance growth [10]-[12]. Thus, operation in such a regime should be avoided.

**linear accelerators:** Since focusing gradients ("tunes") are not limited by imperfection resonances, the intensity can be increased until one hits the structure stopbands. To avoid the largest second-order stopband, the recommendation is typically made to design a zero-current phase-advance below the $90^0$.

**circular accelerators:** If one can compensate emittance growth associated with the crossing of the imperfection resonances then one gets similar limitations. Otherwise, intensity is typically limited by emittance growth due to the machine imperfection resonances.

### Intrabeam Scattering

**linear accelerators:** Since time from time various studies suggest importance of this effect for the high-intensity linacs, a rigorous treatment was needed. From the kinematics of Coulomb collisions one can find the extent of halo around the beam core including the space-charge effect. However, the probability of particle to occupy such a shell in linacs is too small. To confirm this rigorously, some simple scaling formulas were derived

[13]. For the class of the 6D stationary self-consistent distribution, given by:

$$f(r,v) = \left\{ \begin{array}{ll} N(H_0 - H)^n &, H < H_0 \\ 0 &, H > H_0 \end{array} \right\}, \quad (2)$$

where the Hamiltonian includes the space-charge potential and is given by

$$H(r,v) = mv^2/2 + kr^2/2 + e\Phi_{\text{sc}}(r), \quad (3)$$

one finds that the fraction of ions which leave the beam (and form a halo around the bunch) per unit length is

$$\frac{dP}{cdt} \sim \left\{ \begin{array}{ll} r_p^2/\varepsilon_N^3 &, n > 0 \\ (r_p^2/\varepsilon_N^3)\ell n(\varepsilon_N^2/r_p a) &, n = 0 \\ (r_p^2/\varepsilon_N^3)(\varepsilon_N^2/r_p a)^n &, -1 < n < 0 \end{array} \right\}, \quad (4)$$

where, for singular distributions with $0 \geq n > -1$, we assume that the Coulomb force between ions is screened at the Debye length $\lambda_D$. Taking, for example, $r_p = 1.5 \times 10^{-18}$ [m], $\varepsilon_N \cong 10^{-6}$ [m rad] and $a \cong 10^{-2}$ [m], we then have

$$\frac{dP}{cdt} \sim \left\{ \begin{array}{ll} 10^{-15}/\text{km} &, n > 0 \\ 10^{-14}/\text{km} &, n = 0 \\ 10^{-11}/\text{km} &, n = -.5 \\ 10^{-8}/\text{km} &, n = -.9 \end{array} \right\}. \quad (5)$$

which is clearly negligible for linear accelerators.

**circular accelerators:** - In storage rings, the intrabeam scattering is typically one of the dominant mechanisms of emittance growth and is always considered.

### Collective instabilities

Both in linear and circular accelerators collective instabilities lead to halo formation and thus should be carefully avoided. An example of halo growth due to a collective dipole instability, driven by the transverse impedance, is shown in Figs. 4 and 5, for the evolution of the unstable harmonic of the dipole oscillation and associated beam halo, respectively [14].

### Misalignments, magnet errors and noise

Such effects should be considered in both linacs and rings to estimate realistic emittance growth. By itself, very small misalignments, magnet fields errors or magnet noise may not lead to a significant growth of halo but they can enhance other halo mechanisms. For example, the enhancement of parametric halo (due to a mismatch)

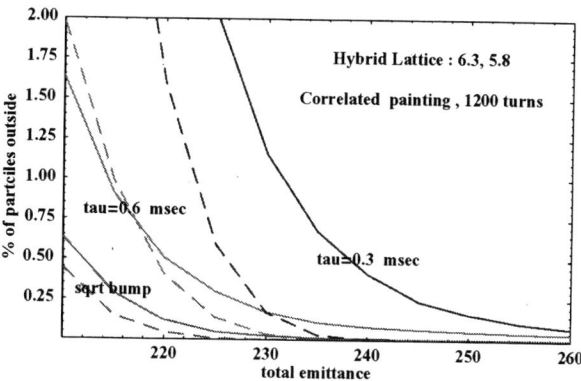

**FIGURE 7.** Beam halo distribution for correlated painting with three different bump functions. Solid and dash lines correspond to simulations with and without space charge, respectively.

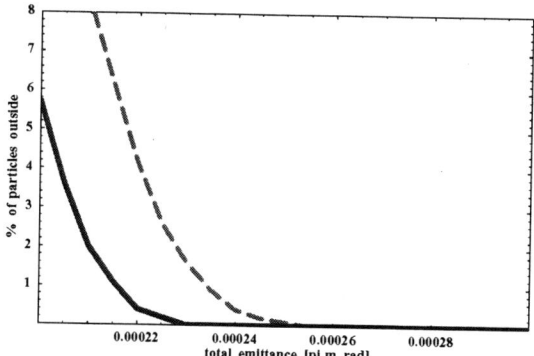

**FIGURE 8.** Beam halo at the end of the 1060-turn SNS injection due the offset of extraction kickers: 1) solid line (blue)- no offset 2) dash line (red)- designed offset.

by misalignments was shown in Ref. [15]. Halo generation due to RF noise (see, for example, [16]) or due to a magnet noise [17], was also recently addressed. Understanding of emittance growth, as a result of many effects combined together, requires realistic computer simulations [18].

## RING SPECIFIC MECHANISMS

Some ring specific mechanisms are related to the fact that characteristic times in rings are much longer than in linacs. Also, there is a possibility to accumulate effects over successive turns. One of the most important mechanisms of halo growth is due to the crossing of magnet imperfection resonances in the tune-space. The high-intensity aspects of halo growth due to various types of resonance was reviewed in Ref. [9]. Most of halo growth can be avoided by choosing the working point appropriately, and by applying the corresponding resonance correction schemes [19]. An example of halo growth due to several nonlinear resonances and its compensation is shown in Fig. 6.

## PROJECT SPECIFIC MECHANISMS

In addition to basic mechanisms described above, each individual project has many other contributions to halo generation. For example, for the SNS ring, the following mechanisms gave significant contribution to beam halo:

### Halo growth due to painting bump function

Beam tail distributions of three different bumps for correlated painting are shown in Fig. 7, where we plot the percentage of particles outside a given emittance (in $\pi$ mm mrad). The bump which collapses as a square-root function (red color) performs better than the other two bumps, which decay exponentially with different time constants $\tau$ ($\tau = 0.6$: pink color, $\tau = 0.3$: blue color) [8].

### Space-charge redistribution of painted beam

In the case of correlated painting, the beam is painted to a square shape; this results in a high density distribution along the diagonals. The inclusion of space charge leads to rapid azimuthal diffusion which was estimated analytically and confirmed numerically [20]. The radial diffusion was also explored numerically.

### "Banana-shape" effect

The extraction kickers of the SNS are not centered with respect to the zero closed orbit. They are offset from the center in order to save on mechanical dimensions when the full-size accumulated beam is extracted. As a result, the longitudinal beam centroid experiences a kick different from the head and tail of the beam, due to the longitudinal current distribution along the bunch. This results in a "banana-shape" distortion along the bunch or oscillation of beam centroid [21]. Figure 8 shows beam halo due to this effect for the case of an old impedance from the extraction kickers, which is a factor of two larger than the present impedance budget.

*Effect of wall image charges*

Increasing the $b/a$ ratio (where $b$ is the beam pipe radius and $a$ is the final radius of the painted beam) decreases the threshold of the instability caused by the extraction kicker impedance, which leads to strong halo. On the other hand, when the ratio $b/a$ becomes too small the damping of instability becomes more effective but one can get significant halo due to the image effects of a quadrupole mode [14].

*Coherent resonances*

The halo due to the crossing of the second-order and high-order coherent resonance was systematically studied. The associated space-charge limit was discussed in Refs. [22]-[23].

## SUMMARY

In summary, we give some general comments: 1) it is important to keep in mind that there are many mechanisms which contribute to halo formation 2) when several mechanisms are present simultaneously, one gets a complex behavior understanding of which requires realistic computer simulations 3) physics of some mechanisms may be the same in linacs and rings but applications of these mechanisms can be very different (for example, due to different regimes of the tune depression, which provides different growth rate for the space-charge driven effects) 4) some generic mechanisms should not be taken as granted and should be always double checked for the parameters of interest 5) there can be some "machine-specific" dominant mechanisms.

## ACKNOWLEDGMENTS

I would like to thank many collaborators with whom I worked on the subject of beam halo in linear and circular accelerators, as well as the SNS Accelerator Physics group for numerous useful discussions on this subject: R.L. Gluckstern, T. Wangler, R. Ryne, S. Kurennoy, N. Pichoff, I. Hofmann, J. Holmes, S. Cousineau, S. Danilov, Y.Y. Lee, J. Wei, D. Raparia, M. Blaskiewicz, G. Parzen, N. Malitsky and many others.

## REFERENCES

1. J.S. O'Connell et al., Proc. of PAC'03 (Washington D.C.), p. 3657; R.A. Jameson, p. 3926 (1993).
2. A.V. Fedotov and R.L. Gluckstern, Proc. of PAC'99 (New York), p. 607, and ref. therein (1999).
3. A.V. Fedotov et al., Phys. Rev. STAB, 2, 014201 (1999).
4. J. Qiang and R. Ryne, Phys. Rev. STAB, 064201 (2000).
5. I. Hofmann et al., Proc. of EPAC'02 (Paris), p.74 (2002).
6. J. M. Lagniel, N. Inst. Meth. A345, p. 46 and p. 405 (1994); A. Riabko et al., Phys. Rev. E 51, p. 3529 (1995).
7. A.V. Fedotov et al., Proc. of Workshop on beam halo and scraping (Wisconsin), p. 27 (1999).
8. A.V. Fedotov et al., Proc. of EPAC'00, p. 1289 (2000).
9. A.V. Fedotov, Resonances and beam loss in high intensity rings, Proc. of PAC'03, Portland (2003).
10. I. Hofmann et al., Part. Accel., 13, p. 145 (1983).
11. J. Struckmeier, M. Reiser, Part. Accel., 14, p.227 (1984).
12. H. Okamoto and K. Yokoya, Nuc. Inst. Meth., A482, p.51 (2002).
13. R.L. Gluckstern and A.V. Fedotov, Phys. Rev. STAB,2, 054201 (1999).
14. A.V. Fedotov et al., Proc. of EPAC'02 (Paris), p. 1350 (2002).
15. P. Chanell, LANL Report, unpublished (1995).
16. P. Ostroumov and K.W. Shepard, Phys. Rev. STAB, 3, 030101 (2000).
17. V. Lebedev, Proc. of PAC'03, Portland (2003).
18. N. Malitsky, A.V. Fedotov, J. Wei, Proc. of EPAC'02, Paris, p. 1646 (2002).
19. A.V. Fedotov and G. Parzen, Compensation of nonlinear resonances in the presence of space charge, Proc. of PAC'03, Portland (2003).
20. A.V. Fedotov et al., Proc. of PAC'01, p. 2851 (2001).
21. A.V. Fedotov, to be published (2003).
22. A.V. Fedotov and I. Hofmann, Phys. Rev. STAB, 5, 024202 (2002).
23. A.V. Fedotov et al., Application of envelope instability to high-intensity rings, Proc. of PAC'03, Portland (2003).

# Halo Diagnostics Overview

Tom Shea[1] and Peter Cameron[2]

[1]*Oak Ridge National Laboraqtory, Oak Ridge, TN, USA*
[2]*Brookhaven National Laboratory, Upton, NY, USA*

## INTRODUCTION

This workshop was unique in two ways. From the perspective of Diagnostics, it was the first workshop devoted exclusively to Halo Diagnostics. From the larger perspective of accelerator design, commissioning, and operation, it was the first workshop to bring together the world's accelerator physicists, instrumentation specialists, and collimation experts to confront the halo-related issues posed by the progressively increasing intensities in existing machines, as well as the order-of-magnitude increases anticipated in the next generation of accelerators.

In both linacs and rings the primary halo issue is the activation of machine components that results from halo scraping on the limiting apertures of the vacuum envelope. The focus of the accelerator physicists is on designing and operating their machines to minimize this halo. The focus of the collimation experts is on cleanly and efficiently disposing of this halo as it appears, a consequence of the clean and efficient disposal being that useful diagnostic information is often lost, buried in the collimators. The focus of the instrumentation specialists is twofold; to provide information useful to the accelerator physicists in their machine tuning efforts to avoid halo formation, and to provide direct measurement of halo.

In storage synchrotrons, serious operational limitations are often encountered as a result of the effect of halo on the experimenters who use the beam. Background due to halo can mask the rare physics processes that the experimenters seek to study. Equally (or more) important, the experiment detectors are often the most radiation sensitive components in the accelerator, as well as being a substantial investment in both money and specialized manpower. A specific example is RHIC, where beam aborts during acceleration ramps due to beam loss at experiment detectors was a major hindrance to maximizing integrated luminosity during the RHIC 2003 run. The beam loss threshold imposed by the most sensitive of the several experiments was far below that imposed by activation of machine components.

With the increasing importance of halo as accelerators evolve, it is not surprising that the agenda of the recent ICFA Workshop on Diagnostics for High-Intensity Hadron Machines [1] was dominated by presentations on profile and halo measurements. It is also not surprising that this workshop was held at Oak Ridge, site of the Spallation Neutron Source, a machine whose specifications for uncontrolled longitudinal and transverse halo are difficult without precedent. A significant output of that workshop was a detailed summary table [2] of diagnostics for the world's various existing and proposed high-intensity hadron machines. The interested reader is encouraged to consult that table.

## OVERVIEW

Efforts to increase our understanding of beam halo are subject to the traditional and complementary division between theory and experiment. Theory can be further subdivided into analytical approaches and simulations. The analytical approaches define the applicable fundamental physics and the scaling laws. While the analytical approaches help provide overall understanding, due to the complexity and scale of the halo problem it is difficult to produce reliable quantitative predictions. Simulations help bridge the gap between theory and experiment. They provide more reliable predictions, and can often include diagnostics elements. Both theory and experiment are required as understanding pushes ahead, with Diagnostics confirming the understanding gained from analysis and simulation, and providing information for further refinement. Diagnostic performance is essential to the overall success of these efforts.

Halo Diagnostics perform within at least three different scenarios. The first scenario is within the normal operation of purpose-built machines like LEDA[3] and UMER[4]. These are not user facilities in the usual sense, but rather are dedicated to the study of specific aspects of machine dynamics, and diagnostics are fundamental to their reason for existence. The second scenario is dedicated beam studies at user facilities. One example is the study of halo growth in the SPS [5]. In this instance life is somewhat more difficult for the instrumentation specialist, as the time available for beam studies is only a small fraction of operations time, and the demands upon diagnostics often extend considerably beyond those imposed by normal operations. The third scenario is normal machine setup and operations. In this case the ingenuity of the instrumentation specialist is challenged by the problem of extracting useful information about halo without perturbing the machine and the users.

## Definitions and Specifications

The first requirement is to have a clear definition of halo. As discussions at this workshop made clear, this proves surprisingly difficult, and a clear consensus did not emerge. During the discussions, the relevant parameters seemed to include; the amount of beam beyond a certain transverse extent, lineshape and deviation of lineshape from Gaussian, the mechanism of halo formation, and one's perspective and purpose as accelerator physicist, instrumentation specialist, or collimation expert. A rigorous set of definitions of halo parameters has appeared in a peer-reviewed journal [6], and might serve both as a stimulus to thought and as a useful common reference.

Given a definition of halo, the next essential element is to have a clear specification of instrument and measurement requirements. The usual specifications include absolute and relative accuracy, resolution, and dynamic range. Beyond that, one often finds specifications given for bandwidth, acquisition time, short and long term stability, invasiveness, the physical envelope, vacuum and baking requirements, the control system interface, and the operator interface. For efficient diagnostics development, it is crucial to have clear specifications as early as possible in the development cycle.

## Halo Challenges

In both theory and experiment, halo is inherently challenging. The primary problem is dynamic range. This is an issue both in simulation and in measurement. In measurement, implicit in the problem of dynamic range is the problem of sensitivity at the lower end of the scale.

A typical specification for profile monitors is to achieve 5% to 10% accuracy in the measurement of RMS beam size. Measurement results are sometimes better than this, and sometimes are not believed even to this level of accuracy. A confounding factor is often the quality of available information regarding beta functions and dispersion at the measurement location. Given that profile measurements are often questioned at the level of a few percent, the difficulty is easily seen in making halo measurements at the level of $10^{-4}$ and beyond.

In the case of traditional wire-scanning profile monitors, the effort to improve dynamic range has followed two paths. In the first, we find that current mode measurements are being supplemented or replaced by counting measurements. While analog front ends and digitizers are improving (the integral nonlinearity of available data acquisition systems is now better than $10^{-5}$), counting experiments still have the edge in linear dynamic range. The second area of improvement is in the utilization of multiple targets, with a wire for beam core measurements and a scraper to increase signal in the beam tail. The remaining issues involve knowledge and control of the beam/target interaction. In general these include systematic effects, the effect of backgrounds (where coincidence techniques are often helpful), and the need for stability for lengthy measurements. Beyond the traditional profile monitors, a variety of new ideas and techniques have appeared and evolved in recent years, driven both by the need for dynamic range and the desire for non-invasive diagnostics. These will be discussed in more detail in the following sections.

The difficulty of halo and profile measurement is compounded by the fact that profile monitors are sparsely distributed. This is most often true within a single machine, where (unlike for instance, with beam position monitors) cross-correlation between different monitors is usually not possible. It is also true across the various machines, where (often varied and exotic) techniques are not available at multiple machines with similar beams.

The challenge to the instrumentation specialist is magnified by the range of measurement time scales. There is a need to extend measurements to shorter time scales and greater bandwidths to permit pulse-by-pulse or turn-by-turn measurements, and even to permit profile (and, were it possible, halo) variations along a single pulse or bunch. At the opposite end of the time scale, there is the need to measure time evolution of halo, which may imply more than one measurement technique and the need to properly correlate measurements from independent devices. In the case of linacs, this implies multiple devices of similar design in a given energy range, and possibly significant variation of device designs for different energy ranges. In the case of rings one is most often limited (by cost, manpower, and the availability of real estate), to a single device of a given type.

How one goes about meeting these challenges is not completely clear. From the overall perspective, the most obvious constraint is the limit on money and manpower. A partial solution is to identify commonalities (both at the intra- and inter-laboratory levels) in existing and proposed devices and applications, and to share and combine data and experience. This approach has been given strong impetus by the inter-laboratory collaborations formed for the construction of the SNS and the LHC, and for the design of the NLC and the VLHC. The globalization of diagnostics is proceeding at an accelerating pace.

## Categories of Halo Diagnostics

Pre-workshop discussions suggested an expansive definition of halo diagnostics, and resulted in their classification into three categories. The first and most obvious includes devices that directly measure halo and halo evolution, and the prime example is the wire scanner. The second category includes devices that contribute to the diagnosis of machine conditions that cause halo formation, and an example would be a tune measurement system. The third category includes devices that measure the effects of halo development, and an example would be the loss monitor system. The table below presents typical parameters for the various types of halo diagnostics, organized by category.

## Profile Measurement Techniques

A compendium of profile measurement techniques was presented by Shafer in 1997 [7]. In the early days, profile measurement was accomplished by the irradiation of foils, photographic film, glass plates, and plastic sheets. The physical analysis was completed offline. This method is still of use for calibration, and also in particularly difficult measurement circumstances, an example being that the SNS will analyze activation profile of the spallation target [8].

Invasive targets still characterized the next step in the evolution of profile measurement techniques. The aperture-collector technique (like modern emittance scanners) is completely invasive. Less invasive methods include harps and multiwires, stepping wires, and flying wires. More exotic methods, like liquid wires and sodium curtains, have enjoyed brief popularity.

Non-invasive (or only slightly invasive) methods have developed more recently. In hadron machines the residual gas ionization monitor has reached a high level of development, and is present in almost all machines with minor variations. Gas fluorescence monitor prototypes have been installed and experimented with at several machines, but none have evolved into baseline operational profile monitors. The two major drawbacks seem to be the need for a pressure bump with a gas whose excited states have fast decay times, and the poor S/N ratio in comparison with other methods. A significant advantage is that the measurement is not affected by space charge, but with the demonstration that the IPM measures accurate profiles in the presence of space charge this becomes less of a consideration. Laser beam probes are promising, both in Compton scattering from highly relativistic beams and electron stripping from H-minus beams. Particle beam probes also merit investigation, particularly the effort to deconvolve profiles from the measurement of the deflection of an electron beam probe [9]. Finally, the quadrupole detector permits measurement of the first moment of the beam profile, useful for matching, as well as for monitoring envelope resonances in linacs.

There have also been recent developments in both invasive and non-invasive two-dimensional imaging diagnostics. These include improved phosphor screens, optical transition radiation, IR imaging, and synchrotron radiation [10]. These devices are usually not specified to measure halo.

| Category | Type | Accuracy (typical) | Resolution (typical) | Dynamic range | Comments |
|---|---|---|---|---|---|
| Profile Measure | emittance scanner | 5% | 1% | ~$10^3$ | Initial distribution for simulations, result after growth [11,12] |
| Profile Measure | Wire Scanner plus scraper | 5% | 1% | $10^5$-$10^6$ | Both SEM and counting mode, also measure mismatch in linacs [13] |
| Profile Measure | Wire Scanner with thin wire | 5% | 1% | $10^7$-$10^8$ | Used HERA-B detector for DAQ [14], also e- beam measurements at JLab [15], injection matching [16] |
| Profile Measure | Vibrating wire | ? | ? | $10^6$-$10^7$ | extremely sensitive (susceptible to ohmic heating?) [17] |
| Profile Measure | Laserwire scanner | 5% | 3% | $10^3$-$10^4$ | Photo-ionization of H$^-$ Non-invasive, no wire burning [18] |
| Profile Measure | IPM | 5% | 100μ | ~$10^3$ | 10MHz BW, pressure bump [19], resolution limit is collector spacing |
| Profile Measure | Luminescence | 5% | 2% | ~$10^3$ | Requires pressure bump [20] |
| Longitudinal Prof Meas | Beam-in-Gap | ~20%? | ~2%? | ~$10^5$? | Kicker/scraper/BLM [21,22] laser/electron detector [23] |
| Machine condition | BPM | ~0.2% x aperture | ~$10^{-5}$ x aperture | ~$10^4$ | Gives machine optics, accuracy is w/o beam-based alignment |
| Machine condition | Fast BPM | 8 bits | 8 bits | Need orbit offset | 1GHz BW scope; fast instability monitor [24] |
| Machine condition | Coherent tune | ~$10^{-3}$ | ~$10^{-3}$ | ~$10^3$ | Beam kick generates halo |
| Machine condition | Schottky tune | ~$10^{-3}$ | ~$10^{-3}$ | ~$10^4$ | Also gives ε, ξ, dp/p,... [25] |
| Machine condition | PLL tune | ~$10^{-5}$ | ~$10^{-6}$ | ~$10^6$ | gives coupling, chromaticity,... [25] |
| Machine condition | Quadrupole monitor | ~equal to wire scan | ~equal to wire scan | ~$10^2$ | matching, emittance, envelope instability, incoherent tune, [26] |
| Machine condition | Electron detector | N/A | N/A | N/A | With retarding field, gives energy spectrum; 80MHz BW [27] |
| Machine condition | Longitudinal RFA | ~$10^{-4}$ | ~$10^{-4}$ | ~$10^2$ | Evolution of beam energy spread |
| Machine condition | AC Dipole | bpm | bpm | ~1mm kick | coherent oscillations without emittance growth [28] |
| Halo effects | Fast loss monitor | Single particle | Single particle | ~$10^6$ | Counting mode [29] |
| Halo effects | Target thermocouples | ? | ? | ? | ISIS beam halo at spallation target, planned for SNS |
| Halo effects | Experiment detectors | Single particle | Single particle | ~$10^8$ | Background in the user's detectors |

## CONCLUSION

General goals were presented for the workshop. These included the formulation of fundamental and operational definitions of halo, clear specification of the requirements for diagnostics, definition of the current state of the art, a statement of goals for halo diagnostics, the identification of promising new halo diagnostic technologies and techniques, the identification of promising beam experiments, and the fostering of continued and improved collaboration in all of the above areas. The workshop was at least partially successful in all of these areas. .

## REFERENCES

1. http://www.sns.gov/icfa/

2. http://www.sns.gov/icfa/DiagnosticMatrix.pdf

3. J.D. Gilpatrick, "Beam Diagnostics Instrumentation for a 6.7-MeV Proton Beam Halo Experiment", Proc. of the Linac2000, Monterey, California (2000).

4. P.G. O'Shea et al, "The University of Maryland Electron Ring (UMER)," NIM A 464, 646-652 (2001).

5. L. Burnod and J.B. Jeanneret, "Transverse Drift Speed Measurement of the Halo in a Hadron Collider", Proc. Workshop on Advanced Beam Instrumentation, Tsukuba, Japan, (1991).

6. C.K. Allen and T.P. Wangler, "Beam halo definitions based upon moments of the particle distribution", PRST-AB 5, 124202 (2002).

7. R. Shafer, "An Overview of Beam Profile Diagnostics for Intense Proton Beams", Proceedings of the Intense Beam Profile Workshop, Los Alamos, NM (1997)

8. J. Haines, private communication.

9. P.K. Roy et al, "Non-Intercepting diagnostic for the HIF Neutralized Transport Experiment", PAC 2003, Portland.

10. Bingxin Yang, "Optical System Design for High-Energy Particle Beam Diagnostics", Proceedings of the 2002 Beam Instrumentation Workshop, Upton, NY (2002).

11. F. Bieniosek et al, "Beam Imaging Diagnostics for Heavy Ion Beam Fusion Experiments", PAC 2003, Portland.

12. L. Groening et al, "Single Shot Measurements of the 4-Dimensional Transverse Phase Space Distribution of Intense Ion Beams at the UNILAC at GSI", DIPAC 2003, Franfurt.

13. J.D. Gilpatrick, "Wide Dynamic-Range Beam-Profile Instrumentation for a Beam Halo Measurement: Description and Operation", these proceedings.

14. S. Arutunian et al, "Beam Tail Measurements using Wire Scanners at DESY", these proceedings.

15. A.P. Freyberger et al, "Large Dynamic Range Beam Profile Measurements", PAC 2003, Portland.

16. S. Igarashi, "Observation of Emittance Growth at KEK PS", these proceedings.

17. S.G. Arutunian, et al, Phys. Rev. ST Accel. Beams 6, 042801 (2003)

18. S. Assadi et al, "The SNS Laser Profile Monitor Design and Implementation", PAC 2003, Portland.

19. R.C. Connolly et al, "The IPM as a Halo Measurement and Prevention Diagnostic", these proceedings.

20. D. Sandoval, et al, "Fluorescence-based Video Profile Beam Diagnostics: Theory and Experience", AIP Conference Proceedings 319 (BIW'93), 1993.

21. A. Drees et al, "Abort Gap Studies and Cleaning during RHIC Heavy Ion Operation", PAC 2003, Portland.

22. D. Gassner et al, "SNS Longitudinal and Transverse Halo Measurement", these proceedings.

23. A. Aleksandrov et al "Beam in Gap Measurements at the SNS Front-End", PAC 2003, Portland.

24. M. Blaskiewicz et al, "Transverse Instabilities in RHIC", PAC 2003, Portland.

25. P. Cameron, "Tune-based Halo Diagnostics", these proceedings.

26. A. Jansson, "The Magnetic Quadrupole Pickups in the CERN SPS", BIW 2002, Brookhaven.

27. R. Macek et al "Electron Cloud Diagnostics in Use at the Los Alamos PSR", PAC 2003, Portland.

28. M. Bai et al, "Measuring Beta Function and Phase Advance in RHIC with an AC Dipole", PAC 2003, Portland.

29. R. Witkover and D. Gassner, "Preliminary Design of the Beam Loss Monitoring System for the SNS", BIW 2002, Brookhaven.

# Beam Collimation at Hadron Colliders[1]

## N. V. Mokhov[2]

*Fermilab, P.O. Box 500, Batavia, IL 60510, USA*

**Abstract.** Operational and accidental beam losses in hadron colliders can have a serious impact on machine and detector performance, resulting in effects ranging from minor to catastrophic. Principles and realization are described for a reliable beam collimation system required to sustain favorable background conditions in the collider detectors, provide quench stability of superconducting magnets, minimize irradiation of accelerator equipment, maintain operational reliability over the life of the machine, and reduce the impact of radiation on personnel and the environment. Based on detailed Monte-Carlo simulations, such a system has been designed and incorporated in the Tevatron collider. Its performance, comparison to measurements and possible ways to further improve the collimation efficiency are described in detail. Specifics of the collimation systems designed for the SSC, LHC, VLHC, and HERA colliders are discussed.

## INTRODUCTION

At hadron colliders, as at any other accelerator, the creation of beam halo is unavoidable. This happens because of beam-gas interactions, intra-beam scattering, proton-proton (antiproton) collisions in the interaction points (IP), and particle diffusion due to RF noise, ground motion and resonances excited by the accelerator magnet nonlinearities and power supplies ripple. As a result of halo interactions with limiting apertures, hadronic and electromagnetic showers are induced in accelerator and detector components causing numerous deleterious effects ranging from minor to severe. An accidental beam loss caused by an unsynchronized abort launched at abort system malfunction can cause catastrophic damage to the collider equipment. Only with a very efficient beam collimation system can one reduce uncontrolled beam losses in the machine to an allowable level [1, 2].

Beam collimation is mandatory at any superconducting hadron collider to protect components against excessive irradiation, minimize backgrounds in the experiments, maintain operational reliability over the life of the machine (quench stability among other things), and reduce the impact of radiation on the environment. It provides:

1. Reduction of beam loss in the vicinity of IPs to sustain favorable experimental conditions during the whole store.
2. Minimization of radiation impact on personnel and the environment by localizing beam loss in the pre-determined regions and using appropriate shielding in these regions.
3. Protection of accelerator components against irradiation caused by operational beam loss and enhancement of reliability of the machine.
4. Prevention of quenching of SC magnets and protection of other machine components from unpredictable abort and injection kicker prefires/misfires and unsynchronized aborts.

## BEAM LOSS AND SCRAPING RATES

Although beam loss and scraping rates depend on the machine specifics, their origin and values have much in common at hadron colliders. They are reliably estimated for the Tevatron [3, 4]. The ultimate Run II parameters include 36 bunches of $2.7 \times 10^{11}$ protons and $1.35 \times 10^{11}$ antiprotons each, with normalized horizontal emittances of 20 mm-mrad and 15 mm-mrad, respectively. The total beam intensities at the beginning of the store are $N_p = 9.72 \times 10^{12}$ and $N_{\bar{p}} = 4.86 \times 10^{12}$. The ultimate luminosity at the beginning of the store would be $3.31 \times 10^{32}$ cm$^{-2}$s$^{-1}$ averaging to $1.43 \times 10^{32}$ cm$^{-2}$s$^{-1}$ over a 13.5-hour store. Estimated evolution of beam loss $\Delta I$ over such a store for three major components is as following:

1. $\bar{p}p$ collisions at two IPs (*collision loss*), $\Delta I = 2.2 \times 10^7$ p/s or $\bar{p}$/s.
2. Particle loss from the RF bucket due to heating of a longitudinal degree of freedom (*longitudinal loss*),

---

[1] This work was supported by the Universities Research Association, Inc., under contract DE-AC02-76CH03000 with the U.S. Department of Energy.
[2] mokhov@fnal.gov

$\Delta I = 2 \times 10^7$ p/s and $6.1 \times 10^6$ $\bar{p}$/s.
3. *Beam-gas scattering*, $\Delta I = 6.5 \times 10^6$ p/s and $2.9 \times 10^6$ $\bar{p}$/s, calculated at a nitrogen equivalent pressure of $10^{-9}$ torr with the following gas content (in nanotorr): $H_2$ (5.7), CO (0.14), $N_2$ (0.07), $C_2H_2$ (0.06), $CH_4$ (0.11), $CO_2$ (0.07), Ar (0.09).

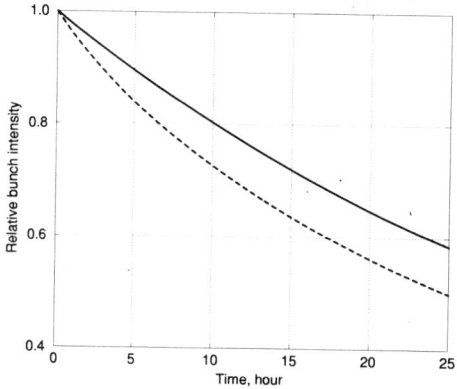

**FIGURE 1.** Relative bunch intensity as evolved in a Tevatron store for protons (solid) and antiprotons (dashed).

Fig. 1 shows beam intensity decay over a store. Inelastic and 60% of elastic events contribute to *collision loss*, because about 40% of protons (antiprotons) elastically scattered at the Tevatron IPs remain in the 3σ core after a bunch-bunch collision. Intensity drops over a 13.5-hour store are 26% and 34% for proton and antiproton beams, respectively. Longitudinal beam loss, beam gas-scattering and elastic part of collision loss are the main mechanisms of the *slow beam halo growth*. The main collimation system is designed to intercept about 99.9% of this halo, with $N_{sp} = 2.93 \times 10^7$ p/s and $N_{s\bar{p}} = 1.15 \times 10^7$ $\bar{p}$/s as the scraping rates for proton and antiproton beams, correspondingly.

## TWO-STAGE COLLIMATION

The most direct way of collimating a beam of particles is to define the physical aperture with a solid block of absorbing material. In the early Tevatron days the first collimation system was designed [5] on the basis of simulations with the MARS and STRUCT codes [6, 7]. The optimized system, which consisted of a set of collimators about 1 m long each, was installed in the Tevatron, that immediately made it possible to raise by a factor of five the efficiency of the fast resonant extraction system and intensity of the extracted 800 GeV proton beam. The data on beam loss rates and their dependence on the collimator jaw positions were in excellent agreement with the calculation predictions.

Depending upon the energy, material and thickness, a certain fraction of the intercepted beam will survive, either be traversing the whole length of the block or by being scattered out of the block. The first component can be reduced by using a longer jaw or a denser material. Suppression of the outscattered particles is much more difficult. For a given material, their yield depends upon the impact parameter $\Delta$ and particle energy. $\Delta$ grows linearly with the halo transverse diffusion velocity v. At Tevatron, v is about 1.5 μm/s and $\Delta = 0.1$-0.5 μm. This results in a probability of outscattering close to 0.5, *i.e.*, low collimation efficiency.

A natural way to catch the outscattering particles is by switching to a *two-stage collimation* system. The whole system consists then of a primary *thin scattering target*, followed by a few *secondary collimators* at the appropriate locations in the lattice. The purpose of a thin target is to increase the amplitude of the betatron oscillations of the halo particles and thus to increase their impact parameter $\Delta$ on the secondary collimators on the next turns, without influencing the unscattered beam. At Tevatron, $\Delta \approx 0.1$-0.3 mm on secondary collimators – almost a factor of 1000 larger than on the primary ones. This results in a significant decrease of the outscattered proton yield, total beam loss in the accelerator and jaw overheating, mitigating requirements to collimator alignment. Besides that, the collimation efficiency becomes almost independent of accelerator tuning. There is only one significant but totally controllable restriction of the accelerator aperture and only the secondary collimator region needs heavy shielding.

In 1995, based on the MARS-STRUCT simulations, the existing scraper in the Tevatron at AØ was replaced with a new one with two 2.5-mm thick L-shape tungsten targets with a 0.3-mm offset relative to the inner surface on either end of the scraper (to eliminate the misalignment problem). This resulted in reduction of beam loss rate upstream of both collider detectors by a factor of five, in agreement with the modeling predictions [8]. The system was further improved for Run II [9] (see below).

Two-stage collimation systems were parts of the original designs in all the superconducting hadron collider projects: $3 \times 3$-TeV UNK at Protvino [10], $20 \times 20$-TeV SSC in Texas [1, 2], $7 \times 7$-TeV LHC at CERN [11, 12], 0.82-TeV proton HERA ring at DESY [13], $20 \times 20$-TeV (Stage-1) and $88 \times 88$-TeV (Stage-2) VLHC in Illinois [14].

## POSITIONING AND DESIGNING COLLIMATORS

Thin movable primary collimators (*scatteres, targets, blades*) are optimized for the beam and heating (scattering, integrity and cooling) and positioned at $x_0 = m\sigma_0$ from the beam axis ($m \approx 5$) in a high-$\beta$ (*betatron cleaning*) and non-zero dispersion (*momentum cleaning*) regions, three in total: horizontal, vertical and off-momentum. Movable secondary collimators (e.g., L-shape jaws) – long enough to absorb showers induced by particles scattered from the primary collimators – are located at the appropriate phase advances $\Delta\phi$ at $x = n\sigma$ from the beam axis. Here, $\sigma_0$ and $\sigma$ are the beam RMS at the entrance to the primary and secondary collimators, respectively, for each plane. The secondary collimator jaws are aligned parallel to the envelope of the circulating beam.

The optimum conditions for the positioning the secondary collimators with respect to the scatterer is determined from [1]

$$\Delta\phi = \pi k \pm \arccos(m/n),$$

$$n - m > |\Delta p/p(\eta_0/\sigma_0 - \eta/\sigma)| + \delta,$$

where $k = 0, 1, 2, ...$, $\eta_0$ and $\eta$ are dispersions at the primary and secondary collimator positions, respectively, and $\delta \sim 1$. The favorable condition is to have the secondary collimator jaw on the same side of the beam as the primary collimator, which results in the optimal phase advances $\Delta\phi = 20\text{-}40°$ and $300\text{-}320°$ for the horizontal scraping, and $\Delta\phi \sim 40°$ and $140\text{-}160°$ for the vertical scraping. For the primary and secondary jaws positioned at the opposite sides of the beam ($n > 0, m < 0$) and for $n - m < 1$, the optimal phase advance is about $160°$. Fortunately, there is no strong dependence of the collimation efficiency on the phase advance in the above ranges of $\Delta\phi$, which leaves freedom to vary collimator positions to match the other requirements [1].

The following design constraints are taken into account while developing and engineering a collimation system at a hadron collider:

- Minimum outscattering from a primary-secondary collimator couple.
- Impedance constraints.
- The apertures do not occlude any beam when in the garage position.
- No quench of downstream superconducting magnets.
- Muon vectors downstream do not create any problem to the experiments and environment.
- Local shielding (if needed) provides protection of ground water and equipment around the unit, and residual dose rate on its outside below 1 mSv/hr.
- Target/jaw material integrity and cooling issues.
- Alignement issues.

# TEVATRON RUN II

The Tevatron Run II collimation system [9] is based on a two-stage approach to localize most of beam losses in the straight sections D49, EØ and F17. Collimator positions in the ring are shown in Fig. 2. Parameters of the scatterers and secondary collimators have been carefully optimized for the 1-TeV proton and antiproton beams. The 5-mm thick tungsten primary collimators are positioned at $5\sigma$ from the beam axis both in vertical and horizontal planes. The 1.5-m long stainless steel secondary collimators consist of L-shape jaws positioned at $6\sigma$ from the beam axis in both planes. Numerical simulations were done for the lattice in the presence of the proton and antiproton orbit separation designed for Run II.

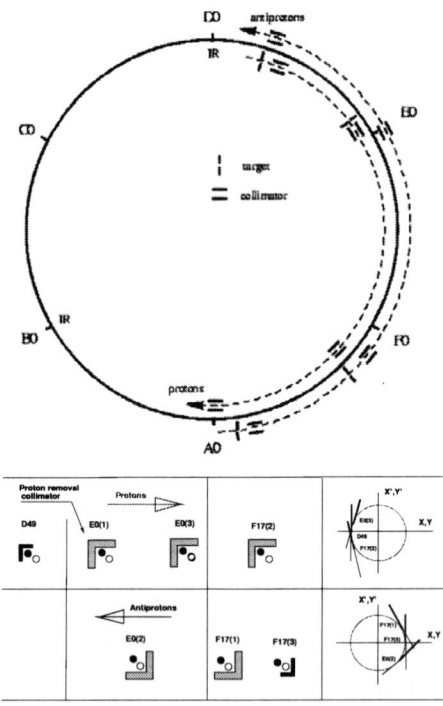

**FIGURE 2.** Tevatron Run II beam collimation system.

Large amplitude protons are intercepted by the secondary collimators during the first turn after interaction with the primary collimator. Protons (antiprotons) with amplitudes smaller than $6\sigma$ survive during several tens of turns until they increase amplitude in the next interactions with primary collimators. These particles produce a secondary halo and occupy the $6\sigma$ envelope. Beam halo particles interact with primary collimators 2.2 times on average. About 0.1% of protons and antiprotons hitting the secondary collimator jaws are scattered back into the beam pipe and later lost on limiting apertures, in most cases upstream of the CDF and DØ collider detectors. Products of beam-gas interactions not intercepted by the collimation system also have a good chance to be lost at the same locations in front of the IPs. The main process of beam-gas interaction, a multiple Coulomb scattering, results in slow diffusion of protons (antiprotons) from the beam core causing emittance growth. These particles increase their betatron amplitudes gradually during many turns and are intercepted by collimators before they reach other limiting apertures. In inelastic nuclear interactions of a beam with residual gas, leading nucleons are generated at angles large enough for them – along with other secondaries – to be lost within tens of meters after such interactions.

Overall, this system provides effective beam cleaning of slowly growing transverse and longitudinal halo, reliably protecting the machine and detectors. It was shown in Ref. [4] that with the Tevatron parameters, nuclear elastic beam-gas scattering can result in a substantial increase of the betatron amplitude. Beam loss distribution due to this process follows the vacuum distribution (Fig. 3). It turns out that a fraction of these particles is not intercepted by the main collimators, and about 25% of them are lost in the vicinity of the IPs adding to the detector background. Moreover, unacceptable beam loss happens in the B0 low-$\beta$ region at the abort kicker prefire, resulting in the SC magnet quench and severe damage to the CDF silicon detectors. To cure this, a 0.5-m long steel mask is installed this summer immediately upstream of the last three dipoles before B0, with its parameters carefully optimized in detailed MARS-STRUCT simulations [15, 16].

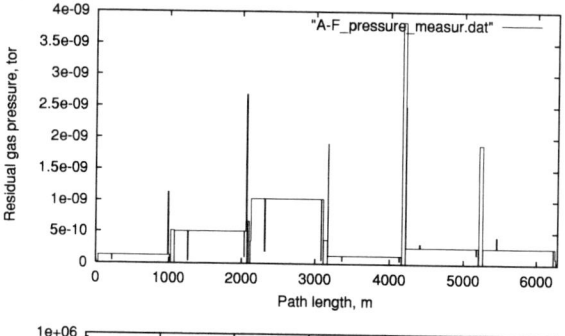
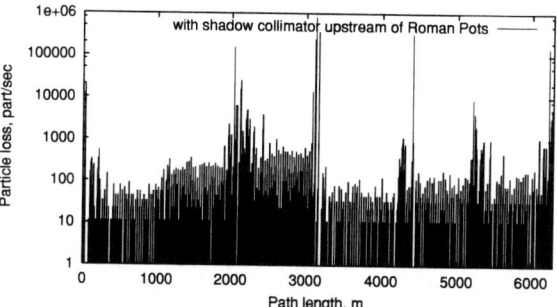

**FIGURE 3.** Measured residual gas pressure (*top*) and STRUCT-calculated beam loss distribution from nuclear elastic beam-gas scattering (*bottom*).

## CRYSTAL COLLIMATION

It was shown for the first time in Ref. [1] that replacing an amorphous primary collimator with a bent crystal can drastically improve the collimation efficiency for TeV beams. A channeling crystal coherently deflects a fraction of the beam halo, directing it, as a whole, deeper into a second collimator body, substantially reducing the outscattering probability. Detailed CATCH-STRUCT calculations [17] have shown that at the Tevatron, beam loss rates in the critical IP locations can be reduced by a factor of ten (see Table 1). Moreover, a number of inelastic nuclear interactions in the optimal crystal is about four times lower compared to that in the optimal amorphous target, that reduces radiation load to the downstream SC magnets by the same factor.

**TABLE 1.** Halo hit rates at the Fermilab D0 and CDF Roman pots and nuclear interaction rates $N$ in target and crystal (in $10^4 p/s$)

|     | With target | With crystal | | |
| --- | --- | --- | --- | --- |
|     |     | Amorphous layer thickness | | |
|     |     | 10 $\mu m$ | 5 $\mu m$ | 2 $\mu m$ |
| D0  | 11.5 | 1.35 | 1.60 | 1.15 |
| CDF | 43.6 | 5.40 | 3.20 | 3.43 |
| $N$ | 270 | 82.4 | 70.6 | 50.3 |

## HERA

It was shown in a detailed study [13] that a single collimator was insufficient in removing the beam halo responsible for the background rates in the H1 and ZEUS detectors at HERA. It has been demonstrated – both by Monte Carlo simulations and experimentally – that these backgrounds can be significantly reduced by installing a two-stage collimation system. Depending on beam lifetime, reduction factors of up to 10 have been observed in dedicated experiments.

Like at Tevatron, the beam-gas induced part of the hadronic background constitutes a constant radiation level that is not affected by the collimators. Recently, another source of the background was discovered at HERA, the C5 mask [18]. The mask's main purpose is to shield from backscattered synchrotron radiation from the lepton beam. However, it is also a scattering source for protons. Reconstruction of the IP location revealed many events coming from this mask located about 0.8 m downstream (as seen by the lepton beam) of the IP. So, it will be made thinner. Another issue was not well pumped Zinc found in the H1's copper-coated tungsten mask. The masks have been replaced and tested to be Zinc-free. Other activities by the machine-detector interface group take place HERA to reduce backgrounds [18].

# LHC

At nominal operation parameters, each of the 7 TeV circulating beams of the LHC contains approximately 334 MJ of energy, which is enough to cause severe damage to the expensive machine and detector equipment. An extremely reliable abort system will use fast extraction to divert the beam to an external graphite absorber at the end of a normal fill or in case of a detected anomaly in beam behavior. There are three collimation systems implemented into the complex: high-luminosity interaction region protection, beam cleaning system and protection at beam accidents.

**The high-luminosity IR protection system** on each side of the IP1 and IP5 has been designed over the years on the basis of comprehensive MARS calculations [19]. It includes:

- The TAS front copper absorber at L=19.45 m from the IP (1.8 m long, 34-mm ID, 500-mm OD).
- A 7-mm thick stainless steel (SS) liner in the Q1 quadrupole.
- The SS absorber TASB at L=45.05 m (1.2-m long, r=33.3-60 mm).
- A ~3-mm thick SS liner in the Q2A through Q3 quadrupoles.
- 40-cm long SS masks at L=23.45 m, r=250-325 mm to protect the Q1 slide bearings.
- The neutral particle 3.5-m copper absorber TAN at 140 m from the IP.
- The 1-m long TCL SS collimator at 191 m from IP.

This system, developed under realistic engineering constraints, will protect the LHC IP1/IP5 region components against luminosity-driven short- and long-term deleterious energy deposition effects with a good safety margin, at least at the design luminosity of $10^{34}$ cm$^{-2}$s$^{-1}$, not compromising the physics both in the main (CMS and ATLAS) and forward (TOTEM) detectors.

**The beam cleaning system** occupies two dedicated insertions for momentum cleaning in IR3 (1 primary and 6 secondary), and betatron cleaning in IR7 (4 primary and 16 secondary), with 54 movable collimators total for two rings. The system layout has been worked out to provide the required cleaning efficiency of 99.998% and integrated into the machine. Open questions still remain [20]: foreseen collimator materials do not withstand the expected beam impact (require a factor of 100-200 better resistance); impedance from collimators is critical; mechanical and operational tolerances are tight; high activation imposes severe restrictions on hands-on maintenance.

**Protection at beam accidents.** A beam loss, caused by an unsynchronized abort launched at abort system malfunction, can cause severe damage to collider inner triplet components and the CMS detector near-beam elements. A set of stationary collimators for the IP5 interaction region has been proposed in [21] to protect its elements and mitigate consequences to the detector. Fig. 4 gives details of the MARS model of the system. The first collimator is positioned at $21\sigma_{collis}=10.3\sigma_{inject}=10$ mm from the beam orbit (11.8 mm from the beam pipe center). Second and third collimators are used to protect magnets from secondary particles emitted from the first one. The collimator configuration, materials and dimensions have been carefully optimized to provide reliable protection of the inner triplet and to ensure collimator survivability. Combined with an unsynchronized abort, such a system reduces peak energy deposition in the IP5 inner triplet quadrupoles by almost six orders of magnitude compared to the disastrous case of a 1-module pre-fire.

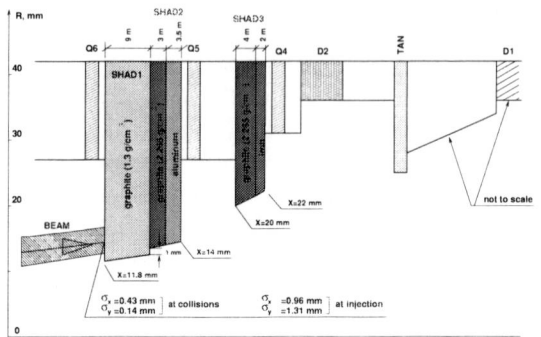

**FIGURE 4.** Stationary collimators in the LHC IP5 outer triplet.

Alternatively, a movable collimator system in the IP6 beam abort straight section, as close to the cause as possible, has been proposed in [22] to protect the entire LHC machine. The configuration of the system is similar to the one shown in Fig. 4. A composite 9.5-m long graphite (8 m) and aluminum (1.5 m) collimator TCDQ is placed at a radial position of 9.1 mm, corresponding to $8\sigma_x$ of the circulating beam at collision energy of 7 TeV, plus orbit deviations. It is movable, i.e. the jaws are retracted at injection to accommodate a larger beam size. The studies revealed that with this system, the entire machine and detector components are reliably protected against any damage at an unsynchronized beam abort. The peak temperature rise in the IP6 components is quite acceptable. If the abort kicker delay time exceeds 1 $\mu$s, several first SC quadrupoles and dipoles in the IP6 can quench. Two additional movable 2-m steel masks are added between the Q5 quadrupole and SC dipoles to reduce the length of the quench region to less than 50% of the first string.

## VLHC

The collimation system [14], designed for the $20\times20$-TeV Stage-1 VLHC, consists of horizontal and vertical primary collimators and a set of secondary collimators placed at optimal phase advances. From the very beginning, the lattice was designed to provide a warm collimation region with enough space to accommodate the system and provide large dispersion for those collimators which intercept the off-momentum protons. The primary collimators are positioned at $7\sigma$ while secondary ones are at $9.2\sigma$ from the beam axis. Eight supplementary collimators are placed in the next long straight section to decrease particle losses in the low-$\beta$ quadrupoles. These collimators are positioned at $14\sigma_{x,y}$ to intercept particles outscattered from the secondary collimators. There are only several SC magnets in the arcs with beam loss rate of 0.3 to 1 W/m, the rest of the arc is clean. Total beam loss in the low-$\beta$ quadrupoles, induced by the tails from collimators, is 61 W. Adding the supplementary collimators, one reduces this by about an order of magnitude.

## CONCLUSIONS

Two-stage collimation, proven to work at Tevatron and HERA, requires further R&D to improve its efficiency, and meet LHC and VLHC challenges – under realistic halo and beam loss scenarios and engineering constraints, and exploring novel techniques.

## REFERENCES

1. Maslov, M. A., Mokhov, N. V., and Yazynin, I. A., The SSC Beam Scraper System, Tech. rep., SSCL-484 (1991).
2. Drozhdin, A. I., Mokhov, N. V., Soundranayagam, R., and Tompkins, J., Toward Design of the Collider Beam Collimation System, Tech. rep., SSCL-Preprint-555 (1994).
3. Mokhov, N. V., and Balbekov, V. I., *Handbook of Accelerator Physics and Engineering, 2nd Printing*, **Ed. A. W. Chao and M. Tigner, World Scientific**, 218–220 (2002).
4. Drozhdin, A. I., Lebedev, V. A., Mokhov, N. V., Nicolas, L. Y., Sidorov, D. V., Striganov, S. I., and Tollestrup, A. V., Beam Loss and Backgrounds in the CDF and DØ Detectors due to Nuclear Elastic Beam-Gas Scattering, Tech. rep., Fermilab-FN-734 (2003).
5. Drozhdin, A. I., Harrison, M., and Mokhov, N. V., Study of Beam Losses During Fast Extraction of 800-GeV Protons from the Tevatron, Tech. rep., Fermilab-FN-418 (1985).
6. Mokhov, N. V., Status of MARS Code, Tech. rep., Fermilab-Conf-03/053 (2003).
7. Baishev, I. S., Drozhdin, A. I., and Mokhov, N. V., STRUCT Program User's Reference Manual, Tech. rep., SSCL-MAN-0034 (1994).
8. Butler, J. M., Denisov, D. S., Diehl, H. T., Drozhdin, A. I., Mokhov, N. V., and Wood, D. R., Reduction of Tevatron and Main Ring Induced Backgrounds in the DØ Detector, Tech. rep., Fermilab-FN-629 (1995).
9. Church, M. D., Drozhdin, A. I., Legan, A., Mokhov, N. V., and Reilly, R. E., "Tevatron Run-II Beam Collimation System," in *1999 Particle Accelerator Conf.*, IEEE Conference Proceedings Fermilab-Conf-99/059, New York, 1999, pp. 56–58.
10. Drozhdin, A. I., Maslov, M. A., and Mokhov, N. V., "Protection of the UNK Superconducting Ring Against Irradiation at Beam Collimation," in *X All-Union Conference on Charged Particle Accelerators*, Proceedings, vol. 2, JINR, Dubna, 1987, pp. 278–285.
11. Burnod, L., and Jeanneret, J. B., Beam Losses and Collimation in the LHC: A Quantative Approach, Tech. rep., CERN SL/91-39 (1991).
12. Trenkler, T., and Jeanneret, J. B., The Principles of the Two-stage Betatron and Momentum Collimation in Circular Accelerators, Tech. rep., CERN SL/95-03 (AP), LHC Note 312 (1995).
13. Seidel, M., The Proton Collimation System of HERA, Tech. rep., DESY 94-103 (1994).
14. Mokhov, N. V., Drozhdin, A. I., and Foster, G. W., "Beam-Induced Energy Deposition Issues in the Very Large Hadron Collider," in *2001 Particle Accelerator Conf.*, IEEE Conference Proceedings Fermilab-Conf-01/135, Chicago, 2001, pp. 3171–3173.
15. Church, M. D., Drozhdin, A. I., Moore, R. S., and Still, D. A., Tevatron Abort Kicker Prefire Simulations, Tech. rep., Fermilab Beams-doc-648 (2003).
16. Nicolas, L. Y., and Mokhov, N. V., Impact of the A48 Collimator on the Tevatron B0 Dipoles, Tech. rep., Fermilab-TM-2214 (2003).
17. Biryukov, V. M., Drozhdin, A. I., and Mokhov, N. V., "On Possible Use of Bent Crystal to Improve Tevatron Beam Scraping," in *1999 Particle Accelerator Conf.*, IEEE Conference Proceedings Fermilab-Conf-99/072, New York, 1999, pp. 1234–1236.
18. Minty, M., *Private Communication*, **DESY** (2003).
19. Mokhov, N. V., Rakhno, I. L., Kerby, J. S., and Strait, J. B., Protecting LHC IP1/IP5 Components Against Radiation Resulting from Colliding Beam Interactions, Tech. rep., Fermilab-FN-732, LHC Project Report 633 (2003).
20. Assmann, R., "Designing and Building a Collimation System for the High-Intensity LHC Beam," in *2003 Particle Accelerator Conf. Proceedings*, Portland, Oregon, 2003.
21. Drozhdin, A. I., Mokhov, N. V., and Huhtinen, M., "Impact of the LHC Beam Abort Kicker Prefire on High Luminosity Insertion and CMS Detector Performance," in *1999 Particle Accelerator Conf.*, IEEE Conference Proceedings Fermilab-Conf-99/060, New York, 1999, pp. 1231–1233.
22. Mokhov, N. V., Drozhdin, A. I., Rakhno, I. L., Gyr, M., and Weisse, E., "Protecting LHC Components Against Radiation Resulting from an Unsynchronized Beam Abort," in *2001 Particle Accelerator Conference*, IEEE Conference Proceedings Fermilab-Conf-01/133, Chicago, 2001, pp. 3168–3170.

# Halo Formation due to Beam-Beam Interaction

## F. Zimmermann

*CERN, AB Division*

**Abstract.** In colliding-beam storage rings transverse and longitudinal beam halo is generated, or depleted, by the effect of head-on and long-range beam-beam collisions. Measurements from several past and present lepton or hadron colliders are compared with simulations and analytical models, highlighting a few halo-specific techniques. This overview concludes with an outlook on the future Large Hadron Collider (LHC), where long-range beam-beam collisions are the dominant perturbation of particle motion and beam-beam compensation schemes appear a promising remedy.

**TABLE 1.** Energy, beam-beam tune shifts, and damping decrement for a few typical colliders

| | beam energy [GeV] | tune shift per IP | total tune shift | damping decr. per IP |
|---|---|---|---|---|
| LEP | 100 | 0.083 | 0.33 | $1.6 \times 10^{-2}$ |
| KEKB | 8, 3.5 | 0.097 | 0.097 | $2 \times 10^{-4}$ |
| DAFNE | 0.51 | 0.03 | 0.03 | $10^{-5}$ |
| LHC | 7000 | 0.003 | 0.01 | $5 \times 10^{-10}$ |
| Tevatron | 980 | 0.01 | 0.02 | $5 \times 10^{-12}$ |

## INTRODUCTION

In storage-ring colliders, the beam-beam interaction can generate beam halo. Different types of colliders may be distinguished according to the particles they bring into collision. Examples for lepton-lepton colliders are, in the order of increasing beam energy, AdA, DAΦNE, BEPC, SPEAR, VEPP-4, CESR, KEKB, PEP-II, PEP, PETRA, TRISTAN, and LEP, all of which provide(d) electron-positron interactions. So far there have been only one lepton-hadron collider, HERA, and 4 hadron-hadron colliders, namely the CERN ISR, the Sp$\bar{\text{p}}$S collider, the Tevatron, and RHIC. A fifth hadron collider, the LHC, is under construction at CERN. Table 1 lists beam energies, the beam-beam tune shift parameter $\xi$ for a single interaction point (IP), the total tune shift parameter over all ($n_{\text{IP}}$) IPs, $\xi_{\text{tot}} = n_{\text{IP}}\xi$, and the damping decrement for a few representative colliders. The latter refers to the relative decrease in oscillation amplitude per turn. Beam energies range from a few GeV to several thousands of GeV. The beam-beam tune shifts per IP are of the order 0.003–0.01 for hadron colliders and about 10 times larger for lepton colliders.

In lepton colliders, radiation damping gives rise to a 'steady-state' equilibrium. Non-Gaussian beam tails may cause background and reduce the beam lifetime. They often limit the luminosity. In 1983, J. Seeman performed a comprehensive review [1] of experimental tails and performance limits in lepton colliders, comparing the experience at SPEAR, CESR, PETRA, and PEP. His observations were as follows. At low beam currents, the luminosity increases as the square of the bunch charge (equal for the two beams), and the beam-beam tune shift increases linearly. From a certain current value, known as the first beam-beam limit, the behavior changes, *i.e.*, now the vertical beam-beam tune shift stays constant, and the luminosity only increases linearly with further increasing beam current. The constant beam-beam tune shift is related to a blow up of the vertical beam size. The ultimate performance is reached at a higher current — the second beam-beam limit —, where the background becomes unbearable and/or the beam lifetime drops to unsustainable values. The background increase can be strongly nonlinear. As the beam current is increased, the tails expand, and the transverse scrapers, which usually ensure an adequate detector background, have to be further and further retracted from the beam, in order to maintain a reasonable lifetime. The second beam-beam limit is reached when the magnet openings near the interaction region cease to be in the shadow of the scrapers. J. Seeman noted that both the beam core and the tails grow in amplitude as a function of current. At the second beam-beam limit and for a 2-hour beam lifetime, the rms beam size of the blown up core corresponds to about 5% of the scraper opening, where the latter roughly equals the translated aperture of the low-beta quadrupoles. The factor of 20 between core size and aperture measures how far the tails extend to larger amplitudes. It was found to be a universal value for all machines, except for one notable exception: When, in a dedicated machine study at CESR, the vertical IP beta function $\beta_y^*$ was reduced from $\sigma_z/\beta_y^* \approx 0.4$ to $\sigma_z/\beta_y^* \approx 0.8$, the ratio of the scraper position to the rms core beam size increased from 20 to 30. Indeed, for this case the luminosity reduction due to the tail growth was three times larger than the purely geometric hourglass effect.

In hadron colliders, halo particles 'never' come back to the core. As in lepton machines, they cause background in the physics experiments. In addition, large tails and losses can destroy the collimators, and they may also lead to quenches of superconducting magnets. Beautiful examples are available from the Tevatron, where at the start of a physics store the halo is carefully removed by transverse scraping. After this beam cleaning, sometimes it takes only 20–30 minutes, before the background signal from proton losses again reaches the original level. The Tevatron losses are sensitive to tune, coupling, and chromaticity, and they can be reduced by a careful control [2]. However, attempts to adjust these parameters may temporarily lead to excessive losses and occasional quenches. Another observation from the Tevatron is that the emittance growth in collision greatly varies from bunch to bunch within a 12-bunch train, as well as from store to store, by up to factors of 4 or 5. Frequently, the emittance blows up quickly, e.g., by 20–30% in the first 30 minutes of a store, for all bunches but the first and last in a train [2]. Due to its appearance on the Tevatron emittance summary display, this is called the 'scallops' effect [2]. The first and last bunches behave differently, because they are missing the closest long-range beam-beam encounters on one or the other side of the head-on IPs, and, thus, their tunes are different.

## GENERATION OF TAILS

In the past, a number of theoretical tail-generating mechanisms have been proposed and studied. These include incoherent scattering processes, e.g., beam-beam bremsstrahlung in LEP [3], stochastic diffusion in the limit of zero memory between subsequent collisions [4], Arnold diffusion [5], resonance trapping [6], phase convection [7], resonance streaming [8, 9], and modulational diffusion [5, 10]. In the following we describe in more detail (1) incoherent scattering, which was the dominant mechanism in LEP and will be for LHC ion-ion collisions, (2) resonance streaming, for generic lepton colliders, and (3) modulational diffusion, for hadron colliders.

In LEP, tails were measured by moving single collimator jaws into the beam and detecting the loss rate downstream using pin diodes and scintillators [3, 11]. The loss rates could be translated into a beam lifetime, which varied with the collimator position. Halo levels corresponding to beam lifetimes of 1000 to 10000 hours could be detected. The observed variation of the halo density with amplitude was reproduced in particle-tracking simulations including beam-beam bremsstrahlung [3, 11]. The main effect was the change in energy of individual particles due to the inelastic scattering at the collision point. The differential probability of incoherent bremsstrahlung for a single collision per particle is [12]

$$\frac{dN}{dE_\gamma} \approx 0.4 \frac{1}{E_\gamma} \alpha \left( \frac{r_0^2 N_2}{\sigma_x \sigma_y} \right) \left[ \ln\left( \frac{4\gamma^3 m_e c^2}{E_\gamma} \right) - \frac{1}{2} \right],$$

where $E_\gamma > E_c \equiv 4\gamma^2 \hbar c/\sigma_z$, and $N_2$ denotes the bunch population of the opposing beam. The LEP halo was sensitive to the dispersion at the IP, which could be varied by symmetric dispersion bumps. Comparisons between simulations and measurements at 46.6 GeV for a beam-beam parameter $\xi \approx 0.025$ showed a reasonable agreement [3, 11]. The remaining small discrepancies are attributed to different physical apertures in the simulation and in the real machine. At higher beam-beam parameters, additional tails were observed, which appeared to be of a dynamic origin. A combination of rare large scattering events due to incoherent beam-beam bremsstrahlung and beam-beam dynamics was also found to be the major source of beam halo in simulations for KEKB [13]. Experimental studies confirming this prediction are not yet available. For the similar PEP-II factory, the background due to beam-beam bremsstrahlung was estimated as tolerable [14].

Coherent bremsstrahlung occurs for $E_\gamma < E_c \equiv 4\gamma^2\hbar c/\sigma_z$ with a probability [15] $dN/dE_\gamma \approx 0.2(1/E_\gamma)\alpha \left(r_0 N_2/(\sigma_x)\right)^2$. This process is relevant only for $E_\gamma < 4\gamma^2 \hbar c/\sigma_z$. Thus, most likely, it does not much contribute to tail formation.

Two other incoherent collision processes, which are expected to limit the maximum luminosity in LHC ion-ion collisions, are pair production followed by electron capture and nuclear excitation with subsequent neutron emission. The approximate event probability for neutron emission is [16]

$$N \approx \left( \frac{N_2}{4\pi \sigma_x \sigma_y} \right) (3.42 \, \mu\text{barn}) \frac{(A-Z)Z^3}{A^{2/3}} \ln(2\gamma^2 - 1).$$

The corresponding probability for electron capture [17] does not reduce to a compact expression, but it scales similarly with energy and roughly as $Z^7$ with the atomic number. Total cross sections for various ion species in the LHC vary between 0.1 and 514 barn [18]. In the case of electron capture and lead-ion collisions, the charge of a lead ion changes by 1 unit, which corresponds to a change in its effective relative momentum error $\Delta\delta = 12 \times 10^{-3}$. This error is so large, that the particle will be lost in the dispersion suppressor behind the IP. In the case of neutron emission, the mass of the ion changes by one atomic unit, which translates into a momentum error of $\Delta\delta = -5 \times 10^{-3}$. Such ions may remain within the momentum aperture, and could survive for a number of turns, forming a beam halo. For other ion species the cross sections and the induced effective momentum errors would be different.

As indicated above for LEP, the generation of tails can be simulated. In standard macroparticle simulations only few particles end up in the tails and the statistical error can accordingly be large. In 1989, J. Irwin proposed an elegant iterative scheme of 'self-generated' boundary conditions' which improves the tail resolution by a factor of 10 in each iteration step [19]. In the first step, a number of macroparticles is tracked for a certain number of turns. A border is then drawn in phase space outside of which 10% of the macroparticles are found. In the next step, the full number of macroparticles are launched outside this boundary, Whenever a particle enters the inner region, it is replaced by another particle that has moved from the inside to the outside during the first step. In the following step again a boundary is determined outside of which 10% of the particles of the second step are found. These now correspond to 1% of the initial distribution. This process can be repeated an arbitrary number of times, until the desired resolution has been achieved. The two key requirements for this recipe to work are, first, randomness (*e.g.*, due to quantum fluctuation) and, second, reaching an equilibrium at each step. This computing scheme was implemented in two independent codes by D. Shatilov ('lifetrac' program) in 1992 [20], and by T. Chen and coworkers around 1993 [21]. These codes do not only include the beam-beam dynamics at the primary IP, but also model some additional effects such as frequent small-angle scattering, and parasitic collisions. They were applied to VEPP-2M, PEP-II, and DAFNE. In 1997, E.-S. Kim and K. Hirata developed a macroparticle algorithm incorporating rare large-angle scattering, which was employed in simulations for KEKB [13]. Rapid advance in computing power may soon render unnecessary the use of sophisticated simulation schemes. Examples for brute force macroparticle tracking are the TRS code by Tennyson [22] and K. Ohmi's code [23]. Typical numbers of 'particle×turns' considered in the simulations range from a few $10^8$ to several $10^9$. All the above codes were written for lepton colliders. Remarkable results were obtained by D. Shatilov [20] and by T. Chen [21], who superposed simulated beam density distributions and resonance lines, calculated in first order perturbation theory, on the transverse action space, and, thereby, revealed the role of individual resonances in the tail formation. Synchro-betatron resonances of typical order 5 to 8 were found to be important in PEP-II. The specific set of resonances strongly varied with the synchrotron tune [21]. Both Chen's and Shatilov's codes were also used to simulate the combined effect of

**TABLE 2.** Tune sensitivity of beam tails, beam lifetime and background, observed at various lepton and hadron colliders

| accelerator | sensitive tune $|\Delta Q|$ | reference |
| --- | --- | --- |
| ISR | 0.002 | Keil et al., 1975 [27] |
| SPS | 0.001 | Meddahi et al., 1991 [28, 29] |
| HERA | 0.002 | Willeke, 1997 [30] |
| VEPP-4 | 0.001 | Temnykh et al., 1989 [31] |
| DAFNE | 0.001 | Boscolo et al., 1999 [32] |
| KEKB | 0.001 | Ohmi et al., 2003 [23] |
| LEP | 0.001 | Burkhardt, 2003 [33] |
| RHIC | 0.001 | Fischer, 2003 [34] |
| Tevatron | 0.001 | Zhang, 2003 [2] |

beam-beam interaction and scattering off the residual gas for PEP-II. In T. Chen's simulations the inverse lifetime was found to be the sum of the inverse lifetimes of the two processes computed separately [24], whereas in the earlier simulations by Shatilov and Zholents a large enhancement was seen [25]. Unfortunately, the PEP-II tail simulations have not been benchmarked against experiments [26]. K. Ohmi's simulations of the present KEKB with 22 mrad crossing angle show tails extending to 30 rms beam sizes in the vertical plane [23]. Without crossing angle (or with crab crossing) the tails are reduced to about $20\sigma$, consistent with Seeman's rule of thumb [1]. For the Super-KEKB upgrade, they will again be similar to the present situation. DAFNE simulations with the lifetrac code show a strong sensitivity of the tails to the working point, easily changing by a factor 2 or 3 in amplitude for tune changes of 0.002 or less. An impressive agreement between simulated and measured optimum working point for maximum luminosity was achieved with a tune precision of about 0.001 [32].

That the tails are sensitive to tune changes of $10^{-3}$, *i.e.*, much smaller than the incoherent tune spread, was observed at many colliders, both with lepton and hadron beams, as is illustrated in Table 2. Tune scans at RHIC parallel to the main diagonal unveiled the different influences of 9th, 13th, 14th, and 17th order resonances [34], causing changes in the background level by a factor of 2 or 3. Surprisingly, the background on the 13th order resonances was better than that on resonances of 17th order [34]. In the Tevatron, two-dimensional scans were conducted, for both protons and antiprotons. Loss rates here vary by a factor 100 depending on the working point [2]. Among the lepton colliders an outstanding example is KEKB, whose automatic tune feedback system continually controls the working point with a precision of a few $10^{-4}$ [23]. At KEKB, the tune-feedback set point depends on the beam current (thus taking care of coherent tune shifts due to impedance and electron-cloud effects) and the target values are different in collision and at injection. The tune feedback uses the signal from non-colliding pilot bunches in either ring, so as to avoid complications from the beam-beam interaction (*e.g.*, strong Landau damping and beam-beam induced tune spread). That it is not always easy to detect the tunes with the necessary accuracy, especially with beams in collision is illustrated by typical Schottky spectra from the Tevatron, which exhibit a multitude of synchrotron sidebands, signals from the other plane or from the opposing beam and a superposition from different bunches, as well as noise 'ghostlines' wandering about in frequency.

The strong sensitivity to the working point hints at the importance of resonances. In general, resonances are of the form $mQ_x + nQ_y + oQ_s = q$, where $m$, $n$, $o$ and $q$ denote integers, and the $Q_i$'s are the betatron and synchrotron tunes. The sum $|m| + |n| + |o|$ is the resonance order $p$. Conversion from transverse phase-space coordinates $x$ and $x'$ to the action-angle variables $I_x$ and $\phi_x$ proceeds, *e.g.*, in the horizontal plane, via $x = \sqrt{2I_x \beta_x} \sin \phi_x$, and $x' = \sqrt{2I_x/\beta_x}(\cos \phi_x - \alpha_x \sin \phi_x)$. Two important parameters characterize the resonance islands: the total island width in the action space $\Delta I_{\text{tot}}$ and the island tune $Q_I$; the latter is the rotation frequency around the island center in units of the revolution frequency. The island tune induced by head-on beam-beam collisions assumes a maximum value at a certain amplitude, that increases with the resonance order. For PEP-II this maximum is reached at about $2\sigma$ for a 4th order resonance and at about $8\sigma$ for a resonance of 16th order [21]. The resonance width increases monotonically with amplitude.

In 1981, J. Tennyson proposed the mechanism of resonance streaming to explain the vertical tails and the beam-beam limit observed in SPEAR [8, 9]. Here an external diffusion or dissipation, *e.g.*, radiation damping or quantum excitation, is enhanced in the vicinity of a resonance. The enhancement is particularly pronounced, if the contour of constant energy intersects the resonance line at a shallow angle $\psi$ in the 2-dimensional action space $(I_x, I_y)$. A small change in energy due to the external diffusion can then translate into a change of the resonance oscillation center that is larger by a factor $1/\sin \psi$. Further denoting the angle between the direction of external diffusion and the energy surface by $\chi$ and the external diffusion coefficient far from the resonance by $D_{\text{ext}}$, the enhanced diffusion coefficient close to the resonance becomes $D = D_{\text{ext}} \sin^2 \chi / \sin^2 \psi$. This mechanism appears to be the dominant transport process in lepton machines, and has surfaced in many tail simulations.

In hadron colliders, tune modulation, *e.g.*, due to power-supply ripple or ground motion, is thought to be responsible for much of the diffusion observed. The tune modulation introduces sideband resonances around the primary resonances. The size of the sidebands depends on the modulation frequency, the modulation amplitude, the size of the primary resonance and the detuning with amplitude. If the modulation frequency is near the island tune, adjacent sidebands overlap and the motion becomes chaotic already for extremely small modulation amplitudes. In the parameter plane spanned by the modulation depth $q$ and the modulation tune $Q_m$, the chaotic region is roughly bordered by the three lines [35] $(q/Q_I)(Q_m/Q_I) = 1/p$, $(q/Q_I)^{1/4}(Q_m/Q_I)^{3/4} = 4/(p\pi)^{1/4}$, and $(q/Q_I)(Q_I/Q_m) = 1/p$, where $p$ is the resonance order. The destruction of resonance islands by tune modulation was verified in the E778 experiment at Fermilab [36]. Studying a simple simulation model of the beam-beam interaction with tune modulation, in particular the chaotic motion near a primary 1-dimensional resonance with weak coupling to a second resonance [37, 38], T. Satogata found an exponential growth in amplitude instead of the diffusive growth predicted by the conventional theory of modulational diffusion [10]. Moreover, normalizing the distance to the coupling resonance as $\alpha \equiv \Delta Q/q$, the simulated growth rate shows steep downward jumps at every integer of $\alpha$, whereas steps are predicted at every second integer only (however, the theoretically predicted integer differs by one unit depending on whether or not the driving resonance is also modulated [10]). More theoretical studies of modulational diffusion are clearly needed.

## DIFFUSION

One of the most important results of the Sp$\bar{\text{p}}$S collider is the necessity of matching the beam sizes of the two colliding beams. If the emittance and intensity of one beam was reduced by scraping, the loss rates of the other beam increased by a factor of 2 [28, 29]. When colliding beams with 20% different beam sizes, the emittance of the larger beam decreased and the emittance of the smaller slightly increased, until the sizes became equal after about 3 hours [28, 29]. These observations indicate fast particle losses at large amplitudes. It is controversial how well the proton tail growth and particle losses in hadron colliders can be parametrized by a diffusion process. While good local fits to a diffusion model have been obtained for collimator experiments at HERA [39], diffusion was found to be inconsistent with various tracking simulations (*e.g.*, survival plot vs. no. of turns [7], or the beam-beam model of S. Peggs & T. Satogata [38]), and also with SPS scraper measurements [7]. A. Gerasimov suggested to employ a generalized 'jump and diffusion' model [7]. The transverse speed of the beam halo was measured at the Sp$\bar{\text{p}}$S collider using collimator retraction by B. Jeanneret and colleagues [40]. At HERA, M. Seidel fitted local loss rates recorded after moving a collimator both into and away from the beam to a diffusion

**TABLE 3.** Transverse diffusion speed at different amplitudes measured in three hadron colliders

| acc. | ampl. | growth [$\mu$m/turn] | growth [$10^{-4}$ $\sigma$/turn] |
|---|---|---|---|
| Sp$\bar{\text{p}}$S [40] | 6, 9 $\sigma$ | 0.13, 0.9 | 1, 7 |
| HERA [39] | 5, 6.5 $\sigma$ | 0.05, 0.25 | 0.6, 3 |
| RHIC [41] | 6, 9$\sigma$ | 0.25, 20 | 2, 200 |

model [39]. Measurements for the inward and outward movements of the HERA collimator resulted in the same fitted diffusion coefficient. R. Fliller III has performed similar studies at RHIC [41]. The measurements in both HERA and RHIC can be described by a diffusion coefficient which steeply increases with transverse action as $B \equiv <\Delta I^2>/\Delta t \approx a J_x^n$ with $a \approx 0.1$ $\mu$m$^{2-n}$s$^{-1}$ and $n \approx 5$. M. Seidel could reproduce the observed diffusion in a simple beam-beam simulation with a random tune drift of amplitude $5 \times 10^{-5}$ (0.1 s correlation time) and a harmonic tune modulation of amplitude $2 \times 10^{-4}$ at 1200 Hz [39]. A few measured data points are compared in Table 3. The transverse speed of escaping particles is nearly the same for all three machines. In the case of HERA and RHIC we estimated the transverse motion per turn as $\Delta x \approx \sqrt{\beta D T_0 / I}$, where $T_0$ is the revolution period, $D$ the diffusion coefficient at action $I$, and $\beta \approx 100$ m.

Tail simulations for the PEP-II B factory revealed a strong effect of a single parasitic collision on either side of the primary collision point [24]. Large tails develop, if the separation is less than 8$\sigma$. Long-range collisions are an even greater concern for the hadron colliders. They perturb the motion at large betatron amplitudes, where particles come close to the opposing beam, thereby causing a 'diffusive' aperture [42], i.e., a sudden steep rise in chaotic diffusion, high background, and poor beam lifetime. This problem increases in importance from the Sp$\bar{\text{p}}$S, over the Tevatron, to the LHC. At the SPS a pretzel scheme separated the two beams over most of the ring, except at three head-on collision points [29]. Most of the 9 long-range encounters had a separation of 6$\sigma$, but the closest was only 3$\sigma$. When the pretzel orbit was reduced by a factor of 2, the background rates increased almost 4 times [28, 29]. At the Tevatron, the number of parasitic encounters is 70. The beams are separated on a helical orbit, created by electrostatic dipoles. The experience is similar to the SPS. The beam lifetime drops and the loss rates increases, if the helix size is decreased from the nominal value [2]. In the LHC, there will be 30 long-range collisions around each of the 4 primary collision points, before the beams are separated into two separate pipes. The total number of long-range encounters is 120. Weak-strong simulations predict the 'diffusive aperture' at about 6$\sigma$ [43]. The diffusion coefficient outside this diffusive aperture can be estimated analytically [44]. A simplified model which combines the diffusion due to intrabeam scattering, gas scattering, and long-range collisions predicts the time evolution of loss rates and the beam profile during a physics store in the LHC [46]. The long-range collisions act as 'dynamic collimators' and deplete the beam halo above 6$\sigma$. An ongoing experiment at the CERN SPS models the combined effect of all LHC long-range beam-beam collisions by a 1 m long longitudinal wire fed with 267 A current [45]. At amplitudes above 8 mm ($\beta \approx 50$ m), the wire should induce a diffusive aperture with a typical transverse diffusion speed well above 1 mm per second. In first experimental tests during the 2002 SPS run, varying the distance between the center of the wire and the beam resulted in a large increase of particle loss and a steep reduction in beam lifetime when the separation decreased below 9$\sigma$ [45]. This is consistent with the simulated effect of changing the crossing angle in the LHC [43].

## COUNTERMEASURES

There are a few basic rules for reducing beam tails, namely the beam sizes should be matched, the collisions should be centered, and the tunes should be optimized. Also the tails are smallest for zero crossing angle. Nonlinear magnetic elements can modify the resonance parameters at large amplitudes [31, 47]. Octupoles were used at VEPP-4/VEPP-2N and at DA$\Phi$NE. The octupoles have two converse effects: they either reduce the tune footprint or they decrease the width of resonances and 'unfold' the detuning with amplitude. Which of these two effects prevails has so far been decided experimentally. Wherever possible, a self-compensation should be introduced, such as a cancellation of long-range beam-beam forces by alternating horizontal and vertical crossing at two interaction points [48, 49], or by an additional cancellation between long-range and near collision effects for superbunches [50]. Lepton tails may be suppressed using quadrupole wigglers [51]. In the future, tails of hadron beams could be cooled by means of optical stochastic cooling or electron cooling.

Tune modulation due to power-supply ripple can be suppressed either by active filters or by a tune modulation feedback on the beam. The second option was demonstrated at HERA in 1996 [52]. A tune monitor detected tune modulation of amplitude $10^{-5}$ in the few 100 Hz frequency range. Additional tune modulations locked to the power supply frequency were introduced, whose phases and amplitudes were adjusted, so as to cancel individual lines in the beam tune spectrum. Cancelling two 300-Hz lines and one 600-Hz line, the proton loss rate in HERA was reduced by 40% [52]. Finally, wires, like the SPS wire mentioned above, can compensate for the effect of long-range collisions [53], or an electron lens may counteract both head-on and long-range beam-beam effects [54]. The wire-based compensation under study for the LHC employs a local correction on either side of each IP, where the beams are already separated. Simulations promise an increase in diffusive aperture by 1 or $2\sigma$ [55]. Understandably, the correction ceases to work close to the opposing beam.

## CONCLUSIONS AND THANKS

Impressive simulations with a high predictive power are available for modern lepton colliders, though a few unsettled discrepancies remain between codes and machines (*e.g.*, concerning the combined effect of gas scattering or coherent bremsstrahlung and beam-beam). For hadrons, the transverse amplitude-dependent diffusion rates in the SppS, HERA, and RHIC are extremely similar. The Tevatron Run-II and the LHC enter a novel regime were long-range collisions are the dominant perturbation. The latter cause fast losses and deplete the beam halo. Examples from LEP and the LHC ion operation have demonstrated the possibility of new and surprising incoherent effects, that in some cases may dominate the halo production. Various methods to manipulate the beam tails are available or under study, for example octupoles, Tevatron electron lens, and the long-range beam-beam compensation using current-carrying wires.

It is a pleasure to thank W. Fischer, R.P. Fliller, A. Drees, and S. Peggs of BNL; T. Sen, X.-L. Zhang, and V. Shiltsev of Fermilab; M. Zobov of INFN; M. Minty, M. Seidel, and F. Willeke of DESY; K. Ohmi and Y. Funakoshi of KEK; H. Burkhardt, J.-P. Koutchouk, J. Jowett, R. Assmann, W. Herr, B. Jeanneret, and F. Schmidt of CERN; M.-P. Zorzano of INTA; Y. Papaphilippou of ESRF; I. Reichel and M. Furman of LBNL; Y. Cai of SLAC; and T. Chen of Teledyne; for kindly providing many helpful informations and material.

## REFERENCES

1. Seeman, J., "Beam-Beam Interaction: Luminosity, Tails and Noise," *Proc. 12th HEACC*, FNAL, 1983.
2. Zhang, X., "Experimental Studies of Beam-Beam Effects in the Tevatron," *Proc. PAC 2003*, Portland, 2003.
3. Burkhardt, H., et al., *PRST-AB*, **3**, 091001 (2000).
4. Cornelis, K., *Proc. 3rd LEP Performance Workshop*, Chamonix, France, January 10–16, 1993, pp. 123–125.
5. Chirikov, B.V., *Physics Reports*, **52**,5, 265–379, 1979.
6. Chao, A.W., Month, M., *NIM*, **121**, 129–138 (1974).
7. Gerasimov, A., FERMILAB-Pub-92/185 (1992).
8. Tennyson, J., *Physica*, **5D**, 123–135 (1982).
9. Tennyson, J., *Beam-Beam Workshop*, Stanford, 1980, pp. 1–20.
10. Chirikov, B.V., et al., *Physica*, **14D**, 289–304 (1985).
11. Reichel, I., Ph.D. thesis, RWTH Aachen, CERN-Thesis-98-017 (1998).
12. Berestetskii, V.B., Lifshitz, E.M., Pitaevski, L.P., *Quantum Electrodynamics*, Pergamon Press, 1982.
13. Kim, E.-S., and Hirata, K., KEK Preprint 97–27 (1997).
14. PEP-II, An Asymmetric B Factory, Conceptual Design Report, LBL-PUB-5379, SLAC-418, pp. 140–142 (1993).
15. Polityko, S.I., and Serbo, V.G., *PRE*, **51**, 3, 2493–2497 (1995), and 'Coherent Bremsstrahlung at pp or pp̄ Colliders,' TPI-MINN-91/51-T (1991).
16. Klein, S., LBL-PUB-45566 (2000).
17. Meier, H., et al., *Phys. Rev. A*, **63**, 032713 (2001).
18. Jowett, J., *Proc. LHC Performance Workshop - Chamonix XII*, CERN-AB-2003-008, 2003, p. 84.
19. Irwin, J., *Proc. ICFA Workshop*, Novosibirsk, 1989, p. 123
20. Shatilov, D., *Part. Acc.*, **52**, 65–93 (1996).
21. Chen, T., et al., *PRE*, **49**, 2323–2330 (1994).
22. Tennyson, J., 'TRS' code, unpublished.
23. Ohmi, K., these proceedings.
24. Chen, T., et al., *Proc. EPAC 96*, Sitges, 1996, p. 1167.
25. Shatilov, D.N., Zholents, A.A., *Proc. IEEE PAC 95*, Dallas, 1995, p. 9671.
26. Seeman, J., private communication, April 2003.
27. Keil, E., et al., CERN/ISR-TH-GE/75-18 (1975).
28. Meddahi, M., Ph.D. thesis, U. de Paris VII, CERN SL/91-30 (BI) (1991).
29. Cornelis, K., *Proc. LHC 99*, Geneva, 1999, pp. 2–5.
30. Willeke, F., "HERA Status and Upgrade Plans," *Proc. IEEE PAC 97*, Vancouver, 1997.
31. Temnykh, A.B., *ICFA Workshop*, Novosibirsk, 1989, p. 5.
32. Biagini, M.E., et al., *Factories'99*, KEK, 1999, p. 181.
33. Burkhardt, H., private communication, 2003.
34. Fischer, W., these proceedings.
35. Chen, T., et al., *ICFA Workshop*, Novosibirsk, 1989, p. 98.
36. Satogata, T., et al., *PRL*, **69**, 1838–1841 (1992).
37. Satogata, T.J., Ph.D. thesis, Northwestern U., FERMILAB-THESIS-1993-52 (1993).
38. Satogata, T.J., Peggs, S., "Hadron Beam-Beam Diffusion in 2.5D," *Proc. LHC 99*, Geneva, 1999, pp. 108–113.
39. Seidel, M., Ph.D. thesis, U. Hamburg, DESY 94–103 (1994).
40. Burnod, L., et al., CERN/SL/90-01 (1990).
41. Fliller, R., "Beam Diffusion Studies at RHIC," *Proc. PAC 2003*, Portland, 2003.
42. Irwin, J., SSC-233 (1989).
43. Papaphilippou, Y., et al., *PRST-AB*, **2**, 104001 (1999).
44. Papaphilippou, Y., et al., *PRSTA-AB*, **5**, 074001 (2002).
45. Koutchouk, J.-P., et al. (2002).
46. Assmann, R., *Proc. EPAC 2002*, Paris, 2002, p. 1326.
47. Zobov, M., Crosstalk between Beam-Beam Effects and Lattice Nonlinearities in DAΦNE, T. Note G-57 (2001).
48. Neuffer, D., Peggs, D., SSC-63 (1986).
49. Herr, W., CERN-SL/90-06-AP (1990).
50. Ruggiero, F., et al., *PRST-AB*, **5**, 061001 (2002).
51. Seeman, J., Cornell CBN 82-32 (1982).
52. Bruening, O.S., Willeke, F., *PRL*, **76**,3719–3722 (1996).
53. Koutchouk, J.-P., CERN LHC-Project-Note-223 (2000).
54. Bishofberger, K., et al., "Tune Shift Compensation using Tevatron Electron Lens," *Proc. PAC 2003*, Portland, 2003.
55. Zimmermann, F., *Proc. Beam-Beam Workshop*, FNAL, and CERN-LHC-Project-Report-502E (2001).

# Dynamic Aperture for Single-Particle Motion: Overview of Theoretical Background, Numerical Predictions and Experimental Results

M. Giovannozzi

*CERN, AB Division, CH 1211 Geneva, Switzerland*

**Abstract.** Higher energies and higher intensities are the necessary conditions for the success of future accelerators. Higher energies need stronger external electromagnetic fields to guide, focus, and accelerate charged particles, while higher intensities result in source of intense self-fields. In both cases, particle motion deviates considerably from a plain linear evolution as described by the classical Hill equation of transverse betatron motion. Particle stability becomes an issue: this problem can be properly tackled using tools from the nonlinear theory of dynamical systems. The concept of dynamic aperture for single-particle motion will be presented underlying links with the fundamental theorems of classical mechanics, such as KAM and Nekhoroshev theorems. Modern numerical techniques to compute the dynamic aperture will be discussed with special emphasis on accuracy analysis. Finally, measurements of particle stability in existing circular accelerators will be reviewed.

## INTRODUCTION

The new generation of high-energy circular machines, such as the CERN LHC [1], will be built using superconducting magnets. The main drawback of superconducting magnets is the presence of intrinsic and unavoidable magnetic field errors, inducing nonlinear effects in the beam dynamics. Therefore, the standard linear theory of betatronic motion [2] is not the best approach to study the new phenomena, and new concepts and tools have to be developed ([3–7] and references therein for an overview of accelerator issues in nonlinear dynamics). The problem of single-particle stability, or equivalently the determination of the dynamic aperture (DA), is the central issue, both from a theoretical and computational point of view (see, e.g. Refs. [8–11]). A number of techniques can be inherited from neighbouring fields (see, e.g. Refs. [12–14]), such as celestial mechanics, which has already tackled similar problems (e.g. the stability of the solar system [15, 16]).

From a conceptual point of view, a well-established definition of dynamic aperture, including error analysis of the proposed numerical approaches is crucial in itself. It becomes even more relevant considering that the problems to be solved, e.g. stability for times comparable with the filling or storage time in a circular machine, are well beyond the capabilities of nowadays computers, and, therefore, clever approximations have to be applied to get a useful answer.

Considerable efforts where devoted to alternative methods, allowing to estimate DA using quantities requiring a reduced computational burden with respect to the direct approach.

Although no final, or optimal, solution was found so far, in the sense that no analytical result is available to compute the DA for a generic dynamical system (a semi-analytical method based on invariant manifolds was shown to be efficient for generic 2D symplectic polynomial maps [17–20], but this cannot be generalised to higher dimensional systems), the progress in the field is certainly remarkable in the recent past.

In this paper the issues related to dynamic aperture computation are discussed, from the very definition to an overview of direct methods [21, 22]. Then, techniques, such as early indicators [22, 23], inverse logarithm interpolation [22, 24], and bounds on invariants [25, 26], are presented as possible alternative solutions to compute and understand the evolution of DA as a function of time. Also, time-dependent effects are mentioned. Finally, the experimental activity related to nonlinear beam dynamics in circular machines is briefly presented [27–33] with a summary of the main results achieved in three selected experiments, i.e. the Fermilab E778 experiment [34, 35], DESY HERA-p [36–40], and CERN Super Proton Synchrotron (SPS) [41–43].

## DYNAMIC APERTURE COMPUTATION

### Definition

Let us consider the phase space volume of the initial conditions that are bounded after $N$ iterations:

$$\int\int\int\int \chi(x,p_x,y,p_y)\,dx\,dp_x\,dy\,dp_y, \quad (1)$$

where $\chi(x,p_x,y,p_y)$ is the characteristic function of the stable domain and $(x,p_x,y,p_y)$ are 4D Courant-Snyder co-ordinates [2]. Since in 4D the invariant curves do not separate different domains of phase space, the concept of last invariant curve, surrounding stable initial conditions is not valid anymore [45, 46]. In principle, the stability domain for a fixed number of iterations could feature holes and very irregular structures. However, numerical simulations seem to indicate [5, 8, 47, 48] that these situations are not typical of weakly nonlinear lattices, where these structures have no practical relevance, as they occupy a negligible fraction of the phase space volume. Therefore, in general, there exists a connected region of initial conditions which are stable for a given number of iterations.

## Direct methods

To exclude the disconnected part of the stability domain in Eq. (1), polar variables $(r_1, \vartheta_1, r_2, \vartheta_2)$ are used, $r_1$ and $r_2$ being the linear invariants. The nonlinear part of the equations of motion adds a coupling between the two planes, the perturbative parameter being the distance to the origin. It is natural to replace $r_1$ and $r_2$ with the polar variables $r\cos\alpha$ and $r\sin\alpha$

$$\begin{cases} x &= r\cos\alpha\cos\vartheta_1 \\ p_x &= r\cos\alpha\sin\vartheta_1 \\ y &= r\sin\alpha\cos\vartheta_2 \\ p_y &= r\sin\alpha\sin\vartheta_2; \end{cases} \quad \begin{array}{l} r \in [0,+\infty[ \\ \vartheta_1,\vartheta_2 \in [0,2\pi[ \\ \alpha \in [0,\pi/2] \end{array} \quad (2)$$

Having fixed $\alpha$, $\vartheta_1$ and $\vartheta_2$, let $r(\alpha,\vartheta_1,\vartheta_2)$ be the last value of $r$ whose orbit is bounded after $N$ iterations. Then, the volume of a connected stability domain is

$$A_{\alpha,\vartheta_1,\vartheta_2} = \frac{1}{8}\int_0^{2\pi}\int_0^{2\pi}\int_0^{\pi/2}[r(\alpha,\vartheta_1,\vartheta_2)]^4\sin 2\alpha\,d\Omega \quad (3)$$

with $d\Omega = d\alpha\,d\vartheta_1\,d\vartheta_2$. Stable islands, not connected to the main stable domain, are disregarded with such a definition (see Fig. 1 for some examples of regular and pathological cases). The DA is defined as the radius of the hypersphere with the same volume as the stability domain:

$$r_{\alpha,\vartheta_1,\vartheta_2} = \left(\frac{2A_{\alpha,\vartheta_1,\vartheta_2}}{\pi^2}\right)^{1/4} \quad (4)$$

The standard approach to evaluate the integral (3) consists in taking $K$ steps in $\alpha$ and $L$ steps in $\vartheta_1$, $\vartheta_2$, and reducing the integral to a sum:

$$r_{\alpha,\vartheta_1,\vartheta_2}^4 = \frac{\pi}{2KL^2}\sum_{\substack{0\le k\le K \\ 1\le l_1,l_2\le L}}[r(\alpha_k,\vartheta_{1l_1},\vartheta_{2l_2})]^4\sin 2\alpha_k. \quad (5)$$

The discretisation condition over the radius $r$ reads

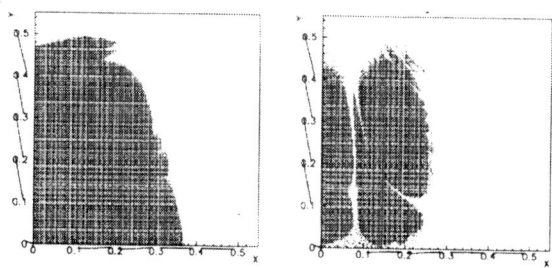

**FIGURE 1.** Example of regular (left) and pathological (right) stability domain for the 4D Hénon map [49] (see Ref. [21] for more details)

$$r(\alpha_k,\vartheta_{1l_1},\vartheta_{2l_2})\frac{I}{R} \in \mathbb{N} \quad (6)$$

where $R$ is the maximum amplitude used in the scan, and $I$ the number of steps in the radial direction.

By direct inspection of Eq. (5) it is clear that the error in the DA computation is optimised by choosing the steps so to produce comparable errors, i.e. $I \propto K \propto L$. Therefore, a DA estimate affected by a relative error of $1/(4I)$ can be obtained by evaluating $I^4$ orbits, i.e. $NI^4$ iterates. The fourth power, due to the phase space dimensionality, makes an accurate estimate of the DA very CPU-time consuming. Nevertheless, some methods to avoid the integration over $\vartheta_1, \vartheta_2$ are available [21]. This can be obtained either by averaging over the angles of the last stable orbit, or by using normal forms [45] to reduce phase space distortions of the last stable orbit.

In addition, a simplified estimate can be given for the case $\vartheta_1 = \vartheta_2 = 0$ and only a scan on $r$ and $\alpha$ is performed [22], namely

$$D(N) = \left(\int_0^{\pi/2}[r(\alpha)]^4\sin 2\alpha\,d\alpha\right)^{1/4}. \quad (7)$$

The advantage of such an approach with respect to the standard technique used in long-term simulation studies (see for instance [8, 9]), where a fixed value of $\alpha$ is considered, is that it provides more accurate results, as $D(N)$ has a smoother dependence on the number of turns allowing to derive interpolating formulae and to extrapolate them to predict long-term particle loss [22, 24]. The estimate of the error $\Delta D$ can be obtained using Gaussian sum in quadrature. An approximated formula reads [22]

$$\Delta D = \sqrt{\frac{(\Delta r)^2}{4}+<|\frac{\partial r}{\partial \alpha}|>^2\frac{(\Delta\alpha)^2}{4}}. \quad (8)$$

In the definitions presented before, the DA is always expressed in terms of an average: this makes it easier to evaluate the associated error and, in addition, the DA estimate is numerically stable. Of course, the minimum DA value is certainly the most relevant quantity for practical applications (see Ref. [7] for a detailed discussion of this point).

## Early indicators

### Lyapunov exponent

The Lyapunov exponent [12, 15] was used as early indicators to predict long-term particle loss with a limited number of turns [23]. The maximal Lyapunov exponent is related to the ratio of divergence of two orbits whose initial conditions are close in phase space. Its estimate after $N$ turns is

$$\lambda(N) = \frac{1}{N} \log \frac{|x^{(N)} - \hat{x}^{(N)}|}{|x^{(0)} - \hat{x}^{(0)}|} \qquad |x^{(0)} - \hat{x}^{(0)}| \ll 1 \quad (9)$$

where $x^{(N)}$ and $\hat{x}^{(N)}$ are the iterates of the initial conditions $x^{(0)}$ and $\hat{x}^{(0)}$ respectively. The distance between the orbits grows linearly with $N$, i.e. $\lambda(N) \to 0$, for regular motion, while it grows exponentially with $N$, i.e. $\lambda(N) > 0 \; \forall \; N$, for chaotic motion. The basic assumption is that particles with regular orbits are stable, those with chaotic orbits are unstable. The threshold is assumed to scale as

$$\sigma_\lambda(N) = \frac{1}{N} \log(nA_\lambda). \quad (10)$$

Such dependence is justified by considering the rate of decay of Lyapunov exponent for stable particles ($A_\lambda$ can be shown to be related to the maximum detuning of regular particles [22]). In Fig. 2 (left) the Lyapunov exponent for a 4D model of the LHC [1] is shown. Stable particles (in white) are rather well separated from unstable ones (in black). The secondary peak in the distribution is an artefact of the algorithm used to compute $\lambda(N)$ [12, 23].

### Tune variation

Another indicator of long-term stability is based on frequency analysis [13, 14, 50–52]. Let $v_x(1:N)$ and $v_y(1:N)$ be the frequencies computed over the first $N$ turns in the phase planes $(x, p_x)$ and $(y, p_y)$ respectively, then the tune difference is

$$\tau(N) = \sqrt{\frac{1}{2} \sum_{i=x,y} [v_i(1:N/2) - v_i(N/2+1:N)]^2}. \quad (11)$$

If the orbit is regular, then $\tau(N) \longrightarrow 0$, otherwise it is bounded away from zero. The threshold can be chosen to scale as

$$\sigma_\tau(N) = \frac{A_\tau}{N} \quad (12)$$

with $A_\tau$ optimised with long-term simulations using the 4D Hénon map [22]. In Fig. 2 (right) the distribution of the tune variation evaluated for the 4D LHC model. Indeed, the distribution of $\tau(N)$ does not feature a sharp drop that could be used for the definition of the threshold, as it was the case for $\lambda(N)$. Furthermore, provided $N$ is large enough, a large fraction of particles has $\tau(N) \approx 0$. Although quite attractive, the DA computation based on

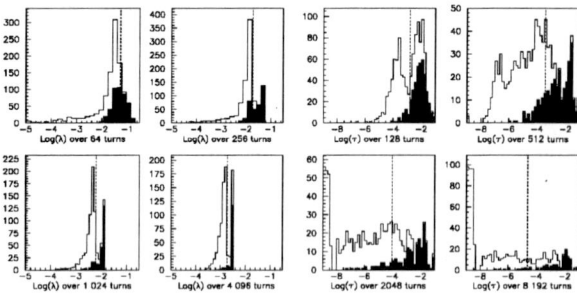

**FIGURE 2.** Distribution of $\lambda(N)$ (left) and $\tau(N)$ (right) for an early model of the LHC. Four turn numbers are used (black area represents initial conditions lost before $10^5$ turns). The dotted lines represent the thresholds (from Ref. [22])

early indicators is affected by a number of difficulties. In the case of Lyapunov exponent the main failures are due to *intermittency* (leading to DA overestimate) and *stable chaos* (leading to DA underestimate, as observed in [38, 43]). In the case of tune estimate, particles locked on a resonance might shown a slow convergence of $\tau(N)$ to zero, thus leading to DA underestimate [22]. This difficulty can be overcome by considering $\tau(N)$ and $\lambda(N)$.

## Inverse logarithm interpolation

A different approach consists in analysing the dependence of $D(N)$ as a function of $N$ (also called survival plots, see Refs. [3, 8]). Thanks to the average over $\alpha$, $D(N)$ can be very well interpolated by the simple law [22] (see Fig. 3 for some examples)

$$D(N) = D_\infty \left(1 + \frac{b}{\log_{10} N}\right). \quad (13)$$

The law (13) deserves some comments. Only two constants are required to fit numerical data, and their physical meaning is transparent, i.e. $D_\infty$ is the DA for infinite times, while $b$ measures the relevance of the long-term losses. Errors on the fitted parameters can be estimated using standard numerical tools [22, 24]. The logarithmic law (13) can be justified in terms of Nekhoroshev theorem [53], and its generalisations to symplectic mappings of arbitrary dimension [54, 55], and KAM theorem [16, 56]. Finally, it allows extrapolation of tracking data with good accuracy (see next Section).

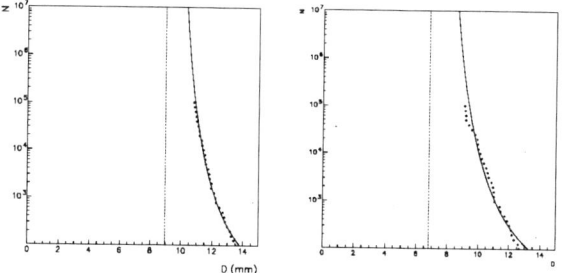

**FIGURE 3.** DA vs. $N$ for the 4D (left) and 6D (right) LHC model: tracking data (dots) and interpolation via Eq. (13) (solid line) (from Ref. [22])

## Numerical results

Detailed analysis of tracking data for realistic 4D model of the LHC [22], are shown in Table 1. The DA estimates are expressed as relative errors with respect to plain tracking at $10^5$ turns. The extrapolation based on Eq. (13) is reliable from 2048 turns onward. The Lyapunov is very pessimistic with respect to plain tracking at $10^5$ turns, but its estimate is consistent with extrapolation of $D_\infty$. Although the predictions with the tune variation are rather pessimistic with respect to tracking at $10^5$ turns, they agree very well with the logarithmic extrapolation for $10^7$ turns. Predictions with only 128 turns already give a clear indication about the relevance of long-term phenomena. Similar conclusions can be drawn from the analysis of 6D data [22].

**TABLE 1.** Relative errors of DA estimates with respect to tracking at $10^5$ turns for the 4D LHC model (from Ref. [22]). A discussion of the errors induced by computer architecture can be found in Ref. [57]

| $N$ | Particle loss | $\lambda(N)$ | $\tau(N)$ | Extrapolation at $10^7$ | at $\infty$ |
|---|---|---|---|---|---|
| 128 | 24% | 4% | -4% | | |
| 512 | 17% | -12% | 0% | 3% | -6% |
| 2048 | 11% | -13% | -4% | -6% | -21% |
| 8192 | 6% | -15% | -6% | -4% | -17% |
| $10^5$ | 0% | | | -5% | -19% |

## Bound on Invariants

The principle is rather straightforward: provided an accurate estimate of the quasi-invariant $J$ in the region of interest for the nonlinear system under consideration is available, and assuming that it is possible to evaluate its maximum variation $\delta J$ over $N_0$ turns (usually performed via Monte Carlo method), then the stability time can be computed as

$$N = \frac{\Delta J}{\delta J} N_0 \qquad (14)$$

where $\Delta J$ is the maximum allowed variation of the quasi-invariant [25, 26]. Although this approach follows from rigorous derivation, its practical implementation could lead to severe underestimate of the stability time, and, what is even more important, its applicability beyond the weakly chaotic region is doubtful.

## Time-dependent effects

Long-term particle losses are drastically enhanced if the betatron tune is modulated by some external causes, such as power supply ripple, or synchro-betatron coupling. The fundamental mechanisms for modulational diffusion have been dealt with in [58, 59]: following this approach, simplified models have been analysed to distinguish between different regimes due to the modulational spectra (see Ref. [60]). Rigorous estimates have been obtained also for the change in adiabatic invariant due to separatrix crossing [61] and these results have been the basis for computing estimates of the diffusion coefficient in limited phase space regions [62, 63].
In this respect, if the underlying mechanism for particle loss is governed by a Fokker-Planck process, which is a questionable hypothesis [64], then the diffusion coefficient has necessarily a strong dependence on both the adiabatic invariants and the local resonances' structure [65].
In Ref. [24], a phenomenological approach was used: the interpolation law (13) was modified by adding a third fit parameter $\kappa$, i.e.

$$D(N) = A + \frac{B}{\log^\kappa N}. \qquad (15)$$

When the modulational amplitude reaches a certain limit, the extrapolation at infinity becomes negative, indicating that all the phase space is unstable. This is in agreement with experiments [41, 43] that show that for large modulations the beam has a finite lifetime. In these cases the exponent $\kappa$ may become negative: a decay of the DA approximately proportional to the logarithm of the number of turns (i.e. $\kappa = -1$) has been observed several years ago in the Superconducting Super Collider simulations. The detailed analysis of numerical stability and accuracy of the proposed fit can be found in Ref. [24], here the numerical results for two different modulation amplitude applied to a 4D modulated Hénon map are shown in Fig. 4 as an example.

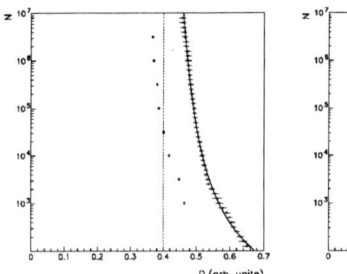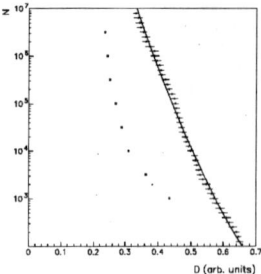

**FIGURE 4.** DA vs. $N$ for the modulated Hénon map (small modulation amplitude - left, large modulation amplitude - right). Tracking data (error bars), interpolation via Eq. (15) (solid line), and extrapolation at infinity (vertical dotted line), prediction through Lyapunov exponent (stars) (from Ref. [24])

## NONLINEAR DYNAMICS EXPERIMENTS

Experimental activities devoted to measuring and understanding nonlinear dynamics effects in circular machines [27] were started in many laboratories (see Refs. [28–44] for an overview, certainly not exhaustive, of results for both hadron and lepton machines). Usually, the main objective consists in measuring single-particle quantities like smear, i.e. the orbit distortion induced by nonlinear forces, detuning with amplitude, islands' parameters, and, in some cases, also DA including its parameteric dependence on relevant physical quantities (e.g. tune ripple parameters). In this paper three experiments at Tevatron (E778, Fermilab), HERA-p ring (DESY), and those carried out at the SPS (CERN) will be considered and summarised.

**Fermilab E778**: the goal consisted in studying the dynamics of charged particles under the influence of nonlinear forces and resonances [34, 35]. To this aim, sextupoles were powered at the injection flat bottom of the Tevatron and the horizontal tune was set to $q_H = 2/5$ to test the behaviour near the fifth-order resonance. Various measurements were attempted, such as smear and detuning with amplitude [34]. A detailed experimental study of islands properties was carried out, with a particular emphasis on the persistence of islands under the influence of artificially generated tune ripple. This point was tested by measuring the coherent signal due to beam trapped inside the islands of the fifth-order resonance as a function of the tune ripple frequency. This allowed probing the complex phase diagram for the beam dynamics under the influence of an external ripple [35]. Beam profile evolution was part of the experimental programme, as well as DA measurements: a reasonable overall agreement with tracking results of about 20 % was found.

**DESY HERA-p**: most experiments were performed at injection energy and after decay of persistent currents to have well defined experimental conditions. No external nonlinear elements were used, the source of nonlinearities being the errors in superconducting magnets. Detuning with amplitude was carefully measured to reconstruct an accurate model of the machine to be used in the numerical simulations. Additionally, two main topics were considered, namely the influence and, possibly, the correction of tune ripple [36], and DA measurements [37–40]. Different techniques were used to measure DA, namely beam profile measurements and beam losses on scrapers. Different regimes were revealed depending on the beam energy: at injection, tune ripple had almost no impact on DA, while at top energy a clear effect was observed. Also, scraper measurements were compatible with a diffusion model only at top energy. The overall agreement between computed and measure DA values is within 20 %.

**CERN SPS**: the experimental programme covered a long period [41–44] during which the objectives evolved from measurements of detuning with amplitude and short-term (few seconds of SPS storage time) DA to long-term (of the order of minutes of storage time) DA, including influence of ripple, such as parametric dependence on ripple frequency and amplitude, as well as two ripple frequencies at a time. In all the cases under considerations, the SPS was operated at 120 GeV to profit from the good reproducibility and linearitiy of the machine at such energy. Six sextupoles were used to generate nonlinear forces without exciting low-order resonances (in particular the third-order resonance).

Short-term DA was found to be in good agreement with numerical simulations, and also with analysitical estimates based on normal forms [66]. As far as long-term DA measurements are concerned, once again a 20 % agreement was found with numerical simulations after a careful tuning and control of the machine parameters. The influence of the ripple parameters on the beam stability was not in quantitative agreement with numerical simulations. A clear signature of the harmful effect of two ripple frequencies was found in the experiment and qualitatively confirmed by the tracking studies [43, 44].

## CONCLUSIONS

Considerable effort was devoted to understanding nonlinear beam dynamics in circular accelerators. In particular, the problem of single-particle beam stability, namely the computation of DA, gained a central role. In this paper, an overview of different approaches, ranging from direct and indirect methods, the latter based either on early indicators, interpolating law, or bound on invariants, were presented. In particular, the issue of accuracy in the DA estimate was considered and results discussed. As far as experimental activities aimed at probing nonlinear effects in beam dynamcs are concerned, a brief review of three experiments performed at Fermilab, DESY, and CERN was presented. Phase space properties (detuning with amplitude, smear, islands' properties) are rather accurately measured and found in good agreement with numerical simulations. As far as the DA is concerned, the definition of the measurement procedure is already a quite delicate point [43]. In all the experiments a qualitative agreement between DA measurements (including also parametric dependence on tune ripple parameters) was found. To reach a quantitative agreement of about 20 % required well-controlled experimental conditions together with extremely accurate numerical simulations.

## ACKNOWLEDGMENTS

Many people contributed to the results presented in this paper. I am particularly indebt with G. Turchetti, E. Todesco and W. Scandale, with whom I worked on the topics of numerical computation of DA. I would like to thank W. Fischer and F. Schmidt with whom I worked on the experiment at the CERN SPS. Finally, I would also like to thank F. Zimmermann for drawing my attention to some references on nonlinear beam dynamics experiments and O. Brüning, M. Martini, E. Todesco for a critical review of the manuscript.

## REFERENCES

1. The LHC Study Group, *CERN–LHC* **95-05** (1995).
2. Courant and H. Snyder, *Ann. Phys.* **3**, (1958) 1.
3. Chao, A., *AIP Conf. Proc.* **230**, (1990) 203.
4. Willeke, F., in *CERN Accelerator School 90-04*, ed. by Turner, S. (CERN, Geneva), 1990, p. 156.
5. Scandale, W., in *EPAC 92*, ed. by Henke, H. (Edition Frontiéres, Gif sur Yvette), 1993, p. 264.
6. Scandale, W., in *CERN Accelerator School 95-06*, ed. by Turner, S. (CERN, Geneva), 1994, p. 109.
7. Todesco, E., *AIP Conf. Proc.* **468**, (1998) 157.
8. Galluccio, F., Schmidt, F., in *EPAC 92*, ed. by Henke, H. (Edition Frontiéres, Gif sur Yvette), 1993, p. 640.
9. Guo, Z., Risselada, T., Scandale W., *AIP Conf. Proc.* **255**, (1992) 50.
10. Böge, M., Schmidt, F., *AIP Conf. Proc.* **405**, (1997) 201.
11. Schmidt, F., Willeke, F., Zimmermann, F., *Part. Accel.* **35**, (1991) 249.
12. Benettin, G., et al., *Meccanica* **15**, (1980) 21.
13. Laskar, J., Froeschlé, C., Celletti, A., *Physica D* **56**, (1992) 253.
14. Laskar, J., *Physica D* **67**, (1992) 257.
15. Hénon, M., Heiles, C., *Astr. J.* **69**, (1964) 73.
16. Siegel, C. L., Moser, J., "Lectures in celestial mechanics", (Springer Verlag, Berlin, 1971).
17. Giovannozzi, M., *Phys. Lett. A* **182**, (1993) 255.
18. Bazzani, A., et al., *Physica D* **64**, (1993) 66.
19. Giovannozzi M., *Phys. Rev. E* **53**, (1996) 6403.
20. Giovannozzi, M., *Cel. Mech.* **68**, (1997) 177.
21. Todesco, E., Giovannozzi, M., *Phys. Rev. E* **53**, (1996) 4067.
22. Giovannozzi, M., Scandale, W., Todesco, E., *Part. Accel.* **56**, (1996) 195.
23. Todesco, E., Giovannozzi, M., Scandale, W., *Part. Accel.* **55**, (1996) 273.
24. Giovannozzi, M., Scandale, W., Todesco, E., *Phys. Rev. E* **57**, (1998) 3432.
25. Warnock, R. L., Ruth, R. D., *Physica D* **56**, (1992) 188.
26. Warnock, R. L., Berg, J. S., *AIP Conf. Proc.* **395**, (1997) 423.
27. Kamada, S., *AIP COnf. Proc.* **344**, (1995) 1.
28. Chen, T., et al., *Phys. Rev. Lett.* **68**, (1992) 33.
29. Zimmermann, F., et al., *SLAC-PUB* **7931** (1998).
30. Kamada, S., et al., *Part. Accel.* **27**, (1990) 221.
31. Tran, P., et al., *SLAC-PUB* **6720** (1995).
32. Lee, S. Y., et al., *AIP Conf. Proc.* **255**, (1992) 370.
33. Wang, Y., et al., *Phys. Rev. E* **49**, (1994) 5697.
34. Chao, A., et al., *Phys. Rev. Lett.* **61**, (1988) 2752.
35. Satogata, T., et al., *Phys. Rev. Lett.* **68**, (1992) 1838.
36. Brüning, O. S., and Willeke, F., *Phys. Rev. Lett.* **76**, (1996) 3719.
37. Zimmermann, F., Willeke, F., *DESY HERA* **91-08** (1991).
38. Brüning, O. S., et al., *DESY HERA* **95-05** (1995).
39. Seidel, M., *DESY-HERA* **93-04** (1993).
40. Brüning, O. S., et al., *DESY-HERA* **94-01** (1994).
41. Brandt, D., et al., in *EPAC 90*, ed. by Marin, P., Mandrillon, P. (Edition Frontières, Gif sur Yvette), 1991, p. 1438.
42. Gareyte, J., Scandale, W., Schmidt, F., *IOP Conference Series* **131**, (1992) 235.
43. Fischer, W., Schmidt, F., *AIP Conf. Proc.* **344**, (1994) 109.
44. Fischer, W., Giovannozzi, M., Schmidt, F., *Phys. Rev. E* **55**, (1997) 3507.
45. Bazzani, A., et al., *CERN* **94-02** (1994).
46. Meiss, J. D., *Rev. Mod. Phys.* **64**, (1992) 795.
47. Zimmermann, F., in *EPAC 94*, ed. by Sueller, V., et al. (World Scientific, Singapore), 1995, p. 327.
48. Galluccio, F., Scandale, W., *CERN SL (AP)* **89-51** (1989).
49. Hénon, M., *Q. Appl. Math.* **27**, (1969) 291.
50. Bartolini, R., et al., *Part. Accel.* **52**, (1996) 147.
51. Bartolini, R., et al., *Part. Accel.* **55**, (1996) 247.
52. Papaphilippou, Y., *CERN-LHC-Project-Report* **299** (1999).
53. Nekhoroshev, N., *Russ. Math. Surv.* **32**, (1977) 1.
54. Bazzani, A., Marmi S., Turchetti, G., *Cel. Mech.* **47**, (1990) 333.
55. Turchetti, G., in *Number theory and physics*, ed. by Luck, J. M., and Moussa, P. (Springer Verlag, Berlin–Heidelberg), 1990, p. 223.
56. Arnol'd, V. I., *Usp. Math. Nauk.* **18**, (1963) 13.
57. Hayes, M., et al., *LHC-PROJECT-NOTE* **309** (2003).
58. Chirikov, B. V., *Phys. Rep.* **52-5**, (1979) 263.
59. Chirikov, B. V., et al., *Physica D* **14**, (1985) 289.
60. Brüning, O. S., *Part. Accel.* **41**, (1993) 133.
61. Neishtadt, A. I., *Sov. J. Plasma Phys.* **12**, (1986) 568.
62. Bazzani, A., Siboni, S., Turchetti, G., *J. Phys. A* **30**, (1997) 27.
63. Bazzani, A., Brini, F., Turchetti, G., *AIP Conf. Proc.* **395**, (1997) 129.
64. Gerasimov, A., *CERN-SL (AP)* **92-30** (1992).
65. Gerasimov, A., *FERMILAB-CONF* **90-250** (1990).
66. Bazzani, A., et al., *Nucl. Instr. & Methods A* **298**, (1990) 102.

# Space Charge Simulation

## Christopher R. Prior

*CCLRC Rutherford Appleton Laboratory,
Chilton, Didcot, Oxfordshire, United Kingdom*

**Abstract.** Based on a recent ICFA mini-workshop held in Oxford, England [1], this paper surveys the computer codes available for simulating the behaviour of charged particle beams under space charge. Modelling tools for both linear and circular accelerating systems are covered. Lists of recent comparisons code v. code and code v. experiment are given and a set of experimental results that might be used to benchmark codes is identified. The Oxford workshop also drew up a detailed spreadsheet of the features of most of the simulation codes in current use, and this is available at [2].

## INTRODUCTION

The revolution in computing technology and the ready availability of fast parallel processors have opened up the path for increasingly detailed and realistic simulations of the behaviour of beams in particle accelerators. The number of macro-particles used for modelling is now typically of the order of $10^6$, compared with only a few thousand ten years ago. Simulations that once were prohibitively slow now routinely run overnight.

Prompted by the need for reliable modelling to complement design work for the coming breed of high intensity proton accelerators, several new codes have recently been developed. With so much invested in the LHC, Tevatron, SNS and J-Parc projects, and ideas for proton drivers for neutrino factories, radioactive ion beams and nuclear waste transmutation under study, it is vital that the underlying theoretical work is comprehensive, accurate and realistic. Such machines reveal new aspects of beam behaviour and include novel techniques in their design. The codes correspondingly need to be able to handle the underlying physics, and advances in recent years now enable such features as halo formation, injection modelling, phase-space painting, resonances, impedances and instabilities to be treated. Optimisation using tracking codes is also now a realistic proposition.

However, in this world of opportunity and promise, caution is advised. Are we sure that our codes are correct? Have we included all the basic physics and do our computer predictions faithfully reproduce what actually happens in a machine? In the case of design work, how can we be sure until the machine is built and running? Are we being consistent in our definitions? Are we correctly interpreting the numbers and pictures that our processors supply?

It was against this background that calls began to be made a few years ago for code comparisons and the development of benchmarking tests. Suggestions were put forward at the Snowmass meeting in Colorado in July 2001, and, following the recent ICFA Beam Dynamics mini-Workshop in Oxford, a renewed attempt is planned in which a larger community will be involved. These tests will be described below, following a survey of the main simulation codes that expect to take part.

## SIMULATION CODES

Simulation codes have traditionally split into two types: those that treat linear systems where a single pass is made through each accelerating element, and circular systems, where repeated passage through periodic elements may permit different modelling techniques.

### Linear Accelerating Systems

Linac codes, such as PARMILA, have been available for a number of years. Quite basic in form in their initial versions with simplified (usually linear) space charge routines, they have recently been subject to some updating, though the underlying structures remain the same.

The main example of a new code that grasps the opportunities for a fast parallel approach offered by linear systems is IMPACT [3], developed by Ryne and co-workers, originally at LANL and now at LBL. This code typically uses $10^6$-$10^7$ simulation particles and runs on the US NERSC computing system. It includes fast map generation capabilities and uses split operator methods so that the Hamiltonian is divided $\mathcal{H} = \mathcal{H}_{ext} + \mathcal{H}_{sc}$, where

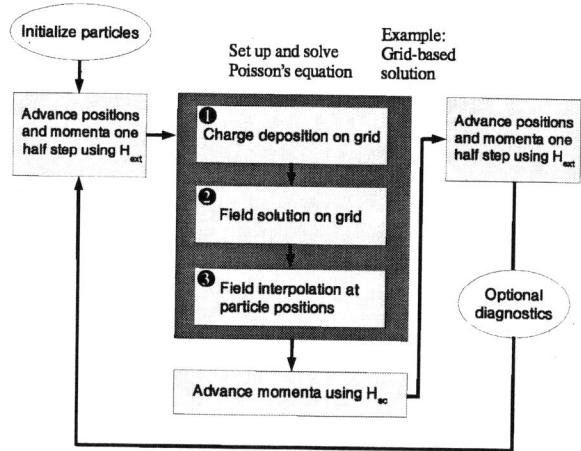

**FIGURE 1.** IMPACT Computational Structure

$\mathcal{H}_{ext}$ corresponds to the magnetic optics and $\mathcal{H}_{sc}$ to the space charge forces. The map for a time step $t$ is then (Figure 1)

$$\mathcal{M} = \mathcal{M}_{ext}\left(\tfrac{1}{2}t\right)\mathcal{M}_{sc}(t)\mathcal{M}_{ext}\left(\tfrac{1}{2}t\right) + \mathrm{O}\left(t^2\right).$$

Space charge is calculated using 3D parallel Poisson solvers and several types of boundary condition are available. The design philosophy is not to take tiny steps to push $\sim 10^7$ particles, but to put the detailed calculation into computing the maps, which are then used to push the particles forward in time.

Systems comprising quadrupoles, dipoles, RF cavities and solenoids can be treated, and 3D constant focussing is included mainly for comparison with analytical predictions. The types of accelerating structures that can be modelled are DTL, CCDTL, CCL and superconducting cavities, and there is also a facility for user-defined elements. Gradient, misalignment and rotation errors can also be taken into account. The code has been used to model the SNS linac, CERN's SPL, LEDA and the J-Parc linac. Studies of the front end of the linac for the European Spallation Source have also been carried out on a system of 16 parallel processors at RAL, where 300 runs with randomly generated errors were easily completed within 3 hrs.

Future developments for IMPACT envisage a new Poisson solver able to handle beams with a range of longitudinal:transverse aspect ratios, a model for high brightness electron bunches and an option for (simultaneous) multiple particle species.

Other codes written mainly for linacs include TRACE/WIN (CEA-Saclay) [4], which is a Windows development of PARMILA with improved space charge and graphics routines. This was used extensively in the CONCERT and ESS studies and works in harness with other codes such as PARTRAN, TRACE3D, TOUTATIS (for RF's) and MONET.

## Circular Accelerating Systems

Given the increasing worldwide use of IMPACT, an extension to cover rings is natural and plans are in place to incorporate its 3D space charge capabilities and RF models into MARYLIE [5]. Based on Lie algebraic ideas using transfer maps, MARYLIE covers both linear beam transport systems and circular storage rings. The aim is for a code with nonlinear symplectic maps for beam line elements, symplectic and non-symplectic tracking, and facilities for optimisation and user-prescribed fitting.

The rings code that appears to be the most versatile is ORBIT [6], developed at ORNL, initially for theoretical studies of the SNS. Written in object oriented $C^{++}$, this can now tackle a wide range of tasks including $H^-$ injection, foil heating, phase space painting, single particle transport through various types of magnets, effects of errors, closed orbit calculations and corrections, longitudinal and transverse impedances, collimation and feedback. Beam transport uses MAD/DIMAD matrices, the Fermilab MXYZPTLK library of differential algebra maps and symplectic "Teapot" style maps. RF cavities are modelled with longitudinal kicks and there is a facility for user-specified harmonics. However plans to develop the code in order to simulate complete cycles of synchrotrons have not yet been implemented.

One aim is to incorporate an electron cloud model to handle the possible e-p instability predicted for many high intensity proton machines under either construction or study. Since existing electron cloud codes tend not to use full lattices in their modelling, it would be an important development if predictions from a full particle tracking code could be achieved.

ORBIT appears to have spawned various offspring. The code has been adopted by Fermilab, where it is being used for Booster studies. Resources have been allocated for in-house development, resulting in incorporation of a Python shell along with other improvements for maintainability and usability. Similarly at Brookhaven, changes to ORBIT feature an interactive Poisson solver for space charge with a conducting or resistive wall boundary condition. One of the more interesting features of this solver is the projection of the beam localized in distance along the closed orbit to a time localized beam distributed over the lattice to perform the space charge calculation. To obtain new maps for acceleration, the parallel BNL-ORBIT can run MAD on a single processor and update the lattice. The space charge solver uses a 3D model, where the particles and longitudinal 2D grid slices are allocated to the processors using a generic algorithm to optimize performance. In application to a 1 MW upgrade of AGS, space charge forces in the longitudinal bunch structure and dynamics have been found to influence the transverse beam dynamics.

At the PSR at Los Alamos, ORBIT has been used in conjunction with experiments on the ring. Good agreement has been found between simulation results and beam profile measurements. In particular, the observation that the beam profile does not depend on the specific painting technique above $3 \times 10^{13}$ protons is understood, and modelling results, showing intensity limitation as the tune is depressed by space charge and the vertical envelope tune approaches the integer 4, have led to a study to produce a compensation scheme.

The Fermilab version of ORBIT has been applied in an attempt to improve Booster performance by comparing simulation with experiment [7]. The motivation for studying the Booster Ring is to obtain the higher intensities necessary for MiniBooNE, NuMI and Tevatron Run 2. Although the linac and main injector can each transmit pulses of $3 \times 10^{13}$ particles, the Booster creates a bottleneck because its losses become unacceptable (> 1 W/m) at intensities above $5 \times 10^{12}$ particles per pulse. The space charge experiment, designed to concentrate on the first 3 ms where losses are about 30%, involved the injection of 11 turns of beam with the RF cavities turned off. During the process, space charge tune shifts of approximately $\sim 0.3$ were reached. Transverse emittances were measured as functions of time and compared with the results from simulations. FNAL-ORBIT modelling used a 2.5D space charge routine with 10000 macroparticles per injection turn (110000 macroparticles total). Tracking was carried out for 2000 turns. Simultaneously, modelling was carried out with a locally written code, SYNERGIA, using a similar $64 \times 64 \times 32$ mesh. The results of the ORBIT calculations showed rapid emittance growth during the 11 turns of injection, followed by a slower growth. The SYNERGIA calculations showed much slower initial growth, followed by a gradual emittance growth to final values not significantly different from those obtained using ORBIT. The emittance measurements by a fluorescent technique were indeterminate, and need refinement. Clearly, at the time of writing, further work is needed on all fronts to resolve the discrepancies.

A rings code of fairly long standing is ACCSIM [8], developed by F. Jones at TRIUMF. This exploits the preexisting codes MAD and DIMAD for lattice preparation and preliminary calculations. Longitudinal space charge forces are calculated from the bunch line density and transverse space charge kicks are calculated from the electric field of the entire ensemble with the local line density as a scaling factor for the force on each particle. The code is therefore effectively 2.5D, though there are plans for a phased upgrade to 3D. The wide range of ACCSIM's utilities is summarised in Figure 2 and a flow chart explaining its integration with MAD/DIMAD is shown in Figure 3.

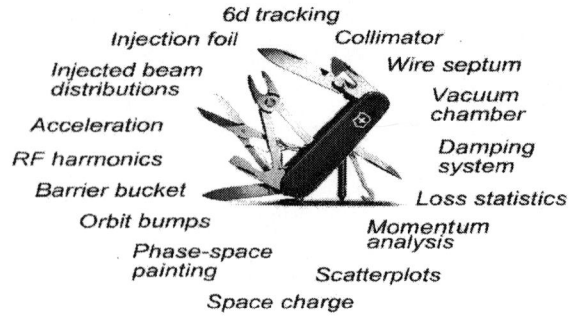

**FIGURE 2.** ACCSIM Capabilities

ACCSIM has been used to model rings such as CERN's PS Booster, the Hitachi medical synchrotron, KEK-PS and the J-Parc 3 GeV ring, generally providing good results (in terms of predictions v. measurements) for rms matching, beam profiles, injection losses and coherent resonance losses. Future study will cover particle redistribution in rms matched beams, space charge resonances, halo formation and synchro-betatron effects from space charge and chromaticity.

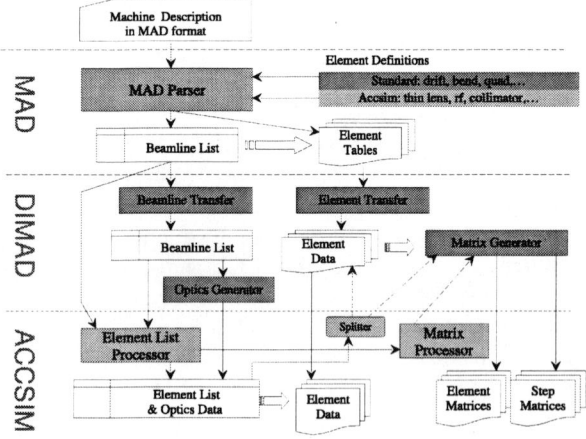

**FIGURE 3.** ACCSIM Data Management

Although in principle very simple, longitudinal (1D) tracking codes can often prove very useful for initial system design. Into this category fall ESME (FNAL), LONG1D (TRIUMF) and Track1D (RAL). From knowledge of only a few global properties of the ring, these codes model bunches by sequentially updating values of phase and energy, with space charge calculated from the longitudinal line density. In recent years, phase space painting, RF manipulations, feedback loops, time-dependent external parameter variation and interactive graphics have been added. ESME has been used to model many machines, including the Fermilab Booster and

LANL-AHF; LONG1D has been used on ISIS, as a basis for the original longitudinal elements of ACCSIM and in the development of tomography codes at CERN; Track1D was at the heart of the ESS injection design, for improvements to the ISIS cycle and as part of the modelling carried out for the present ISIS 240 kW upgrade.

Also at RAL, Track2D [9] is a transverse tracking code for either linear or circular machines with several Poisson solvers based on finite element techniques approach allows irregular shaped boundaries to be treated, both perfectly conducting and lossy. The mesh "breathe" with the beam to ensure good resolution and to handle rapid field changes as particles near the wall. A triangular serendipity scheme is used (up to order 6) but a simple adaptation to a rectangular grid converts the space charge solver to r-z for use in an axisymmetric 2.5D code. The transverse code was one of the first to model multiturn injection under non-linear space charge (in connection with inertia confinement fusion studies) and has been used to model ISIS, CERN-PS and PSB, ESS and the Fermilab proton drivers in Studies I and II.

Extending the triangular mesh to one based on tetrahedra gives a 3D Poisson solver and is built into another RAL code, Track3D. Many techniques developed over the years for data management, efficient storage and computational speed are incorporated into these codes. Track3D is currently undergoing major re-development to convert to a fast parallel processing system. Additional features of Track1D, Track2D and Track3D include the treatment of machine errors, variable charge macro-particles and the ability to track simultaneously particles of more than one species.

Another important code with similar facilities is KEK's SIMPSONS. This has both 2D and 3D options. Although it deals only with circular perfectly conducting boundaries and is applicable to long bunches only, it is able to model acceleration, multiturn injection, painting, apertures, collimation and errors; it also has a variable charge facility. Recently it has been tested at Femilab and compared with FNAL-ORBIT.

## Other Codes

An interesting new code is GPT (General Particle Tracer) [10] developed by Pulsar Physics, partly under contract from TESLA. Working in the time domain, this package is very efficient, with cpu scaling as $N^{1.1}$ ($N$ = number of simulation particles) and is able to handle the difficult problem of space charge calculations for bunch aspect ratios from 0.01 to 100. A large effort has been put into developing a suitable methodology for 3D space charge routines, based on multigrid and pre-conditioned conjugate gradient techniques. Integration is via a 5th order Runge Kutta technique and there are interfaces to other codes such as Superfish. Recent applications of GPT include the design of electron beam optics and radiation yield simulations for the FZR Free Electron Laser, a study of a high-brightness photo-injector, and an energy recovery system for a 12 A, 50-375 keV electron beam. GPT can also be used for designing and optimising collimation systems.

Another code, GenTrackE, is under development by Adelmann (LBL/PSI). Aiming to model large scale machines with complicated 3D geometries, this code has been making full use of the opportunities offered by the US NERSC Seaborg parallel system, showing good performance and results. There are plans to incorporate a full model for electron cloud studies.

At Princeton, a group has been working on a code called BEST (Beam Equilibrium Stability and Transport) [11], which models both linacs and rings in 3D using nonlinear Vlasov-Maxwell equations and the $\delta f$-method. The advantage of this approach is that simulation noise is considerably reduced. The code's main use has been for two-stream instability and beam echo studies, and progress has been made in understanding the beam loss, which is possibly electron cloud related, seen in the LANL-PSR.

Of longer standing is the WARP suite of codes [12] developed at Livermore National Laboratory specifically for the study of space charge dominated ion beams with applications to fusion driver concepts. The package contains 2D ($(x,y)$ and $(r,z)$) and 3D options and uses electrostatic particle-in-cell techniques. There is a wide range of possibilities for specifying the lattice of external fields, including bent beam pipes, which are treated using the "warped" coordinate system from which is derived the code's name. In a parallelised form, it has been used for high resolution simulations of the LBL electrostatic quadrupole injector and multi-lap simulations of emittance growth in a small recirculator experiment.

In a separate category fall codes such as VADOR [13], based on Vlasov techniques for studying the evolution of particle density in phase space. The approach has some advantages, such as effective use in regions of low phase space density. However, while their provision as a tool for cross-checking PIC co it is clear that extension to cover even 2D (transverse) space with reasonable cpu times remains a development for the future.

## CODE COMPARISON

This list of codes, though far from comprehensive, covers the main space charge accelerator modelling tools identified in the Oxford workshop [1]. Questions then arise as to their accuracy and their uses in modelling real ma-

**TABLE 1.** Code v. Code Comparison

| Codes | Test | Result |
| --- | --- | --- |
| ACCSIM, ORBIT, SIMPSONS | KV beam, rms emittance, PSR model | Good |
| ORBIT, ESME | 1D longitudinal | Good |
| ORBIT, SYNERGIA | FNAL booster, multiturn injection, emittance blow-up | Discrepancy |
| Track1D, LONG1D, ACCSIM | 1D long., ISIS, SNS, ESS models | Good |
| Track2D, SIMPSONS, ORBIT | SNS ring modelling; FNAL proton driver injection | Good |
| Micromap, IMPACT | Octupole resonance with space charge | Good |

**TABLE 2.** Code v. Experiment Comparison

| Code v. Machine | Measurement | Result |
| --- | --- | --- |
| ORBIT v. PSR | Profile measurements | Good |
| ESME v. FNAL | 1D longitudinal | Good/fair |
| ESME v. CERN-PS | 1D longitudinal | Good/fair |
| Micromap v. CERN-PS | Montague resonance | In progress |
| ACCSIM v. CERN-PSB | 1D profile | Fair |
| ACCSIM v. KEK-PS | 1D profile | Good |
| IMPACT v. LANL-LEDA | Halo studies | Matched agreement; mismatched discrepancy |
| ORBIT v. FNAL-Booster | Emittance growth from multiturn injection | Inconclusive |
| GPT v. Felix | Emittance, radiation, profiles | Good |
| BEST v. PSR | Electron cloud effect | Fair |
| Track1D v ISIS | Long. profiles, beam loss, injection studies | Good |
| Track2D v. CERN-PSB | Instability, emittance effect | Good |

chines. However, as a prelude to a formal benchmarking programme, it is useful to identify first those comparisons that have been made to date. Table 1 shows instances of simulations where codes have been compared with each other and Table 2 provides examples where codes have been tested against results from experiments.

Apart from the FNAL-ORBIT/Synergia discrepancy mentioned above, there is reasonably good agreement in code v. code comparisons using the same piece of analytical data. However, the agreement is more variable when codes are tested against experiments and this is to be expected as it is almost impossible, without taking very special measures, to cover every aspect of particular machines.

## BENCHMARKING

Two benchmarking tests have been specifically identified based on experiments recently carried out at CERN. The aim of the first exercise was to measure transverse emittance increase due to space charge in the PS as a function of such parameters as tune, bunching factor and bunch intensity. The variations considered were sufficient to cause the beam to cross integer and half-integer resonances under space charge tune depression. Emittance increases up to a factor 3 on a time scale of 10-100 ms were recorded. The results, including all the necessary machine parameters, are published at [14] with the intention that interested parties should attempt to simulate the experiments as a means of testing their modelling codes.

A second benchmarking exercise involves a comparison of simulations and measurements on the CERN PS of emittance exchange in crossing the Montague resonance $2Q_h - 2Q_v = 0$. In tests using the Micromap code (GSI) [15], agreement has been achieved on the level of emittance exchange but not on the width of the stopband, which experimentally is much wider than predicted. Though theoretical work is needed to resolve the discrepancy, this study also provides scope for benchmarking and details are also included at [14].

It is expected that the following codes will take part (names of participants in brackets):

- Micromap (I. Hofmann, G. Franchetti)
- FNAL-ORBIT (W. Chou, J-F. Ostiguy, P. Lucas)
- BEST (H. Qin)
- ACCSIM (F.W. Jones)
- BNL-ORBIT (A. Luccio)
- ORBIT (J. Holmes, S. Cousineau)
- GenTrackE (A. Adelmann)
- IMPACT (R. Ryne, J. Qiang)
- SIMPSONS (D. Johnson, F. Neri)

Based on work using BEST, Hong Qin has also suggested a number of theoretical predictions for benchmarking tests: 1D thermal equilibrium beam profiles, stable beam propagation, and eigenmodes in a space charge dominated beam. Modelling the two-stream instability and beam echo are other possibilities.

Plans for an experimental study of space charge using the ISIS synchrotron at RAL are also under way and the results could provide further material for code comparison. With a tune shift of about -0.4, several resonant lines are crossed during ISIS injection, including the half-integer line (3.5) in the vertical plane and the integer line (4.0) in the horizontal plane, and space charge plays a significant role in the total beam loss. Measurements have been made using a residual gas beam profile monitor and it is expected that parallel simulations with Track2D and Track3D will start soon.

## CODES SPREADSHEET

The Space Charge Simulation workshop [1] concluded by drawing up a spreadsheet containing information about the main codes described above. Available at [2], each code is categorised under language, platform, GUI, whether parallelised or not, the type of transport systems that can be modelled and whether in 1D, 2D or 3D. Details of space charge solvers and their associated boundaries are included, along with the nature of the tracking and whether impedances and field maps are options. Nonlinear modelling relies heavily on graphical output and, particularly if included as an integral part of the package, this can often have a bearing on a code's portability. Such aspects are included in the spreadsheet, in addition to availability of user manuals and standard test cases. The final columns cover special features, limitations and, most importantly, contact details of each code's owner.

## REFERENCES

1. XIIth ICFA Beam Dynamics Mini Workshop on Space Charge Simulation, Trinity College, Oxford, April 2003. A summary is published in the August 2003 edition of the ICFA Beam Dynamics Newsletter.
2. `http://www.isis.rl.ac.uk/AcceleratorTheory/`
3. Qiang, J., Ryne, R., Habib, S., Decyk, V., *An Object-Oriented Parallel Particle-in-Cell Code for Beam Dynamics Simulation in Linear Accelerators*, J. Comp. Phys. 163, 434 (2000)
4. Uriot, D., Pichoff, N., *New Implementation in TRACEWIN/PARTRAN Codes: Integration in External Field Map*, Proceedings of 2003 Particle Accelerator Conference, Portland, Oregon, May 2003
5. `http://www.physics.umd.edu/dsat/dsatmarylie.html`
6. Galambos, J., Danilov, S., Jeon, D., Holmes, J., Olsen, D., Beebe-Wang, J., Luccio, A., *ORBIT - A Ring Injection Code with Space Charge*, Proceedings of the 1999 Particle Accelerator Conference, PAC'99, New York, May 1999.
7. `http://www-bd.fnal.gov/pdriver/booster`
8. Jones, F.W., *A Hybrid Fast-Multipole Technique for Space-Charge Tracking with Halos*, Proceedings of Workshop on Space Charge Physics in High Intensity Hadron Rings, Shelter Island, New York, May 1998.
9. Prior, C.R., *Simulation with Space Charge*, Proceedings of Workshop on Space Charge Physics in High Intensity Hadron Rings, Shelter Island, New York, May 1998.
10. `http://www.pulsar.nl/gpt/`
11. Qin, H., Davidson, R.C., Lee, W.W., Physical Review Special Topics on Acceleration and Beams 3, 084401 (2000); 3, 109901 (2000)
12. `http://hif.lbl.gov/theory/WARP_summary.html`
13. Filbet, F., Sonnendrucker, E., Lemaire, J-L., *Direct Axisymmetric Vlasov Simulations of Space Charge Dominated Beams*, Lecture Notes in Computer Sciences, ICCS 2002, part 3, pp 305-314.
14. `http://www-wnt.gsi.de/ihofmann/Benchmarking/Benchmarking.htm`
15. Franchetti, G., Hofmann, I., Turchetti, G., AIP Conf. Proc. 448, 223 (1998)

# Beam Cleaning in High Power Proton Accelerators

Jie Wei

*Collider-Accelerator Department, Brookhaven National Laboratory, Upton, New York 11973, USA* [1]

**Abstract.** One of the primary concerns in the operation of high-power accelerators is machine component radio-activation caused by uncontrolled beam loss. Beam collimation and halo cleaning play a crucial role in minimizing uncontrolled beam loss. This paper reviews past experience, and discusses design principle, operational strategy, and challenging issues in beam cleaning in high-power accelerators like the Spallation Neutron Source (SNS) and the Japan Proton Accelerator Research Complex (J-PARC).

## INTRODUCTION

During the past decades, the development of accelerator science and technology has sustained exponential growth in both the intensity and power of proton beams. The high proton beam power has extended its use from nuclear and high-energy physics to modern applications including spallation neutron production, kaon factories, nuclear transmutation, energy amplification, neutrino factories and muon collider drivers [1]. Table 1 lists high-intensity applications, including some existing proton and neutron facilities, the two spallation-neutron projects (SNS and J-PARC) presently under construction, and some proposed neutrino-factory proton drivers (NFPD) and nuclear transmutation projects.

The primary concern in designing and operating high-power proton facilities is that machine component radio-activation caused by uncontrolled beam-loss can limit a machine's availability and maintainability. Based on operational experience, hands-on maintenance (1 – 2 mSv/hour at 30 cm from the surface, 4 hours after shutdown) demands an average uncontrolled beam loss not exceeding about 1 Watt of beam power per tunnel-meter [2, 3]. Beam collimation plays a crucial role in localizing beam loss to well-shielded, controlled locations where special practice (e.g. remote handling) can be applied, thus minimizing uncontrolled beam loss and leaving the rest of the facility accessible for hands-on maintenance.

Existing proton synchrotrons have beam losses as high as several tens of percent, mostly occurring when the beam is injected, during its initial capture by the accelerating system, at the start of acceleration ramping, and at the time of transition-energy crossing when the particle motion is non-adiabatic [1]. Experience with beam collimation and halo cleaning is mixed. At the ISIS neutron facility at Rutherfold-Appleton Laboratory, the proton beam is accelerated from 70 to 800 MeV energy at 50 Hz repetition rate to reach 160 kW beam power. Collimation is provided in the transverse and longitudinal planes, with limited two stage systems where possible [4]. The copper and graphite jaws used are adjustable and water cooled. Control of the dominant trapping loss (8%), below ~100 MeV, is crucial for machine operation. About 90% of the 2 kW loss is localized within ~180° in betatron phase from the primary jaw. Uncontrolled losses at high energy are at 0.01% levels.

Among existing accelerators, the lowest fractional beam loss is about $3 \times 10^{-3}$, achieved at a beam intensity of about $3.5 \times 10^{13}$ protons per pulse at the Proton Storage Ring (PSR) accumulator at the Los Alamos National Laboratory [5]. Single-stage beam scraping was attempted but abandoned due to an excessive amount of out-scattering. Pre-injection $H^-$ beam-halo collimation was also attempted but abandoned since beam loss is dominated by foil-scattering of the stored beam. Instead,

**TABLE 1.** Beam parameters of some existing and proposed proton accelerator facilities.

| Machine | $N$ ($10^{13}$) | $f_{rep}$ (Hz) | $E_k$ (GeV) | $P_{ave}$ (MW) | Type |
|---|---|---|---|---|---|
| Existing: | | | | | |
| ISIS (RAL) | 2.5 | 50 | 0.8 | 0.16 | RCS |
| AGS (BNL) | 7 | 0.5 | 24 | 0.13 | RCS |
| PSR (LANL) | 3.5 | 20 | 0.8 | 0.09 | AR |
| MiniBooNE (FNAL)[a] | 0.5 | 7.5 | 8 | 0.05 | RCS |
| NuMI (FNAL) | 3 | 0.5 | 120 | 0.3 | RCS |
| CNGS (CERN) | 4.8 | 0.17 | 400 | 0.5 | RCS |
| In construction: | | | | | |
| SNS | 14 | 60 | 1 | 1.4 | AR |
| J-PARC 3 GeV | 8 | 25 | 3 | 1 | RCS |
| J-PARC 50 GeV | 32 | 0.3 | 50 | 0.75 | RCS |
| Proposed: | | | | | |
| ESS | 46.8 | 50 | 1.334 | 5 | AR (2) |
| AAA (LANL) | – | cw | 1 | 100 | Linac |
| AHF (LANL) | 3 | 0.04 | 50 | 0.003 | RCS |
| PD (FNAL) I | 3 | 15 | 16 | 1.2 | RCS |
| PD (BNL) I | 10 | 2.5 | 24 | 1 | RCS |
| PD/SPL (CERN) | 23 | 50 | 2.2 | 4 | AR (2) |

[a] Including planned improvements.

---

[1] SNS is managed by UT-Battelle, LLC, under contract DE-AC05-00OR22725 for the U.S. Department of Energy. SNS is a partnership of six national laboratories: Argonne, Brookhaven, Jefferson, Lawrence Berkeley, Los Alamos, and Oak Ridge.

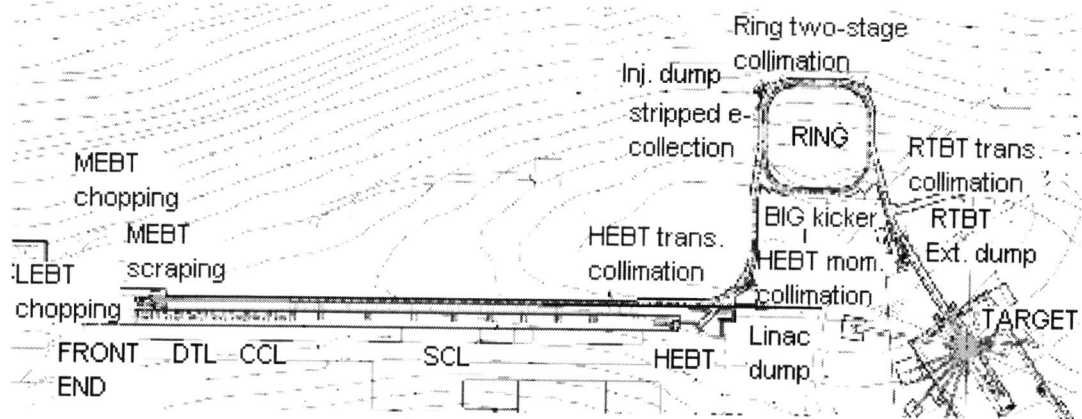

**FIGURE 1.** Layout of beam cleaning devices and beam dumps at the SNS accelerator complex.

measures were made to reduce the beam loss during accumulation at 800 MeV energy (direct $H^-$ injection, control of electron cloud, etc.).

Dedicated collimation experiments were performed at CERN's SPS at a proton energy of 120 GeV to benchmark K2 computer simulation of the collimation process taking into account energy loss, multiple Coulomb scattering, and nuclear elastic and inelastic scatterings [6]. An agreement was found (within a 20% error) on the proton absorption rate at the secondary collector [7]. Similar experiments were also performed at IHEP's U-70 ring at Protvino to validate the two-stage collimation principle at a proton energy of 1.3 GeV[8]. Experience with collimation of TeV beams is described in Ref. [9]. Other computer simulation codes used to design collimation systems include ACCSIM [10], MARS [38], ORBIT [11], UAL [12], and STRUCT [13].

In newly designed high-power accelerators, dedicated beam collimation and halo cleaning is mandatory. In both the 3-GeV and the 50-GeV synchrotrons of the J-PARC project, two-stage systems are used for both the transverse and momentum beam collimation. With the collimation systems' expected high efficiency, the amount of tolerable beam loss is about 1% [14].

Fig. 1 shows the layout of beam cleaning systems of the SNS accelerator complex. Tables 2 and 3 list the expected controlled and uncontrolled beam losses, respectively. The design goal is an average level of uncontrolled beam loss below 1 W/m [15]. The expected collimation efficiency is above 99% for the $H^-$ beam-halo, and above 90% for the proton beam-halo.

## GUIDING PRINCIPLES

In order to reduce machine component radio-activation, beam losses at high energies need to be minimized. For facilities involving a circular accelerator, the beam is usually chopped near the ion source at a low energy to prepare the beam gap for a low-loss extraction from the

**TABLE 2.** Estimated controlled beam loss at the SNS with a 2 MW $H^-$/proton beam.

| Mechanism | Location | Fraction |
|---|---|---|
| Front end: | | |
| gap chopping | LEBT chopper | 0.277 |
| gap chopping | MEBT chopper | 0.043 |
| HEBT: | | |
| $H^0$ from linac | linac dump | $10^{-5}$ |
| linac transverse tail | H/V-collimator | $10^{-3}$ |
| linac energy jitter/spread | L-collimator | $10^{-3}$ |
| Ring: | | |
| beam-in-gap | BIG kicker | $10^{-4}$ |
| excited $H^0$ at foil | collimator | $1.3 \times 10^{-5}$ |
| partial ionization at foil | injection dump | $10^{-2}$ |
| foil miss | injection dump | $10^{-2}$ |
| ring beam-halo | collimator | $1.9 \times 10^{-3}$ |
| energy straggling at foil | collimator | $3 \times 10^{-6}$ |
| RTBT: | | |
| kicker misfiring | collimator | $10^{-5}$ |

ring. The beam is intentionally scraped at a low energy to reduce the beam loss at higher energies. The SNS beam will be chopped twice, first in the Low-Energy Beam Transport (LEBT) at 65 keV with a 25 ns rise/fall time and then in the Medium-Energy Beam Transport (MEBT) at 2.5 MeV with a 10 ns rise/fall time. Transverse scrapers are also designed to clean front-end beam-halo at 2.5 MeV [16]. Low-energy scraping has also been practiced in existing machines: reactor-grade graphite sleeves were fitted in the ISIS drift-tube linac (DTL) [17], and tungsten sleeves were used in the BNL linac to scrape about 5% of the beam at 750 keV energy [18].

Major sources of beam loss need to be identified. Such sources include front-end optical abberations [19]; mismatches across the linac due to changes in accelerating structure, frequency, and focusing strength, and space-charge resonances [20]; physical and momentum aperture limitations; ring-specific items including injection loss, resonances due to space charge and magnetic errors,

**TABLE 3.** Estimated uncontrolled beam loss at the SNS with a 2 MW $H^-$/proton beam.

| Mechanism | Location | Fraction | Power [W/m] |
|---|---|---|---|
| Front end: | | | |
| RFQ transmission | RFQ | 0.2 | 80.7 |
| DTL linac: | | | |
| emit. growth | end of tank 1 | $6.6 \times 10^{-5}$ | 0.24 |
| emit. growth | end of tank 2 | $2.4 \times 10^{-5}$ | 0.18 |
| CCL linac: | | | |
| double $H^-$ stripping | module 1 | $1.9 \times 10^{-5}$ | 0.35 |
| emit. growth | module 1 | $4.7 \times 10^{-5}$ | 0.86 |
| SRF linac: | | | |
| emit. growth | warm sections | $3.7 \times 10^{-6}$ | $\leq 0.2$ |
| emit. growth | supl. 9 periods | $1.2 \times 10^{-5}$ | 0.35 |
| HEBT: | | | |
| $H^-$ magnetic strip. | all HEBT | $1.7 \times 10^{-6}$ | 0.02 |
| collimator out-scatt. | achromat | $7.5 \times 10^{-6}$ | 0.1 |
| Ring: | | | |
| $H^-$ magnetic strip. | inj. dipole | $1.3 \times 10^{-7}$ | 0.3 |
| nucl. scatt. at foil | foil | $3.7 \times 10^{-5}$ | 2.5 |
| collimation ineff. | all ring | $10^{-4}$ | 0.9 |
| RTBT: | | | |
| nucl. scatt. at window | target window | $4 \times 10^{-2}$ | |

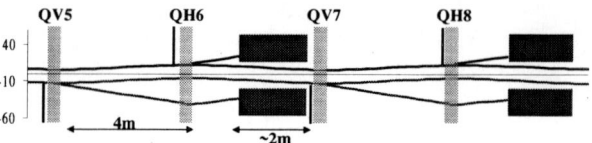

**FIGURE 2.** Layout of $H^-$ beam collimation in the SNS. Top (bottom) half corresponds to the horizontal (vertical) plane.

and collective instabilities [1].

A beam collimation system needs to be designed not only for normal operation but also for fault conditions. For SNS (Fig. 1), the HEBT system protects the ring components from accidental ion-source and linac malfunction. The RTBT system protects the target components from accidental ring extraction misfires.

Two-stage collimation is essential to reduce beam out-scattering from the collimator surface for enhanced collimation efficiency. The primary scraper excites the motion of the particles and increases their impact parameters when they encounter the secondary collectors [21].

Since collimators usually impose the limiting aperture to the beam, their coupling impedance to the beam, especially in the transverse directions, needs to be studied. Steps in vacuum chambers need to be tapered, and sharp bellows need to be avoided to reduce field emission and beam-coupling complications. Inner surface exposed to the beam often needs to be treated to reduce the secondary-emission yield of the electron cloud.

System integrity and maintainability are important design issues. Due to exposure to the high beam power, collimation components need to be resistant to heat, mechanical shock, and prompt radiation. Remote handling is often preferred.

## BEAM CLEANING STRATEGIES

This section discusses the cleaning of $H^-$ beam-halo, proton beam-halo, and the electron cloud in high-power proton facilities.

### $H^-$ beam-halo cleaning

In ring applications, charge-exchange injection is routinely used to accumulate high beam intensity. In order to increase injection efficiency and to reduce radioactivation due to scattering on the injection stripping foil, the pre-injection $H^-$ beam needs to be collimated. The collimation system usually consists of movable stripping foils as scrapers followed by magnetic fields that separate the stripped beam-halo from the primary $H^-$ beam, and then collectors or beam dumps.

Collimation of the transverse phase space requires at least two sets of scrapers/collectors in each direction to minimize the escaping radius for a single-pass collimation. As shown in Fig. 2, SNS uses FODO lattice with 90° betatron phase advance per cell [22]. With two scrapers/collectors in each direction located one cell apart in a non-dispersive region, the escaping radius is $\sqrt{2}$ times the scraper half-aperture.

Collimation in the longitudinal direction requires a high-dispersion location where off-momentum particle's orbit deviation is much larger than the betatron oscillation amplitude, $D \frac{\Delta p}{p}|_{scrape} \gg \sqrt{\beta_{x,y} \varepsilon_{x,y,full}}$, where $D$ is the dispersion, $\beta_{x,y}$ are betatron functions, $\varepsilon_{x,y,full}$ is the full transverse emittance, and $\frac{\Delta p}{p}|_{scrape}$ is the value of momentum scraping. As shown in Fig. 1, a bending achromat is used in SNS to create high dispersion within a localized region. Movable foils are positioned near the maximum-dispersion location to scrape both positive and negative momentum tails. The scraped beam are guided to a shielded collector.

### Proton beam-halo cleaning

Cleaning in the transverse dimensions uses two-stage scraper/collector systems located in a dispersion-free region. The multiturn collimation process requires an adequate acceptance ratio between the collimators (scraper and collectors) and the rest of the vacuum pipe (Table 4). Optimization of the relative phase-advance between the primary scraper and secondary collectors further improves the efficiency of collimation and minimizes beam loss on unprotected elements [23, 24, 32]. As shown in Fig. 3, ideally the secondary collectors are located at phase advances of $\mu_1$ and $\mu_2$ from the primary scraper, where $\mu_1 = \cos^{-1}\left(\frac{A}{A+H}\right)$, $\mu_2 = \pi - \mu_1$, $\sqrt{\beta_s} A$

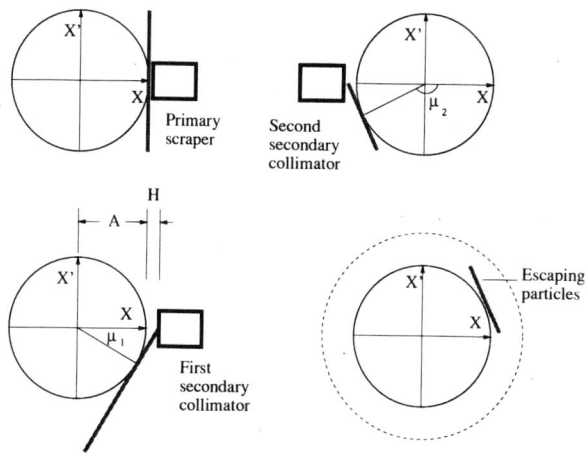
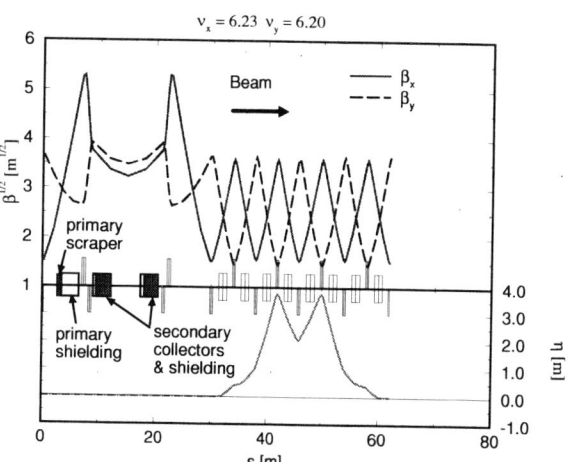

**FIGURE 3.** Schematics in normalized phase space of a two-stage collimation with one scraper and two collectors.

**FIGURE 4.** SNS ring FODO/doublet superperiod containing collimation components. The lattice periodicity is 4.

is the scraping aperture, $\sqrt{\beta_c}(A+H)$ is the collecting aperture, and $\beta_s$ and $\beta_c$ are the beta functions. Actual placement of the collimation elements is restricted by the machine lattice. Detailed simulation is needed considering the actual geometry to minimize activation on nearby magnets. In the SNS ring, the lattice with FODO-cell arcs and doublet-structure straights provides long, uninterrupted section for the collimation system (Fig. 4), significantly increasing the collimation efficiency in comparison with the previous all-FODO lattice [25].

Cleaning in the longitudinal dimension can be accomplished in the momentum and in the azimuthal-phase space. The momentum halo can be cleaned in several ways: (1) Injecting in a high-dispersion region and collecting at 180° phase-advance downstream (ESS) as shown in Fig. 6 [26]; (2) scraping at a high-dispersion lattice location (ISIS, J-PARC Fig. 5) [14, 4]; and, (3) using a beam-in-gap (BIG) kicker (SNS) [27, 28]. The latter two multiturn cleaning methods require an adequate momentum acceptance (Table 5). Azimuthal-phase cleaning, or cleaning of the "beam gap", can be done with a wide-band stripline kicker to resonantly excite coherent betatron-oscillations, driving the residual beam in the gap into the transverse collimators. The BIG cleaning efficiency depends on the kick strength, pulser rise/fall time, beam tune-spread, and resonance kick sequence.

**TABLE 4.** SNS ring transverse acceptances and beam emittance (unnormalized).

| | |
|---|---|
| Full total beam emittance, $\varepsilon_x + \varepsilon_y$ | 240 $\pi\mu$m |
| Ring primary scraper acceptance | 240 ~ 300 $\pi\mu$m |
| Ring secondary collector acceptance | 300 $\pi\mu$m |
| Ring acceptance (w/o collimation) | $\geq$ 480 $\pi\mu$m |
| Extraction channel acceptance | $\geq$ 400 $\pi\mu$m |
| RTBT acceptance (w/o collimation) | $\geq$ 480 $\pi\mu$m |
| RTBT collimator acceptance | 300 $\pi\mu$m |

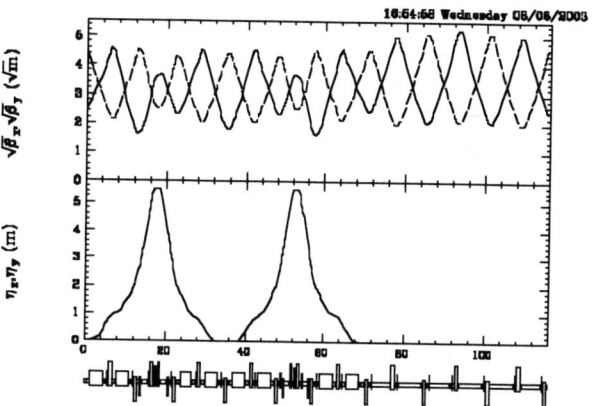

**FIGURE 5.** J-PARC Project 3-GeV ring lattice superperiod (courtesy S. Machida) of FODO structure. The lattice periodicity is 3. The split quadrupole creates a high-dispersion drift for momentum halo scraping and chromatic adjustment.

SNS simulation shows that with a pulser rise/fall time of $+/-30$ ns (10% of the 300 ns gap width) and kicking strength of 0.6 mr, the gap can be cleaned in about 60 turns at about 90% efficiency [28].

To reduce activation at ring extraction, the beam in the gap must be cleaned either during the initial ramping for synchrotrons, or with beam-in-gap kickers for accumulators. Additional collimators are placed in the post-extraction transport to protect experimental and application targets.

**TABLE 5.** SNS ring momentum ($\Delta p/p$) acceptance and beam momentum spread.

| | |
|---|---|
| Full beam momentum spread | $\pm 0.7\%$ |
| RF bucket admittance at 40 kV voltage ($h=1$) | $\pm 1\%$ |
| Ring acceptance for a nominal emittance beam | $\pm 2\%$ |

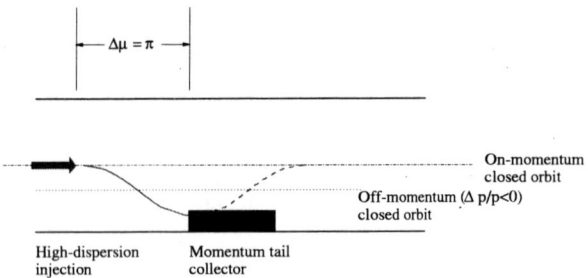

**FIGURE 6.** Collection of injection off-momentum tail by injecting at a high-dispersion region.

## Electron-cloud cleaning

Electrons present in high-intensity machines can cause trailing-edge tune-shift and resonance crossing, electron-proton instability, emittance growth and beam loss, increases in vacuum pressure, heating of the vacuum pipe, and interference with beam diagnostics. We classify electron production into the following categories: (1) electrons generated at the stripping foil in the injection region; (2) electrons generated at the surfaces of collimators and vacuum pipe due to the impact of lost protons; (3) electrons produced around the ring from residual-gas ionization, and, (4) electrons further generated by beam-induced multipacting from the vacuum-pipe wall [29]. Mitigation methods include collection of stripped electrons at injection, surface treatment of the beam vacuum chamber using TiN or NEG coating or beam scrubbing for suppression of the secondary-emission yield, clearing electrodes to suppress electron production, and solenoid windings to suppress multipacting.

Fig. 7 shows the layout of electron cleaning at SNS ring injection. About 2 kW of 545 keV electrons stripped from the 1 GeV $H^-$ beam are guided by a special magnetic field towards the collector made of carbon-carbon material attached to a water-cooled copper plate [30, 31]. Selecting a low-charge-state material for the collector reduces the effects of back-scattered electrons. The inner surfaces are coated with 100 nm thick TiN. A clearing electrode is installed capable of applying 10 kV voltage.

## COLLIMATOR DESIGN & HANDLING

Primary scrapers typically are adjustable, thin blades, the material and thickness of which are optimized for scattering, heating, and other engineering properties. High charge-state materials are preferred to achieve large impact parameters at the secondary collectors with minimum energy losses. The primary scraper of the SNS ring consists of four tantalum blades, each 5 mm thick. They are spaced in 45 degree angles, adjustable to the varying needs of collimation aperture. The scraper assembly is shielded for radiation containment (Fig. 8) [33, 34].

Secondary collectors designed for high-power machines typically are long, complex elements. Unlike col-

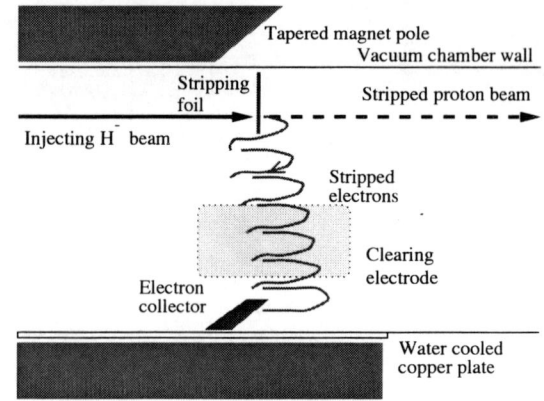

**FIGURE 7.** Collection of stripped electrons during the injection of $H^-$ beam at the SNS ring.

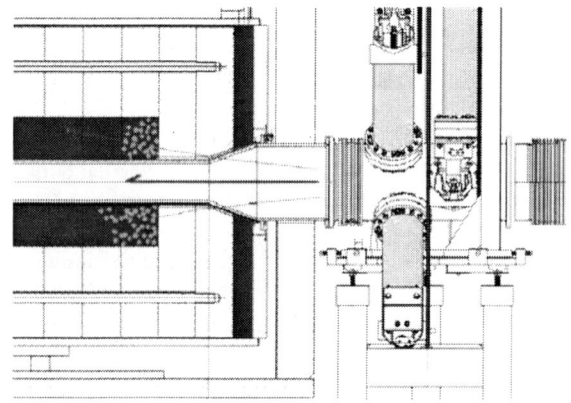

**FIGURE 8.** Schematics of SNS ring primary scraper and shielding (courtesy H. Ludewig and N. Simos).

limators meant for high-energy accelerators that often consist of adjustable solid-metal jaws [9, 35], due to stringent engineering requirements for stress- and heat-tolerance, thermal contraction, and shielding, for operational reliability these units are usually not adjustable in cross-section [33]. Their lengths are designed to stop the primary beam-halo. Anticipating a relatively high beam power (typically 10 kW) and radiation level, they are cooled by closed-loop, de-ionized water. Fig. 9 shows the collector and shielding design for the SNS ring. The vacuum chamber is made of double-layered Inconel-718 with helium gas in the 1 mm gap to detect leaks and copper winding for heat transfer between the two walls [34]. With layers of stainless-steel blocks, stainless-steel particle bed submerged in cooling water, borated aluminum on the outside wall of the collimator vessel, and steel shielding, it contains radiation [33].

Prompt and residual radiation in the accelerator area and nearby air, soil, and ground water activation are important concerns [36, 33]. Beam dump, collimator, and nearby devices must withstand the effects of the beam's high power, power density, and repetition rate.

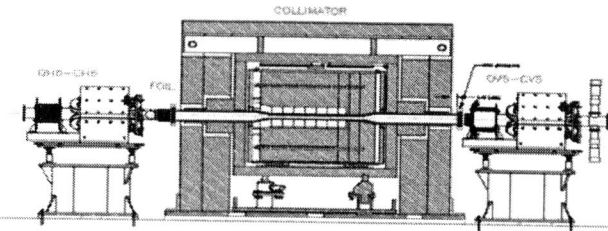

**FIGURE 9.** Schematics of a SNS ring collector showing layers of material for radio-activation containment. The effective length is about 1.5 m. The collimator is designed to withstand an average beam power of up to 10 kW at 1 GeV kinetic energy.

Computer simulations with programs like MCNPX and MARS are often used to determine the transport of the primary and secondary particles through the surrounding material, and to estimate the level of radioactivation and needed protections [37, 38]. Mechanical analyses with computer simulation programs like ANSYS and performance tests of prototypes are essential, including their resistance to thermal-, radiation-, and mechanical-stresses, vacuum out-gassing, vacuum leakage, and long-term fatigue [39].

Maintenance of collimation systems often requires remote handlings. Quick-disconnect vacuum flanges and remote water fittings allow easy access to the systems [40]. For example, the SNS collectors are carefully designed for easy assembly and disassembly (Fig. 9). The tunnel is equipped with a 25-ton crane for transportation. Separate, moveable shielding is further used for personnel protection.

Machine protection is an integral part of the system. The SNS collimation system is designed to withstand at least two full pulses ($\sim$24 kJ per pulse) under fault conditions. Diagnostics devices including fast beam-loss monitors with tens of ns response are linked with machine-protection systems to immediately stop subsequent beam pulses.

## SUMMARY

In high-power proton accelerators, beam-halo cleaning and collimation are key to localizing beam loss for overall machine maintainability. Based on the experience at existing machines, two-stage collimation is essential in achieving a high efficiency. A comprehensive system includes the cleaning in both the transverse and longitudinal directions. Cleaning of electron cloud and suppression of electron generation are also important aspects in the design of new rings. Engineering challenges include the design of devices for robust and yet reliable operations, selection of materials tolerable to high beam power and radio-activation, machine protection under normal and fault conditions, and remote handling and maintenance. These issues are also shared by high-energy accelerators and linear colliders [9, 41].

I thank J. Brodowski, N. Catalan-Lasheras, S. Cousineau, J. Galambos, D. Jeon, H. Ludewig, R. Macek, S. Machida, N. Mokhov, Y. Mori, G. Murdoch, C. Prior, D. Raparia, G. Rees, N. Simos, E. Tanke, J. Tuozzolo, C. Warsop for many useful discussions.

## REFERENCES

1. Wei, J., Rev. Mod. Phys., (in press, 2003)
2. *RF Linear Accelerators*, Wangler, T., Wiley & Sons, New York (1998)
3. *Workshop on Beam Halo and Scraping*, ed. Mokhov, N.V., Chou, W., (FNAL, 1999)
4. Warsop, C., AIP Conf. Proc. 448, p. 104; these proceedings
5. Macek, R., AIP Conf. Proc. 448, (1998) p. 116
6. Jeanneret, J.B., Trenkler, T., CERN SL 94-105 AP (1994)
7. Catalan-Lasheras, N. et al, EPAC (Stockholm 1998) p. 242
8. Catalan-Lasheras, N. et al, EPAC (Vienna 2000) p. 1474; PAC (Chicago, 2001) p. 424
9. Mokhov, N.V., these proceedings
10. *Users' Guide to ACCSIM*, Jones, F.W., TRIUMF Design Notes, TRI-DN-90-17 (1990)
11. *ORBIT User's Manual*, Galambos, J. et al, ORNL (1999)
12. *UAL User Guide*, Malitsky, N., Talman, R., BNL Report 71010 (2002)
13. *STRUCT Program User's Reference Manual*, Drozhdin, A.I. et al., SSCL-MAN-0034 (1994)
14. *Accelerator Technical Design Report for High-Intensity Proton Accelerator Facility Project*, JAERI/KEK (2002)
15. Catalan-Lasheras, N. et al, SNS/AP/7 (ORNL, 2001)
16. D. Jeon et al, Phys. Rev. ST-AB **5**, 094201 (2002)
17. *Spallation Neutron Source: Description of Accelerator and Target*, ed. Boardman, B., RL-82-006 (RAL, 1982)
18. Alessi, J., Raparia, D., private communications
19. Keller, R., et al, EPAC (Paris, 2002) p. 1025
20. Hofmann, I. et al, EPAC (Paris, 2002) p. 74
21. Teng, L.C., Report FN-196/0400 (FNAL, 1969)
22. Raparia, D. et al, SNS/52 (BNL, 1998)
23. Bryant, P.J. et al, CERN Report SL/92-40 (AP)
24. Trenkler, L.C., CERN Report SL/92-50 (EA)
25. Wei, J. et al, Phys. Rev. ST-AB **3**, 080101 (2000)
26. *The European Spallation Source Project Technical Report*, ESS Council (2002)
27. Witkover, R. et al, PAC (New York, 1999) p. 2250
28. Cousineau, S. et al, PAC (Chicago, 2001) p. 1723
29. Wei, J., Macek, R., ECLOUD'02 (CERN, 2002) p. 29
30. Abell, D. et al, EPAC (Vienna 2000) p. 2107
31. Brodowski, J., private communications (2002)
32. Ludewig, H. et al, PAC (New York, 1999) p. 548; Catalan-Lasheras, N. et al, Phys. Rev. ST-AB **4**, 010101 (2001)
33. Ludewig, H. et al, these proceedings
34. Simos, N. et al, these proceedings
35. Kaltchev, D.I. et al, PAC (Vancouver 1997) p. 153
36. Mori, Y., these proceedings
37. *MCNPX Users Manual - Version 2.1.5*, Waters, L.S., ed., TPO-E83-G-UG-X-00001 (LANL, 1999)
38. *The MARS Code System User's Guide*, Mokhov, N.V., FN-628 (FNAL, 1995)
39. ANSYS Engineering Analysis of Systems, USA
40. Murdoch, G. et al, these proceedings
41. Raubenheimer, T., these proceedings

# Working Group I:
# Halo Dynamics

# Summary of the beam dynamics working group

I. Hofmann and A.V. Fedotov

## TOPICS OF DISCUSSION

1. Mechanisms of halo formation
2. Analytic developments and theories
3. Observation from existing machines
4. Measurements

## HOW COMPLETE IS OUR UNDERSTANDING OF HALO?

A good deal of contributions to this working group has explicitly or implicitly addressed the mechanism of halo production, and the diversity of phenomena encountered in simulation work, as well as the few experiments on halo presented at the workshop. In spite of the practical importance of halo, with its impact on machine design and protection as well as diagnostic, the imbalance between the relatively large number of theoretical-numerical studies presented over the years on the hand, and the scarcity of experimental data on the other hand is still striking.

In the joint discussion with the diagnostics working group the need was emphasized to understand better the actual mechanism of halo formation, and to have some kind of "definition of a halo", either by the geometrical characteristics, or by the nature of mechanism. From the discussions during the working group meetings the latter appeared to be the more unambiguous choice. Nonetheless it was suggested as sometimes useful to also bring into the discussion geometrical features of halo, like "whatever exceeds a Gaussian" or "whatever is beyond $3\sigma$".

## Resonant halo

Here, the idea is that particles are pushed out of the core – which may have a uniform or Gaussian profile or anything in between – by a resonantly acting force. This force is in most a periodic or nearly periodic one. In principle, the origin of the force may be from the coherent motion of the beam itself (dipole, envelope, etc.), or from a driving lattice harmonic term; combinations of various contributions are also possible.

1. Coherent motion of beam core.
   The "energy" for driving particles into a halo stems from a periodic core coherent oscillation, which gets damped by "pumping" energy into single particles provided that their oscillation frequencies match the core frequency. Several possibilities exist:

   - *envelope mismatch* (parametric 2:1 halo), as the most well-known case reflected in the experimental papers by T. Wangler and N. Pichoff. Both succeeded to relate the observed rms emittance growth to the imposed initial mismatch strength. In this context it was brought to the attention of the working group that similar halo properties were measured at LBL for heavy ion fusion beam transport already in 1985 by M. Tiefenbach et al. Promising first results of the University of Maryland (UMER) Ring have demonstrated flexibility and unique diagnostics possibilities.

   - *mismatch by dipole offset* (commonly understood as coherent oscillation) was shown for the SNS ring to be a potential candidate of significant halo, which in turn is damping the coherent motion (V. Danilov); more study is needed including full impedances and distributions other than Lorentzian.

   - *error driven envelope mismatch*, where random lattice errors drive the envelope mismatch, although the role of resonant behavior is not quite clear; evidence for halo formation was given by F. Gerigk (CERN sc p-Linac study), while P. Ostroumov (sc Linac for the RIA project) found that such errors can be kept sufficiently low by a proper design.

   - *anisotropy effects*, where the papers by J.-M. Lagniel, J. Qiang and D. Jeon suggest that the expected emittance exchange should not lead to any significant halo.

2. Lattice harmonics as driving source.

   In circular accelerators lattice errors may have a resonant effect on halo formation by driving single particles as well as coherent modes.

   - Measurements at the KEK synchrotron (S. Igarashi et al.) during few ms of foil stripping injection exhibit shoulders, which are identified by simulation as a result of fourth order space charge resonance. S. Machida's talk discusses loss in 3D bunch simulations with yet unclear origin.
   - An experiment at the CERN-PS on a 1.2 s long injection flat-top using a strong octupole has been simulated in 3D bunches over $10^6$ turns (tune modulation by space charge and synchrotron motion causing island trapping) with good agreement (G. Franchetti et al.).
   - In the discussion the role of tune modulation studied earlier (in the 1970's) for beam cleaning was pointed out (see note by A. Chao); islands moving in and out as a result of the tune modulation by whatever reason (like chromaticity in the earlier work) lead to trapping of particles.
   - Related to the CERN octupole experiment coherent modes were shown (in 2D) to be practically damped for Gaussian beams in contrast with KV or waterbag beams (I. Hofmann et al.), which simplifies the picture as one of single particle crossing.

### Non-resonant halo

Several papers discussed halo, which was not related to resonant motion.

- Halo by intrabeam scattering – in practice relevant to rings only – was considered by N. Pichoff as well as by G. Turchetti et al. with respect to numerical noise in linacs.
- Also, halos due to $H^-$-stripping or due to interaction with the residual gas (talk by W. Weng) were addressed.

### Machine specific halos

The working group discussed the following cases of halo for specific machines, which were not part of the above groups.

- For LANSCE, by far the largest measured contribution to emittance growth was reported to be a result of strong rf bunching mismatch.
- In both the SNS and J-PARC MEBT lines "spherical aberration" (octupolar force from space charge or RF leading to S-shaped emittance projections) are believed to cause the measured halo profiles (simulations carried out by D. Jeon and M. Ikegami).

## WHAT DETERMINES THE SIZE OF HALO?

For the parametric 2:1 halo, the $10^{-4}$ level of beam intensity is shifted from the 3 $\sigma$ level of the matched beam to typically $5 - 7\sigma$ for a mismatch as large as 30%. F. Gerigk suggested a similar number for lattice error driven halo, but more work needs to be done in this context.

In the KEK experiment the effect was mainly found in the development of shoulders caused by resonance islands pushed away from the core during stacking – in principle, no upper bound to the halo size.

In the CERN PS-experiment the halo size is also predicted to be unbounded for working points sufficiently close to the resonance condition, which is confirmed by the observed beam loss.

## DOMINANT MECHANISMS OF HALO GENERATION

In linacs, the following conclusions can be attempted:

- Initial mismatch from poor matching between sections - in principle it can be kept low if sufficient diagnostics is available.
- Are random errors (gradients, misalignment etc.) large enough to justify worries about a sizeable halo? Clearly more systematic work is needed here, which should also address sufficiently large error sets.

For rings, the picture is somewhat different:

- Initial mismatch halo (parametric resonance) may be observable for poorly matched bunch transfer, but it should be possible to minimize this effect.
- The effect of nonlinear resonances is suggested to depend on the time scale (1 ms vs. 100-1000 ms); experiments and simulations should be extended to a larger range of parameters. Systematic comparison of codes and experiments is needed.

In summary, for lattice error driven halos in linacs as well as for nonlinear resonance driven halos in rings work is just at the beginning. Significantly more efforts are needed in both experiments and interpreting simulations.

# CEA Studies on Halo Formation

N. Pichoff[1], P-Y. Beauvais[2], R. Duperrier[2], G. Haouat[1], J-M Lagniel[1], D. Uriot[2],

[1] *CEA Bruyères-le-Châtel, BP 12, 91680 Bruyères-le-Châtel, France.*
[2] *CEA Saclay, 91191 Gif sur Yvette cedex, France,*

**Abstract.** Beginning with the TRISPAL project, halo formation has been extensively studied at CEA last 10 years. Effect of mismatching, non-linear forces, resonances, longitudinal-transverse coupling, intrabeam scattering, and interaction with the residual gas have been explored. They have been studied theoretically from both analytical models and dedicated simulation codes and, for some of them, experimentally from proton beam profile measurements over a high dynamic range in a 26 periods FODO channel. Our knowledge, strongly improved through collaborations with our worldwide colleagues, has been applied to the design of several linac projects, whose last are SPIRAL2 and RX2. The goal of this presentation is to summarise the contribution of the CEA teams to the understanding of the halo formation.

## INTRODUCTION

About 10 years ago, a great interest for Halo production in high intensity proton linac arises at CEA with the TRISPAL Project aiming to produce Tritium using a 40 mA, 600 MeV, cw proton accelerator. At that time, a large program of halo studies (experimental as well as theoretical) began. At the end of the TRISPAL project (1998), the interest was kept alive with the arrival of new projects where CEA was involved : ASH, aiming to produce energy and destroy nuclear waste by coupling a reactor with an accelerator, CONCERT, proposing a multipurpose installation, ESS, the European Spallation Source, and the last one, SPIRAL2, a cw Deuteron-Heavy ions linac for nuclear physics studies. For all these projects, beam loss control is crucial to avoid large activation of the machine. Many suspected sources of halo have then been studied theoretically and some even experimentally. A short summary of these studies with relevant references is presented here.

### Experimental Setup

Two machines have been used for diagnostics development and experimental measurements.

◆ **ELSA** is an electron RF accelerator located at Bruyères-le-châtel. It was essentially used to develop profile measurement diagnostics with a high dynamics range. Two methods based on RTO or scintillating screens observed by an intensified camera were used [1]. They allowed to reach beam profile measurement over up to 7 decades.

Because of the specificity of the tails of a beam created in a RF photo-injector it was difficult to extrapolate the measurements made on ELSA to long linacs.

◆ The **FODO experiment** took place in Saclay on the former Saturne DTL injector [2]. It is a 52-quadrupole FODO channel. The beam profiles as well as beam transverse phase-space distribution have been measured in front of and behind the channel. Unfortunately, no diagnostic was possible in the channel. A strong effort was made on the measurements of initial beam characteristics and on the matching of the beam (comparison between simulations and experiments). Some experimental results are presented here.

## THEORETICAL STUDIES

### Equilibrium

Before "exciting" the beam, we spent some time studying the equilibrium for transverse distribution in a focusing channel. We have learned that a beam tries always to tend to an equilibrium in a infinitely long channel. There is infinite number of equilibriums. They are characterised by a distribution function which is only a function of the motion Hamiltonian. We have proven that, whatever the equilibrium, the profile of a space-charge dominated beam tends to the profile of a opposite charges distribution whose space-charge would produce the focusing force (for example, a homogenous distribution producing a linear space-charge force ⇨ the beam profile tends to an homogenous profile in linear forces). In these conditions, particles trajectories in phase-space looks like rectangles (whatever the external force) instead of ellipses in linear external force, for example.

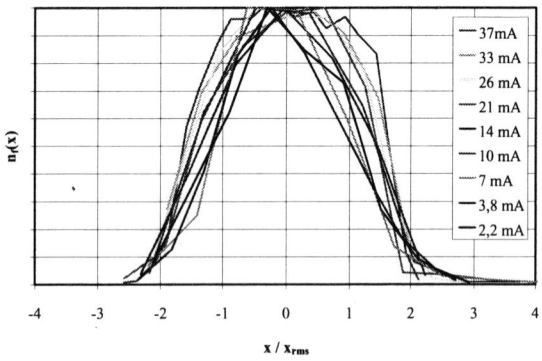

**Figure 1** : Experimental normalised beam profiles for different currents

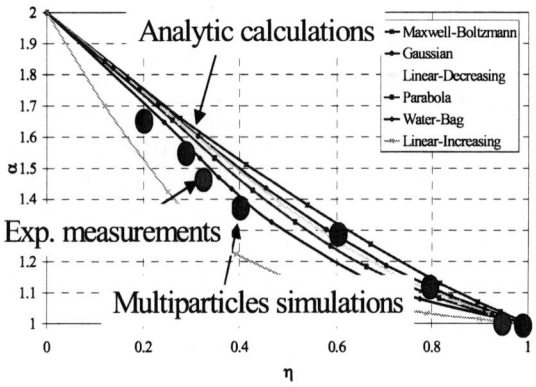

**Figure 2** : Comparison between experience, analytic model and simulations of the beam "shape parameter" in phase-space as a function of tune depression.

Experimental measurements of beam transverse profile at the channel exit (Figure 1) and phase-space distribution (Figure 2) for different tune-depression confirmed the theoretical models. The phase-space "shape parameter" α used in Figure 2 and defined in [3] is 1 for tune depression $\eta = 1$ and goes to 2 (same amplitude for both particles) when $\eta \to 0$.

## Mismatch

Mismatch is known as the main source of halo observed in the PIC simulations. It has been studied by many accelerator physicists worldwide. This halo is the result of the contribution of 2 phenomenon :
- a filamentation due to non linear forces,
- a parametric resonance of particle trajectories with the core oscillation.

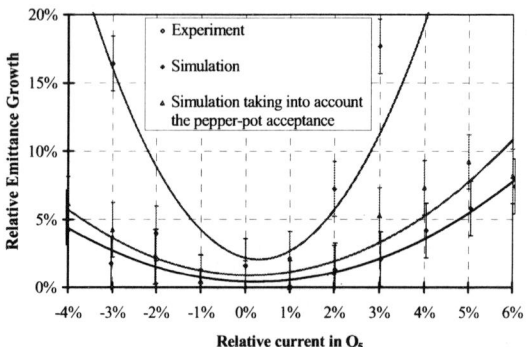

**Figure 3** : Emittance growth through the FODO channel by mismatching the beam with an upstream quadrupole

Using Poincaré plots, we have shown that the half integer resonance between particles and core motions is always excited when the beam is mismatched [4]. This resonance is responsible of large amplitude particle oscillations. As many resonances can be excited, chaotic motion is observed in phase-space when these resonances are overlapping. Depending on the working point (coupling resonances can be also excited), the impact of the mismatch is more or less important. To find safe working point, we had to calculate the 3 mismatched mode frequencies of bunched beam in a continuous focusing channel with no approximation in the case where the forces along x and y are the same [5]. Measurements of emittance growth in the FODO channel as a function of the beam mismatch have been compared to simulation (Figure 3). By taking into account the acceptance of the emittance measurement, the agreement is quite good.

## Coupling resonances

In the CEA-CNRS-INFN collaboration for the ASH design (1999) an emittance exchange has been observed when the longitudinal and transverse phase advances were the same. Following I. Hofmann advices, we have studied the effect of the coupling resonance in these conditions with the help of S. Nath from LANL. Some results of these studies are presented in this workshop in a dedicated talk [6].

We have seen that for reasonable average tune depression $\eta$ the time needed for a maximum emittance exchange depends linearly on a space charge factor $(1/(1-\eta))$. The resonance width seems to be almost insensitive to the initial beam distribution. Emittance exchange is not necessary accompanied with halo production.

## Intrabeam scattering

PIC codes usually do not simulate properly intrabeam scattering. In the Vlasov equation dealing with the motion of particles without collision, the collision process must be modelled by adding a collision term. The denser the beam, the lower the two-body collision influences on the beam dynamics. Space-charge routines of PIC codes generally smooth the space-charge forces. This smoothing suppresses or reduces the low range collisions in the beam compared to a simple particle-particle interaction (PPI) routine. Nevertheless, the remaining collision process in the PIC code is generally far from reality.

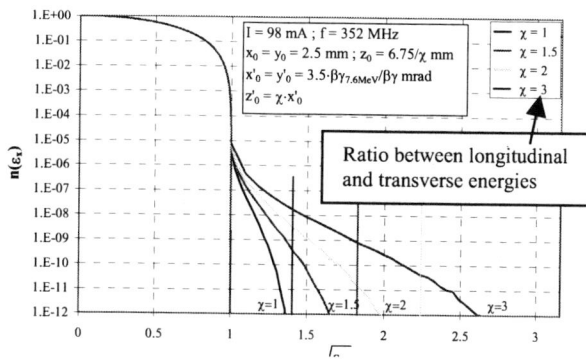

**Figure 4**: Beam profile and tails from intrabeam scattering at the end of a 1GeV linac

We have developed a model to estimate the contribution of the two-body intrabeam scattering to halo formation. The extend of the induced halo depends on the beam equipartionning factor (ratio between longitudinal and transverse average energies), but its level is low ($<10^{-8}$ m$^{-1}$) for 100 mA proton beams [7], (Figure 4).

## Interaction with residual gas

The beam interaction with residual gas is usually not taken into account in PIC simulations. Many kinds of interactions can happen. For the moment, we have studied only 3 of them:

◆ The **stripping of H$^-$ ions**, electron exchange of H$^-$ ions with residual gas atoms, is probably the major source of losses in linacs like SNS or ESS in the normal conducting sections (even if it is not really a source of halo). The loss rate knowing the residual gas pressure and composition, or the vacuum needed to keep beam losses lower than a given limit can easily be estimated from the reaction cross section. For ESS linac losses around 1W/m have been estimated in the end of the room-temperature linac [8] (Figure 5).

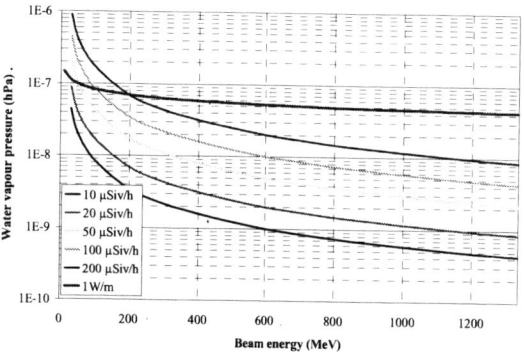

**Figure 5**: H$_2$O maximum pressure needed to stay below a given rate of H$^-$ stripping beam losses

◆ The **space-charge compensation** plays an important role in the beam transport at low energy. The beam ionises a part of the residual gas, trapping species with opposite charge and expulsing those with same sign. The average space-charge force acting on the beam is then reduced. This phenomenon, happening mainly at low energy, should be completely understood to predict the correct matching of the beam to the linac. Its time evolution should be modelled and, why not, compensated to avoid strong mismatch of a pulsed beam front end. A 10 µs time constant is actually small, but represents 1% of a 1ms pulse, much more than what one can afford to lose! A PhD student, A. Ben Ismaïl, is now working on this subject at Saclay [9].

◆ The **elastic scattering** of the beam on the residual gas can give large angles to some particles. Once scattered, the particles can either be lost after a fourth of betatron period (called direct loss) or populate the beam halo. The fraction of direct losses and of halo particle can be estimated from the scattering cross section [10]. In a Nitrogen 10$^{-6}$ hPa gas pressure, a beam with 0.2 π.mm.mrad normalised transverse emittance and an average rms size of 3 mm would accumulate a 1.3 10$^{-6}$ beam fraction out of 4 RMS size from 5 MeV to 1 GeV in a 340 m linac. This is not a lot (only 13 W from a 10 MW beam at 1 GeV) but can contribute to large local losses in case of errors or reduction of the acceptance.

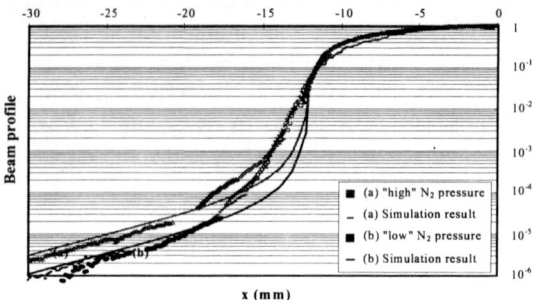

**Figure 6**: Experimental and simulated beam profile and tails at the end of a drift for 2 Nitrogen gas pressures

## Error studies

Errors on quadrupoles, and on cavities can cause localised losses. These errors are responsible of beam c.o.g motion (transverse and longitudinal) and mismatch. As these are statistical errors, a statistical treatment is thus the most appropriate. Having chosen the amplitude and the distribution law of each error, beam transport is simulated through many (1000, for example) linacs with different sets of errors. The different results are used to define the probability to find a particle at a given position.

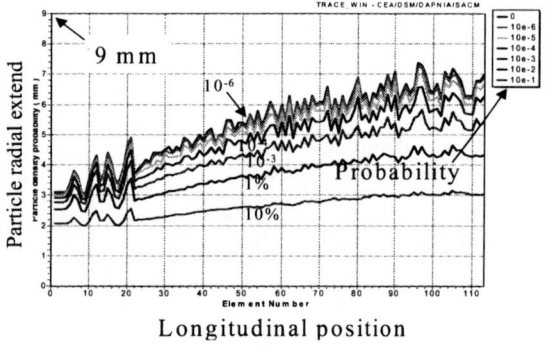

**Figure 7**: Probability to find a particle to above given radial extend in the IPHI DTL including errors.

At the same time, a correction strategy with diagnostics and corrections can be included in the calculation. Some effects can be compensated by the correction scheme (misalignment, ...), some cannot (fast vibrations, ...). These error studies are very design-dependent and must be done once the design is done, at the end of a design process. They have been done in Saclay for IPHI, IFMIF, ESS and SPIRAL2 projects [11].

## Conclusion

A small CEA team has worked for 10 years on halo formation. Its work has always been supported by projects of high intensity proton or H- beam. A large effort has been put on effects that are not simulated by PIC codes. In the same time, codes like TraceWIN (envelope code), PARTRAN and TOUTATIS (PIC codes) have been developed and used to design and simulate the linacs. Our strategy is to use the increasing power of computers to put more linac physics in our codes. Our participation to strong collaborations has been a great help to understand the physics of these kinds of accelerators. We would like to thank our colleagues from LANL, ORNL, RAL, INFN, CERN, FZJ as well as physicists of other labs we met in the conferences, workshops and through their books (M. Reiser, R. Gluckstern, I. Hoffman, R. Ryne, Ji Qiang, K. Crandall, and many others ...).

## References

[1] G. Haouat et al., *Halo of a high brightness electron beam*, PAC95.

[2] N. Pichoff et al., *Measurement of space-charge dynamics effects in a FODO channel*, EPAC98.

[3] N. Pichoff et al., *Transverse profile equilibrium in a space-charge dominated beam*, APAC98.

[4] J-M. Lagniel, *On Halo formation from space-charge dominated beams*, NIMA, 345, 1994.

[5] N. Pichoff, *Envelope Modes of a Mismatched Bunched Beam*, internal note, DAPNIA/SEA 98/44.[6] J-M. Lagniel et al., *Equipartition, emittance and halo exchanges*, this workshop.

[7] N. Pichoff, *Intrabeam scattering on Halo formation*, PAC99.

[8] N. Pichoff et al., *Beam losses from H- stripping on residual gas*, ESSLIN-TN-0202-01.

[9] A. Ben Ismail., *Compensation de charge d'espace dans les sources de haute intensité*, JA-SFP, 2003.

[10] N. Pichoff et al., *Halo from Coulomb Scattering of beam particles on residual gas*, Particles Accelerators, vol.63, 2000.

[11] D. Uriot, *Le DTL du projet IPHI*, DSM/DAPNIA/SEA 99/20, Internal note.

# Sources of Beam Halo Formation In Heavy-Ion Superconducting Linac And Development Of Halo Cleaning Methods[†]

P.N. Ostroumov

*Physics Division, Argonne National Laboratory, 9700 S. Cass Avenue, Argonne, IL, 60439*

**Abstract.** The proposed Rare Isotope Accelerator (RIA) Facility, an innovative exotic-beam facility for production of high-quality energetic beams of short-lived isotopes, will contain two superconducting linacs. To produce sufficient intensities of secondary beams the driver linac will provide 400 kW accelerated beams of any ion from hydrogen to uranium. A detailed design has been developed for the focusing-accelerating lattice of the RIA driver linac which is configured as an array of short SC cavities, each with independently controllable rf phase. To obtain high-power heavy-ion beams the driver linac uses simultaneous acceleration of multiple charge states and two strippers. End-to-end beam dynamics simulations in six-dimensional phase space were applied to study all possible sources of beam halo formation in the driver linac. The concept of a "beam-loss-free" linac is developed and implies beam halo collimation in designated areas.

## DRIVER LINAC LAYOUT

The RIA is being considered as a major nuclear science facility for the near future. A cw SC 1.4 GV driver linac and 123 MV post-accelerator are being designed for the RIA Facility. A conceptual design of the driver linac has been developed [1,2], the major elements of which are shown in Fig. 1. Except for the injector radio frequency quadrupole (RFQ), the entire linac is based on SC accelerating structures. The "baseline" driver linac design consists of ~400 SC cavities of 9 different types. The majority of the cavities (98%) falls into 7 different types. Recently the baseline design of the driver linac has been modified: 1) peak surface electric field in all drift-tube SC resonators is assumed to be 20 MV/m except the first seven 4-gap quarter wave resonators; 2) the high-β section of the driver linac contains two types of triple-spoke resonators [3] instead of three types of elliptical resonators operating at a peak electric field 27.5 MV/m. This option reduces total number of cavities to ~350 with fewer types of the SC resonators.

The driver linac can be tuned to provide a uranium beam at an energy of 400 MeV/u and can be re-tuned to provide a proton beam at 900 MeV. To obtain 400 MeV/u uranium beams the driver linac uses two strippers. The low-β section which is the linac part upstream of the first stripper can accept simultaneously two charge states of heaviest ions and accelerate up to ~10 MeV/u. The medium-β and high-β sections are designed for acceleration of multi-q beams. After each stripper, there is a magnetic transport system (MTS) [4] which provides six-dimensional matching of multiple-q beams into the following accelerating structure. In addition, the MTS is designed to separate and dump the low-intensity unwanted charge states in order to avoid beam losses in the high-energy section of the driver linac.

The driver linac is a high intensity machine and beam losses in the high-energy section must be kept below $10^{-4}$. Acceleration of multi-q uranium beams places stringent requirements on the linac design. Any other lighter ion beam with much smaller emittances can be accelerated with no losses. Recently we have

---

[†] Work supported by the U. S. Department of Energy under contract W-31-109-ENG-38.

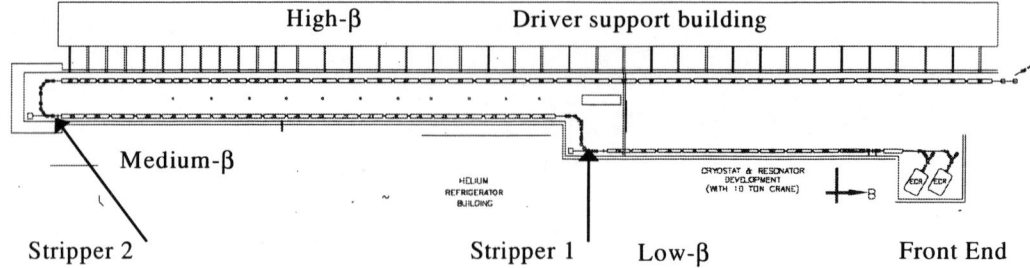

**FIGURE 1.** Elements of the proposed RIA driver linac.

performed detailed beam dynamics studies with the goal of optimization of the linac structure in order to reduce possible emittance growth of multi-q uranium beams [5,6]. End-to-end beam dynamics simulations in six-dimensional phase space have been applied to study all possible sources of beam halo formation and possible beam loss in the driver linac. Major contributors to the effective emittance growth were identified as: a) multiplicity of charge states in the accelerator based on resonator groups operating on higher harmonics of the fundamental frequency, b) passage through the stripping foils, and c) random errors of rf fields and misalignments of accelerator elements. Due to the cw mode of linac operation space charge is negligible after the low-energy beam transport (LEBT) and there is no source of uncontrolled beam halo formation in the high-energy section of the linac. There are several other minor "single-particle" mechanisms that can develop beam halo in the driver linac: higher-order distortions in the LEBT, coupling of r-z motion in the RFQ, mismatch of multi-q beam, nonlinear motion in the longitudinal phase space, higher-order distortions in the beam transport systems due to the charge spread $\Delta q/q$, effect of the dipole component of magnetic field in some types of SC resonators, effect of the quadrupole component of defocusing electric field in some types of SC resonators.

## STRIPPER SECTION

Passage of a heavy ion beam through the stripping film or foil results in a decrease of the average energy due to the ionization losses and both transverse and longitudinal emittance growth due to scattering and energy straggling. The most significant emittance growth and halo formation in the driver linac is associated with the second stripper. There are no experimental data of detailed particle energy and angle distributions after the passage of ~85 MeV/u uranium beam through a stripping foil. We have used the code SRIM [7] for the Monte Carlo simulation of the transport of incident monochromatic beam containing $10^6$ uranium ions through a stripping foil. Basic results of the simulations are summarized in Table I. Beam intensity distribution as a function of particle energy at the exit of the second stripper is shown in Fig. 2. The energy acceptance of the high-$\beta$ section of the linac is $\pm 0.6$ MeV/u and it should be noted that the low-energy halo particles do not fit into the acceptance. Figure 3 shows the distribution of 99.8% of particles that experienced elastic scattering in the ($\alpha$,W) plane, where $\alpha = \sqrt{x'^2 + y'^2}$, and W is the ion energy per nucleon. About 0.2% of ions will experience nuclear reactions. Currently our studies do not include the dynamics of nuclear reaction products. According to preliminary estimations these products can result to ~$10^{-6}$ relative beam losses in high energy section of the

**TABLE 1.** Stripper effect on the uranium beam.

| Parameter | 1st st. | 2nd st. |
|---|---|---|
| Beam energy (MeV/u) | 10.5 | 85.0 |
| Medium | Be, C | C |
| Thickness (mg/cm$^2$), d | 0.05Be+0.22C | 15.0 |
| Energy loss (MeV/u), $\Delta W$ | 0.137 | 3.29 |
| rms angle of scat. (mrad), $\delta\theta$ | 0.43 | 0.5 |
| rms energy spread (keV/u), $\delta W$ | 2.0 | 17.6 |

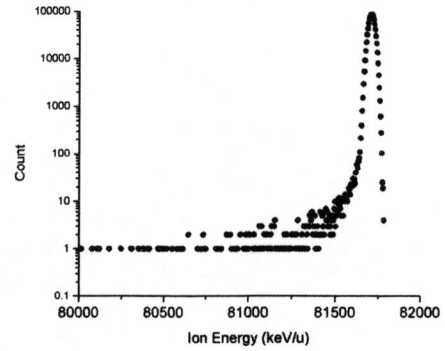

**FIGURE 2.** Uranium beam energy distribution at the exit of the stripper.

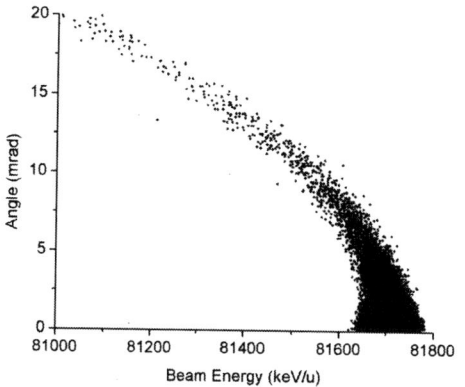

**FIGURE 3.** Distribution of uranium ions in the angle-energy plane.

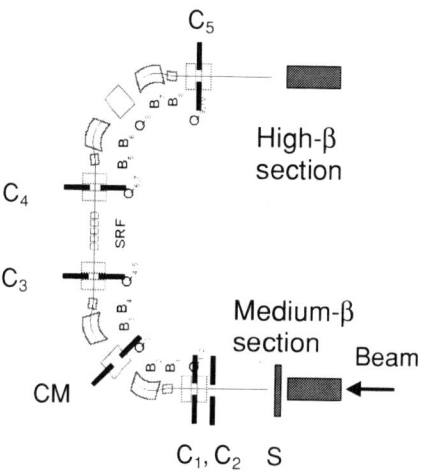

**FIGURE 4.** Post-stripper magnetic transport system with collimators. Legend: S is the stripping foil; $Q_1$, $Q_2$, $Q_4$-$Q_7$, $Q_9$-$Q_{10}$ are the quadrupoles; $Q_3$, $Q_8$ are the multipoles; CM is the main collimator, $C_1$-$C_5$ are the halo collimators, SRF is the superconducting rebuncher cavity, $B_1$-$B_8$ are the bending magnets.

driver linac. A strong correlation between the large scattered angles and ion energy is clearly seen from this figure. This correlation suggests a simple way to remove the low-energy halo by the system of collimators along the MTS, as is shown in Fig. 4. The main collimator CM is located in a highly dispersive area and dumps all unwanted charge states. The collimators $C_1$-$C_5$ are designed to clean beam halo in the transverse phase planes. The transverse acceptance of the MTS with the collimator openings of ±10 mm is shown in Fig. 5. The acceptance of the MTS with collimators is ~10 times smaller than the acceptance of the high-β section. As shown in the beam dynamics simulations, the collimators clean beam in four-dimensional phase space. Because in the driver linac there are no uncontrollable mechanisms for halo formation, beam collimation in the MTS creates the possibility of avoiding any beam losses associated with the beam dynamics. To avoid beam losses completely, the longitudinal acceptance of the high-β section must accept the full beam emittance, including halo particles. As was mentioned in ref. [3] the longitudinal acceptance can be increased by the factor of four if the triple-spoke resonators operating at 345 MHz are used in the high-β section of the linac.

One of the most uncertain parameters of the foil is thickness fluctuation $\eta = \Delta d/d$ across the beam spot. There are several causes of thickness fluctuations: an initial manufacturing error, thermal deformations due to the beam power absorption, and sputtering and radiation damage. Due to the large energy loss in the second stripper the non-uniform foil thickness can result in an additional appreciable beam energy spread. For detailed studies of possible beam loss, including the stripper effect, the code TRACK [8] has been modified in order to regenerate the distribution of particle ensembles after the stripper, on the basis of SRIM calculations taking into account incident beam distribution, and the foil thickness fluctuation. The two-dimensional angle-energy distribution obtained by the code SRIM for the input monochromatic beam has been used for generation of the particles' distribution in 6-dimensional phase space. The foil fluctuation is included as a uniformly distributed change of particle energy within $\eta\Delta W$.

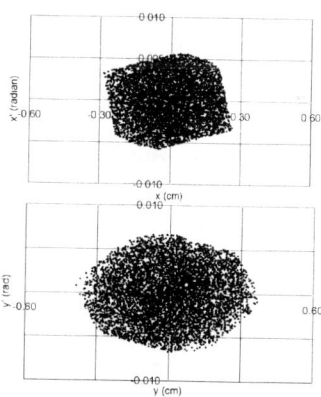

**FIGURE 5.** Transverse acceptance of the MTS with collimators.

## BEAM DYNAMICS SIMULATIONS

The end-to-end simulation of beam dynamics in the driver linac is being performed by the TRACK code which traces particle trajectories in 6D phase space, and generally represents the dynamics of the multi-component heavy-ion beams with good spatial resolution. After recent modifications the TRACK code supports all elements of the driver linac such as RFQ, multi-harmonic buncher, bending magnets, and magnetic and electrostatic focusing devices. All TRACK elements are represented with realistic three-dimensional field distributions and naturally include fringe fields. The detailed beam dynamics studies in the driver linac including errors of electromagnetic fields and misalignments of the linac elements were reported recently [6]. Using the TRACK code we have studied particle losses in high-$\beta$ section of the driver linac in the presence of all errors, misalignments and stripper effects. For these studies the simulation starts at the entrance of the second stripper. The initial beam parameters at this point have been taken from detailed simulation of beam dynamics in the upstream linac section. The beam core within ±0.3 MeV/u is accelerated in high-$\beta$ section without any losses if the collimation system is implemented in the MTS. Therefore the tracking has been performed only for halo particles outside the energy "window" of ±0.3 MeV/u. This part of beam halo is represented by $2 \cdot 10^4$ particles of the multiple-charge state beam which corresponds to a small fraction of the total beam intensity. These particles have been tracked through the MTS with collimators and high-$\beta$ section and the particle losses along the linac have been noted in 200 randomly seeded linacs.

The available experimental data after the stripper shows significantly wider beam energy distribution [9] that has been obtained by the SRIM code. Therefore, the post-stripper distribution has been generated for three different values for the standard deviation of the energy straggling $\delta W$, as is presented in Table 2. We have assumed ±5% stripping foil thickness fluctuation. Two designs of the high-$\beta$ section have been studied: 1) elliptical cavity linac (ECL) which is the baseline design for the RIA driver linac and 2) triple-spoke linac (TSL). The results of the simulation are presented in Table 2. If the energy distribution after the stripper is taken from the SRIM code there are no losses in the high-$\beta$ section for the both linac designs. The TSL is much less sensitive to the energy distribution after the stripper as well as to the stripper thickness fluctuations.

**TABLE 2. Relative beam losses in high-$\beta$ section of the driver linac**

| $\delta W$ (keV/u) | Rf errors | Beam loss in ECL | Beam loss in TSL |
|---|---|---|---|
| 17.6 (SRIM) | ±0.5°, | no | no |
| 53 | ±0.5% | $6 \cdot 10^{-5}$ | no |
| 88 | | $2 \cdot 10^{-4}$ | no |

# CONCLUSION

The Monte Carlo simulations by the SRIM code show strong correlation between the low energy particles and scattering for large angles at the exit of the stripper. The concept of a "beam-loss-free" linac which implies beam halo collimation in designated areas at the post-stripper MTS has been developed. Preliminary data of the beam energy distribution after the stripping foil shows appreciable discrepancy between the measurements and SRIM calculations. More careful measurements of the beam energy spread and transverse emittance (or scattering angles) are required for uranium beam at 85 MeV/u.

# ACKNOWLEDGMENTS

The author thanks his colleagues E. Kanter, J. Nolen, R. Pardo and K.W. Shepard for valuable discussions.

# REFERENCES

1. K. W. Shepard et al., in *the Proc. of the 9th Inter. Workshop on RF Superconductivity*, edited by B. Rusnak, Santa Fe, New Mexico, p. 345-351.

2. J.A. Nolen, Paper MO302 presented at LINAC 2002, Gyeongju, Korea, August 2002.

3. K. W. Shepard, P. N. Ostroumov and J. R. Delayen, "High-Energy Ion Linacs Based On Superconducting Spoke Cavities", submitted to Phys. Rev. STAB, June 10, 2003.

4. M. Portillo et al., in *the Proc. of the 2001 PAC*, edited by P.W. Lucas and S. Webber, June 2001, Chicago, IL, p. 3012.

5. P. N. Ostroumov. Phys. Rev. ST. Accel. Beams 5, 030101 (2002).

6. P.N. Ostroumov, Paper WOAB007 presented at *the 2003 PAC*, May 2003, Portland, Oregon..

7. J.F. Ziegler, J.P. Biersack and U. Littmark, *The Stopping and Range of Ions and Solids*, Pergamon Press, NY, 1996.

8. P. N. Ostroumov and K. W. Shepard, Phys. Rev. ST. Accel. Beams 11, 030101 (2001).

9. J. Nolen and E. Kanter, ANL, Physics Division, Private communication.

# Halo formation and its mitigation in the SNS linac

Dong-o Jeon

*Representing the SNS Project, ORNL, Oak Ridge, TN 37830, USA*

**Abstract.** A halo generation mechanism in the non-periodic lattices such as the SNS linac MEBT (Medium-Energy Beam-Transport between RFQ and DTL) is reported. We find that the nonlinear space charge force resulting from large transverse beam eccentricity ~2:1 in the ~1.6-m-long MEBT chopper section is responsible for halo formation. As a result, the beam distribution, based on the Front End (FE) emittance measurements and multiparticle simulation studies, develops halo that leads to beam loss and radio activation of the SNS linac. Designing lattices with transverse beam eccentricity close to 1:1 suppresses this kind of halo generation. Modifying the MEBT optics and introducing adjustable collimators in the MEBT significantly reduced beam losses in the CCL, which is a preferred scheme for mitigating halo.

## INTRODUCTION

Beam dynamics simulations of the SNS linac show that the beam halo develops at low energy, and some halo particles survive acceleration to higher energies before being lost primarily on the CCL bore as shown in Fig. 1. This particle loss at higher energies results in radio activation of the CCL. In order to find ways to mitigate this halo related beam loss, we conducted studies to identify the sources and mechanism of halo formation. It turns out that the MEBT is the largest contributor to FE halo generation in the SNS linac [1].

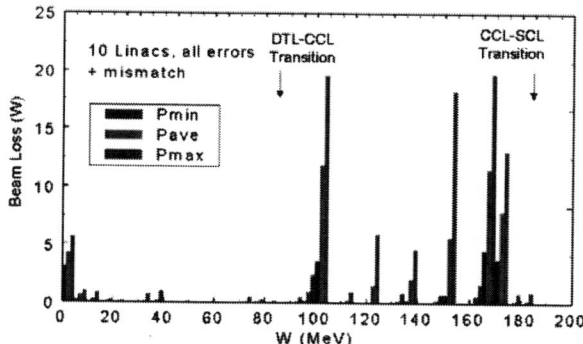

**FIGURE 1.** Plot of the beam loss along the linac.

## HALO FORMATION MECHANISM

To better understand MEBT optics, the horizontal (x curve) and vertical (y curve) envelope profiles of the sqrt(5) * rms beam size in the MEBT are plotted in Fig. 2. The top curve is longitudinal envelope profile. The beam is squeezed vertically to clear the vertical deflection plates of both the chopper and anti-chopper and relaxed horizontally. This arrangement is necessary to have 90° zero-current betatron phase advance between the chopper box and the chopper target in the middle (between the chopper target and the anti-chopper box as well). However, this 1.6-m-long chopper section with a large beam eccentricity is the source of halo formation.

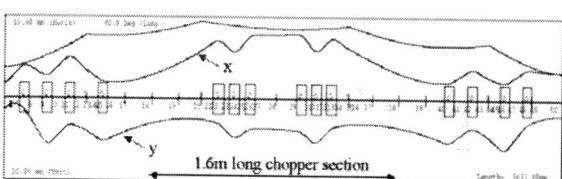

**FIGURE 2.** Trace3D beam envelope profiles of MEBT.

Figure 3 shows the electric field (in arbitrary unit) on top of real space projections of beam distribution at the chopper target (in the middle of the MEBT) where the beam eccentricity is ~2:1. The beam is wide in x and narrow in y. The Ex becomes nonlinear beyond x=0.5 cm, which is well inside the core. This means that the outer part of core with |x| > 0.5 cm (marked as "potential halo") is subject to nonlinear space charge force and their phase advance is quite different from the inner part of the core seeing linear space charge force. The phase advance difference over the 1.6-m chopper section leads to severe beam distortion in horizontal phase space. In the case of Ey, only small fraction of halo particles sees nonlinear space charge force. This is why the tail develops mainly in x phase space by the end of the MEBT.

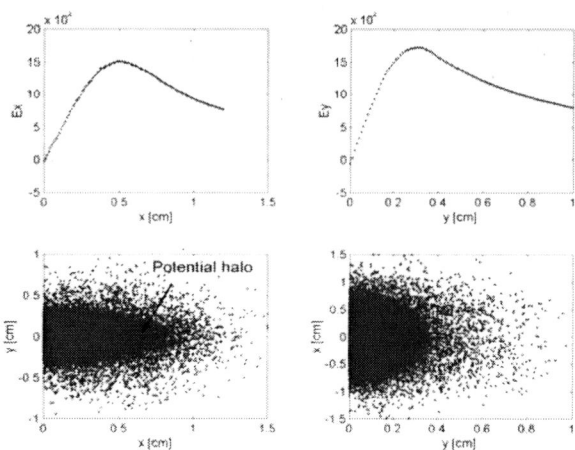

**FIGURE 3.** Plots of E field and real space projections of beam distribution. The unit of E field is arbitrary. x rms beam size is 3.40mm and y rms beam size 1.71 mm.

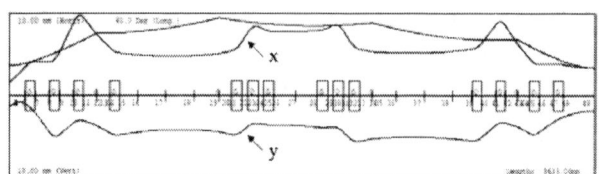

**FIGURE 4.** Trace3D envelope profiles of round beam optics.

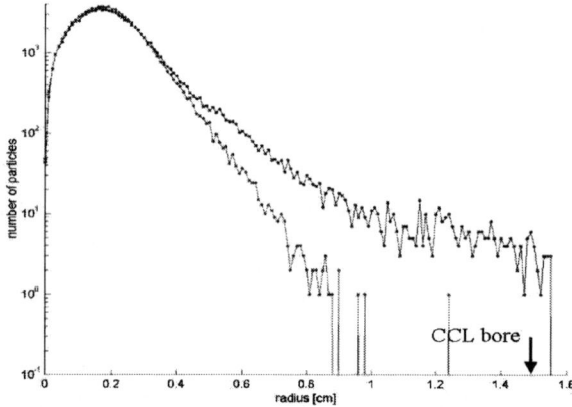

**FIGURE 5.** radial beam profile at 171MeV before (the upper curve, baseline case) and after the MEBT optics change alone (the lower curve).

To suppress halo formation induced by large transverse beam eccentricity, the optics of lattices should make the beam as round as possible. For the purpose of studying, we modified the entire MEBT optics to reduce beam eccentricity as shown in Fig. 4 (compare with Fig. 2).

Making the beam round indeed suppresses the halo formation as shown in Fig. 5 that depicts the beam profiles at 171 MeV before and after optics modification. However, modification of the entire MEBT optics is not viable to facilitate the beam chopping for ring injection. At least the first half of the MEBT should not be modified, while the second half can be modified.

## HALO MITIGATION SCHEME

A hybrid scheme is adopted for halo mitigation that is a combination of an alternative MEBT optics and adjustable collimators at the MEBT chopper target.

### Alternative MEBT optics

In an alternative design, the upstream half of MEBT optics is preserved to preserve MEBT chopping while the downstream half of MEBT optics is modified for round beam. The resulting beam cross section is more circular as shown in Fig. 6. The beam now has a larger vertical extent and approaches the anti-chopper plates as designed.

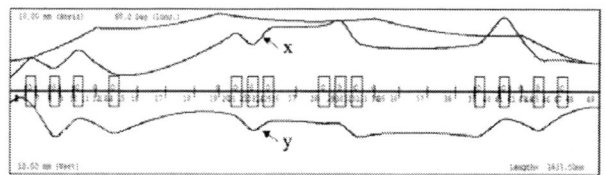

**FIGURE 6.** The proposed alternative MEBT optics.

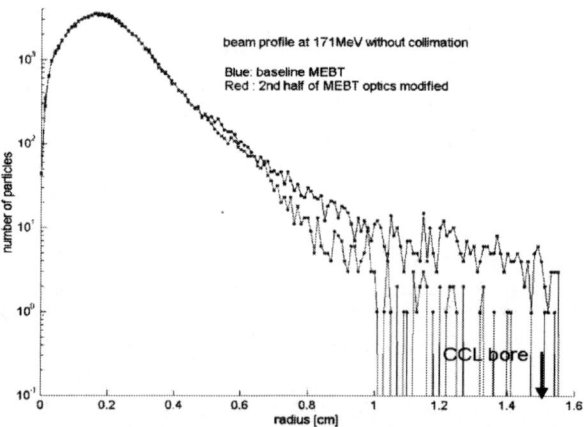

**FIGURE 7.** Radial beam distribution at 171MeV for the baseline MEBT optics without collimation (the upper curve) and for the alternative MEBT optics without collimation (the lower curve). 87% of the beam tail with r > 9 mm is removed just due to optics change.

This simple modification to the optics alone reduces the formation of transverse tails substantially and improves the beam quality in the downstream linac. Figure 7 shows that 87% of the beam tails with r > 9 mm at 171MeV is removed. The halo reduction is comparable to the effect of MEBT collimation with the baseline MEBT optics.

## MEBT collimation

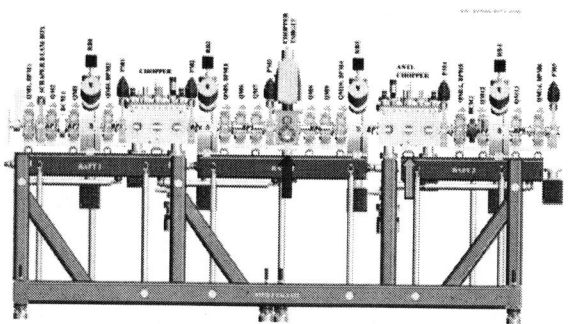

**FIGURE 8.** Schematic layout of MEBT indicating the location of adjustable collimators.

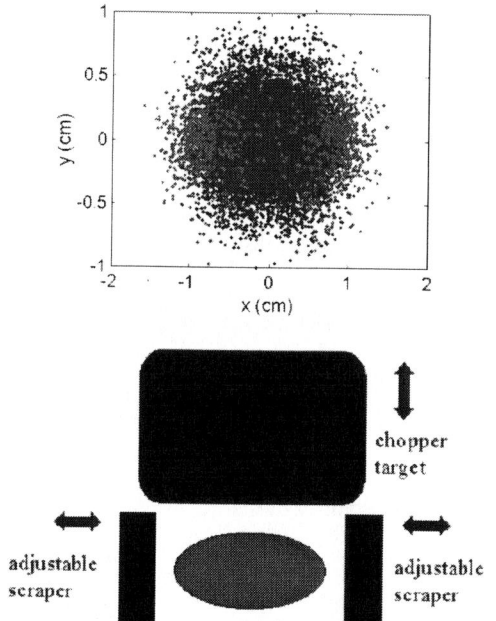

**FIGURE 9.** Top plot shows particles with r > 9 mm at 171MeV mostly populated at both ends. The bottom plot shows the schematic drawing of adjustable horizontal collimators and the chopper target.

There are only a few places where collimators will fit in the MEBT. One convenient place is at the chopper target. Figure 8 shows the layout of the MEBT with the chopper target and anti-chopper box indicated by arrows. A pair of adjustable horizontal collimators would be installed in the chopper target box (at the red arrow). The chopper target itself is located above the mid-plane to intercept beam that is deflected upward. Collimators mounted on horizontal actuators will not interfere with the function of the target.

The particles on both ends in the top plot of Fig. 9 are those with r > 9 mm at 171 MeV. So horizontal collimators will be effective and its assembly is shown schematically at the bottom. This collimator implementation has the advantage that it is readily adjustable to accommodate the actual beam conditions, which are expected to vary with different operating conditions. The other advantage is that the proposed collimators can be cooled easily. The adjustable collimators are designed to scrape up to about 20% of beam power when they are made of Carbon/Carbon composite [2], in other words, up to about 10% by each of the two adjustable horizontal collimators.

Combining MEBT optics modification and MEBT collimation, we expect to reduce uncontrolled beam loss associated with halo to a manageable level.

Figure 10 shows the radial beam distribution at 171 MeV resulting from this hybrid solution, which combines the alternative MEBT optics and the MEBT scraping at the chopper target. 97% of the halo with r > 9 mm is removed compared with the baseline case (in blue). Even for the increased peak current of 54mA rather than 38mA, there is also an enough safety margin even for this case.

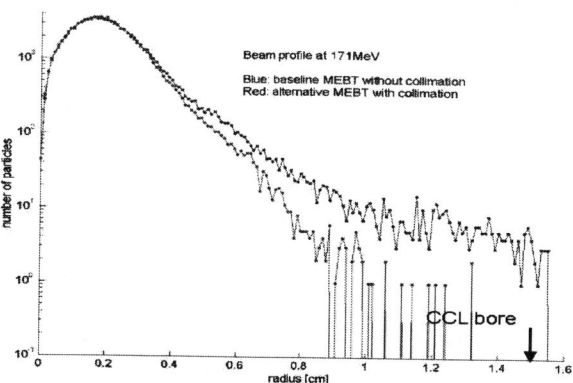

**FIGURE 10.** Radial beam profile at 171MeV for the adopted hybrid scheme mitigating halo (lower curve).

## DTL collimation

We also explored the possibility of DTL collimation of SNS linac. The focusing lattice in the DTL is FFODDO, where O means empty drift tubes. We considered inserting circular collimators in the first 11 empty drift tubes. The bore radius of drift tubes is 12.5 mm. By using only empty drift tubes, we avoid the possibility of overheating and possibly approaching the Curie point of the permanent-magnet quadrupole lenses (PMQs).

The projections in Fig. 11 clearly show halo particles inside the core of the beam. In fact, the halo is completely unobserved at some points (e.g. drift tubes 1 and 16). The tails created in the MEBT have become so well integrated with the core of the beam that drift-tube collimation is not effective.

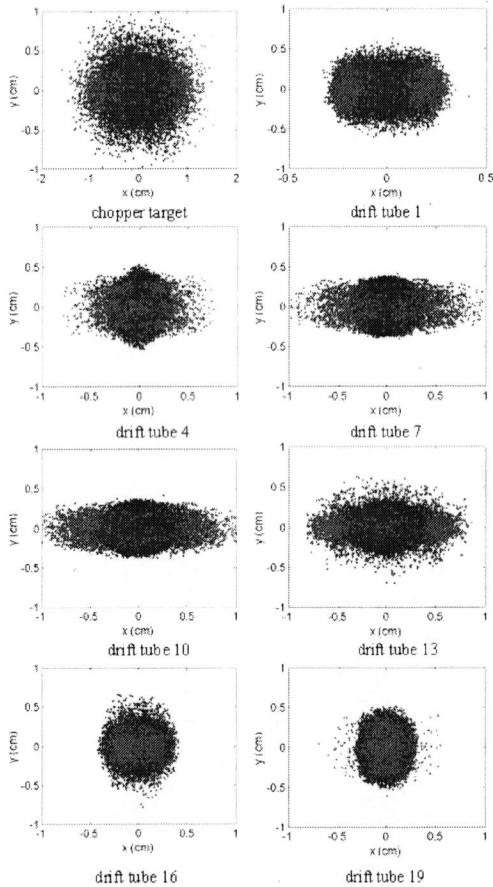

**FIGURE 11.** Real space projections of the beam distribution at the MEBT chopper target and at the first 7 of the proposed DT collimators. Halo particles "at risk" are plotted lighter.

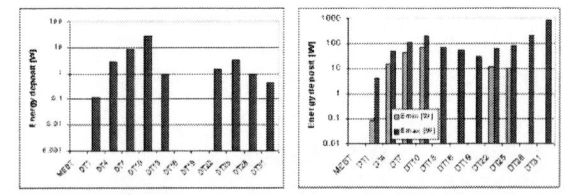

**FIGURE 12.** Expected power deposited in 6 mm radius DTL collimators. The left-hand plot assumes no errors. The right-hand plot summarizes the results of 100 linac runs with errors. The drift tube bore radius is 12.5mm.

The left-hand plot of Fig. 12 shows that, excluding machine imperfections, the beam power deposited in drift tube 10 would double the design thermal load. Including machine imperfections (right-hand plot), the maximum expected power deposited in drift tube 22 is 444 W, which is ~6 times the design cooling capacity of this drift tube.

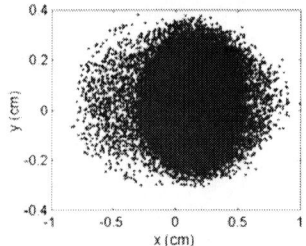

**FIGURE 13.** Real-space beam distribution at the end of DTL tank 1 for one set of random machine imperfections. The resulting asymmetric scraping leaves a significant halo.

The random misalignment of the PMQs steers the beam, so the beam centroid is seldom on axis. Without beam-position monitors (BPMs) in the low-energy end of tank 1 we have no way to steer the MEBT beam onto the DTL axis. As a consequence, the beam will scrape the bore asymmetrically as shown in Fig. 13.

We conclude that placing collimators in drift tubes is not so an effective approach to removing halo particles that are expected to be lost in an uncontrolled way.

## CONCLUSION

A new halo generation mechanism in a non-periodic lattices is presented together with a mitigation scheme. The nonlinear space charge force resulting from large transverse beam eccentricity is responsible for the halo formation in the FE of the SNS linac.

## ACKNOWLEDGEMENT

This work is supported by the DOE, under contract No. DE-AC05-00OR22725 with UT-Battelle, LLC for ORNL.

## REFERENCES

1. D. Jeon et al, Phys. Rev. ST Accel. Beams **5**, 094201 (2002).

2. Private communication with S. Kim.

# Halo Studies for the ESS and Linac4 Front-Ends

## Frank Gerigk

*ISIS Accelerator Theory & Future Projects Group*
*CCLRC Rutherford Appleton Laboratory, Chilton, Didcot, Oxon, U.K.*

**Abstract.** In high intensity proton linacs, beam halo can be observed as early as the very first stages of acceleration. The inital population of halo particles found in the output distribution of the RFQ, increases in chopper lines, which interrupt the regular focusing pattern of RFQ and DTL. The resulting emittance growth and halo development contributes to beam loss in the low-energy accelerator sections and partly defines the beam quality in the high-energy sections. Using the examples of the ESS and Linac4 (CERN) front ends (RFQ to 20 MeV) we investigate emittance growth and beam halo due to statistically distributed field and gradient errors.

## INTRODUCTION

The analysis of statistical errors represents the final stage in every linac design. Emittance growth, phase & energy jitter and beam loss due to these errors define the tolerances in the machine and thus have a direct impact on the project costs.

This paper reports on the progress in simulating statistical errors in the ESS [1] and Linac4 [2] front-ends, both high-intensity $H^-$ linacs with a beam chopper in their Medium Energy Beam Transport (MEBT) line. The simulation approach using IMPACT [3] is outlined, and results from the simulation of quadrupole gradient errors as well as RF phase and gradient errors are reported. Finally the results are interpreted with respect to the current understanding of halo development.

## SIMULATION

*Lattices.* Both front-ends start with a simulated RFQ output distribution [4] followed by a MEBT containing a beam chopper with several focusing elements. Two Alvarez Drift Tube Linac (DTL) tanks then raise the energy from 2.5 to 20 MeV in the ESS case and from 3.0 to 25 MeV for Linac4. Long term effects will be visualized using the full Linac4 lattice up to an energy of 120 MeV. The lattice parameters are summarized in Table 1.

*Conventions.* In order to quantify the amount of halo in a distribution snapshot, the fraction of particles outside $n$ times its r.m.s. emittance (in each plane) is plotted. Thus the development of beam halo can be clearly separated from $\varepsilon_{r.m.s.}$ growth and/or density oscillations in

**TABLE 1.** Lattice parameters

|  | ESS | Linac4 |
|---|---|---|
| **RFQ output** | 2.5 MeV | 3.0 MeV |
| peak current | 57 mA | 30 mA |
| RF frequency | 280 MHz | 352 MHz |
| **MEBT** | 3.11 m | 3.73 m |
| no. of chopper plates | 4 | 2 |
| no. of buncher cav. | 6 | 3 |
| no. of quadrupoles | 13 | 11 |
| **DTL** | 11 m | 9 m + 6.1 m* |
| no. of RF sources | 2 | 2+1 |
| no. of gaps | 77 | 82+29 |
| no. of quadrupoles | 78 | 83+29 |
| **output energy** | 20.3 MeV | 24.9 MeV |
| **CCDTL**[†] |  | 47.6 m |
| no. of RF sources |  | 10 |
| no. of 3/4-gap tanks |  | 37 |
| no. of quadrupoles |  | 37 |
| **output energy** |  | 120 MeV |

\* + refers to the full Linac4
† for full Linac4

real space. An example is shown in Fig. 1 where the two input distributions are plotted. Both distributions carry particles with core radii of up to 5 times the r.m.s. value. A uniform error distribution is used for all simulations, meaning that the r.m.s. errors can be obtained by multiplying the quoted max. error amplitudes by $\approx 0.58$.

*Simulation details.* A preprocessor creates $m$ random error sets, applies them to the original lattice and stores the new input files in scratch directories. With a simple shell script the jobs are then submitted to a Linux

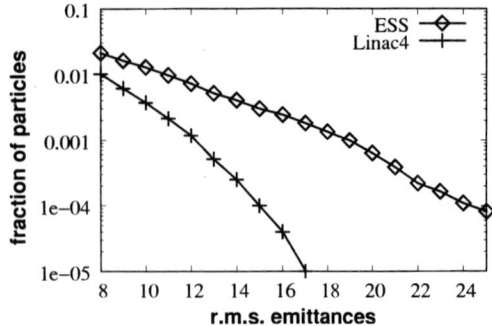

**FIGURE 1.** Transverse halo: particles outside $n \times \varepsilon_{r.m.s.}$ in the input distributions for the ESS and Linac4 front-ends.

cluster. For the full Linac4, 500 error sets with 50000 particles are simulated for each error amplitude, using $\approx$ 14 hours of CPU time on 30 processors. For the shorter front-end simulations 300 error sets are created. The process is then repeated for each error amplitude and evaluated by a post-processor module.

## STATIC & DYNAMIC ERRORS

In the following we use the classic distinction between 'dynamic' errors that change from pulse to pulse or within single RF pulses and 'static' errors that change very slowly (seasonal changes) or remain completely unchanged during operation [5], [6]. Static errors are generally a residue of the initial adjustment of the lattice elements, e.g. random (gap to gap) RF amplitude errors from the field adjustment with bead pull ($\approx 1\%$), or random quadrupole gradient errors ($\approx 1\%$). The same category applies for mismatch and alignment errors which are not covered in this paper. While the static errors usually change from element to element the dynamic errors tend to be grouped, originating from pulse to pulse variations of RF ($< 1\%$, $1°$) or quadrupole ($< 0.5\%$) power supplies.

Figure 2 shows the average $\varepsilon_{r.m.s.}$ growth rates with respect to maximum transverse and longitudinal error amplitudes for grouped or completely random errors. In the transverse plane the grouped and ungrouped errors result in the same average $\varepsilon_{r.m.s.}$ growth rates, cutting in half the number of simulations needed to evaluate both error sources. In the longitudinal plane, however, the grouped errors yield larger $\varepsilon_{r.m.s.}$ growth (and larger phase & energy jitter) than the ungrouped errors. Although modern RF power sources with temperature controlled cables provide a margin of 0.5% field error and 0.5° phase error during the pulse, the error margins for the first bunches of a pulse may be considerably higher [5] and thus become a major concern.

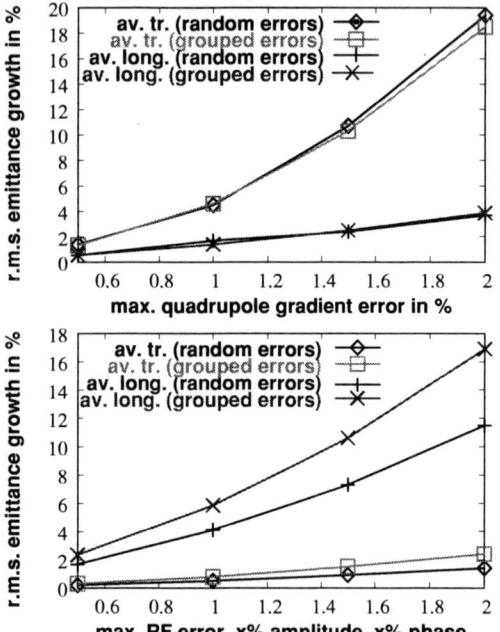

**FIGURE 2.** Upper graph: average $\varepsilon_{r.m.s.}$ increase for dynamic (grouped) and static (random) quadrupole gradient errors in the full Linac4; lower graph: dynamic and static RF field and phase errors

The average $\varepsilon_{r.m.s.}$ growth for realistic dynamic and static error margins seems to be very small and therefore of no concern for low-loss machine operation. Nevertheless, a combination of all static and dynamic errors including alignment errors and mismatch may easily yield unacceptable losses and has to be evaluated for each particular machine.

In both planes, the maximum $\varepsilon_{r.m.s.}$ growth rates for the worst cases are $\approx$ 3 to 4 times higher than for the plotted average cases. For dynamic errors the worst cases are of little importance since normal machine operation will automatically yield the average. The same reasoning applies for certain static errors (e.g. quadrupole gradient errors) which can be reduced with automated tuning systems. Some static errors, however, cannot be compensated (e.g. residual field adjustment errors in multi-gap RF tanks) and for those the worst case results should be taken into account.

## PHASE & ENERGY JITTER

The dynamic errors in the longitudinal plane not only result in $\varepsilon_{r.m.s.}$ growth but also in phase and energy deviation (jitter) from the nominal trajectory. If the linac beam is injected into a subsequent accelerator or a transport line with RF cavities, the phase and energy jitter has to be limited to fit into the RF bucket of the following system.

We find that the largest phase & energy deviations usually originate from transitions between sections that are powered by different amplifiers and/or have different focusing structures. Opposite error amplitudes at these transitions can provide large kicks which are then amplified in subsequent drifts. An example is shown in Fig. 3, where the four worst case energy deviations are plotted for the two front-ends.

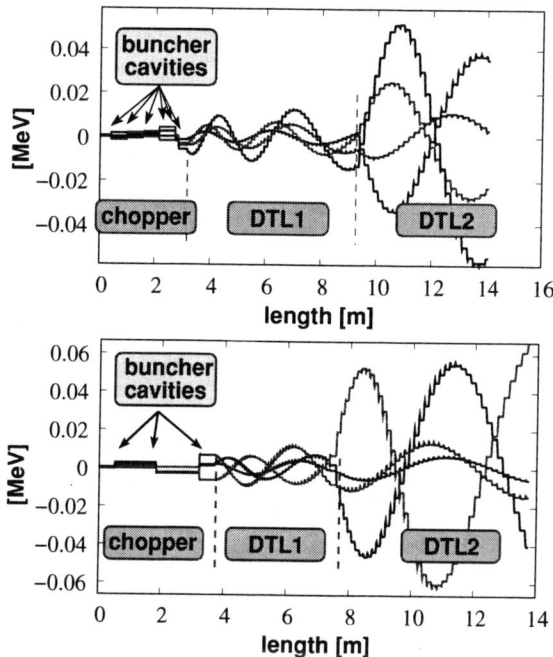

**FIGURE 3.** The four worst case energy deviations for a maximum of 1% field error and 1° phase error in the ESS (upper graph) and Linac4 (lower graph) front-end.

The average and maximum phase and energy jitter at the end of Linac4 are between 15 and 30% higher than for the ESS. This can probably be explained by the more compact design of the ESS MEBT (see Table 1) and the 'missing gap' at the transition of DTL1 and DTL2 in Linac4. In the ESS front-end, the two DTL tanks have a 'seamless' transition.

Considering Fig. 2 and 3 it is obvious that the partitioning of RF power supplies and cavities plays a crucial role in the development of phase & energy jitter. In the following example the jitter at the end of Linac4 is plotted for its original partitioning of power supplies and RF cavities, and for a partitioning with a reduced number of amplifiers (5 instead of 16, Fig. 4).

Contrary to the results with grouped and ungrouped RF errors in Fig. 2, we find that the lower number of power supplies results in a much lower phase & energy jitter and a much reduced average longitudinal $\varepsilon_{r.m.s.}$ growth (2.6% instead of 5.8%). This suggests that the power

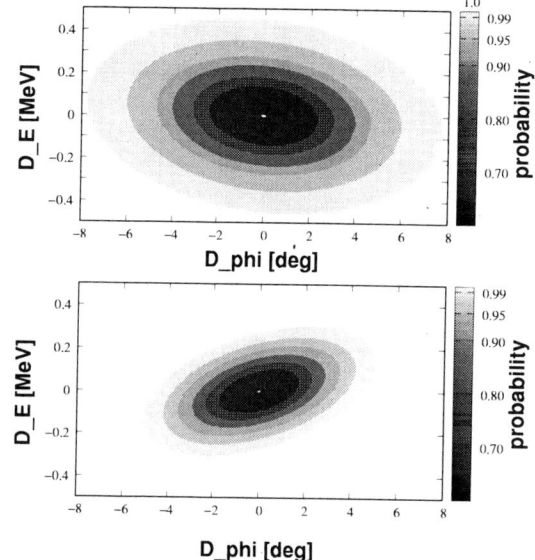

**FIGURE 4.** Phase & energy jitter at the Linac4 output for a maximum of 1% field error and 1° phase error using **16** (upper graph) or **5** (lower graph) RF amplifiers.

splitting should be carefully optimized for each machine and that the optimum number of RF gaps per power source may be different for different lattice types and different energies.

## HALO FROM STATISTICAL ERRORS?

Parametric beam halo develops for integer ratios (usually 2:1) between the oscillations of a mismatched beam core and the oscillations of single particles. This type of coherent core oscillation can be excited by initial mismatch and can remain remarkably stable throughout an entire linac. Figure 5 shows one such example for a fast-mode

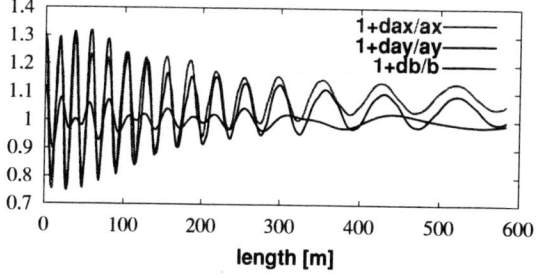

**FIGURE 5.** 30% fast mode excitation in the superconducting section of the SPL.

excitation in the 600 m long superconducting section of the SPL [7]. Energy transfer from the core oscillations to the single particle orbits eventually damps the oscillations of the core and thus creates halo.

Similar oscillations, though much more irregular, can be observed in simulations with statistical errors, e.g. in Fig. 6 where the radial deviations for an average case of quadrupole errors (max. ±1%) reach amplitudes between 5 and 10%. For the worst case they increase to 40%, reaching similar levels to those generated by strong initial mismatch. Figure 6 also shows that the worst ra-

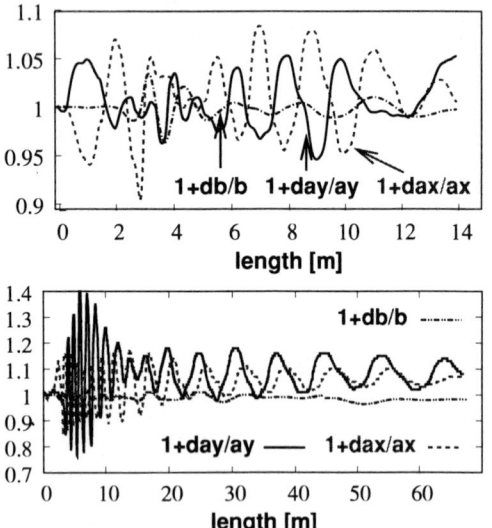

**FIGURE 6.** **Upper graph:** average case of radial deviations for ±1% quadrupole gradient errors in the ESS front-end ($\approx$ 5% $\varepsilon_{r.m.s.}$ growth), **lower graph:** worst case for Linac4 ($\approx$ 100% $\varepsilon_{r.m.s.}$ growth).

dial deviations occur after the MEBT, in the first two DTL tanks. Unless these tanks are designed with large bore radii they are likely to suffer from high beam loss. The corresponding beam halo for the ESS front-end (Fig. 7) shows an almost invisible increase for the average case and roughly a doubling of large amplitude particles for the worst case. For the full Linac4, Fig. 8 shows a more visible effect for the average error case.

Despite the difficult interpretation of these curves (beam halo already for the matched case, particle loss, low number of particles $\leq$ 50000, etc.) it seems as if statistical errors might have the potential to create beam halo. Fig-

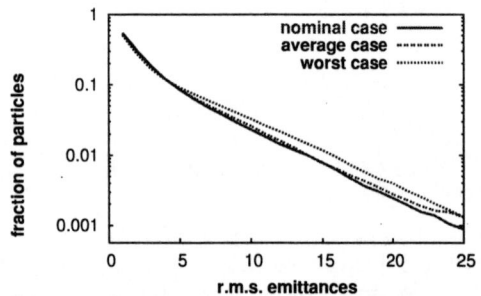

**FIGURE 7.** Transverse halo due to ungrouped 1% (max.) quadrupole gradient errors in the ESS front-end.

ures 7 and 8 both show an increased number of particles beyond the core emittance ($> 5 \cdot \varepsilon_{r.m.s.}$) even though the amount of the observed halo is less dramatic than for initial mismatch.

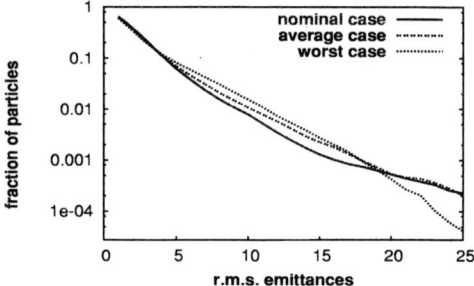

**FIGURE 8.** Transverse halo due to ungrouped 1% (max.) quadrupole gradient errors in Linac4.

## SUMMARY AND OUTLOOK

First results of IMPACT simulations of dynamic and static gradient & phase errors were presented for Linac4 and the ESS front-end. Quadrupole gradient variations for grouped and ungrouped errors resulted in equal $\varepsilon_{r.m.s.}$ growth rates and seem to be of little concern. Longitudinally the partitioning of power sources and RF gaps had a significant influence on the resulting $\varepsilon_{r.m.s.}$ growth and energy & phase jitter. Additional concerns are the possibility of an unfortunate combination of residual field adjustment errors, and the enhancment of RF error bars for the first bunches of an RF pulse. The simulations indicate the possibility of halo development due to statistical errors, although more systematic work is needed to understand the process fully. A combination of all error sources and an evaluation of alignment errors has yet to be done and might add significance to the effects of single error sources.

## REFERENCES

1. Gerigk, F., Revised ESS Front-End (2.5 - 20 MeV), Tech.Rep. (2003).
2. Gerigk, F., and Vretenar, M., "Design of a 120 MeV H$^-$ Linac for CERN High-Intensity Applications," in *Proceedings LINAC02*, Kyongiu, Korea, 2002.
3. Qiang, J., Ryne, R. D., Habib, S., and Decyk, V., *Journal of Computational Physics*, **163**, 1–18 (2000).
4. *The ESS Project Volume III, Technical Report*, ISBN 3-89336-303-3, 2002.
5. Vretenar, M., private communications (2003).
6. Findlay, D., private communications (2003).
7. Vretenar, M., editor, *Conceptual Design of the SPL, a High-Power Superconducting H$^-$ Linac at CERN*, CERN 2000-012, 2000.

# Self-consistency and coherent effects in nonlinear resonances

I. Hofmann*, G. Franchetti*, J. Qiang† and R.D. Ryne†

*GSI, 64291 Darmstadt, Germany
†LBNL, 94720 Berkeley, USA

**Abstract.**
The influence of space charge on emittance growth is studied in simulations of a coasting beam exposed to a strong octupolar perturbation in an otherwise linear lattice, and under stationary parameters. We explore the importance of self-consistency by comparing results with a non-self-consistent model, where the space charge electric field is kept "frozen-in" to its initial values. For Gaussian distribution functions we find that the "frozen-in" model results in a good approximation of the self-consistent model, hence coherent response is practically absent and the emittance growth is self-limiting due to space charge de-tuning. For KV or waterbag distributions, instead, strong coherent response is found, which we explain in terms of absence of Landau damping.

## INTRODUCTION

The effect of space charge on nonlinear resonances has received increasing attention with the need of optimizing the performance of storage rings or synchrotrons with high intensity or high phase space density. In a recent experiment carried out at the CERN proton synchrotron (see companion paper to this conference, Ref. [1]) the 3D aspects of the resonance effect of a single octupole on an 1 s long injection flat-bottom have been explored by simulation as part of a code bench-marking effort by abandoning self-consistency and using an analytical solution of Poisson's equation for the "frozen-in" density profile. The present study is carried out in order to advance our understanding of the importance of self-consistency as well as the differences in resonance behavior of coasting and bunched beams. We adopt similar parameters as in Ref. [1], like a fixed working point varying over an interval above the fourth order resonance at $Q_x = 6.25$, as well as comparable space charge tune shifts.

The most obvious effect of space charge is the incoherent tune shift and spread, which leads to an extended foot-print of single particle tunes in the tune diagram. Less obvious is the observation that the response on a resonance may be modified by the coherent motion of all, or of a large fraction of particles, which is the main expression of self-consistency. It leads to an additional time-dependent force, which must be added to the external forces, and which may cause a coherent shift of the resonance condition. This phenomenon was studied in some detail for second order resonances in connection with the Spallation Neutron Source Ring, where it was found that this coherent shift has a favorable effect on the tolerable intensity [2]. The present study, however, shows that for Gaussian beams and fourth order resonances such a coherent effect is practically absent.

## SIMULATION

We use the self-consistent 2D particle-in-cell (PIC) version of the MICROMAP code [3] with $10^5$ simulation particles and employing different distribution functions on a $64 \times 64$ grid filling a rectangular boundary of $70 \times 70$ mm size. In most examples intensity is normalized to a maximum space charge tune shift $\Delta Q_x = 0.09$ for the Gaussian distribution, and rms equivalence for all others, which implies $\Delta Q_x = 0.045$ for the equivalent KV-beam. Before discussing the effect of space charge we consider the octupole effect alone. While the experiment was done with 40 A octupole excitation, we find it convenient in this 2D study to raise the octupole strength by a factor 2.5 (corresponding to 100 A) in order to enhance the otherwise weak effects. The normalized strength (in $m^{-3}$) is given here as

$$K_3 = 1.215 \cdot I, \quad (1)$$

where $I[A]$ is the octupole excitation current.

For a Gaussian distribution we find that the response in terms of rms emittance growth is 24% for not too strong octupoles, independent of the octupole strength. This is due related to the fact that the stabilizing de-tuning by the octupole increases simultaneously with the resonance driving term. The time required to reach the maximum emittance growth is, however, about inversely proportional to the octupole strength ($\approx 10^3$ betatron periods for 100 A). The response curve as a function of machine tune with clear de-tuning shift is shown in Fig. 1 for a Gaussian distribution. Due to the strong octupole there is

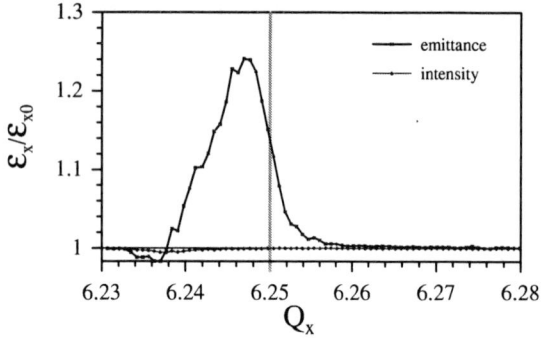

**FIGURE 1.** 2D simulation without space charge showing relative rms emittance growth and intensity from octupole (100 A).

a visible (< 1%) beam loss effect for $6.235 < Q_x < 6.24$. This is accompanied by some emittance reduction due to extraction of the larger amplitude particles, which is absent for a 40 A octupole. In order to interpret loss effects we have first carried out a simplified numerical study on the dynamic aperture by searching the maximum stable radius of test particles placed into 20 different directions in the upper half of the $x - y$ plane. Assuming a 200 A octupole current we have calculated, in the absence of space charge, that the dynamic aperture shrinks to a radius of $2.5\sigma$ near $Q_x = 6.25$ for $10^3$ turns, and to about $2.2\sigma$ for $10^5$ turns. Comparing several octupole strengths we have found the approximate scaling

$$DA \approx 35/\sqrt{K_3}, \quad (2)$$

for the dynamic aperture $DA$ expressed in units of $\sigma$.

## Distribution function and Landau damping

Including space charge we obtain results, which depend sensitively on the distribution function. The KV-beam response after 1000 turns is shown in Fig. 2. In order to appreciate the coherent nature of this response we first discuss the rms emittance growth for a "frozen-in" space charge electric field, where the initial values are not updated, hence the response is entirely incoherent. The response is non-zero for a distance of the bare machine tune from the resonance line less than 0.045, which equals the incoherent space charge tune shift - common to all particles - of the initial KV-beam. Similar to Fig. 1 we observe in the "frozen" response a stop-band width of $\approx 0.01$ and a shift from the ideal resonance condition, which is now at $Q_x = 6.295$. Both effects are a result of the resonance broadening due to the

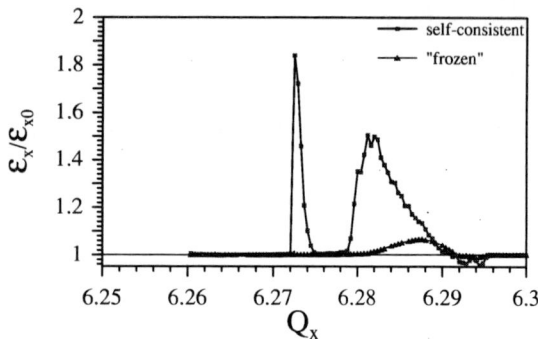

**FIGURE 2.** 2D simulation of KV distribution with $\Delta Q_x = 0.045$ and octupole (100 A) as function of bare machine tune, showing self-consistent and "frozen-in" models (after 1000 turns).

strong octupole force. There is, however, a significant additional de-tuning by space charge, which limits the maximum emittance growth to < 6% compared with the 25% growth without space charge.

The self-consistent simulation, instead, shows two high separate peaks split and shifted by coherent space charge effects. The broader peak is interpreted as direct result of the fourth order resonance, but with a coherent tune shift, which suggest that the resonance can be approached roughly 20% more than suggested by the single-particle response of the "frozen-in" model. The height of this peak exceeds significantly (more than five times) the maximum "frozen-in" response - a pronounced coherent effect. The striking and unexpected large peak at $Q_x = 6.273$ cannot be explained as direct result of the octupole, but is associated with an envelope instability. Such an envelope instability requires a fractional phase advance of the envelope of half an integer relative to the lattice periodicity as was shown in Refs. [2, 4]. This condition is analogous to the envelope instability in linear accelerators, where a single-particle phase advance above 90° per focusing period may induce a half-integer unstable envelope. In the present case the "structure period" cannot stem from the smooth first order lattice, but only from the local perturbation induced by the relatively strong octupole. The latter occurs at only one position on the circumference, hence the total phase advance of particles per turn is exceeding $6 \times 360° + 90°$.

To distinguish between second and fourth order resonance effects the coherent coupled mode coefficients $C_{mk}$ introduced in Ref. [5], and elaborated in more detail in Ref. [6], are useful. In a pure single-particle picture it is assumed that the bare machine tune should stay above the resonance by at least the incoherent tune shift. From the point of view of coherent resonance crossing this critical distance is multiplied by $C_{mk}$, which is smaller than unity in most cases. The $C_{mk}$ thus reflect the fact that the additional coherent space charge force partly compensates the stationary space charge force and - depending on the order of the resonant mode - allows to bring the bare machine tune closer to the resonance than could be inferred from the single-particle point of view. From the position of the left peak of Fig. 2 we calculate that $C = 0.023/0.045 \approx 0.5$, which is typical for the envelope breathing mode [6].

For the rms equivalent waterbag distribution in Fig. 3 we find that the envelope instability peak is unchanged. The fourth order resonance effect is visibly reduced, however. We explain this as a weakening of the coherence induced by the finite tune spread. Note that the "frozen-in" emittance response reflects the distribution of single-particle tunes, which is the origin of Landau damping. From Fig. 3 we note that the "frozen-in" curve has no overlap with the coherent tune of the envelope instability (marked by the location of the strong peak) similar to Fig. 2, which therefore cannot be "Landau-damped"; the overlap with the fourth order resonance visibly weakens the coherent response.

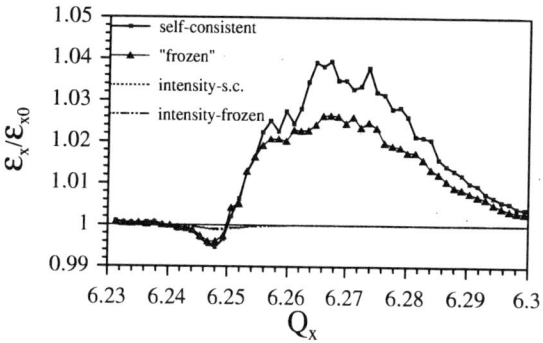

**FIGURE 4.** 2D simulation of Gaussian distribution (same parameters as Fig. 2, note changed scales).

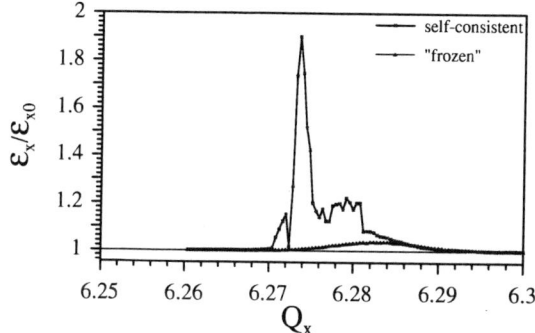

**FIGURE 3.** 2D simulation of waterbag distribution (same parameters as Fig. 2).

For the Gaussian distribution the picture is essentially different. The much broadened single-particle spectrum of the Gaussian fully overlaps the position of the expected envelope instability frequency, which is therefore effectively "Landau-damped". The much broadened direct response curve is only slightly enhanced compared with the "frozen-in" model (Fig. 4), hence there is an almost complete suppression of the coherent resonance effect. There is also a region of about 0.2% loss, which is practically identical for the self-consistent and "frozen-in" model, and is caused by the extended Gaussian tails.

### Dependence on octupole strength

The strong reduction of emittance growth, caused by the suppression of coherent resonance response for the Gaussian beam, is studied in Fig. 5 as function of octupole strength. The graph shows the relative rms emittance growth and intensity for $Q_x = 6.27$ obtained from a simulation, where the frozen-in space charge was calculated by an analytical formula valid for round beams (for faster parameter scans, and justified by section ), i.e. equal emittances. Note the drastic suppression of the rms emittance growth compared with the 24% of the zero-space-charge level. The suppression is most visible for small octupole strength, where the space charge detuning is strongly dominant over the resonance excitation. For octupole strengths above 100 A the shrinking of the dynamic aperture - consistent with Eq. 2 - causes loss from the Gaussian tails, which makes the effective rms emittance shrink even below the starting value.

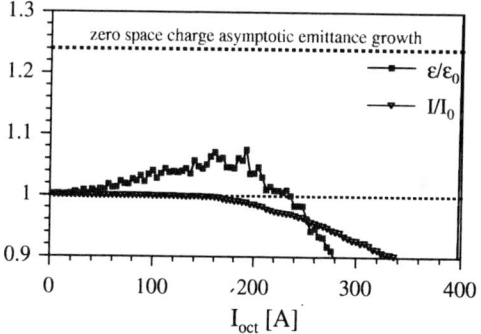

**FIGURE 5.** 2D simulation of dependence of rms emittance growth and intensity on octupole strength for Gaussian distribution (same space charge shift as Fig. 2).

## CODE COMPARISON AND LONG-TERM EFFECTS

We have compared the output from the MICROMAP code using $10^5$ macro particles with a simulation applying the 3D IMPACT code [7] with $10^6$ macro particles. Note that IMPACT with a 3D Poisson solver applied to the essentially 2D coasting beam problem should employ a larger number of simulation particles than is used in MICROMAP with a 2D Poisson solver. The scope of the comparison with two sufficiently different numerical codes is to gain confidence in the particle-in-cell simulation for large turn number. Here we particularly worry about code-specific noise effects, which may be negligible on the time-scale of $10^3$ turns, but could have a significant effect over much longer times.

We first compare the output of simulation at the end of 1000 turns for the case of 200 A octupole current. At this level significant beam loss is expected, and we can thus test both, the emittance growth and dynamic aperture effect caused by the octupole. Note that we have chosen the slightly smaller space charge tune shift of $\Delta Q_x = 0.075$ for this comparison. The result of Fig. 6 shows excellent agreement even in the dependence on $Q_x$. The detailed

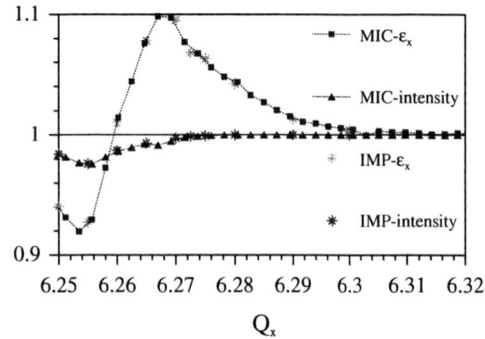

**FIGURE 6.** Comparison of MICROMAP and IMPACT simulation for 200 A octupole and $\Delta Q_x = 0.075$ at 1000 turns.

evolution for $Q_x = 6.27$ and 3200 turns (for MICROMAP 3400 turns) is shown in Fig. 7. It is noted that the rms emittances show a slow but steady growth after about 1000 turns, with at the same time a continuing small rate of loss. This steady growth cannot be attributed to just

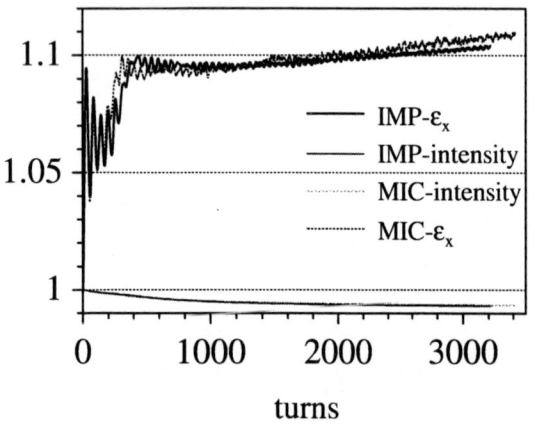

**FIGURE 7.** Evolution of MICROMAP and IMPACT simulations for $Q_x = 6.27$.

simulation noise as it is completely absent if the octupole is turned off, in which case the relative emittance growth is less than $10^{-4}$. It may be associated with slow diffusion of particles into the dynamic aperture, hence these particles contribute to the rms emittance before they hit the simulation boundary. It must not be assumed that this process continues for ever on the same rate, since tails get depleted, and eventually the process slows down. This was supported by an extended run of this case up to $10^4$ turns with MICROMAP, where the growth was found to saturate at 13% beyond 8000 turns.

## CONCLUSION

This study demonstrates that for Gaussian coasting beams, and under stationary external parameters, there is practically absence of coherent resonance effects, which justifies the use of non-self-consistent space charge calculation. The main reasoning behind this 2D finding is the effect of Landau damping due to the broad frequency spread of the Gaussian, which is even enhanced in bunched beams due to the additional spread from longitudinal current variation. This encourages the use of analytical space charge models for 3D studies, particularly in the realm of $10^5 - 10^6$ turns as use din Ref. [1], where fully self-consistent simulation appears impractical, except for special model tests on a reduced time scale. Obviously, such conclusions cannot be applied to a situation, where changing parameters (stacking injection, peak intensity increase during bunch rotation, shifting working point) enforce a significant dynamical change of the distribution function.

## ACKNOWLEDGMENTS

This work was performed in part using resources of the NERSC scientific computing center of the US DOE.

## REFERENCES

1. G. Franchetti et al. this conference.
2. A.V. Fedotov and I. Hofmann, *Phys. Rev. ST Accel. Beams* **5**, 024202-1 (2002).
3. G. Franchetti, I. Hofmann, and G. Turchetti, *AIP Conference Proceedings* **448**, 233 (1998).
4. A.V. Fedotov . R.L. Gluckstern and I. Hofmann, in *Proceedings of the Particle Accelerator Conference*, Portland, Oregon, 2003.
5. R. Baartman, AIP Conf. Proc. 448 (New York, 1998), p. 56.
6. I. Hofmann, G. Franchetti, O. Boine-Frankenheim, J. Qiang, R.D. Ryne, Phys.Rev.ST Accel.Beams 6:024202 (2003).
7. J. Qiang *et al.*, *J. Comp. Phys.* **163**, 434 (2000).

# Longitudinal Mismatch in SCL as a Source of Beam Halo*

Alessandro G. Ruggiero
(E-mail: agr@bnl.gov)
*Brookhaven National Laboratory, Upton, NY 11973*

**Abstract.** An advantage of a proton Super-Conducting Linac (SCL) is that RF cavities can be operated independently, allowing easier beam transport and acceleration. But cavities are to be separated by drifts long enough to avoid they couple to each other. Moreover, cavities are placed in cryostats that include inactive insertions for cold-warm transitions; and interspersed are warm insertions for magnets and other devices. The SCL is then an alternating sequence of accelerating elements and drifts. No periodicity is present, and the longitudinal motion is not adiabatic. This has the consequence that the beam bunch ellipse will tumble, dilute and create a halo in the momentum plane because of inherent nonlinearities. When this is coupled to longitudinal space-charge forces, it may cause beam loss with latent activation of the accelerator components.

## THE AGS SUPERCONDUCTING LINAC

At BNL we have few projects that require Super-Conducting Linacs (SCL). One of them is the AGS Upgrade for a proton average power of 1 MWatt (Ref. 1). The location of the AGS-SCL is shown in Figure 1. It follows the 200-MeV Room Temperature Linac and accelerates negative ions to 1.2 GeV for multi-turn, charge-exchange injection into the AGS.

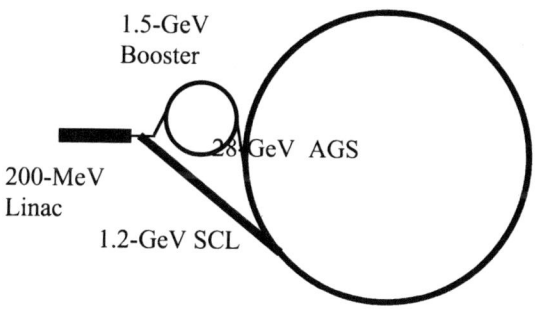

Figure 1. The AGS complex with the 1.2-GeV SCL

The SCL is made of three sections, as shown in Figure 2, that are: the Low-Energy (LE) section from 200 to 400 MeV, the Medium-Energy (ME) section from 400 to 800 MeV, and the High-Energy (HE) section to the final 1.2 GeV. The first section operates at 805 MHz, the last two at 1,610 MHz.

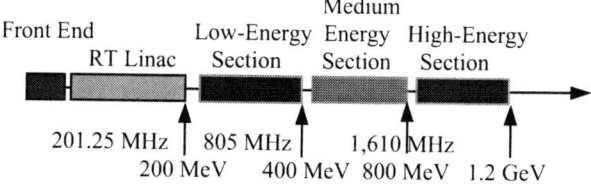

Figure 2. Layout of the AGS-SCL with three Sections

---

* Work performed under the auspices of the U.S. Department of Energy

Any of the three sections is made of a sequence of identical Periods, each made of a Warm Insertion for quadrupoles and other beam devices, and a Cryomodule inside which cavities are located, as shown in Figure 3. Each Cryomodule is made of $M$ Cavities, and each cavity is made of $N$ cells. Cavities are separated from each other by a distance $d$ large enough for the insertion of RF couplers and for decoupling. Each end of a Cryomodule has in addition a transition from the cold to the warm region that takes space. Finally Cryomodules are separated by the Warm Insertions that take an extensive length.

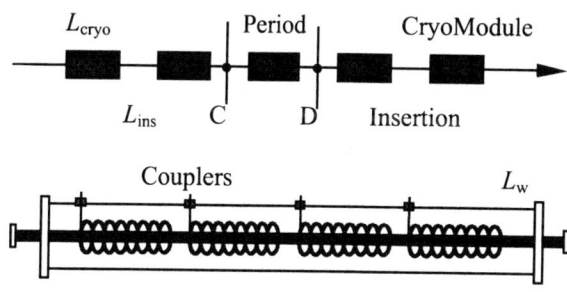

Figure 3. Sequence of Periods, Cavities, Cells and Warm Insertions in one SCL Section

Thus a section of SCL is made of a alternating sequence of Drifts of different length ($g$ and $d$) and Accelerating elements, namely Cavities with a number of RF Cells, as shown in Figure 4. The summary of the RF and geometry parameters for the AGS Upgrade SCL is given in Table 1.

## LONGITUDINAL EQUATIONS OF MOTION

Introduce the Time Delay $\tau = t - t_s$ and the Energy Difference $\varepsilon = E - E_s$. Let a prime denote derivative with respect to the path length. After linearization

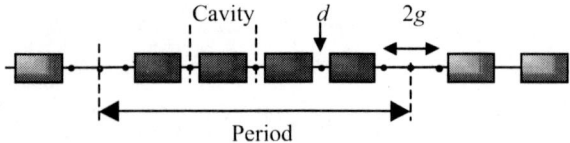

Figure 4. Arrangement of Periods, Cavities and Drifts

Table 1. AGS-Upgrade SCL

| Linac Section | LE | ME | HE |
|---|---|---|---|
| Injection Energy, MeV | 200 | 400 | 800 |
| Final Energy, MeV | 400 | 800 | 1,200 |
| RF Frequency, MHz | 805 | 1,610 | 1,610 |
| Ave. Axial Field, MV/m | 13.4 | 29.1 | 29.0 |
| No. of Periods | 6 | 9 | 8 |
| No. of Cavities/Period, $M$ | 4 | 4 | 4 |
| No. of Cells/Cavity, $N$ | 8 | 8 | 8 |
| Cell Length, cm | 11.45 | 7.03 | 7.92 |
| Cavity Length, cm | 91.60 | 56.24 | 63.36 |
| Cavity Separation, $d$, cm | 32 | 16 | 16 |
| Warm-Cold Transition, cm | 30 | 30 | 30 |
| Warm Insertion, cm | 107.9 | 137.9 | 137.9 |
| Period Drift, $g$, cm | 67.95 | 90.95 | 90.95 |

$$\tau' = -\varepsilon / c \beta_s^3 \gamma_s^3 E_0 \qquad (1)$$

$$\varepsilon' = -eE_{acc} (\sin \phi_s) \omega \tau \qquad (2)$$

where $\phi_s$ is the reference RF phase. For a segment of the accelerating structure, short enough to neglect the variation of $\beta_s^3 \gamma_s^3$, by combining Eq.s (1) and (2)

$$\tau'' + K^2 \tau = 0 \qquad (3)$$

$$K = [eE_{acc} (-\sin \phi_s) \omega / c \beta_s^3 \gamma_s^3 E_0]^{1/2} \qquad (4)$$

## MATRIX METHOD

The particle longitudinal motion can also be described with matrices (Ref. 2), in analogy to the method used to study transverse betatron oscillations. Each transport matrix applies to a column vector of components $\tau$ and $\tau'$. The transfer matrix for a Drift of length $\ell$ is

$$M_{drift}(\ell) = \begin{vmatrix} 1 & 1 \\ 0 & 1 \end{vmatrix} \qquad (5)$$

and for a Cell of length $L$

$$M_{cavity} = \begin{vmatrix} \cos\theta & (\sin\theta)/K \\ -K\sin\theta & \cos\theta \end{vmatrix} \qquad (6)$$

Define $\theta = KL$ and $\eta = d/L$. When the various elements are multiplied together, we obtain the transfer matrix for a Cavity

$$M_c = M_{drift}(d/2) \, M_{cavity} \, M_{drift}(d/2) \qquad (7)$$

$$M_c = \begin{vmatrix} \cos\mu_c & \beta_c \sin\mu_c \\ -(\sin\mu_c)/\beta_c & \cos\mu_c \end{vmatrix} \qquad (8)$$

where

$$K\beta_c = (1 - \theta^2 \eta^2 / 4 + \theta\eta \cot\theta)^{1/2} \qquad (9)$$

$$L \, tg \, \mu_c = \theta \beta_c / (\cot\theta - \theta\eta/2) \qquad (10)$$

Similarly the transfer matrix for a Period

$$M_p = M_{drift}(g) \, M_c^M \, M_{drift}(g) \qquad (11)$$

$$M_p = \begin{vmatrix} \cos\mu_p & \beta_p \sin\mu_p \\ -(\sin\mu_p)/\beta_p & \cos\mu_p \end{vmatrix} \qquad (12)$$

where

$$\beta_p = [\beta_c^2 - g^2 + 2g\beta_c \cot(M\mu_c)]^{1/2} \qquad (13)$$

$$L \, tg \, \mu_c = \beta_p / [\beta_c \cot(M\mu_c) - g] \qquad (14)$$

It is to be observed that the use of the transfer matrices $M_c$ and $M_p$ gives a good approximation when the energy gain across one Cavity section and, eventually, one Period is a small fraction of the energy in entrance.

Define the Bunch Area $S = \pi \Delta\tau \Delta\varepsilon$, the Bunch (Half) Length $\Delta\tau = (S_n \beta_{c,p} / \pi)^{1/2}$, and the normalized bunch area $S_n = S / c \beta_s^3 \gamma_s^3 E_0$.

## STABILITY OF MOTION

The condition for stability is that | Trace of $M_c$ and $M_p$ | < 2, that is | $\cos \mu_{c,p}$ | is real and less than unit, and also $\beta_{c,p}$ is real and positive. The stability Diagrams are shown in Figures 5 and 6. It is seen that when the major drifts $d = g = 0$, the motion is always stable. Increasing the length of the drifts causes a reduction of the range of stability. Figure 7 gives the maximum value $\theta_{max}$ of the rotation angle, that is the accelerating gradient, versus the ratio $\eta$.

## MISMATCH OF MOTION

We have assumed that $K$ does not change across a *period*. This is justified if the acceleration rate is not too high, and the energy change is only a small fraction of the total kinetic energy. Exact matching is achieved by requiring that the amplitude value $\beta_p$ remains constant from one period to the next. In turn that requires that also $\beta_c$ and $\mu_c$ per cavity interval remain unchanged. That is the rotation angle $\theta$ and the restoring parameter $K$ are also constant. For given RF angular frequency $\omega$ and phase $\phi_s$, we derive the following condition for exact matching

$$E_{acc} / \beta_s^3 \gamma_s^3 = \text{constant} \quad (15)$$

A difficult condition to satisfy!

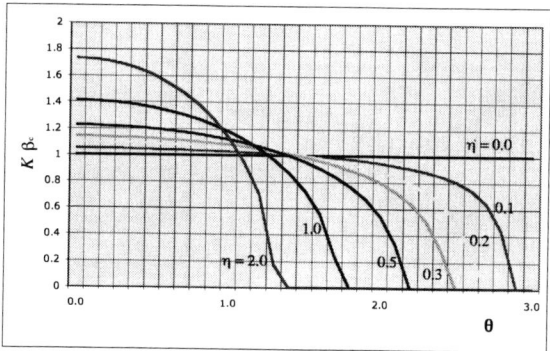

Figure 5. Plot of $K\beta_c$ versus $\theta$ for various values of $\eta$

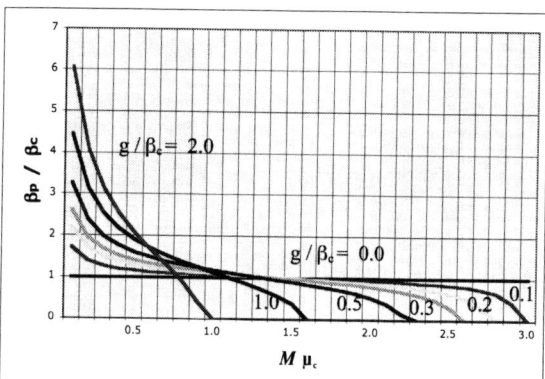

Figure 6. Plot of $\beta_p / \beta_c$ vs. $M\mu_c$ for several values of $g/\beta_c$

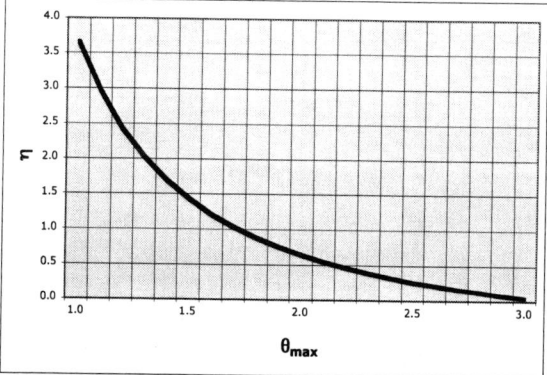

Figure 7. Plot of max value of $\theta$ versus $\eta$

A simpler mode of operation assumes a constant energy gain per *period*, so that the restoring parameter $K$ will decrease with the beam energy. This will cause continuous mismatch of the particle motion, resulting in a beam bunch rotation (with consequent possible dilution) of an amount that depends on the acceleration rate, and on the length of the drift insertions ($d$ and $g$). In analogy to the conventional approach used to describe the transverse betatron motion, a beam bunch can be made to correspond to an ellipse in the phase space ($\tau$, $\tau'$). Between periods, the ellipse is described by the amplitude $\beta_p$ and the inclination $\alpha_p$.

To estimate the amount of mismatch as the motion progress, we assume that the beam bunch is exactly matched at the entrance of the SCL section, where $\alpha_p = 0$. It is well know then how to estimate the bunch ellipse rotation, dilation or contraction from one period to the next with the transformation

$$\left| \begin{array}{c} \beta_p \\ \alpha_p \\ \gamma_p \end{array} \right|_2 = \left| \begin{array}{ccc} m_{11}^2 & -2 m_{11} m_{12} & m_{12}^2 \\ -m_{21} m_{11} & 1 + 2 m_{12} m_{21} & -m_{12} m_{22} \\ m_{21}^2 & -2 m_{22} m_{21} & m_{22}^2 \end{array} \right| \left| \begin{array}{c} \beta_p \\ \alpha_p \\ \gamma_p \end{array} \right|_1 \quad (16)$$

$$\beta_p \gamma_p = 1 + \alpha_p^2 \quad (17)$$

Figure 8 shows some typical parameters in the Low-Energy section of the AGS-SCL that indicate how close is the stability limit (Ref. 3). The number of phase oscillations over one period is large when compared to that of circular accelerators. The inclination parameter $\alpha_p$, a measure of the mismatch of motion, and the envelope amplitude function $\beta_p$ are also shown, together to the bunch dimension.

Let then the equation of the longitudinal bunch ellipses

$$S_n/\pi = \gamma_p \tau^2 + 2\alpha_p \tau \tau' + \beta_p \tau'^2 \quad (18)$$
$$\tau'/\tau_0' = -\alpha \tau/\tau_0 + [1 - (\tau/\tau_0)^2]^{1/2} \quad (19)$$

where $\tau_0$, $\tau_0'$ are the extensions in the ($\tau,\tau'$) phase space. The evolution of the bunch ellipse is shown in Figure 9 at the end of each period for the AGS Low-Energy section in actual coordinates, whereas the same ellipses are also shown in Figure 10 using normalized coordinates (19). The continuous mismatch is noticeable.

## CONCLUSIONS

Usually, our perception of a beam moving down an accelerator is made of bunches having the longitudinal shape of unchanging upright ellipses. This may be indeed a good approximation in circular accelerators where the energy gain per turn is small, but is not correct in the case of linear accelerators where the acceleration rate is considerably higher and

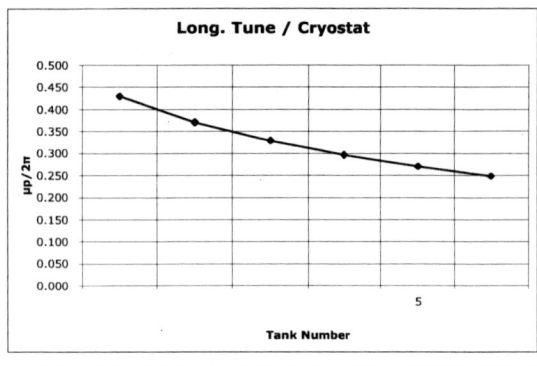
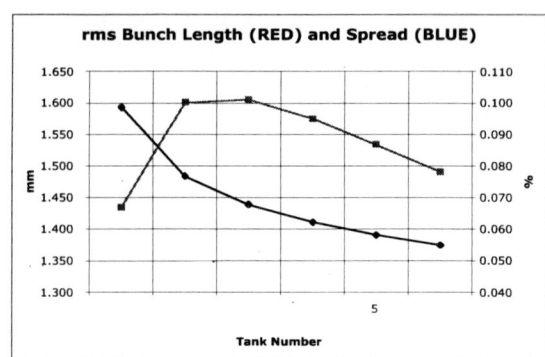
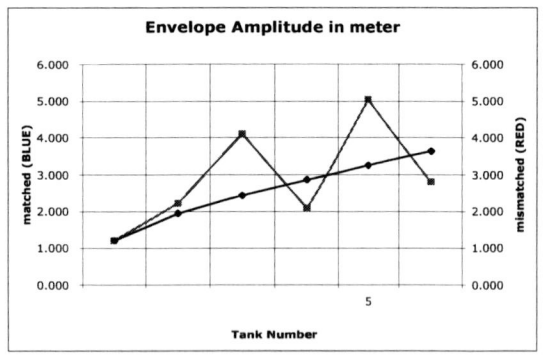
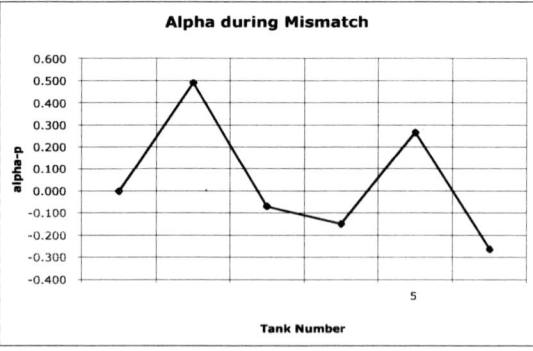

Figure 8. Variation of some bunch parameters with the period number.

the beam energy changes over short periods of length. Moreover, active accelerating cavities are separated by drifts of various lengths that introduce an intrinsic mismatch by which beam bunches actually rotate, elongate and contract over considerably short periods of length of only few meters. This has an analog with the transverse motion where also the betatron emittance ellipse changes continuously but periodically. The longitudinal motion nevertheless has no intrinsic periodicity and it is continuously mismatched from one location to the next. Our concern is that the continuous tumbling rapidly changing in a SCL of the bunches may lead to the creation of longitudinal halos accompanied by latent beam losses. Operation of high-power SCL in the GeV range needs to be demonstrated. There are several proton SCL projects being proposed, but only one is presently under construction (SNS) and will be soon, hopefully, in operation.

**REFERENCES**

1. A.G. Ruggiero et al., "AGS Upgrade to 1-MW with a Super-Conducting Linac Injector", presented at PAC03, Portland, Oregon, May 2003.
2. A.G. Ruggiero, "Design Considerations on a Proton Superconducting Linac", BNL-62312, Aug. 1995.
3. A Visual Basic Program for the Design of SCL, available by request to the Author.

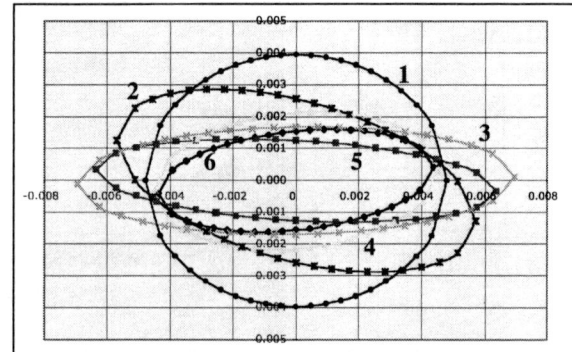

Figure 9. Bunch ellipses at the exit of each period of the SCL Low-Energy section.

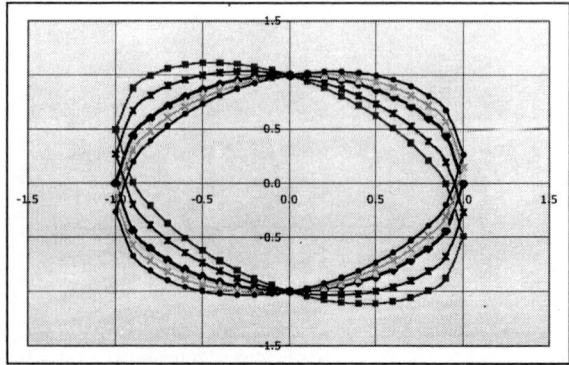

Figure 10. The same bunch ellipses in the variables $\tau'/\tau_0'$ and $\tau/\tau_0$

# Observations and simulation of a fourth order resonance with space charge

G. Franchetti*, I. Hofmann*, M. Giovannozzi[†], M. Martini[†] and E. Metral[†]

*GSI, 64291 Darmstadt, Germany
[†]CERN, 1211 Geneva, Switzerland

**Abstract.** A benchmarking experiment with a high intensity bunched beam stored for 1.2 s in a nonlinear lattice has been performed at the CERN Proton Synchrotron (PS). Beam emittance and beam intensity have been measured for several working points at different distances from a lattice-induced resonance. We found a regime of emittance growth for machine tunes far from the resonance and a regime of emittance shrink/beam loss for working point close the resonance. We compare the observations with 3D simulations and in the blow-up regime we find good agreement. We interpret these results in terms of space charge induced trapping and de-trapping on the lattice resonance. We show that this mechanism is responsible of halo formation which depends on the distance from the resonance and discuss the interplay of dynamic aperture and halo size as main mechanism to explain beam loss in the experiment.

## INTRODUCTION

Up to now space charge effects in beams and single-particle nonlinear dynamics have always been two distinct areas of interest. Typically, space charge effects have been studied in rings with both simulation and experimental work on the range of thousands of turns [1, 2]. Resonances in this regime are dominated by space charge. On the contrary, single-particle nonlinear dynamics studies have been carried out in the range of hundred thousand turns (see for instance Ref. [3]). There, nonlinearities are intrinsic to the machine lattice, while self-consistent effects are absent. For several projects this distinction is natural, as storage times for high-energy machines are long (typically hours), but space charge effects are negligible. In high-intensity machines, however, the beam is stored for much shorter times. This situation allowed an independent development of space charge beam physics and single-particle nonlinear dynamics. However, in new projects such as the SIS100 synchrotron at GSI [4, 5] this distinction is no more justified. Storage for 1 s of a high intensity beam in a nonlinear lattice with loss levels not exceeding 1% is requested. The interplay between pure single-particle lattice driven nonlinear effects (long-term) and pure space charge effects (short-term) becomes critical. In the simplified simulation model of Ref. [6] it was recently suggested that synchrotron motion induces, because of space charge, a single-particle tune modulation. Consequently, single-particle periodic crossing of a resonance eventually leads to particle trapping into and de-trapping from resonance islands. The synchrotron-space-charge motion induced on resonance islands may take particles in and out the beam core. This process is related to the trapping during a single passage through a higher order nonlinear resonance described in Refs. [7, 8] as a results of a varying tune. In this work we present the results of an experiment performed in a regime of high-intensity beam stored for 1 s in presence of well-controlled lattice nonlinearity and attempt an interpretation of these results through numerical simulations.

## MEASUREMENTS

The measurements were carried out as part of a high intensity machine development time at the PS in October 2002. The number of protons in the single-bunch (200 ns long at $4\sigma$) was $1.1 \times 10^{12}$, small enough to avoid overlap with other resonances. A vertical maximum space charge tuneshift of 0.12, and a horizontal one of 0.075 (for minimum amplitude particles) were achieved with relatively small emittances of $\varepsilon_x = 9$ mm mrad and $\varepsilon_y = 4.5$ mm mrad (unnormalized at $2\sigma$). The bunch profiles measured 10 ms after injection were found to be Gaussian in all directions in the absence of the octupole. The vertical machine tune was set to $Q_y = 6.12$, and the horizontal tune was varied in the interval $6.25 < Q_x < 6.32$. The chromaticity was close to the natural one, hence the small momentum spread of $10^{-3}$ (at $2\sigma$) allows ignoring chromatic effects. The kinetic energy was kept at the injection value of 1.4 GeV with a measurement window of 1.2 s ($4.4 \times 10^5$ turns) over which the bunch intensity was monitored with a current transformer. The calibrated octupole (here $K_3 = 1.215 \times I$ m$^{-3}$) was powered to 40 A at 110 ms after injection to excite the resonance $4Q_x = 25$. We used the transverse profiles measured with the flying wire (20 m/s), fitted them with a Gaussian profile, and determined the corresponding rms emittances. Initial and

- in most cases - final profiles were actually found quite close to Gaussian. In Fig. 1 results of final measurements 1.2 s after injection are plotted as a function of the machine working point.

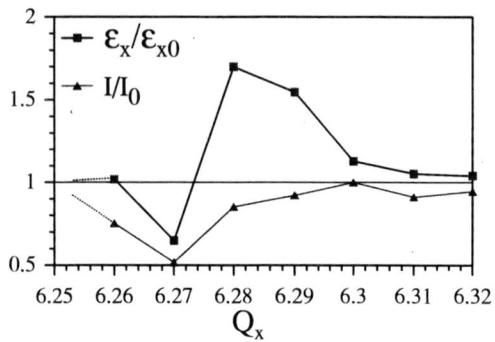

**FIGURE 1.** Experimental results on final rms emittance (of Gaussian fit) and beam current relative to initial values.

## SIMULATION

Interpretation of these measurements must rely on adequate computer simulations. The comparison is also a necessary basis for code benchmarking on such a long time scale, which has not been undertaken so far to our knowledge. To initiate such a process we have carried out a series of simulation runs in 3D. We have replaced, for simplicity, the linear PS focusing lattice by constant focusing and ignored lattice nonlinearities besides the contribution from the well-defined octupole. We have also ignored the smaller vertical beam emittance of the experiment and assumed a circular cross section to match with the limitation in the analytical 3D space charge model, which is based on a rotational ellipsoid. The horizontal emittance has been re-defined accordingly such as to reproduce accurately the maximum horizontal space charge tuneshift extracted from the measurements, which we believe is the crucial issue here, since we are not dealing with a coupling resonance.

### Dynamic aperture

The loss observed in the experiment must be related to the shrinking of the dynamic aperture, since the beam was too small to hit a physical aperture. To roughly explore the effects on the dynamic aperture of the octupole alone we have carried out a numerical test by searching the maximum stable radius of test particles placed into 20 different directions in the upper half of the $x-y$ plane, and ignoring space charge at this point. We have found that the nominal octupole (40 A) leads to a dynamic aperture ($10^5$ turns) of about $5\sigma$ near $Q_x = 6.25$, where $\sigma$ is the horizontal rms beam size of the injected beam. This value is not small enough to explain the observed extraction of particles as will be seen in the subsequent simulations. Hence, a more complete knowledge of machine nonlinearities at the working points used here may be needed to explain the observed loss. Assuming 200 A octupole current we have calculated that the dynamic aperture shrinks to a radius of $2.5\sigma$ near $Q_x = 6.25$ for $10^3$ turns, and about $2.2\sigma$ for $10^5$ turns.

### 3D simulations with synchrotron motion

In [9] it has been found that for a Gaussian beam coherent effects are "Landau-damped". This fact allows to use frozen models for space charge calculation. It was also found in a 2D simulation for $Q_x = 6.27$ using an octupole powered to 200 A that over 3000 turns the emittance increases by $\sim 10\%$. The failure of 2D simulation to describe the experiment justifies the need for including the longitudinal motion in a 3D bunch model, which induces a tune modulation primarily via space charge. Space charge is computed using a fully analytical method (see Ref. [10] for more details). It only uses integrals of algebraic expressions in $x, y, z$ and is therefore sufficiently fast and noise-free. We employ a density distribution of the kind $(1-x^2/a^2)^3$ (in $x$, and similar in $y$ and $z$). Its core is sufficiently close to that of a Gaussian, but it has a finite beam edge at $3\sigma$. Using 2000 test particles we generate the corresponding (initially) consistent distribution in 6D phase space, assuming the same bunch length (200 ns at $4\sigma$) and synchrotron period (645 turns) as in the experiment.

The dependence of emittance growth on the working point is seen (Fig. 2) to have a similar trend as in the experiment for $Q_x > 6.28$, but no loss for smaller tunes. For better comparison with the experiment we have applied a Gaussian fit to the simulation data and determined the rms emittance from this fit, which puts the emphasis on the core emittance rather than rms emittances. Note that the relatively large emittance growth without accompanying loss reflects the large physical aperture in both experiment and simulation, if compared with the initial beam size. The resulting maximum halo increases for $Q_x \to 6.25$ (Fig. 3). This is due to the fact that larger betatron amplitudes must be adopted to compensate the increasing space charge, while the particle moves longitudinally to the bunch center and trapping on the resonance island is maintained. Note that for $Q_x \to 6.32$, where the resonance loses its effect, the halo shrinks to the initial beam edge of $3\sigma$. This halo formation by island trapping is the reason for beam loss in the experiment, where apparently further nonlinearities cause a smaller dynamic aperture than in the simulation and lead to halo extraction. An example of a simulated transverse beam density cut with pronounced halo determined after $5 \times 10^5$ turns is shown in Fig. 4 on a logarithmic vertical scale. The to-

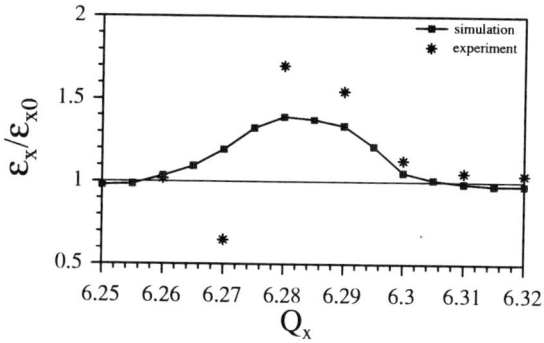

**FIGURE 2.** 3D simulation using analytical space charge (40 A octupole). Shown are simulated rms emittances (Gaussian fit) after $5 \times 10^5$ turns, and experimental values.

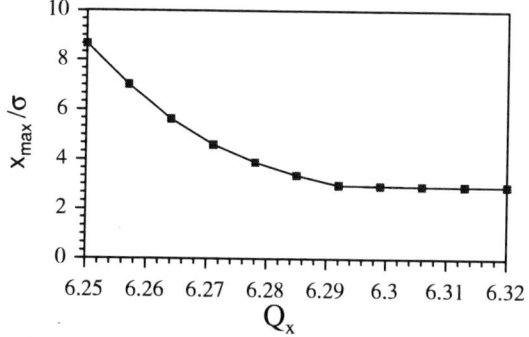

**FIGURE 3.** Halo radius (in units of initial $\sigma$).

tal number of particles in the halo at this instant is about 1%. This quantity does not include the particles, which have been temporarily in the halo at some earlier time, and which would determine the total loss, if a physical aperture would intercept the halo.

## Interpretation

Our interpretation of the significant difference between 2D (only 10% emittance growth for I=200A, see Ref. [9]) and 3D emittance growth relates this to syn-

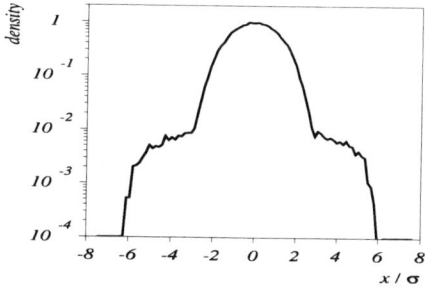

**FIGURE 4.** Normalized density in simulation for $Q_x = 6.26$ (radial units in initial $\sigma$).

chrotron motion: in 2D particles have static tunes and are on resonance for basically *one* value of the betatron amplitude. Once on resonance they get easily de-tuned again after small amplitude increase due to the dominant space charge detuning; in 3D the synchrotron motion makes the particles oscillate between high and low space charge, which induces an efficient periodic tune modulation. Eventually such particles are locked to the resonance islands, which implies that they may be carried to larger transverse amplitude to compensate the enhanced space charge when moving back to the bunch center. As a single-particle process, hence in the absence of space charge, such amplitude growth by island trapping was studied in detail in Ref. [8], and proposed as beam cleaning method in Ref. [11]. A relatively large number of particles is thus able to cross periodically the resonance at various betatron amplitudes, until trapping occurs. As it was shown in Ref. [6], such trapping may be followed by de-trapping after some time, unless the particle hits the aperture before. This process is shown in Fig. 5 for a single test-particle arbitrarily chosen with maximum synchrotron amplitude, hence initiated at the bunch end. Units on the abscissa are single particle emittances relative to the initial (transverse) beam edge emittance corresponding to $3\sigma$. The initial relative single-particle emittance is chosen to be 0.8 and it is plotted at each turn. The tune of $Q_x = 6.257$ is sufficiently close to the resonance that a large halo radius may be expected, which is confirmed by the simulation. This growth up to 5.5-times the beam edge emittance is consistent with Fig. 3 as it corresponds to $x_{max}/\sigma = 7$. From Fig. 5 (top), we find that the single-particle normalized emittance follows mainly two behaviours: in one it has almost stationary oscillations which last for about half a synchrotron oscillation. The other feature is characterized by emittance jumps. According to our previous description, we attribute such a jump to resonance trapping. In this respect, Fig. 5 (top) suggests that this test particle occasionally gets and remains trapped while going through the bunch center. This in fact corresponds to one oscillation of the single-particle emittance which is a rare event found before and after turn 2000. The most common pattern of the test-particle, however, seems to reveal that trapping and de-trapping happens before the particle passes through the bunch center. For this particular example with $Q_x = 6.257$ the resonance crossing happens relatively close to the bunch end, justifying therefore the frequency of emittance jump of half synchrotron frequency. In Fig. 5 (top) it is also shown that between two jumps the emittance exhibits almost stationary emittance oscillations. This stems from the combined effect of space charge and octupole nonlinearity. Hence, the larger the particle amplitude the larger the single-particle emittance oscillations. For small emittances, instead, the effect of the octupole is reduced and the emittance os-

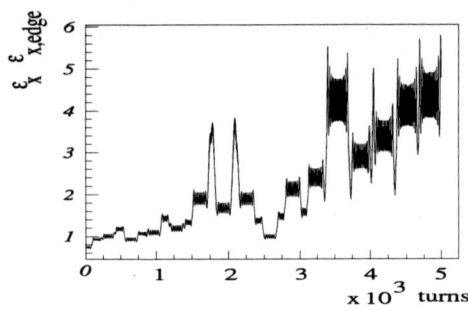

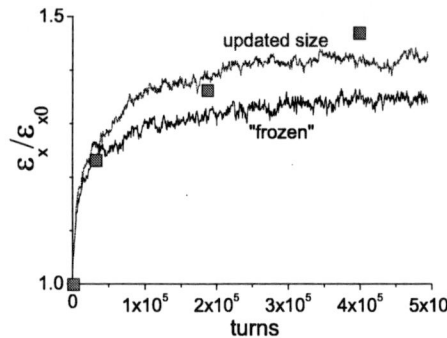

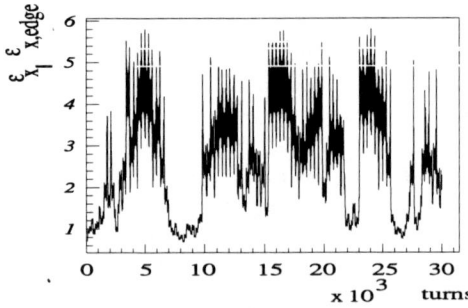

**FIGURE 5.** Trapping and de-trapping of a single test-particle and $Q_x = 6.257$ (top frame zoom of bottom frame).

cillations become much smaller. In Fig. 5 (bottom), the single-particle emittance is shown over a longer time scale.

## Rms self-consistency

The simulation rms emittance evolution for $Q_x = 6.29$ using the fully frozen analytical space charge is shown in Fig. 6. The time profile compares well with the measured data, but saturates somewhat below the experimental values. This saturation is due to the fact that an equilibrium occurs between trapping and de-trapping among the particles with sufficiently large synchrotron amplitude (depending on $Q_x$) to cross the resonance. We compare this result with a modification, where the growing rms emittance is used to update the horizontal rms size. As a result of this rms self-consistency space charge gets weaker with progressing emittance growth. This allows even particles cross the resonance, which have initially had too small synchrotron amplitudes - correspondingly large tuneshift - to be able to reach the resonance condition. With this increasing number of potentially resonant particles, further emittance growth takes place, and better agreement with the measurement is achieved. While this simple modification works well in the emittance growth regime, it does not help to improve the agreement in the loss regime, where better knowledge of the dynamic aperture is needed.

**FIGURE 6.** 3D simulation emittance evolution for $Q_x = 6.29$ comparing analytical fully frozen space charge with results obtained by using a continuously updated rms size (squares: measured values).

## CONCLUSION

Synchrotron oscillations have been shown to enhance significantly the effect of the octupole. In the emittance growth regime quite good agreement is achieved with the measurements over half a million turns, which supports our 3D space charge model and the interpretation in terms of tune modulation. We predict the formation of a halo increasing in radius for $Q_x \to 6.25$ and claim this is the source of the measured loss. Future measurements should consider weaker octupoles, where the predicted halo might be entirely inside the dynamic aperture and observable by scrapers. Cross-checks with fully self-consistent 3D simulation over some $10^4$ turns (stronger octupole case) are planned for the near future.

## REFERENCES

1. S. Cousineau *et al.*, Phys. Rev. ST Accel. Beams **6**, 034205 (2003).
2. K. Takayama *et al.*, EPAC02, p. 1413 (2002).
3. M. Giovannozzi, this conference.
4. W. Henning, Proceedings of the 2003 Particle Accelerator Conference, Portland, Or., USA (2003).
5. P. Spiller *et al.*, ibid.
6. G. Franchetti and I. Hofmann, AIP Conf. Proc. **642**, 260 (2002).
7. A. Schoch, CERN Report, CERN 57-23 (1958).
8. A.W. Chao and M. Month, Nucl. Instr. and Meth. **121**, 129 (1974).
9. I. Hofmann *et al.*, this conference.
10. G. Franchetti, I. Hofmann, M. Giovannozzi, M. Martini, E. Metral, to be submitted to Phys. Rev. ST Accel. Beams.
11. A.W. Chao and M. Month, Nucl. Instr. and Meth. **133**, 405 (1976).

# Halo and RMS Beam Growth due to Transverse Impedance

V. V. Danilov and J. A. Holmes

*SNS Project, Oak Ridge National Laboratory, Oak Ridge, TN 37831-6473*

**Abstract.** Collective beam dynamics will play a major role in determining losses in high intensity rings. We demonstrate here, using both analytic and computational models, that beam halo can form under the influence of transverse impedances, even for stable cases well below the instability threshold. It is shown for cases above the instability threshold that rms beam size and halo develop more rapidly than the beam centroid.

## I. INTRODUCTION

We consider the well known case of a coasting beam with energy spread in the presence of a transverse impedance interacting with dipole moments of the beam. In our analysis, we look beyond the simple dipole moment point of view to examine other characteristics of the beam. These characteristics include the rms beam size and the maximum extent of the beam halo.

It is shown that, close to the instability threshold, the rms size of the beam grows at a more rapid pace than the centroid. The resonant particles, in addition to providing Landau damping for the beam below the instability threshold, themselves undergo maximal betatron amplitude growth, thus generating beam halo. Section 3 presents the comparison of analytic models with simulations using the ORBIT Code[1]. The simulations confirm all the analytically calculated effects.

## II. EVOLUTION OF A PENCIL BEAM WITH TRANSVERSE IMPEDANCE: ANALYTIC EVALUATION

We now consider an exactly solvable model and compare the results calculated analytically with those obtained using ORBIT's transverse impedance module [2]. We consider a coasting beam with a Lorentz energy distribution in a constant focusing periodic beam channel containing one localized harmonic impedance element, which is taken to be real. In this problem, we ignore the direct space charge force and take the beam line to be straight so that we can ignore dispersion effects. Also, throughout this paper we will neglect chromaticity. There is special interest in this problem. First, we can solve it exactly and study the evolution of the distribution, the emittance, and the beam halo for different beam intensities. These are the main concerns for systems like the Spallation Neutron Source ring because they can lead directly related to its activation. Second, exactly solvable models give an excellent opportunity to check the accuracy and predictive capability of numerical methods.

To begin the analysis, let us consider the dipole moment $D(z,\delta,\tau)$, now expressed as a function of $z, \delta$, and $\tau$, which are, respectively, the longitudinal coordinate in the bunch, the energy deviation $\delta = \frac{\delta E}{E_{Tot}} = \beta^2 \frac{\delta p}{p_0}$, and the commoving position coordinate $\tau = \frac{s}{\Pi}$, where $\Pi$ is the periodic length. We assume that the beam receives an impedance kick once each lattice period and undergoes betatron oscillations between kicks. Thus, the equation for $D(z,\delta,\tau)$ is

$$\frac{d^2D}{d\tau^2} + K(s)\Pi^2 D = \frac{F}{\gamma m}(\frac{\Pi}{\beta c})^2 \qquad (1)$$

where $K(s)$ is the transverse focusing function and $F$ is the force of the impedance kick. By defining a normalized dipole moment $D_N = \frac{D}{\sqrt{\beta(s)\varepsilon}}$ and a phase coordinate $\phi = \frac{1}{2\pi v_b} \int_{s_0}^{s} \frac{ds}{\beta(s)}$, where $v_b$ is the betatron tune, $\varepsilon$ is the emittance of the beam, and $s_0$ is some chosen reference position in the lattice, the equation for the dipole moment for a general periodic lattice can be written $\frac{d^2 D_N}{d\phi^2} + (2\pi v_b)^2 D_N = \frac{F}{\gamma m}(\frac{2\pi v_b}{\beta c})^2 \sqrt{\frac{\beta^3(s)}{\varepsilon}}$.

For the uniform focusing case considered here, it can be written $\frac{d^2 D}{d\tau^2} + (2\pi v_b)^2 D = \frac{F}{\gamma m}(\frac{\Pi}{\beta c})^2$.

To analyze this equation, some further assumptions are required. For the localized impedance, we assume a single harmonic, so that $Z_\perp = Z_\perp(n\omega_0 + \omega_b)$ for integer $n$. The force with this single harmonic is then given by the equation $F = -\text{Re}\{iqID(z,\delta,\tau)Z_\perp(n\omega_0 + \omega_b)\delta_\Pi(s - s_0)\}$, where $q$ is the charge, $I$ is the current, $\delta_\Pi(s - s_0)$ is the periodic delta function with period $\Pi$, and $s_0$ is the location of the impedance. The dipole moment can then be factored into rapidly and slowly varying parts:

$$D(z,\delta,\tau) = d_s(\delta,\tau)\exp\{i(2\pi n \frac{z}{\Pi} + \omega_b t)\} = d_s(\delta,\tau)\exp\{i[2\pi(n + v_b)\frac{z}{\Pi} + 2\pi v_b \tau]\}$$

where $d_s(\delta,\tau)$ varies slowly and is independent of $z$. We also note that, as a total derivative, $\frac{d}{d\tau} = \frac{\partial}{\partial\tau} + \frac{dz}{d\tau}\frac{\partial}{\partial z} + \frac{d\delta}{d\tau}\frac{\partial}{\partial\delta}$, where $\frac{dz}{d\tau} = \frac{\Pi \eta \delta}{\beta^2}$, $\frac{d\delta}{d\tau} = 0$, and $\eta = \frac{1}{\gamma^2} - \frac{1}{\gamma_T^2}$ is the phase slip factor. For the straight lattice considered here, $\frac{1}{\gamma_T} = 0$. We also assume that $\delta$ and $\frac{dd_s}{d\tau}$ are small quantities and that $\frac{d^2 d_s}{d\tau^2}$ is negligible. Substitution of $D(z,\delta,\tau)$ into the uniform focusing equation and application of the above assumptions yields a first order equation:

$$\frac{\partial d_s(\delta,\tau)}{\partial\tau} + i\frac{\eta\delta}{\beta^2}2\pi(n + v_b)d_s(\delta,\tau) = i\frac{F}{2\gamma m}(\frac{\Pi}{\beta c})^2 \frac{1}{2\pi v_b}\exp\{-i[2\pi(n + v_b)\frac{z}{\Pi} + 2\pi v_b \tau]\}.$$

To place this equation in a form for further analysis, we now make use of the fact that the collective force $F$ is small and apply the method of averaging to take the average of the force term over $K$ turns, where $K \gg 1$. Finally, we include the effect of the energy distribution in the force term by replacing the dipole moment by its integral over the unit normalized energy distribution $g(\delta)$. In the present calculation we assume $g(\delta) = \frac{\delta_0}{\pi(\delta^2 + \delta_0^2)}$ is the Lorentz distribution with width $\delta_0$. The resulting equation is:

$$\frac{\partial d_s(\delta,\tau)}{\partial\tau} + i\Delta(\delta)d_s(\delta,\tau) = \chi \int_{-\infty}^{\infty} g(\delta)d_s(\delta,\tau)d\delta, \quad (2)$$

where $\Delta(\delta) = \frac{\eta\delta}{\beta^2}2\pi(n + v_b)$, $\chi = \frac{-qIZ_\perp(n\omega_0 + \omega_b)}{2\gamma m(\beta c)^2}\beta_{s_0}$, and $\beta_{s_0}$ is the lattice beta function at $s_0$, the location of the impedance. For solutions with instability, we concentrate on the slow mode, and the signs of both $Z_\perp$ and $n$ are negative. We note that in this approximation we have neglected the dependence of $v_b$ on energy, consistent with the assumption that chromaticity is zero. The factor of $v_b$ appears in the equation for $\Delta(\delta)$ because the betatron phases of the particles change due to difference in velocities. Assuming an initial condition $d_s(\delta,\tau = 0) = 1$, the solution of Eq. (2) is:

$$d_s(\delta,\tau) = \exp(-i\Delta\tau) - \frac{\chi(\exp(-(\Delta_0 - \chi)\tau) - \exp(-i\Delta\tau))}{\Delta_0 - \chi - i\Delta}, \quad (3)$$

where $\Delta_0 = |\Delta(\delta_0)|$. Equation (3) provides one of the principle points of comparison between the analytic and computational solutions to this problem. Before carrying out this comparison, we complete the analytic evaluations.

Equation (3) shows that the threshold condition for instability of the slow dipole moment occurs when $\operatorname{Re} \chi = \Delta_0$. This can happen only when either $Z_\perp(n\omega_0 + \omega_b) < 0$ or $n + \nu_b < 0$. These conditions correspond to so-called slow waves (see the details in [3]). Below, for simplicity, we deal only with slow waves and real impedance $Z_T = -Z_\perp(n\omega_0 + \omega_b) > 0$. Solving the threshold condition for a coasting beam containing $N$ particles in a periodic length in a straight uniform focusing lattice, we obtain for the threshold number of particles:

$$N_{th} = \frac{2\pi\delta_0}{\gamma\beta}\frac{Z_0}{Z_T}\frac{\nu_b(n+\nu_b)}{r_c}, \quad (4)$$

where $Z_0 = 377\,\Omega$ and $r_c$ is the classical radius of the particle.

Because of the energy spread and the impedance, the size of the coasting beam changes in time. It is possible to evaluate the beam size using Eq. (3) and the definitions $\bar{d}_s = \int_{-\infty}^{\infty} \operatorname{Re}(d_s) g(\delta) d\delta$ and $\sigma^2 = \int_{-\infty}^{\infty} \{\operatorname{Re}(d_s) - \bar{d}_s\}^2 g(\delta) d\delta$ to obtain, asymptotically for large times:

$$\frac{\sigma^2}{d^2(0)} = \frac{\Delta_0}{2(\Delta_0 - \chi)} = \frac{\delta_0}{2(\delta_0 - \zeta)} = \frac{N_{th}}{2(N_{th} - N)} \quad (5)$$

for stable cases (intensities below threshold), and

$$\frac{\sigma^2}{|\bar{d}_s|^2} = \frac{\Delta_0}{2(\chi - \Delta_0)} = \frac{\delta_0}{2(\zeta - \delta_0)} = \frac{N_{th}}{2(N - N_{th})} \quad (6)$$

for unstable cases (intensities above threshold). In Eqs. (5,6), we use the threshold condition to define and obtain $\zeta = \frac{\chi\delta_0}{\Delta_0} = \frac{N\delta_0}{N_{th}}$.

Now we determine a maximal halo extent $h$, which we define to be the maximum amplitude of the beam dipole moment for asymptotically large time. From (3) it is evident that the maximum dipole moment occurs for particles with zero energy offset, $\Delta = 0$. Therefore, below the threshold we have:

$$\frac{h}{d(0)} = \frac{\Delta_0}{\Delta_0 - \chi} = \frac{\delta_0}{\delta_0 - \zeta} = \frac{N_{th}}{N_{th} - N}. \quad (7)$$

Above the threshold the halo extent is infinite but, as in the case of the rms size, its ratio to the average dipole moment of the beam approaches a finite number:

$$\frac{h}{|\bar{d}|} = \frac{\chi}{\chi - \Delta_0} = \frac{\zeta}{\zeta - \delta_0} = \frac{N}{N - N_{th}}. \quad (8)$$

We now compare for a specific case the analytically obtained results of Eqs. (3-8) with results calculated using the ORBIT code.

## III. EVOLUTION OF A PENCIL BEAM WITH TRANSVERSE IMPEDANCE: ANALYTIC - COMPUTATIONAL COMPARISON

The case we consider uses a coasting proton beam having kinetic energy 1 GeV ($E_{Tot} = 1.938\,\text{GeV}$) and a Lorentz energy distribution of width $\delta E = 10\,\text{MeV}$. We use a straight periodic uniform focusing channel of length $\Pi = 40$ meters with horizontal tune $Q_x = 1.10$ and vertical tune $Q_y = 1.05$. The single local impedance harmonic is taken at $n = -2$ in the vertical direction, and the value is set to $Z_\perp = 2 \cdot 10^5 \frac{\Omega}{\text{meter}}$. Thus, this is a one-dimensional problem. Finally, we assume a pencil beam having a "slow" $n = -2$ harmonic in the vertical plane of initial size 1 mm. The computer calculations presented here are carried out with $2 \cdot 10^5$ macroparticles. Results from the same calculations, when done with $1 \cdot 10^5$ macroparticles, show almost no difference.

Figure 1 compares the evolution of the simulated and analytic vertical beam phase space for particles from -5 to +5 of the Lorentz energy width for $N = 3 \cdot 10^{13}$. The simulation shows somewhat more rapid growth than the analytical prediction.

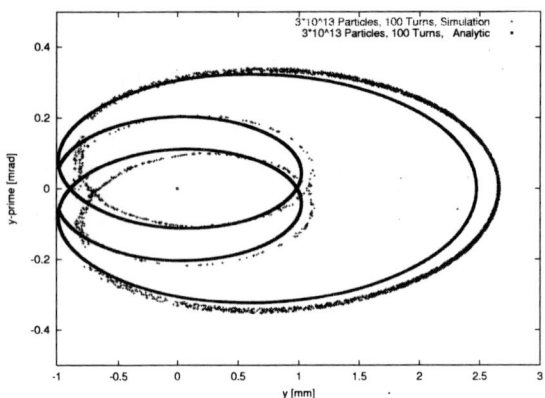

**FIGURE 1** Comparison of evolution of simulated and analytic phase space distributions for $N=3*10^{13}$.

Figure 2 shows the fraction of particles, calculated both numerically and analytically, with the vertical coordinate greater than 1.1 times the initial displacement for the stable case. The numerical results for the maximum and the ultimate fraction of particles in the halo exceed the analytic values by about 10%. This may be related to the finite longitudinal size of the slice (contrary to the infinitesimally short slice in the analytical solution). It is interesting that even the relaxation of stable beams will lead to halo growth.

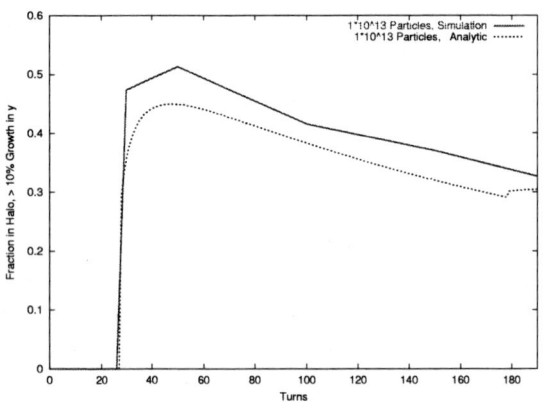

**FIGURE 2** Simulated and analytic fractions of particles with positions exceeding 1 mm by at least 10% for $N = 1 \cdot 10^{13}$ particles.

In the case of nonzero initial beam size $\sigma_0$, a simple calculation for the size $\sigma$ of the beam distribution shows that

$$\sigma^2 = \sigma_{Pencil}^2 + \sigma_0^2, \qquad (9)$$

where $\sigma_{Pencil}$ is the size of the pencil beam from previous calculations.

## IV. CONCLUSIONS

In this paper we have described the development of beam halo as part of the relaxation process, even for stable cases. The results presented are valid for coasting beams or for long bunches with high frequency instability, since the latter can be described approximately by the coasting beam model. We considered only Lorentz energy distributions, but the phenomenon of halo growth is related to the Landau damping process and is expected to appear for all types distributions having particles in resonance with collective modes.

More complicated instabilities, such as weak head-tail or transverse mode coupling instabilities, may behave similarly regarding rms size and halo. The derivation of particular formulas for these cases, similar to Eq. (5-8), is the subject of future work.

## V. ACKNOWLEDGMENTS

The authors wish to thank Mike Blaskiewicz, John Galambos, Stuart Henderson, Andrey Shishlo, and Jie Wei for many useful discussions and suggestions during this investigation.

*Research on the Spallation Neutron Source is sponsored by the Division of Materials Science, U.S. Department of Energy, under contract number DE-AC05-96OR22464 with UT-Battelle Corporation for Oak Ridge National Laboratory.

## REFERENCES

1. J. Galambos, J. Holmes, D. Olsen, A. Luccio, and J. Beebe-Wang, ORBIT Users Manual, http://www.sns.gov//APGroup/Codes/Codes.htm

2. V. Danilov, J. Galambos, and J. Holmes, in *Proceedings of the 2001 Particle Accelerator Conference*, (Chicago, 2001).

3. A. W. Chao, "Physics of Collective Beam Instabilities", John Wiley & Sons, Inc. (New York: 1993) p.247.

# Collective effects and collisions in halo genesis and growth

G. Turchetti*, C. Benedetti*, A. Bazzani* and S. Rambaldi*

*Dipartimento di Fisica Universitá di Bologna, INFN sezione di Bologna*

**Abstract.** For a KV beam the halo is created by test particles, initially out of the core, via nonlinear resonances with the betatron motion. The number of resonances and the related diffusive processes are enhanced by the presence of mismatch oscillations of the core. Numerical (PIC) or physical (collisions) noise modifies the core of a KV beam creating tails on time scales relevant for storage rings. We consider a 2D model defined by $N$ pseudo line-charges per meter, whose limit $N \to \infty$, at fixed perveance, is the 2D mean field theory for a coasting beam in a constant focusing channel. Using an algorithm of optimal computational complexity we find that the relaxation time scales linearly with $N$ in agreement with Landau's kinetic theory we have developed for the 2D model.

## INTRODUCTION

Space charge effects are relevant for high intensity beams at moderate energies which are being considered for neutron spectroscopy, wastes transmutation and research on nuclear structure. The presence of a halo around the core is a serious concern since very low losses can be tolerated. For a linac the beam transits once through the focusing line but the short bunches and the acceleration require a 3D treatment of the problem. The test particles (particle-core) models for a uniformly charged ellipsoidal core (an analytic self consistent phase space distribution like KV is not known) provide a picture of halo dynamics [1] in rather good agreement with the 3D PIC simulations [2]. In this case the intra-beam scattering effects are not relevant [3]. Linearly unstable perturbations of a KV beam were also proposed as possible mechanisms for particles escape from the core [4]. In the case of a storage ring the bunches are long, there is no acceleration, and the number of visited FODO cells is high $10^5 \sim 10^7$. The use of a 2D model is justified and the effects of any small amplitude noise must be examined. The numerical noise in a PIC code, whose amplitude depends on the discretization parameters, causes an emittance increase [5]. A white noise in the equations of motion has a similar effect [6]. The effect of frequent small angle collisions was examined in the framework of Landau's kinetic theory by introducing a drift and a noise in the equations of motion of a particle which moves in the self consistent field generated by all the others [7]. We plan to explore the effect of soft and hard Coulomb collisions by direct integration of Hamilton's equations in order to check the validity of the basic assumptions of kinetic theory and to find a scaling law for the relaxation time to the Maxwell-Boltzmann equilibrium. The 2D model, even though the coherence of the longitudinal motion for long times can be questioned, is a good laboratory. Indeed it is numerically softer and the related kinetic theory easier to handle once a good approximation for the scattering cross section is found. We achieved this goal obtaining the asymptotic scaling for the drift and diffusion coefficients. After developing an efficient integration scheme and a field evaluator of optimal ($N \log N$) computational complexity we have shown, by integrating Hamilton's equations, that the basic phenomena (thermalization of mismatch oscillations, equipartition for an asymmetric focusing) were reproduced. Moreover the relaxation times to the self consistent Maxwell-Boltzmann distribution proved to scale linearly with the pseudo line-charges number $N$ in agreement with our theoretical predictions. This enables us to estimate the relaxation for a beam with $10^{11}$ protons per meter and emittance 1 mm mrad in a focusing channel with a bare phase advance of $\omega_0 = 1\,\mathrm{m}^{-1}$ and depressed one of $\omega = 0.6\,\mathrm{m}^{-1}$ may occur after $10^6$ m.

## THE MODEL

We consider a (non relativistic) a coasting beam of $N_p$ protons per meter whose mass and charge we denote by $e$ and $m_p$. Letting $v_0$ be the longitudinal velocity and $k_0 = eB_1 v_0/c$ be the coefficient of the linear restoring force, where $B_1$ is the average quadrupolar gradient, the equations of motion, where we choose $s = v_0 t$ as independent variable read

$$\frac{d^2 x_i}{ds^2} + \omega_0^2 x_i = \frac{\xi}{N} \sum_{j \neq i}^{N} \frac{x_i - x_j}{r_{ij}^2} \qquad \frac{d^2 y_i}{ds^2} + \omega_0^2 y_i = \frac{\xi}{N} \sum_{j \neq i}^{N} \frac{y_i - y_j}{r_{ij}^2}$$

The only parameters are the bare phase advance and the perveance

$$\omega_0^2 = \frac{k_0}{mv_0^2} = \frac{q}{m}\frac{B_1}{cv_0} \qquad \xi = \frac{q}{m}\frac{2Nq}{v_0^2}$$

where $q$ is the charge per unit length of a wire, $m$ is its mass and $N$ is the number of wires. The ratio $q/m$ is constant and equal to $e/m_p$, the total charge per unit length $qN$ is also constant and equal to $eN_p$, where $N_p$ is the number of protons per meter. Since the proton density in a cylindrical beam of radius $R$ is $\rho = N_p/(\pi R^2)$ and the specific length is $\ell = \rho^{-1/3}$ to each proton in a cylinder of height $\ell$ we associate a wire. Their number is

$$N_{\text{phys}} = \rho \ell \pi R^2 = N_p \ell = N_p^{2/3}(\pi R^2)^{1/3}$$

It is convenient to scale the longitudinal coordinates with the cell length $L$ we assume to be 1 m and the transverse coordinates with a typical length $a$ comparable with the *beam* radius, we choose 1 mm. After the scaling $\omega_0$, multiplied by $L$ becomes dimensionless and $\xi$ being multiplied by $L^2/a^2$, becomes of order 1. The typical values we choose are $\xi = 1$, $\omega_0 = 1$. For emittance $\varepsilon = 1$ (mm mrad) the depressed phase advance is $\omega = 0.618$, the core radius is $R = 1.27$ and to $N_p = 10^{11}$ corresponds $N_{\text{phys}} = 3 \times 10^5$. The Hamiltonian of the system is

$$H = \sum_{i=1}^{N}\left(\frac{p_{xi}^2 + p_{yi}^2}{2} + \omega_0^2 \frac{x_i^2 + y_i^2}{2} + \frac{\xi}{2}V^{(i)}\right)$$

where the potential of the force acting on particle i splits into 2 components

$$V^{(i)} = -\frac{2}{N}\sum_{k, r_{ik} > R_D} \log r_{ik} - \frac{2}{N}\sum_{k, r_{ik} < R_D} \log r_{ik}$$

The first component is the mean field potential of far charges, the second component is the potential of nearby charges (within the Debye cylinder). The Debye potential and Debye radius are given by

$$V_D(r) = 2QK_0\left(\frac{r}{\lambda_D}\right) \qquad \lambda_D^2 = \frac{k_BT}{4\pi Ne^2\rho_{0S}} = \frac{2\omega\varepsilon}{\xi}\left(\frac{R}{4}\right)^2$$

where $-Q$ is the charge per unit length in a Debye cylinder, $K_0$ a Bessel function. The asymptotic behavior is

$$V_D \underset{r \to \infty}{\simeq} 2Q\frac{e^{-r/\lambda_D}}{\sqrt{r/\lambda_D}} \qquad V_D \underset{r \to 0}{\simeq} -2Q\log\frac{r}{\lambda_D}$$

Even though the Debye radius is independent of $N$ in the limit $N \to \infty$ the contribution of the collisional part vanishes and the mean field theory is recovered since the charge fluctuations vanish. In section 4 we show that the drift and diffusion matrix vanish in this limit.

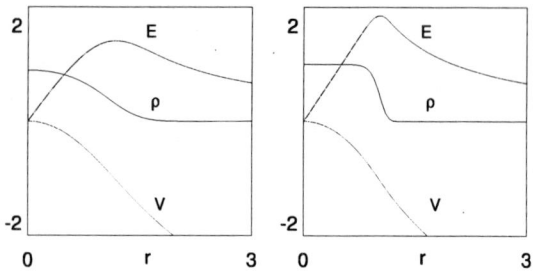

**FIGURE 1.** Self consistent MB distribution $\rho$, its electric field $E$ and potential $V$ for $\xi = 1$, $\omega_0 = 1$: (left) $\varepsilon = 1$ so that $R = 1.27$, $k_BT = 0.154$ (right) $\varepsilon = 0.2$ so that $R = 1.02$, $k_BT = 0.0096$

## MEAN FIELD EQUATIONS

Neglecting the collisional component the potential of the mean field is a solution of the Poisson equation

$$\Delta V = -4\pi\rho_s \qquad \rho_s(x,y) = \int \rho(x,p_x,y,p_y)\,dp_x dp_y$$

where $\rho_s$ is the space density. The phase space density $\rho$ is a solution of the Liouville equation with the mean electric field

$$\frac{\partial \rho}{\partial s} + [\rho, H] = 0 \qquad \int \rho\, dxdydp_xdp_y = 1$$

The Hamiltonian $N$ in the continuum limit has a self consistent solution corresponding to a uniform charge distribution within a disc of radius $R$ with total potential

$$V_{\text{tot}} = \left(\omega_0^2 - \frac{\xi}{R^2}\right)\frac{x^2 + y^2}{2}$$

The static equilibrium condition occurs when the the potential is identically zero namely for $R = \xi^{1/2}/\omega_0$. The dynamic equilibrium condition occurs when

$$H = \frac{p_x^2 + p_y^2}{2} + \frac{\omega^2}{2R^2}(x^2+y^2) = \frac{\omega\varepsilon}{2} \qquad \omega^2 = \omega_0^2 - \frac{\xi}{R^2}$$

where $\pi\varepsilon$ is the area of the orbit (emittance) and $R$ is self consistently core radius fixed bye

$$E = \frac{\omega^2 R^2}{2} = \frac{\omega\varepsilon}{2} \qquad \frac{\varepsilon^2}{R^4} = \omega_0^2 - \frac{\xi}{R^2}$$

Letting $k_B$ be the Boltzmann constant divided by $mv_0^2$ we have $E = 2k_BT$. The core radius increases with the perveance and with the emittance: $R(\xi,\varepsilon) > R(\xi,0) \equiv \xi^{1/2}/\omega_0$ and $R(\xi,\varepsilon) > R(0,\varepsilon) \equiv (\varepsilon/\omega_0)^{1/2}$. Any function $\rho = f(H)$ is a solution of the Vlasov equation provided that $V$ is determined self consistently. The KV distribution which gives a uniform space and momentum distribution is

$$\rho = \frac{\omega}{2\pi^2\varepsilon}\delta(H-E) \qquad E = \frac{\varepsilon\omega}{2}$$

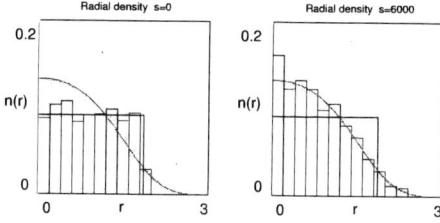

**FIGURE 2.** Space density for an initial KV distribution (left) and after $s = 6 \times 10^3$ m (right). The histogram refers to the simulation with $N = 10^3$ the continuous line to KV and MB.

The self consistent Maxwell-Boltzmann distribution corresponding to the canonical ensemble is

$$\rho = Z^{-1} \exp\left(-\frac{p_x^2 + p_y^2 + \omega_0^2(x^2 + y^2) + \xi V}{2k_B T}\right)$$

the potential $V$ being solution of the Poisson equation with initial condition $V(0) = V'(0) = 0$

$$\frac{1}{r}\frac{d}{dr}r\frac{dV}{dr} = -4\pi\rho_s(0)\exp\left(-\frac{\omega_0^2 r^2 + \xi V(r)}{2k_B T}\right)$$

## SIMULATIONS AND COLLISIONS

In order to integrate Hamilton's equations we have considered three different methods:
i) fourth order Runge Kutta with variable time step (4+2 evaluations)
ii) fourth order symplectic Runge Kutta with constant time step (4 evaluations)
iii) Stoerm method with variable time step (1 evaluation)
The computational complexity $C(N) \propto N^2$ of the field evaluation was reduced to $C(N) \propto N \log N$ by multipolar expansion of far field, continued fraction reconstruction, hierarchical space splitting. Requiring $10^{-8}$ accuracy the direct computations is convenient up to $N = 10^3$ and the CPU time on a single 2.5 Ghz processor is 0.02 s. For higher $N$ the proposed method gives a linear rise for some selected values: for instance with $N = 10^5$ the CPU time time is 2 s. We have considered a *low resolution*: integration with $N = 100$ steps/cell as in a PIC integration of Vlasov equations ($\Delta s = 1$ cm).
The hard collisions are not resolved. The CPU time for $N = 10^3$ is 2 s per cell which amounts to 1 h to reach relaxation ($\sim 1800$ cells). With $N = 10^4$ the CPU time per cell is about 20 s and the time to reach relaxation is about 4 days. Indeed the both computational complexity and the relaxation time both increase linearly with $N$ so that the CPU time to reach the relaxation increases quadratically with $N$.

The *high resolution* integration is carried out with $N = 10^4$ steps/cell ($\Delta s = 0.1$ mm). The hard collisions are resolved and the CPU time grows by a factor 100. As Initial distribution a matched or sightly mismatched KV was chosen. We have checked that the total energy and angular momentum are conserved an that the relaxation to a self consistent Maxwell-Boltzmann distribution occurs with a time scale compatible Landau's kinetic theory. In the mismatched case the energy of oscillations is damped and in the case of asymmetric focusing we observe equalization of of horizontal and vertical temperatures during the relaxation process. In figure 2 we show the space density after relaxation for an initial KV distribution.

## KINETIC THEORY

Assuming the binary collisions are small angle frequent and instantaneous, we can treat them as a random process. Let the changes of position and momentum in the interval $\Delta s$ be given by

$$\Delta \mathbf{x} = \frac{\partial H}{\partial \mathbf{p}}\Delta s \qquad \Delta \mathbf{p} = -\frac{\partial H}{\partial \mathbf{r}}\Delta s + (\Delta \mathbf{p})_{\text{coll}}$$

The master equation reads

$$\frac{\partial \rho}{\partial s} + [\rho, H] = -\sum_i \frac{\partial}{\partial p_i}(K_i \rho) + \frac{1}{2}\sum_{i,j=1}^2 \frac{\partial^2}{\partial p_i \partial p_j}(D_{ij}\rho)$$

where the r.h.s. is the contribution of collisions

$$\mathbf{K} = \left\langle \frac{\Delta \mathbf{p}}{\Delta s} \right\rangle \qquad D_{ik} = \left\langle \frac{\Delta p_i \Delta p_k}{\Delta s} \right\rangle$$

and $H$ is the mean field Hamiltonian

$$H = \frac{p_x^2 + p_y^2}{2} + \omega_0^2 \frac{x^2 + y^2}{2} + \frac{\xi}{2}V \qquad \Delta V = -4\pi \int \rho \, d\mathbf{p}$$

For a binary collision the kinematics is defined by figure 3 and, denoting by $\mathbf{P}$ the total momentum, the relation between the momenta in the laboratory and center of mass frames is

$$\mathbf{p}_1^{\text{out}} = \mathbf{P} + \frac{\mathbf{p}^{\text{out}}}{2} \qquad \mathbf{p}_2^{\text{out}} = \mathbf{P} - \frac{\mathbf{p}^{\text{out}}}{2}$$

From the definition of drift $\mathbf{K}$ where $\Delta \mathbf{p} = \mathbf{p}_2^{\text{out}} - \mathbf{p}_2$ we obtain

$$\mathbf{K} = N \int d\mathbf{p}_1 \rho(\mathbf{p}_1, \mathbf{r}, s) \|\mathbf{p}_1 - \mathbf{p}_2\| \int d\theta \frac{d\sigma}{d\theta}(\mathbf{p}_2^{\text{out}} - \mathbf{p}_2)$$

Going from the lab. to the c.m. system and integrating over the scattering angle $\Theta$ the result is

$$\mathbf{K} = \frac{N}{2}\int d\mathbf{p}_1 \rho(\mathbf{p}_1, \mathbf{r}, s)\|\mathbf{p}_1 - \mathbf{p}_2\|(\sigma_0 - \sigma_1)(\mathbf{p}_1 - \mathbf{p}_2)$$

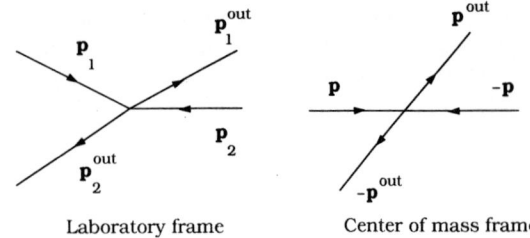

**FIGURE 3.** Momenta in the laboratory and c.m. frames

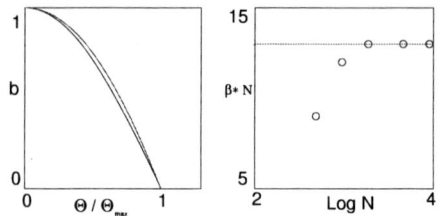

**FIGURE 4.** Left: comparison of exact and approximate impact parameter $b$. Right: plot of $N\beta$ to show the asymptotics.

where $\sigma_k = \int \frac{d\sigma}{d\Theta} \cos^k \Theta \, d\Theta$ are the moments of the cross section which depend on $p \equiv \|\mathbf{p}_1 - \mathbf{p}_2\|$. The final result is

$$\mathbf{K}(\mathbf{p}_2) = -\beta \mathbf{p}_2$$

where $\beta$ is a positive function defined by

$$\beta = \frac{N}{2} \int_0^\infty dp_1^2 \frac{\rho}{2} \int_0^{2\pi} d\phi \, p \, (\sigma_0 - \sigma_1) \left(1 - \frac{p_1}{p_2} \cos\phi\right)$$

where $\rho$, which depends on $p_1^2$, $r^2$ and $s$, is equal KV and to Maxwell-Boltzmann for $s=0$ and $s \to \infty$ respectively. We discovered that that the differential cross section $\frac{d\sigma}{d\Theta} = -\frac{db}{d\Theta}$, computed for the cut-off potential

$$\frac{\xi}{2} V(r) = -\frac{\xi}{N} \log \frac{r}{\Lambda} \vartheta(\Lambda - r)$$

has a simple asymptotic expression as $\xi/NE \ll 1$, see figure 4

$$\frac{d\sigma}{d\Theta} = 2\Lambda \frac{\Theta}{\Theta_{max}^2} \vartheta(\Theta_{max} - \Theta) \qquad \Theta_{max} = \frac{\xi}{NE} \frac{1}{\sqrt{c}}$$

where $c^{-1/2} = 1.5$ which implies that asymptotically

$$\sigma_0 - \sigma_1 \simeq \frac{\Lambda}{4} \Theta_{max}^2$$

The consequence is that for $N$ large we have

$$\beta \simeq \beta_* \frac{\xi^2}{N}$$

Numerical results obtained with $\xi = \omega_0 = \varepsilon = 1$, show that the relaxation length is $s_{rel} \simeq 2N$ for a number of pseudo-particles $N$ in the range $[10^3, 10^4]$ to be compared with the estimate $s_{rel} = \beta^{-1}$. For $N_p = 10^{11}$ protons/m the number of pseudo-wires is $N_{phys} \sim 3 \times 10^5$ and the relaxation length $6 \times 10^5$ m.

## CONCLUSIONS

We have considered a 2D model for a beam moving in a constant focusing channel by developing an efficient integration scheme and procedure to evaluate the field with computational complexity $N \log N$. For a cell of unit length (1 m) and bare phase advance $\omega_0 = 1$ Hamilton's equations are integrated with a large integration step $\Delta s = 10^{-2}$ which does not resolve hard collisions and with a small one $\Delta s = 10^{-4}$, which does resolve them. The relaxation process is similar within the statistical errors involved. The relaxation time $t_{rel} = s_{rel}/v_0$ appears to scale linearly with $N$ for $N \geq 2 \times 10^3$. The relaxation time obtained from kinetic theory, using an asymptotic approximation to the scattering cross section, exhibits this scaling law which becomes exact for $N \to \infty$. As a consequence we can estimate with a high confidence the relaxation time for a physical number of particles. The reliability of our scheme is confirmed by the behavior of rms radii, emittances and temperatures for a symmetric and asymmetric focusing channel. The CPU times for the low resolution integrator are comparable with a PIC method and after implementation on a parallel architecture can be used to investigate the long time behavior of an intense beam in a ring taking into account the soft collisions.

## REFERENCES

1. M. Communian, A. Pisent, A. Bazzani, G. Turchetti, S. Rambaldi, Physical Review Special Topics - Accelerators and Beams **4**, 124201 (2001)
2. G. Turchett, S. Rambaldi, A. Bazzani, M. Comunian, A. Pisent *3D solutions of the Poisson-Vlasov equations for a charged plasma and particle core model in a line of FODO cells* European Journal of Physics in press (2003)
3. N. Uhlmann, G. Zwicknagel, M. Communian, A. Pisent NIM **A488**, 1-10 (2002)
4. L. Gluckstern, W. H. Cheng, H. Ye Phys. Rev. Letters **75**, 2835 (1995)
5. J. Struckmeier Phys. Rev. **E 54**, 830 (1996)
6. A. Bazzani, G. Turchetti, C. Benedetti, A. Franchi, S. Rambaldi *Accuracy analysis of a 2D Poisson-Vlasov PIC solver and estimates of the collisional effects in space charge dynamics* ICAP-02 Conference proceedings Institute of Physics in press (2003)
7. Ji Qiang, R. D. Ryne, S. Habib *Self consistent simulations of Coulomb collisions in charged particle beams*

# Effects of Halo on the AGS Injection from 1.2Gev Linac*

W.T. Weng**, J. Beebe-Wang, D. Raparia, A. G. Ruggiero, N. Tsoupas

*Brookhaven National Laboratory, Upton, NY 11766*

**Abstract.** BNL is conducting a design study of a 1.0 MW super neutrino beam facility. It requires 230 turns charge exchange injection from a 1.2 GeV superconducting linac with 28 mA current for 0.72 msec. This report studies the impact of halo distribution of the linac beam on the efficiency of injection and the final beam distribution in the AGS as functions of the injection orbit bump and the foil thickness. Another important consideration is the residual radiation generated on the accelerator components near the injection area. If necessary, radiation hardened components and local shielding have to be provided.

## INTRODUCTION

We have examined [1] possible upgrades to the AGS complex that would meet the requirements of the proton beam for a 1.0 MW neutrino superbeam facility. We are proposing to build a superconducting upgrade to the existing 200 MeV linac to an energy of 1.2 GeV for direct H- injection into the AGS.

The requirements of the proton beam for the super neutrino beam are summarized in Table 1 and a layout of upgraded AGS is shown in Figure 1. Since the present number of protons per fill is already close to the required number, the upgrade focuses on increasing the repetition rate and reducing beam losses (to avoid excessive shielding requirements and to maintain activation of the machine components at workable level). It is also important to preserve all the present capabilities of the AGS, in particular its role as injector to RHIC.

**TABLE 1. AGS Proton Driver Parameters.**

| | |
|---|---|
| Total beam power | 1 MW |
| Beam energy | 28 GeV |
| Average beam current | 42 µA |
| Cycle time | 400 msec |
| Number of protons per fill | $0.9 \times 10^{14}$ |
| Number of bunches per fill | 24 |
| Protons per bunch | $0.4 \times 10^{13}$ |
| Injection turns | 230 |
| Repetition rate | 2.5 Hz |
| Pulse length | 0.72 msec |
| Chopping rate | 0.75 |
| Linac average/peak current | 20 / 30 mA |

Present injection into the AGS requires the accumulation of four Booster loads in the AGS which takes about 0.6 sec, and is therefore not suited for high average beam power operation. To minimize the injection time to about 1 msec, a 1.2GeV linac will be used instead. The injection Linac consists of the existing warm linac of 200 MeV and a new superconducting linac of 1.0 GeV. The multi-turn injection from a source of 28 mA and 720 µsec pulse width is sufficient to accumulate $0.9 \times 10^{14}$ particle per pulse in the AGS. The minimum ramp time of the AGS to full energy is presently 0.5 sec. This must be reduced down to 0.2 sec to reach the required repetition rate of 2.5 Hz to deliver the required 1 MW beam to the garget.

## SUPERCONDUCTING LINAC (SCL)

The superconducting linacs accelerate the proton beam from 200 MeV to 1.2 GeV. The presented configuration follows a design described in detail in [2]. All three linacs are built up from a sequence of similar periods. The major parameters of the three sections of the SCL are given in Table 2. The low energy section operates at 805 MHz and accelerates proton from 200 to 400 MeV. The two sections, accelerating to 800 MeV and 1.2 GeV, operate at

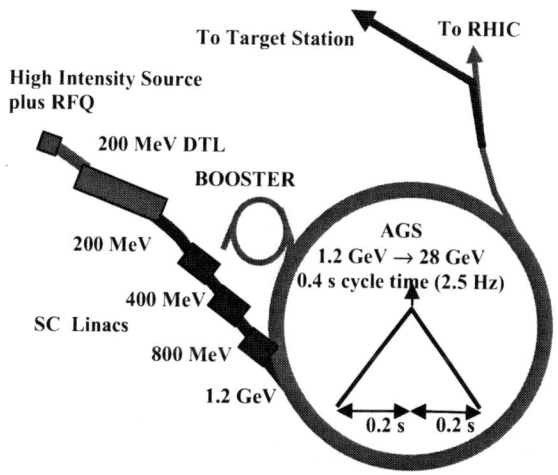

**FIGURE 1.** Schematic diagram of the accelerators for the "neutrino production".

---

* Work performed under the auspices of the US Department of Energy
** E-mail weng@bnl.gov

1.61 GHz. A higher frequency is desirable for obtaining a larger accelerating gradient with a more compact structure and reduced cost. The SCL will be operated at 2 °K for reaching the desired gradients.

**TABLE 2. General parameters of the SCL.**

| Linac section | LE | ME | HE |
|---|---|---|---|
| Average beam power, kW | 7.14 | 14 | 14 |
| Average beam current, μA | 35.7 | 35.7 | 35.7 |
| Initial kinetic energy, MeV | 200 | 400 | 800 |
| Final kinetic energy, MeV | 400 | 800 | 120 |
| Cell reference $\beta_0$ | 0.615 | 0.755 | 0.887 |
| Frequency, MHz | 805 | 1610 | 1610 |
| Cells/cavity | 8 | 8 | 8 |
| Cavities/cryo-module | 4 | 4 | 4 |
| Cavity internal diameter, cm | 10 | 5 | 5 |
| Total length, m | 37.82 | 41.4 | 38.32 |
| Accelerating gradient, MeV/m | 10.8 | 23.5 | 23.4 |
| Cavities/Klystron | 1 | 1 | 1 |
| Norm. rms emittance, π mm mrad | 2 | 2 | 2 |
| rms bunch area, $\pi^\circ$ MeV (805 MHz) | 0.5 | 0.5 | 0.5 |

## H⁻ BEAM INJECTION INTO AGS

The H⁻ injection region has been chosen to be the location of the B20 straight section of the AGS. A schematic diagram of the injection region is shown in Figure 2. In this diagram the following components of the injection region are shown: (a) three of the main magnets of the AGS; (b) the stripping foil; (c) the closed beam orbits; (d) The trajectory of the injected H⁻ beam; (e) the trajectories of the H⁻ beam which is not stripped by the stripping foil; (f) the trajectory of the partially stripped H⁻ beam (H⁰) and (g) the trajectories of the electrons emanating from the stripping foil.

All secondary particles must be collected downstream of the stripping foil into various "beam-dumps". In order to make the circulating proton beam collinear with the H⁻ injected beam at the injection point, the circulating proton beam is "locally bumped" by using two horizontal "bump-magnets" [3].

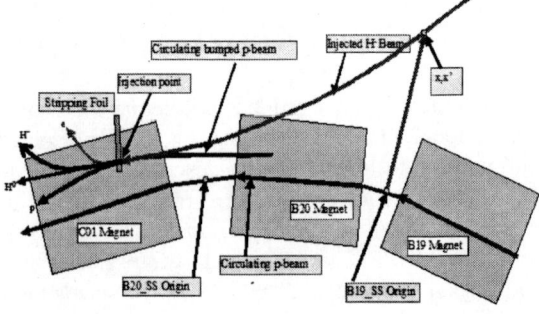

**FIGURE 2.** Schematic diagram of the injection region.

## HALO/TAIL GENERATION VS. LINAC EMITTANCE

For high intensity proton accelerators, such as the upgraded AGS, there are very stringent limitations on uncontrolled beam losses. In this paper, we present the estimate of emittance growth and uncontrolled beam losses as function of linac emittance by computer simulations.

All of the physical quantities used in the simulations (Table 1 and 3) are chosen according to the design specifications [1]. Correlated Painting [4] is chosen for injection into AGS, considering the available aperture at injection and beam halo/tail control. A significant effort has been made to optimize injection painting [4]. The optimized injection bump collapses as an exponential function of time with a time-constant of 0.1 msec. The initial foil-hit by each incident H⁻ is counted as thrice to include the effects of two stripped electrons. The average foil thickness is assumed to be 300 μg/cm². In order to separate the effects of linac emittance from the other issues, the effects of space charge and magnet errors are not included in this study.

**TABLE 3 Simulation parameters.**

| | |
|---|---|
| Horizontal beta at the injection | 28.0 m |
| Vertical beta at the injection | 8.0 m |
| Horizontal emittance of injected beam | 2π mm-mrad |
| Vertical emittance of injected beam | 2π mm-mrad |
| Horizontal beam size at injection, $\sigma_x$ | 5.2293 mm |
| Vertical beam size at injection, $\sigma_y$ | 2.7952 mm |
| Horizontal Foil size (2.5 $\sigma_x$) | 13.0731 mm |
| Vertical foil size (2.5 $\sigma_y$) | 6.9878 mm |

A direct effect of linac beam emittance is the halo/tail generation in the circulating beam. Figure 3 shows the estimated halo/tail generation in the beam [4] as a function of normalized RMS emittance of linac beam. Here, the Halo/tail generation is defined as the ratio of number of particles with emittance larger than the designed acceptance of 49π mm-mrad to the total number of particles in the circulating beam.

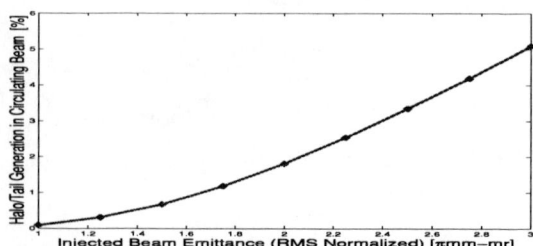

**FIGURE 3** The estimated halo/tail generation in the circulating beam as functions of normalized RMS emittance of injected beam.

# INJECTION EFFICIENCY VS. LINAC EMMITANCE

Injection efficiency is associated with two kinds of beam losses at the injection stripping foil: (1) controlled losses which can be directed to the injection beam dumps located down stream of the injection foil; and (2) uncontrolled losses which are scattered in all directions and generate radiation in the injection area [5].

As a consequence of particle traversal in the stripping foil, there are beam losses associated to: (1) nuclear scattering, (2) energy straggling, and (3) multiple scattering [5].

## Nuclear Scattering

The beam loss due to nuclear scattering in the foil is a function of the foil traversal rate, foil thickness and Linac beam emittance. Figure 4(a) shows the fractional nuclear scattering losses as a function of the normalized RMS Linac beam emittance. If all the losses are located in the injection straight section (~10m), the resulting estimated radiation level, at 1 foot from the beam line after a 100 day run followed by 4 hours of shutdown, is shown in Figure 4(b) as a function of normalized RMS emittance of injected beam.

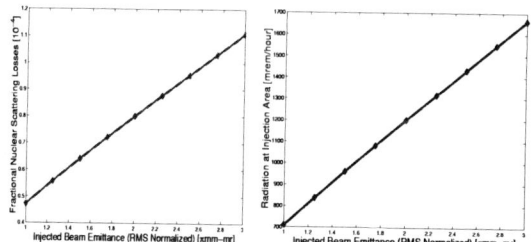

**FIGURE 4** (a) Fractional beam losses at injection area and (b) radiation level due to nuclear scattering in the stripping foil as functions of normalized RMS emittance of injected beam.

## Energy Straggling

As particles traverse through the injection foil, a fraction of their energies are lost and deposited in the foil. Some of the circulating protons, which lie in the tail of Landau distribution, will go through large synchrotron oscillations and may end up in the beam gap. These protons will be lost at the time of extraction or at a dispersive location in the ring. This beam loss is proportional to foil hitting rate, which is associated to the emittance of injected beam when the foil size of $5\sigma_{inj}$ is kept. Figure 5(a) shows the estimated fractional beam loss in the gap due to energy straggling as functions of normalized RMS emittance of injected beam.

## Multiple Scattering

The major effect of particle multiple scattering in the foil is to increase the transverse beam emittance. Figure 5(b) shows the estimated transverse emittance growth due to multiple scattering as functions of normalized RMS emittance of injected beam.

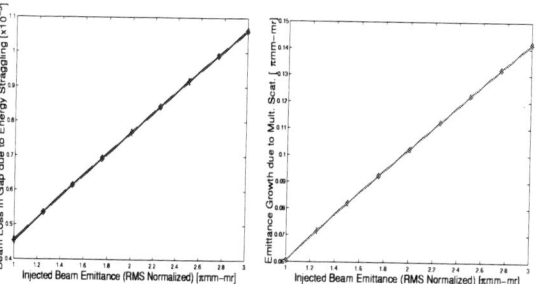

**FIGURE 5** (a) The estimated fractional beam loss in the gap due to energy straggling; (b) the estimated transverse emittance growth due to multiple scattering as functions of normalized RMS emittance of injected beam.

# MAXIMUM FOIL TEMPERATURE VS. LINAC EMMITANCE

The foil temperature distribution is governed by the current density distribution on the foil during injection. Different painting schemes [4] produce not only different final particle distributions, but also generate different foil traversal patterns and therefore different current density distributions in the foil. In this study the current density distributions are simulated through turn-by-turn beam tracking during the correlated painting with the optimized injection bump. The initial foil-hit by each incident H- is counted as thrice to include the effects of two stripped electrons. The foil temperature distributions are then converted from the current density distributions through the relationship presented in Figure 6. This is calculated from the model described in reference [6] with a foil thickness of 300 µg/cm$^2$. The model includes:
1. the radiation heat transfer between the carbon foil and the stainless steel beam pipe,
2. the heat conduction through the foil to its base,
3. a natural convection condition on the outer surface of the beam pipe, and
4. a Gaussian distribution of injected beam.

Figure 7 shows the maximum foil temperature distribution with the injected beam of 2π mm-mrad normalized RMS emittance of injected beam.

The maximum foil temperature is a function of the linac beam emittance. For a given foil thickness, the maximum foil temperature decreases while the linac beam emittance increases. However, with a foil size of $5\sigma_{inj}$, the foil-hitting rate also increases with $\sigma_{inj}$. It will induce more uncontrolled beam loss.

The foil lifetime tests at BNL Linac has indicated that the maximum single foil lifetime is ~78 hours and decrease sharply when the foil temperature exceeds 2500 °K [7]. In general, the injection foil temperature of upgraded AGS can be kept under 2500 °K. Therefore, upgraded AGS injection foil will have a long lifetime.

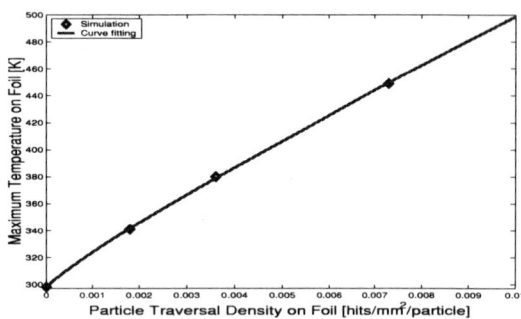

**FIGURE 6** Maximum temperature on foil vs. beam current density (blue points) deduced from the model [6] and the curve fit (red line) used in foil temperature distribution studies.

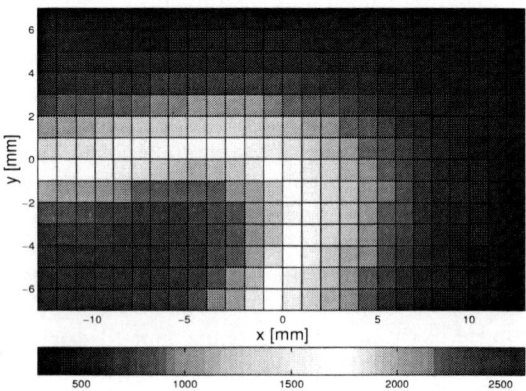

**FIGURE 7** Simulation result of the foil temperature distribution. The entire bottom portion of H⁻ injection foil is shown with horizontal coordinate $x$ and vertical coordinate $y$.

## CONCLUSIONS AND DISCUSSIONS

From the results shown above, it is clear that the correlated painting with the optimized injection bump, collapsing as an exponential function of time with a time-constant of 0.1 msec, gives the best final proton beam distribution in the upgraded AGS. The preferred linac emittance is about $1.5\pi$ mm-mrad for acceptable injection losses.

Two methods have been considered to improve the H⁻ injection from linac. One is to bring the RFQ immediately adjacent to the output of the ion source to reduce the emittance growth. In our estimate, this can reduce the linac output emittance from $3\pi$ mm-mrad to about $1.5\pi$ mm-mrad. Well-designed low level feedback system has to be provided to prevent further emittance growth in the high-energy section of the linac.

Another possible improvement is to introduce a second harmonics cavity for the AGS during injection time. It can effectively reduce the space charge tune shift by 30%. This will either reduce resonance losses for same intensity, or allow for 30% more intensity to be accepted by the AGS.

## ACKNOWLEDGMENTS

The authors would like to acknowledge the contributions by the members of the C-AD accelerator physics group. The assistances of C. J. Liaw in the foil temperature simulation and foil lifetime experiments are also greatly appreciated.

## REFERENCES

1. Diwan, M, et al., "AGS Super Neutrino Beam Facility Accelerator and Target System Design", BNL-71228-2003-IR, April 15, 2003.

2. Ruggiero, A. G., "AGS Upgrade to 1 MW with a Super-Conducting Linac Injector" in Proceedings of PAC2003 (2003).

3. Tsoupas,N.,et al."Injection Acceleration and Extraction of High Intensity Proton Beam For The Neutrino Facility Prodject at BNL", in Proceedings of PAC2003.

4. Beebe-Wang, J., et al., "Beam Properties in the SNS Accumulator Ring due to Transverse Phase Space Painting", Proceedings of the EPAC 2000, Vienna, Austria, 26-30 June 2000, p.1465-1467.

5. Beebe-Wang, J., et al., "Injection Carbon Stripping Foil Issues in the SNS Accumulator Ring", Proceedings of the PAC2001, Vienna, Austria, 26-30 June 2000, p.1508-1510.

6. Liaw, C. J., et al., "Calculation of Maximum Temperature on the Carbon Stripping Foil of the Spallation Neutron Source", Proceedings of the PAC99, New York, USA, March 27- April 2, 1999.

7. Liaw, C. J., et al., "Life Time of Carbon Stripping Foils For the Spallation Neutron Source" PAC2001, Vienna, Austria, 26-30 June 2000.

# Beam Halo from Quadrupole Rotation Errors

R.A. Kishek, S. Bernal, I. Haber, H. Li, P.G. O'Shea, B. Quinn, M. Reiser, and M. Walter

*Institute for Research in Electronics and Applied Physics,
University of Maryland, College Park, MD 20742*

**Abstract.** Systems where the beam has a rotation angle relative to the plane of the quadrupoles can support a larger number of envelope modes. Since beam halos can be produced by parametric resonance between the oscillations of the core and individual particle orbits, many more classes of such resonances are to be expected in rotated beams. Early simulation studies are reviewed in which halos are formed by random quadrupole rotation errors. The injection of a rotated beam into a perfect lattice is also explored. The results are compared to a preliminary experiment on the University of Maryland Electron Ring (UMER).

## INTRODUCTION

Early studies indicated that beam halos can pose a severe restriction on beam current or aperture due to the fear of activation of the accelerator structure by lost halo particles [1]. Much has been accomplished in understanding halo dynamics in charged particle beams. The emerging perspective is that halos are created by parametric resonances in beams whose potential has strong time-dependence, where particles orbiting at frequencies close to the frequency of the core pulsations will be "kicked out" of the beam [2]. This picture has been largely substantiated by studies using the particle-core model [3], which typically tracks particles in the potential of a core undergoing rms mismatch oscillations, an idea that was tested recently in the LEDA experiment at Los Alamos [4].

A more general, and potentially much more serious situation is the case when the core contains a non-zero angular momentum, as can happen in systems with skew quadrupole errors. In such a system, the beam can support a larger number of envelope modes [5-6], and the particle orbits are also more complicated, opening up wider possibilities for parametric resonances that can result in halo. This paper reviews the results of particle-in-cell simulations of FODO quadrupole lattices with skew errors [7] which led in some cases to halo formation. The injection of rotated beams, as can arise from quadrupole rotation errors in the injector, is also explored. These results are compared to preliminary experiments on the University of Maryland Electron Ring (UMER) [8]. UMER is a small, low-cost machine which uses low-energy nonrelativistic electrons to access the scaled dynamics of ion accelerators. It thus follows in the footsteps of the pioneering scaled experiments at the University of Maryland (c. 1989) which led to one of the first experimental observations of beam halos [9-11]. It is one of the goals of UMER to be able to do a controlled halo experiment.

## MODEL

We base the simulations on the UMER beam, which presently drifts at 10 keV, so the $\beta$ of 0.2 implies the beam is nonrelativistic and emission of synchrotron radiation by the electrons is not a factor. Three induction gaps will be used to provide longitudinal focusing and eventual acceleration to 50 keV. Transverse focusing is provided by 36 FODO cells around the 11.52 meter circumference ring. With the design beam current of 100 mA, an initial effective emittance of $\varepsilon = 50$ μm and a zero-current phase advance per period of $\sigma_0 = 76°$ (giving a beam radius a = 1 cm), UMER can achieve a space charge intensity of $\chi = 0.97$, which is near the extreme intensity. Here,

$\chi$ [defined in Ref. 8] represents the ratio of space charge forces to external focusing forces. The intensity of the UMER beam can be reduced over a wide range (down to $\chi \sim 0.35$) by changing the anode-cathode spacing, the potential at the cathode grid, or by changing the aperture size in the beam collimator at the exit of the gun [12].

For the simulations here we use the particle-in-cell code WARP [13] in the 2-1/2-D "Slice" mode of operation, which evolves an infinitely-long beam along the direction of propagation, s. Numerical parameters have been thoroughly tested to ensure good resolution and minimal numerical collisionality over the equivalent of 10 turns of propagation. In the work covered by this paper, the dipoles are ignored, but in some cases the superperiodicity of the lattice is imposed on the quadrupole errors to simulate recirculation.

## RANDOM SKEW QUAD ERRORS

The coupling introduced by quadrupole rotation errors means that one no longer expects the x and y normalized emittances to be conserved, even for a beam with an initial Kapchinskij-Vladimirskij (K-V) distribution with a linear space charge force profile propagating under linear external forces. However, if the equations of motion result from linear forces and are derivable from a Hamiltonian system, constants of the motion may be obtained analogous to the normalized x and y emittances [14]. Further, the K-V distribution has been generalized [15] to distributions in which the principal axes do not align with the x and y axes, and moment equations have been derived [16] that assume the space-charge profile remains linear, consistent with the assumption of the KV-like distribution of ref. [15]. In Refs. 16 and 7, these conservation constraints were derived and will henceforth be labeled the "generalized emittances", $\varepsilon_g$ and $\varepsilon_h$.

Ref. 7 published results of testing these generalized emittances, in which the WARP code was used to evolve the UMER beam in a lattice with skew quad errors. Using random errors of the order of 0.1°-0.2° (~ 2-4 mrad) rms rotation, a moderate growth of the generalized emittances was observed (about a factor of 2 over 10 turns), indicating thermalization of free energy [17] and irreversible mixing. When the superperiodicity of the lattice was removed to simulate a linac, however, the same moderate growth persisted for only a few turns, at which point it gave way to explosive growth [Fig. 1]. A halo was observed to form at that point. The conclusion was that the superperiodicity of the lattice imposed some periodicity on the beam rotation, constraining the beam angles attainable, hence also the emittance growth. Where no superperiodicity exists, on the other hand, the beam rotation angles can grow beyond the threshold which would result in a halo [Fig. 2].

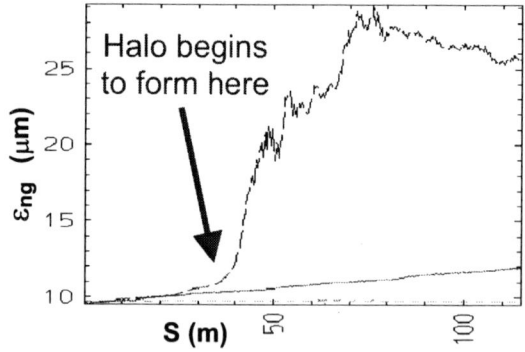

**Fig. 1**: Generalized emittance $\varepsilon_{ng}$ along 10 turns in UMER; low current (lower); nominal current, straight (upper) and periodic (middle curve) (adapted from Ref. 7).

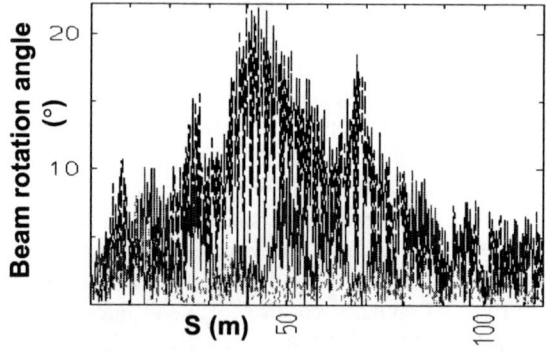

**Fig. 2**: Beam rotation angles from the simulations of Fig. 1. Note that the angles are constrained in the ring lattice (red), but grow substantially until the halo forms at s ~ 40 m in a straight lattice (blue).

In order to mitigate the effects of rotation errors, we have explored correction schemes using quadrupole/skew-quadrupole pairs, corresponding to an electronically rotatable quadrupole. Preliminary simulation results were encouraging, indicating that the irreversible growth in generalized emittance can be reduced by a factor of two if the same (DC) correction is applied once per turn [18]. An experiment on the UMER injector was carried out to test the concept of this skew-corrector [19]. One of the injector matching section quadrupoles was replaced by this device, and the skew angle was varied while beam images indicating beam size and rotation angle were taken at various locations downstream. The results showed excellent agreement between experiment and simulation, and demonstrated the experimental viability of this device.

More significantly, the experiment revealed the extreme sensitivity to skew errors in the injector matching section. Whereas a 1.5° rotation error in a single magnet in an otherwise perfect periodic lattice does not affect the beam much after several turns, an error of the same magnitude in the matching section was observed to lead to as much as 40° beam rotation angle just a few magnets downstream [19]. It is not clear at this point what causes this extreme sensitivity, but it obviously poses a grave problem to machines with high intensity beams. This higher sensitivity in the injector is particularly problematic for UMER since a sizeable length of beamline at the end of the matching section is completely blind due to the injection Y.

## INJECTING ROTATED BEAMS

The logical next step is to assess the impact of injecting a rotated beam into a perfect lattice. This was done for the same parameters as the previous simulations described here. For ease of setup, the beam was injected upright relative to the x-y plane, but all the magnets were rotated by the same angle. As the beam encounters the first quadrupole, it receives a rotational kick which imparts a nonzero angular momentum to it. The beam continues oscillating ("wobbling") in coherent rotational motion for the rest of the simulation, with the angular momentum changing periodically inside each quadrupole. The amplitude of these oscillations is dependent upon the initial rotation angle (injection error).

It is observed that these periodic beam rotations gradually damp, as the coherent motion is phase-mixed. For small injected beam rotation angles (~1°), there is very little effect on the beam envelope or emittance. For larger angles (~5°), a halo begins to form after a few plasma periods and the generalized emittances therefore increase. This halo is even more pronounced for larger injection angles, generally developing earlier and containing a larger fraction of particles [Fig 3]. Figure 4 shows the generalized emittance, beam envelope and x and y emittances associated with a 10° injection error. Notice the substantial growth of generalized emittance accompanying the halo formation.

The halo in Fig. 3 appears strikingly similar to halos seen in early experiments on UMER [20]. Fig. 5

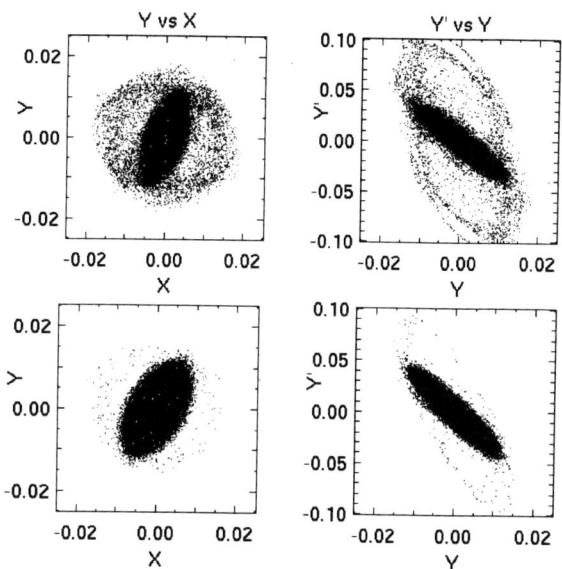

**Fig. 3**: X-Y and Y-Y' phase space snapshots at s = 8.64 m for 20° (top) and 10° (bottom) injection rotation errors. Note that the principal planes of the quadrupoles are rotated relative to the simulation coordinates shown.

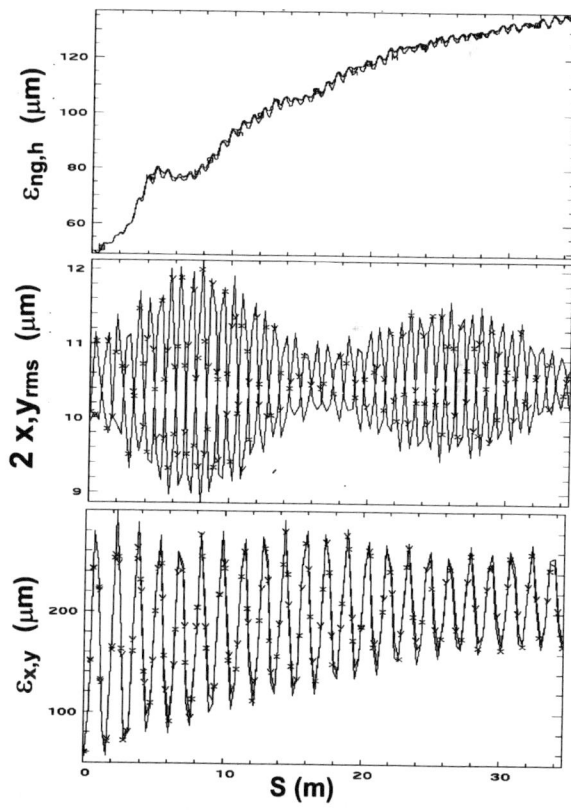

**Fig. 4**: Generalized emittances, beam envelope, and x, y emittances as functions of s for a 10° injection rotation error.

shows a pseudo-color image of the UMER beam taken about 1.1 m downstream from the injection point. This image clearly shows a beam core rotated by about

15° in addition to a well-defined halo (the combined rotation angle of core and halo is reduced to 7°). Before a systematic comparison between experiment and simulation can be undertaken, a baseline halo-free operating point needs to be established in the experiment. This requires very careful alignment and matching, and perhaps changes to the matching section to reduce its sensitivity to rotational errors. After that, it will be possible to use a skew-corrector magnet in the injector to induce a controlled halo that can be studied experimentally.

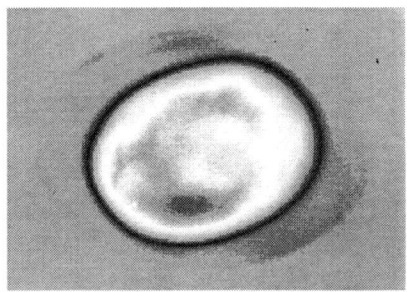

**Fig. 5**: Pseudo-color image of beam from experiment with somewhat different parameters (adapted from Ref. 20).

Much remains to be done in terms of theoretical understanding of skew-produced halos. As apparent from this brief presentation, this problem is quite challenging and at the same time extremely rich in terms of physical phenomena that deserve study.

## ACKNOWLEDGMENTS

Many thanks to John Barnard who originally inspired this work, and to David Grote who suggested skewed injection and whose exemplary support for the WARP code is much appreciated. Part of the simulations here were run on supercomputers provided by the National Energy Research Scientific Computing Center (NERSC). This work is performed under the auspices of the US Dept. of Energy under contracts no. DEFG02-94ER40855 and DEFG02-92ER54178.

## REFERENCES

1. Jameson, R.A., *Fus. Eng. & Des.* **32-33**, 149 (1996).
2. Gluckstern, R.L., *Phys. Rev. Lett.* **73**, 1247 (1994).
3. Wangler, T.P., Crandall, K.R., Ryne, R., and Wang, T.S., "Particle-core model for transverse dynamics of beam halo," *Phys. Rev. ST-AB* **1**, 084201 (1998).
4. Allen, C. K. *et al.*, *Phys. Rev. Lett.*, **89**, 214802 (2002).
5. Chernin, D., "Evolution of RMS Beam Envelopes in Transport Systems with Linear x-y Coupling," *Particle Accelerators*, **24**, 29-44 (1988).
6. Barnard, John J., and Losic, Bojan, "Envelope Modes of Beams with Angular Momentum," *Proc. 2000 Linac Conference*, Piscataway, NJ: IEEE Press, 2001.
7. Kishek, R.A., Barnard, J.J., and Grote, D.P., "Effects of Quadrupole Rotations on the Transport of Space-Charge-Dominated Beams," *Proc. PAC 1999*, Piscataway, NJ: IEEE Press, 1999, pp. 1761.
8. Reiser, M., *et al.*, "The Maryland Electron Ring for Investigating Space-Charge Dominated Beams in a Circular FODO System", *Proc. PAC 1999*, Piscataway, NJ: IEEE Press, 1999, pp. 234.
9. Kehne, D., Reiser, M., and Rudd, H., in *High-Brightness Beams for Advanced Accelerator Applications*, AIP Conf Proc **253**, eds. William W. Destler and Samar K. Guharay, New York: AIP, 1992, pp. 47-56.
10. Reiser, Martin, *Theory and Design of Charged Particle Beams*, New York: John Wiley & Sons, 1994, sec. 6.2.
11. Haber, I., Kehne, D., Reiser, M. and Rudd, H. *Phys. Rev. A* **44**, 5194 (1991).
12. Kishek, R.A., Bernal, S., Bohn, C.L., *et al.*, "Simulations and experiments with space-charge-dominated beams," *Physics of Plasmas*, **10** (5), 2016 (2003).
13. Grote, D.P., Friedman, A., Haber, I., and Yu, S., *Fus. Eng. & Des.* **32-33**, 193-200 (1996).
14. Dragt, A.J., Neri, F., and Rangarajan, G., *Phys. Rev. A*, **45**, 2572 (1992).
15. Sacherer, F.J., Ph.D. Thesis, Univ. of California, Berkeley, UCRL-18454 (1968).
16. Barnard, John J., Proc. 1995 PAC Conf., Piscataway, NJ: IEEE Press, 1996, pp. 3241.
17. Reiser, M., *J. Appl. Phys.* **70**, 1919 (1991).
18. Kishek, R.A., Reiser, M., O'Shea, P., Venturini, M., and Zhang, W.W., in *The Physics of High Brightness Beams*, edited by J. Rosenzweig & L. Serafini, Singapore: World Scientific, 2000, pp. 297 – 308.
19. Li, H., Bernal, S., Kishek, R.A., *et al.*, "Printed-Circuit Magnets for the University of Maryland Election Ring (UMER)-New Developments", *Proc. PAC 2001*, Piscataway, NJ: IEEE Press, 2001, pp. 1802.
20. Bernal, S., Beaudoin, B., Cui, Y., *et al.*, "Beam Transport Experiments over Half-Turn at the University of Maryland Election Ring (UMER)", *Proc. PAC 2003*, Piscataway, NJ: IEEE Press, 2003, to appear.

# Simulation of halo particles with Simpsons

Shinji Machida

*KEK, Accelerator Division*
*1-1 Oho, Tsukuba-shi, Ibaraki-ken, 305-0801, JAPAN*

**Abstract.** Recent code improvements and some simulation results of halo particles with Simpsons will be presented. We tried to identify resonance behavior of halo particles by looking at tune evolution of individual macro particle.

## INTRODUCTION

It is now clear that coherent oscillations tell us the stability of a space charge detuned beam as a whole [1]. Since the coherent tune shift is smaller than the maximum incoherent tune shift, the threshold of coherent resonance is always higher than that of incoherent one. On the other hand, only a few percent or even less beam loss is allowed in a high current machine such as J-PARC and SNS. It seems that the stability criterion based on coherent resonance is not enough to predict such beam loss. We want to understand particle behavior of those few particles. In this paper, we will show some particle simulation results with Simpsons [2] and discuss the behavior of those particles.

## CODE IMPROVEMENTS

In multi-particle tracking, quantities such as rms emittance and higher order moments are calculated to present beam behavior. Coherent oscillations are characterized by the oscillation frequency of moments, namely dipole coherent oscillations by frequency of the first moment, quadrupole oscillations by the second moment, and so on.

In order to observe a few percent or less beam loss, trajectories of individual macro particle becomes important. It was, however, presented last year that single particle orbit strongly depends on grid size [3], although quantities such as rms emittance and 99% emittance do not.

We have improved algorism of the tracking code and increased the number of macro particles so that now it gives a stable single particle trajectories, namely, trajectories do not depend tracking parameters such as grid size and time step. Figure 1 shows the single particle amplitude as a function of grid size and number of macro particles. When we take more than 1E5 macro particles, results are independent of grid size. Here, 3D space charge force is applied.

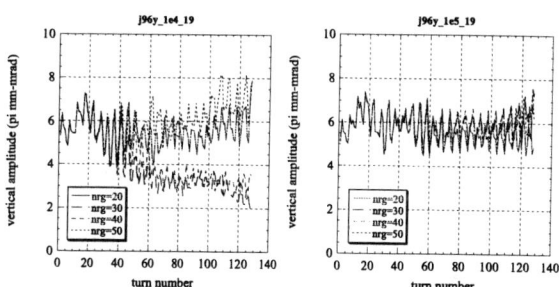

**FIGURE 1.** Evolution of single particle amplitude. 1) number of macro particles is 1E4 (left). 2) number of macro particles is 1E5 (right). Four lines correspond to longitudinal grid from 20 to 50.

The other improvement of the code is to model aperture of each element. Previously, particle loss is checked at each time step by calculating transverse amplitude, sqrt($x_i^2+y_i^2$), if that is larger than a certain value. The vacuum chamber with a constant radius was assumed. Now, at each magnet, transverse coordinates of $x_i$ and $y_i$ are checked if they are within a cross section of vacuum chamber. As a shape of cross section, five different kinds are modeled such as racetrack, diamond, etc. As a matter of fact, a collimator has the minimum aperture and most of the particle loss occurs there.

# SIMULATIONS OF JPARC RCS

## Tune shift and operating point

The new code was applied to the tracking of JPARC RCS [4]. JPARC RCS is a rapid cycling synchrotron that accelerates protons from 181 MeV to 3 GeV with 25 Hz repetition rate. Figure 2 shows bunching factor and incoherent tune shift of the first 8 ms (a whole cycle is 20 ms). Except injection period that lasts about 500 μs, the bunching factor is –0.15 or less. The operating point is chosen at (6.72, 6.35). Sextupoles for chromaticity correction are excited so that chromaticity is zero.

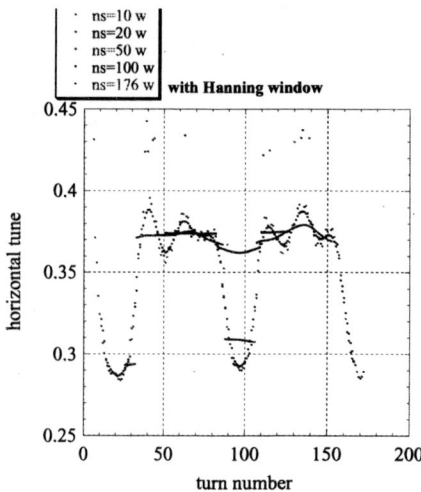

**FIGURE 3**. Tune measurement with data window and interpolation. Too much sampling (turns) loses details of tune change. Too less sampling does not calculate reasonable tune.

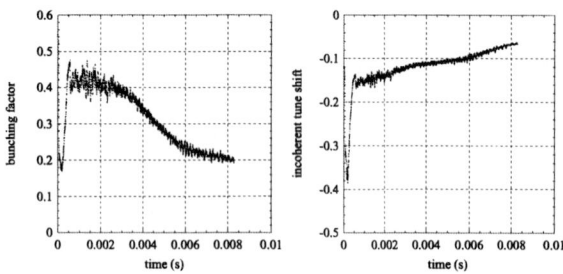

**FIGURE 2**. Bunching factor and tune shift for the first 8ms. JPARC RCS accelerates protons from 181 MeV to 3 GeV in 20 ms. It is around 1 GeV at 8ms.

## Tune measurement

Tune of an individual particle is measured by the technique of Fourier Transform (FT) with data windowing and interpolation [5]. Since incoherent tune changes rapidly due to space charge force with synchrotron oscillations, tune should be determined with as small turns as possible. Figure 3 shows tune as a function of turn number. When many turns such as more than 100 are used to determine tune, rapid change is averaged and detailed information is lost. On the other hand, when only 10 turns are used, turn by turn tune variation becomes large and it is not reasonable. In the following calculations, we take FT of 20 or 32 turns.

## Tracking procedure

First, we track all macro particles (~1E5) and identify the initial coordinates of some particles that are eventually lost. Secondly, we track the same macro particles with recording the coordinates of lost particles turn by turn. The idea is to trace the trajectory of lost particles from the beginning until lost.

Figure 4 shows phase space plot of a lost particle and incoherent tune as a function of turn until it gets lost. Because of large amplitude in longitudinal plane, incoherent tune is modulated according to synchrotron oscillations with an amplitude of 0.1. Distorted ellipse of the longitudinal trajectory is due to second harmonic RF that creates two fixed points near the head and tail.

It is, however, not necessarily true that particle loss is associated with tune modulation by synchrotron oscillations. There is another particle that is lost with very small synchrotron oscillations and almost no tune modulations as shown in Fig. 5.

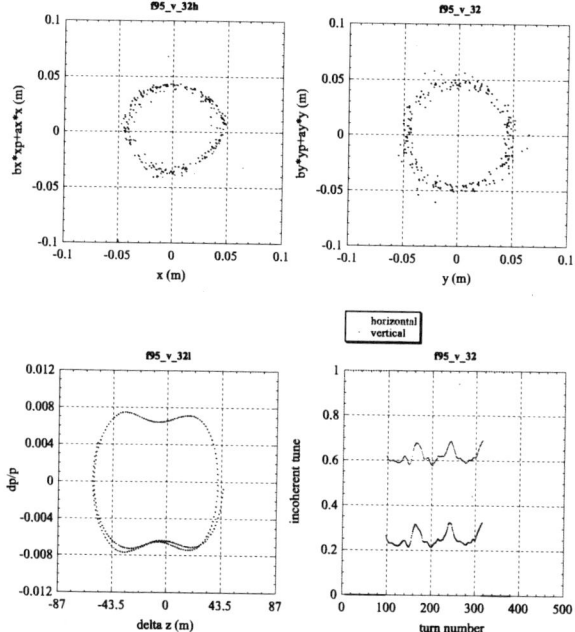

**FIGURE 4.** Particle trajectory of lost particle #32. Top left is horizontal phase space, top right is vertical, bottom left is longitudinal, and bottom right is incoherent tune evolution.

We then plot tune of all lost particles in Fig. 6 when they are lost. Although there is a structure resonance of $vx-2vy=-6$ just above the operating point, correlation with the tune of lost particles is not clear.

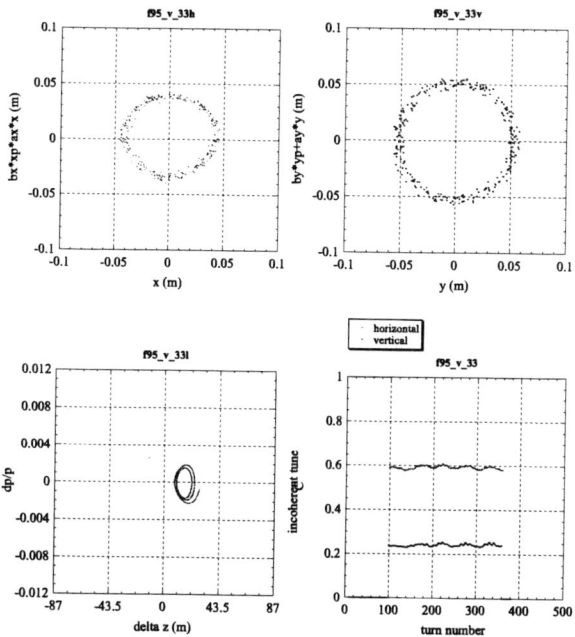

**FIGURE 5.** Particle trajectory of lost particle #33. Top left is horizontal phase space, top right is vertical, bottom left is longitudinal, and bottom right is incoherent tune evolution.

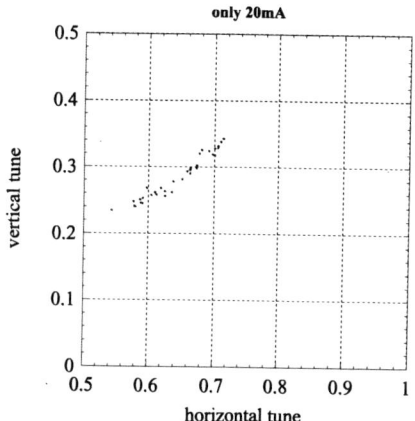

**FIGURE 6.** Tune of lost particles when it is lost. There is a structure resonance of $vx-2vy=-6$ although the correlation is not clear.

## SUMMARY

Simpsons has been improved to give stable single particle trajectory in a bunched beam with 3D space charge force. We studied a behavior of a few percent particles that are lost during a first few 100 turns. It was not successful to correlate tune of a particle and any resonance nearby. More detailed study is continued.

## REFERENCES

1. For example, S. Machida, "Space-charge effects in circular accelerators", AIP Conference Proceedings 592, New York, American Institute of Physics, 2000, pp. 405-434.

2. S. Machida, "Simulation of space charge effects in a synchrotron", AIP Conference Proceedings 448, New York, American Institute of Physics, 1998, pp. 73-84.

3. S. Machida, "Behavior of intense beams simulated with Simpsons", AIP Conference Proceedings 642, New York, American Institute of Physics, 2002, pp. 245-247.

4. Y. Yamazaki (eds.), "Accelerator technical design report for JPARC", KEK Report 2002-13, March 2003.

5. R. Bartolini, A. Bazzani, M. Giovannozzi, W. Scandale, and E. Todesco, "Tune evaluation in simulations and experiments", CERN SL/95-84 (AP), November 1995.

# Comparison of Particle Simulation with J-PARC Linac MEBT Beam Test Results

Masanori Ikegami*, Takao Kato*, Zenei Igarashi*, Akira Ueno*, Yasuhiro Kondo[†], Robert Ryne** and Ji Qiang**

*KEK, High Energy Accelerator Research Organization, 1-1 Oho, Tsukuba, Ibaraki 305-0801, Japan
[†]JAERI, Japan Atomic Energy Research Institute, Tokai, Naka, Ibaraki 319-1195, Japan
**LBNL, Lawrence Berkeley National Laboratory, 1 Cyclotron Road, Berkeley, CA 94720, USA

**Abstract.** The construction of the initial part of the J-PARC linac has been started at KEK for beam tests before moving to the JAERI Tokai campus, where J-PARC facility is finally to be constructed. The RFQ and MEBT (Medium Energy Beam Transport) has already been installed at KEK, and the beam test has been performed successfully. In this paper, the experimental results of the beam test are compared with simulation results with a 3D PIC (Particle-In-Cell) code, IMPACT.

## INTRODUCTION

The J-PARC (Japan Proton Accelerator Research Complex) accelerator consists of a 400-MeV linac, a 3-GeV RCS (Rapid Cycling Synchrotron), and a 50-GeV synchrotron [1, 2]. The linac is comprised of a 50-keV negative hydrogen ion source, a 3-MeV RFQ, a 50-MeV DTL, a 190-MeV SDTL (Separate-type DTL), and a 400-MeV ACS (Annular Coupled Structure linac). The construction of the initial part of the J-PARC linac has been started at KEK to develop and establish the linac system before moving to the JAERI Tokai campus, where the J-PARC facility is finally to be constructed. The 324-MHz RFQ and the MEBT (Medium Energy Beam Transport) has already been installed at KEK, and the beam test has been performed. For the details of the beam test, refer to the reference [3, 4]. In this paper, we focus on the comparison of 3D PIC (Particle-In-Cell) simulations with the experimental results obtained with a transverse emittance monitor and wire scanners. As a simulation code, we use IMPACT [5] which is developed at LBNL.

Before moving to the description of experimental set up, we briefly review the layout of the MEBT. The MEBT has two main roles, namely, to perform transverse and longitudinal matching to the succeeding 324-MHz DTL, and to chop beams to minimize the beam loss at the injection into the RCS. The schematic layout of the MEBT is shown in Fig.1. The MEBT includes eight quadrupole magnets (Q1 to Q8) for transverse matching, two 324-MHz buncher cavities for longitudinal matching, two rf deflection cavities (RFD's) and a scraper for beam chopping, and various instrumentation for beam

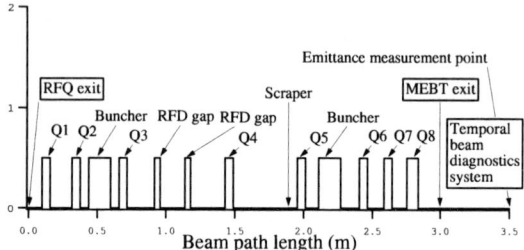

**FIGURE 1.** Schematic layout of the MEBT.

diagnosis. We also have five two-plane steering magnets for beam steering which are built-in to quadrupole magnets. In the measurements described in this paper, we do not perform beam chopping. The peak current of 28.7 mA has been achieved at the exit of the RFQ, and the transmission ratio through the MEBT reaches 99.3% without using steering magnets.

## EXPERIMENTAL SETUP

In the beam test, a TBD (Temporal Beam Diagnostic system) is placed at the exit of the MEBT, which will be removed when installing the DTL. The TBD includes a transverse emittance monitor and a Faraday cup. The emittance monitor is double-slit type, and its first slit is located about 0.5 m downstream from the exit of the MEBT. The slit width and slit interval of the emittance monitor are 0.1 mm and 205 mm, respectively.

In the actual operation, the beam should be strongly

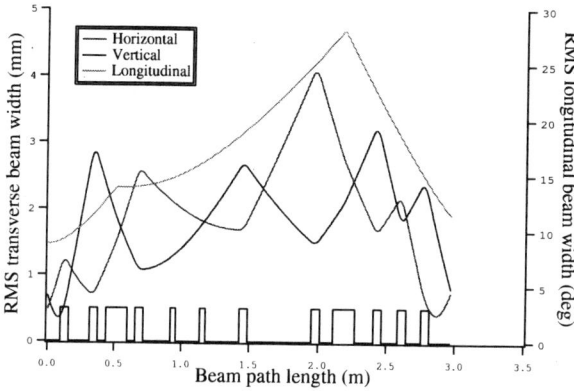

**FIGURE 2.** Beam envelope along the MEBT for a matched case.

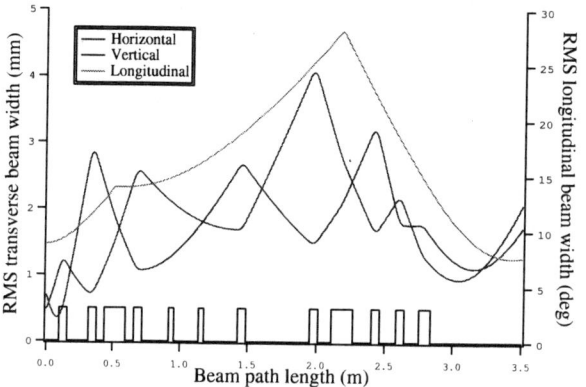

**FIGURE 3.** A typical beam envelope along the MEBT in the experiment.

focused at the exit of the MEBT to satisfy the matching condition to the DTL. However, the strengths of the last two quadrupoles are weakened in the experiment to enable the emittance measurement at the downstream beam diagnostic system. Figure 2 shows a beam envelope for a typical quadrupole setting which satisfies the matching condition to the DTL, and Fig.3 shows a typical beam envelope for the experiment, in which only the last two quadrupoles are weakened. In Fig.3, the downstream end of the plot corresponds to the first slit position of the emittance monitor in the TBD. The quadrupole and buncher setting in Fig.3 corresponds to those in the emittance measurement discussed in the next section.

In the MEBT, we have four WS's (Wire Scanners) for beam profile measurement. Each WS has horizontal, vertical and oblique (45 deg) carbon wires with the diameter of $7\mu m$. In the profile measurement, beam width is shortened to about $50\mu sec$ to reduce the heat load of the carbon wire. In the emittance measurement, we usually use the repetition ratio of 12.5 Hz or 25 Hz to shorten the measurement time, which is typically about 15 min for one plane.

## EXPERIMENTAL RESULTS

The measured normalized rms emittances are 0.252 $\pi$mm·mrad and 0.214 $\pi$mm·mrad in the horizontal and vertical planes, respectively. The emittance is measured with the emittance monitor at the TBD. The peak current in the measurement is 28.7 mA at the exit of the RFQ. The emittances at the exit of the RFQ was measured to be 0.173 $\pi$mm·mrad and 0.194 $\pi$mm·mrad in the horizontal and vertical planes, respectively, before installing the MEBT. The available beam current was, however, limited to about 10 mA at the time of the measurement, because the ion source has been developed in parallel with the construction. These emittances are measured with the same TBD which was placed just after the RFQ. Accordingly, we don't have measurement data for the emittance at the exit of the RFQ with the present maximum available beam current of around 29 mA.

Figure 4 shows the phase space distribution obtained in the measurement with the TBD after the MEBT. In Fig.4, $x$ and $y$ denote the horizontal and the vertical positions, and $s$ is the path length of the design particle. Measured phase space density is represented by 100k dots (particles) in Fig.4 for comparison with particle simulations.

Figure 5 shows a typical beam profile measured with WS3, which is located 81 mm upstream from Q4. In the measurement, quadrupole setting is the same with the emittance measurement, while the buncher cavities are turned off.

## COMPARISON WITH SIMULATION

As a preliminary test on the agreement between experiments and simulations, we have performed 3D PIC (Particle-In-Cell) simulations with IMPACT assuming a 6D Gaussian distribution at the exit of the RFQ. In the simulations, we assume transverse Twiss parameters at the exit of the RFQ which was obtained with the emittance measurement just after the RFQ, and the initial transverse emittances are adjusted to reproduce measured ones at the TBD after the MEBT. We also assume initial longitudinal parameters obtained with PARMTEQ[6] simulations for the RFQ. Figure 6 shows obtained phase-space distribution at the emittance monitor after the MEBT, in which we consider the same lattice setting and beam conditions with the MEBT emittance measurement. In the simulation, 1M simulation particles and $64 \times 64 \times 64$ meshes are employed, and

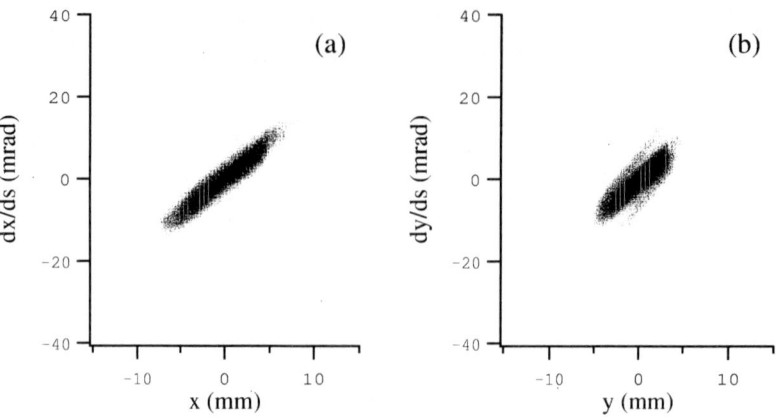

**FIGURE 4.** Phase-space distribution measured with the emittance monitor after the MEBT. (a) Horizontal phase plane. (b) Vertical phase plane.

the integration step width is set to $\beta\lambda/10$ with $\beta$ and $\lambda$ being the particle velocity scaled by the speed of light and the rf wave length, respectively. The assumed initial normalized rms emittances are 0.234 $\pi$mm·mrad, 0.193 $\pi$mm·mrad, and 0.0822 $\pi$MeV·deg in the horizontal, vertical, and longitudinal directions, respectively. In Fig.6, 100k particles out of 1M particles are displayed. Comparing Fig.6 with Fig.4, it is seen that the qualitative agreement between the simulation and the experiment is reasonable, while the shape of the tail portion is slightly different. In Fig.5, we show the beam profile obtained in a similar IMPACT simulation. While the simulated rms beam width is slightly wider in the horizontal direction, the agreement in the vertical direction is excellent. These agreements indicate that the tail portion is already developed to some extent at the exit of the RFQ. For comparison, we show in Fig.7 a result for waterbag case, in which we assume a 6D waterbag distribution as the initial distribution. The tail portion in the waterbag case is obviously less pronounced than in the measurement. Efforts to obtain more realistic initial distribution at the exit of the RFQ is now underway, with which the agreement between experiments and simulations is expected to be improved.

## SUMMARY

The beam tests of the RFQ and the MEBT for the J-PARC have been performed at KEK. The measured normalized rms emittances are 0.252 $\pi$mm·mrad and 0.214 $\pi$mm·mrad in the horizontal and the vertical directions. The peak current is 28.7 mA at the exit of the RFQ, and the transmission ratio through the MEBT is 99.3 %. Preliminary simulation studies are performed with a 3D PIC code, IMPACT, and the agreement between experiment and simulation results is found reasonable in the case where a 6D Gaussian distribution is assumed at the exit of the RFQ. The agreement indicates that the tail portion is developed to some extent in the RFQ. Further study is needed to obtain more realistic distribution at the exit of the RFQ to enable quantitative prediction of tail or halo development in the downstream linac structures.

## ACKNOWLEDGMENTS

The beam test has been performed by the members of J-PARC linac group. Especially, the authors would like to thank K. Nigorikawa of KEK for developing data acquisition software for the wire scanners and other beam monitors in the MEBT.

## REFERENCES

1. Yamazaki, Y., "The JAERI/KEK Joint Project (the J-PARC Project) for the High Intensity Proton Accelerator," in *Procs. of PAC 2003*, 2003.
2. Yamazaki, Y. (ed.), Accelerator Technical Design Report for J-PARC, Tech. rep. KEK Report 2002-13; JAERI-Tech 2003-044, KEK/JAERI, (2003).
3. Ikegami, M., et. al., "Beam Commissioning of the J-PARC Linac Medium Energy Beam Transport at KEK," in *Procs. of PAC 2003*, 2003.
4. Kato, T., et. al., "Beam Studies with RF Choppers in the MEBT of the J-PARC Proton Linac," in *Procs. of PAC 2003*, 2003.
5. Qiang, J., et. al., *J. Compt. Phys.*, **163**, 434 (2000).
6. Crandall, K. R., et. al., RFQ Design Codes, Tech. Rep. LA-UR-96-1836, LANL (1998).

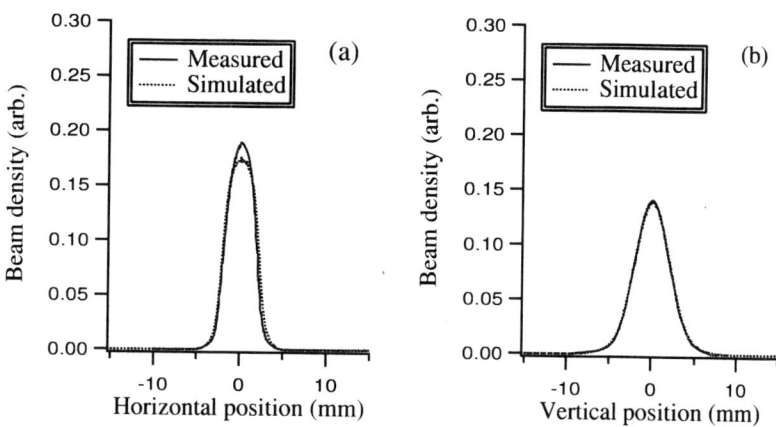

**FIGURE 5.** Beam profile measured with WS3 located before Q4. The quadrupole setting is the same with the emittance measurement, while the bunchers are turned off. (a) Horizontal beam profile. (b) Vertical beam profile. The beam profile obtained in an IMPACT simulation (Gaussian case) is also shown.

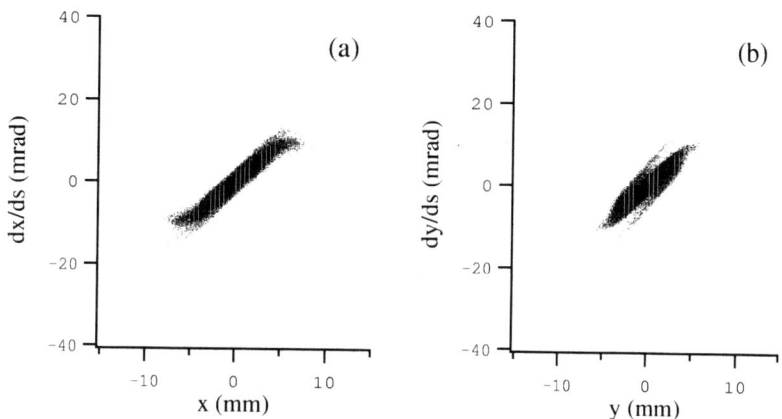

**FIGURE 6.** Phase-space distribution at the emittance monitor obtained with an IMPACT simulation for the emittance measurement (Gaussian case). (a) Horizontal phase plane. (b) Vertical phase plane.

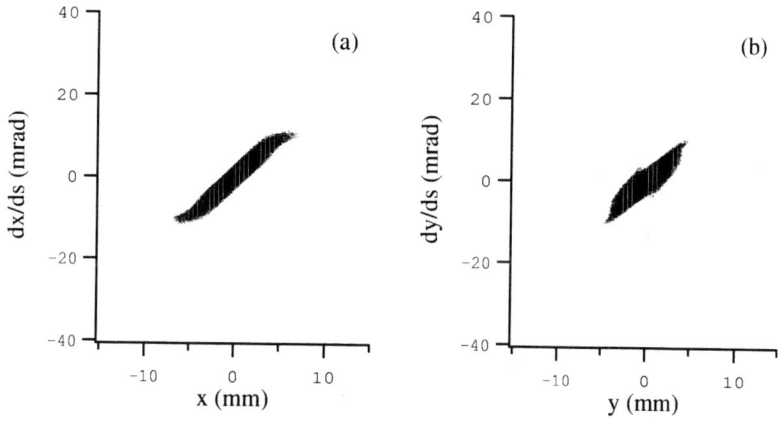

**FIGURE 7.** Phase-space distribution at the emittance monitor obtained with an IMPACT simulation for the emittance measurement (waterbag case). (a) Horizontal phase plane. (b) Vertical phase plane.

# Working Group II:
# Halo Diagnostics

# Halo Diagnostics Summary

Peter Cameron[1] and Kay Wittenburg[2]

[1] *Brookhaven National Laboraqtory, Upton, NY, USA*
[2] *Deutsches Elektronen Synchrotron, DESY Germany*

## INTRODUCTION

The plenary presentation on Halo Diagnostics (T. Shea) offered the participants several different approaches to organizing their thinking on this diverse and complex subject. In terms of the fundamental goal of understanding halo, the involvement of diagnostics was considered in analytical approaches, in simulations, and in experiments. In terms of the priority given to diagnostics, the range goes from purpose built experiments, to dedicated beam experiments at user facilities, to measurements parasitic to normal operations. In terms of challenges, the task begins with a simple definition of halo, requires clear specification of diagnostics requirements as a variety of next generation machines take shape, confronts the dynamic range and sensitivity problems, and finally addresses two omnipresent challenges to the diagnostics specialist – cost, and building trust in the measurements. With the above considerations in mind, and after a survey of the current state of the art, goals for the workshop were presented.

As the working group efforts got underway, it became clear that even at this workshop a general definition of "Beam Halo" could not be given, because of the very different requirements in different machines, and because of the differing perspectives of instrumentation specialists and accelerator physicists. Definitions were offered both from a geometric perspective (point of departure of beam from Gaussian profile, portion of the density distribution beyond n sigma,...) and from the perspective of the formation mechanism (space charge halo, parametric resonance halo,...). From the Diagnostics point of view, one thing is certainly clear – by definition halo is low density and therefore difficult to measure, due to both low count rates in the tails and the large dynamic range (it is desirable to measure the profile of the core 'simultaneous' with the tails).

At this workshop diagnostics were approximately separated into two subgroups – diagnostics for halo measurement, and diagnostics for halo prevention. Diagnostics for halo measurement presented at this workshop include wire scanners (both solid and laser wires), scrapers, the ionization profile monitor, and (for measurement of longitudinal halo) kickers with gated detectors. Presented diagnostics for halo prevention include electron cloud monitors, instability monitors, the quadrupole moment monitor, energy analyzers, tune monitors, and the AC dipole. The above two categories are somewhat artificial – it is straightforward to imagine using halo measurement in conjunction with machine tuning as a halo prevention diagnostic, though somewhat more difficult to imagine using the diagnostics in the halo prevention group for halo measurement. In the following, short summaries of the presented devices and their results will be given. The halo measurement devices are summarized in a table.

## HALO MEASUREMENT

Many of the contributions presented refinements to conventional wire scanners to permit tail measurements. Several schemes were described to achieve the required large dynamic range.

In an extracted beamline at the PSR (R. Macek), detection of secondary emission currents from $100\mu$ SiC wires led to a dynamic range up to $10^6$. The emphasis was on noise suppression, which was achieved by locating the electronics on the beamline, minimizing integration time, limiting bandwidth, performing background subtraction, and auto-zeroing leakage currents and offset voltages. A logarithmic amplifier was used to make possible the required dynamic range with 12 bit digitizers. Measured profiles were approximately Gaussian to almost $10^{-4}$, with broad shoulders beyond that. The demarcation between the central Gaussian and the shoulders was strikingly clean and abrupt. The mechanism of halo production is not understood. Future plans include adding a −5KV bias to the wire to minimize electron cloud effects in the measurement.

At LEDA (D. Gilpatrick) a combination of a 33 µ carbon wire for core measurements and a 1.5mm thick graphite scraper for the tails are mounted on a single scanning actuator, and make possible a $10^5$ dynamic range. Detection was by measurement of secondary emission current for the wire and stopped protons for the scraper. Considerable effort was given to determining optimum bias for both wire (-12V) and scraper (+25V). Scraper insertion was limited to about 2σ by beam heating. The differentiated scraper data was merged with the wire data on-line. Measured profiles appeared to have three distinct regions – a sharp peak extending to $\sim 10^{-1}$, a broader approximately Gaussian shape extending to $10^{-3}$, and broad shoulders beyond that. The mechanism of formation of this structure was not explained.

Both counting and secondary emission techniques were used with good agreement in the slow extracted beam at the AGS (D. Gassner). In the counting measurement, three-fold coincidence horizontal and vertical scintillator telescopes effectively minimized background. Acceptance was about $10^{-4}$ steradian. The scattering targets were 2.5mm diameter tungsten rods. Singles peak rates of ~5M/sec were a factor of 10 below saturation level. Dynamic range was $10^{-2}$ to $10^{-3}$ for secondary emission and $10^{-3}$ to $10^{-4}$ for counting. The purpose of this installation was to diagnose emittance growth during slow extraction from the AGS. Measurements showed an asymmetric halo. The problem was solved by positioning the extraction kicker and ejection septum further from the beam, causing the slow extracted beam to spend less time in the fringe fields of the AGS main magnets. The wire scanner was then de-commissioned.

Counting techniques were also used with a 7µ wire at HERA (K. Wittenburg). The very clean background conditions at HERA led to a dynamic range of $>10^8$, using the fast scan technique in the beam core and a very slow scan in the beam tail. The efficiency (counts per proton intersecting the wire) was about $10^{-7}$. Measurements clearly showed the effectiveness of the scraper in removing beam tails.

Scintillation counting techniques were supplemented by local silicon detectors as well as the experiment detectors in the HERA-B experiment (K. Wittenburg), which uses an internal target inserted into the halo of the beam. Detection efficiency was (remarkably) greater than 50%. The large bandwidth of the counting method permitted beam-in-gap measurements between the ~10MHz bunches. With improved machine performance beam halo practically vanished, a circumstance that made it necessary to artificially generate halo without disturbing the core by means of tune modulation together with beam-beam interaction (C. Montag). With this forced diffusion it was possible to stabilize interactions at the desired rate. It was also shown that the core of the beam and the luminosity at the colliding experiments were not deteriorated by this method, a result of possible significance for the PLL tune measurements mentioned in the next section.

A new readout scheme for wire scanners was successfully tested in Yerevan (reported by K. Wittenburg), where the change of the natural oscillation frequency caused by the heating of the 90µ beryllium-bronze wire was detected. Temperature resolution of this method is of the order of $10^{-4}$°C. Beams with intensities down to 3.4nA were successfully scanned, which shows the very high sensitivity of this technique. The authors expect a dynamic range of up to $10^7$. Concern was expressed about the effect of ohmic heating of the wire by the beam fields. A vibrating wire scanner has been installed at PETRA at DESY, and is waiting for beam.

In an effort where the focus was not so much on dynamic range as the measurement of fast profile changes, counting techniques were utilized in the fast (20 m/s) 7µ carbon flying wire at KEK (S. Igarashi). The flying wire was used to measure the beam profile during injection with the goal of minimizing losses. A series of profiles were acquired by changing the trigger setting in 0.2ms increments (each scan needs 4ms). Full beam profiles with good time resolution were assembled by reconstructing the data from successively injected pulses at different sweep trigger settings/wire positions but at the same time after injection. Tune and intensity dependent profile modulations were observed, and were attributed to the space charge driven fourth order resonance. There was good qualitative agreement between measurements and simulations.

Proof-of-principle for a laser wire scanner for the SNS linac was demonstrated at Brookhaven, with further development work and installation taking place at Oak Ridge (S. Assadi). Short (~10ns) pulses from a 1032nm Nd-YAG laser strip electrons from the H-minus beam particles. In the Brookhaven prototype profile measurements were accomplished by using differential current measurements to observe the resulting current notch. The sensitivity of the laser wire at Oak Ridge was improved by collecting and measuring the stripped electrons, allowing a dynamic range of up to $10^4$. The laser wire method permits measurement of full power beam during normal operations without burning wires, and reduces the danger of expensive contamination of RF cavities in the superconducting portion of the linac.

The RHIC ionization profile monitors (R. Connolly) have been steadily and significantly improved since prototype testing in a transfer line in 1996, and since the acquisition of the first single bunch turn-by-turn injection profiles in the RHIC ring in 1999 (which clearly showed quadrupole oscillations due to injection mismatch). Modifications include improved RF shielding to minimize coupling to beam fields, improved detector geometry to minimize sensitivity to radiation from beam losses, improved geometry of adjacent electric and magnetic fields to prevent migration of non-signal electrons into the detector, improved sweep field geometry and higher sweep voltage to minimize measured profile sensitivity to sweep voltage, and the addition of electron sources to permit calibration. The resulting background reduction has made it possible to begin to probe into the beam tails. A dynamic range approaching $10^3$ has been achieved. A similar (but considerably larger) version of this IPM will be built for the SNS Ring.

To meet the stringent $10^{-4}$ loss requirement of the SNS (losses due to beam-in-gap are estimated to be second only to those due to space charge driven halo), longitudinal halo measurement in the SNS Ring (D. Gassner) will be accomplished by a beam-in-gap kicker driving the gap beam onto a scraper, with detection by a fast gated micro-channel plate. In addition, this system will accomplish gap cleaning before extraction, with the scraper retracted and the gap beam landing in the collimators. Without the kicker, the scraper and data acquisition system will also be used for transverse halo measurements.

| Machine | Type | Signal | Dynamic range | Status |
|---|---|---|---|---|
| LEDA (LANL) (6.7MeV p) | Scanner+ Scraper | SEM | $10^5$-$10^6$ | Working in control-system |
| AGS slow extraction line (2GeV p) | Scanning Target | Counting mode + SEM | $10^4$-$10^5$ $10^2$-$10^3$ | De-commissioned |
| PSR extraction line (LANL) (800MeV p) | Wire Scanner with thin wire | SEM Log amp | $10^6$ | In regular operation |
| SNS LINAC (2.5MeV to 1GeV H⁻) | Laserwire scanner | Photo-neutralization, electron detection | $10^3$-$10^4$ | In operation |
| DESY HERA (40 – 920GeV p) | Wire Scanner with thin wire | Counting mode | $10^7$-$10^8$ | In operation, Readout prototype |
| Yerevan (20 MeV e⁻) DESY PETRA (40GeV p) | Wire Scanner with thin wire | Vibrating wire; natural frequency | $10^6$-$10^7$ | Preliminary tests; More tests planed |
| KEK PS (12GeV p) | Wire Scanner with thin wire | Scintillators | ~$10^3$ | In operation |
| RHIC (polarized p, ions) | IPM | Current | $10^2$-$10^3$ | In operation |

**TABLE 1.** Presented instruments for beam profile measurements, their dynamic range and operational status

# HALO PREVENTION

The summary of the Beam Dynamics working group in these proceedings (A. Fedotov, I. Hoffman) separates halo formation mechanisms into the categories of 'non-resonant' and 'resonant'. The diagnostics presentations for non-resonant mechanisms addressed tools for the observation and analysis of instabilities. The presentations for resonant mechanisms dwelt primarily upon measurement of tune-related parameters.

An overall view of instabilities in both time and frequency domains (M. Blaskiewicz) presented data showing recent measurement of the head-tail instability in RHIC as well as the electron cloud instability in both the AGS Booster and the LANL PSR. The usefulness of principal component analysis in understanding this data was demonstrated. In the effort to understand the conditions for onset of instability, data was presented showing the measurement of longitudinal impedance in RHIC from Schottky spectra. The agreement between theory and measurement in this data was remarkably good. Particularly interesting in this presentation was the observation of long-lived (~1hr) longitudinal 'hot spots' (solitons?) in the proton beam at store in RHIC, and their possible role in triggering instabilities during acceleration. These hot spots were not observed with gold beams, probably as a result of the damping/diffusion effect of IBS.

The electron cloud instability results from a form of multipacting driven by time gradients of the beam space charge field in the longitudinal tail of the bunch. Different instruments were developed and studied at the PSR (R. Macek) for the measurement of electron cloud parameters. Historically, the presence of electron cloud was observed from pressure rise due to multipacting electrons causing desorption at the beampipe wall, and looking at ion pump currents with sufficient bandwidth provided early time domain information. Early studies of direct electrons utilized simple biased electrodes to collect electrons. Data from these electrons was often puzzling and difficult to interpret, particularly the copious presence of electrons at the wall in bending dipoles in the plane normal to the dipole field (the IPM has clearly demonstrated the practical efficiency of magnetic field in confining electron trajectories). This motivated the development of the Retarding Field Analyzer (RFA), where the presence of a repeller grid permits measurement of the electron energy spectrum at the beampipe wall. The 80MHz bandwidth of the RFA makes possible detailed study of the time evolution of the electron cloud, and clearly shows the 'trailing edge multipactor' behavior. This multipactor is seeded by electrons released from the beam space charge potential well at the tail of the bunch. A key factor in buildup of the instability is how many seed electrons survive the gap to be captured by the following bunch. To this end, an Electron Sweeping Detector (ESD) was developed. The ESD is basically an RFA with the added capability of applying a fast high voltage (~1KV) pulse at the beginning of the following bunch. Successful measurements were made of surviving electrons.

In an application of similar techniques to the longitudinal, a high-resolution retarding field energy analyzer (Y. Zou) was developed at UMER (a 10keV, 100mA, electron accelerator) to study the mechanism of beam energy spread and its evolution in the electron beam. Detailed analysis of the device parameters had led to an energy resolution of better than 2eV. First experimental results show excellent agreement between the experiments and the theory (Boersch effect and longitudinal-longitudinal effect). More experiments and further improvements are in preparation.

In the case of resonant mechanisms of halo formation, measurement of tune-related parameters (coherent tune, incoherent tune, beam-beam tune shift, coupling, chromaticity, non-linear tune spread, tune shift due to electron cloud,...) is essential. The application of Phase-locked Loop (PLL) and Schottky techniques to these measurements in RHIC was presented (P. Cameron), as well as plans for utilizing these methods in the SNS Ring. Concern was expressed by Professor Hoffman regarding the reliability of these methods in space-charge dominated beams, and this subject is under investigation. PLL data was also presented showing lock to the large amplitude oscillations of beam in islands of the 2/9 resonance in RHIC, data which might be used in resonance compensation. Preliminary data from a resonant quadrupole monitor was presented, in which the quadruple mode was resonated to improve sensitivity and diminish the dynamic range problem characteristic of this type monitor.

The techniques of electron cooling and stochastic cooling are complementary – electron cooling effectively cools the beam core, whereas stochastic cooling is efficient on the beam tails. In the case of Optical Stochastic Cooling (OSC), undulator magnets would be used for both the pickup and the kicker, and cooling time is power limited rather than bandwidth limited (V. Yakimenko). Present plans on the path to eventual OSC in RHIC include summer 2003 testing of an optical parametric amplifier, in which a frequency doubled $CO_2$ laser pumps a nonlinear crystal to amplify the signal (ultimately from the undulator pickup). Gains of over 100dB are anticipated. Pickup and kicker should be installed in regions of non-zero dispersion to permit simultaneous longitudinal and transverse cooling. In the case of halo cooling, the power limitation can be overcome by adjusting timing of the pump laser to apply power predominantly to the longitudinal beam tails, reducing the practical cooling time from ~1hr for the entire beam to ~<1min for the beam tails. A concern is the requirement for an isochronous lattice between pickup and kicker, the requirement being a few microns (~$10^{-14}$ sec) at the 12 micron cooling wavelength.

With the AC Dipole (M. Bai), adiabatic turn on and off of a 1m long 0.01T kicker driving the beam at a frequency close to the betatron tune permits excitation of large amplitude coherent oscillations without measurable emittance growth. Data was presented showing 2mm p-p coherent oscillations in RHIC. Measurements accomplished to date include beta functions, phase advance, and linear coupling. By driving the beam at the spin tune rather than the betatron tune, this kicker is also used as a spin flipper for the polarized beam program. Possible applications of this method to halo measurements are under study.

RHIC boasts an active and well-organized Beam Experiments program focused on improvements in machine performance. Essentially all experiments in this program have direct bearing on halo formation. In a summary of beam experiments results (F. Pilat), data was presented on measurements of dynamic aperture, diffusion coefficients (both at injection and store), beam transfer functions in the 4-8GHz range (in preparation for initial stochastic cooling efforts in 2004), IR triplet nonlinearities and corrections, beam-beam tune shift vs. crossing angle, nonlinear chromaticity, resonance compensation, and pressure rise/electron cloud. Plans for 2004 include addressing the requirements for luminosity increases and RHIC upgrade plans, improved machine modeling, and increased collaboration with FNAL and CERN.

## CONCLUSION

Wire scanners and scrapers define the state of the art in halo measurements, with high sensitivity and huge dynamic range. With the variable-delay trigger method, and given the possibility of repeated replication of beam conditions, wire scanners can measure fast profile changes. With continuous steady improvements, the IPM is starting to make contributions to halo measurement, including fast profile changes without the requirement of repeated replication of beam conditions. Beyond these bread-and-butter tools, there are evolving a variety of nice (and sometimes even fancy) instruments to measure the parameters which may drive beam tails. A difficult challenge posed by the accelerator physicists is the turn-by-turn (or better yet, bunch-by-bunch) measurement of beam parameters, both in static and ramped machines.

# Physics Results from the Los Alamos Beam-Halo Experiment

## Thomas P. Wangler

*Los Alamos National Laboratory, Los Alamos, NM 87544*

**Abstract.** We present physics results from an experimental study of beam halo in a high-current 6.7-MeV proton beam propagating through a 52-quadrupole periodic-focusing channel. The gradients of the first four quadrupoles were independently adjusted to match or mismatch the injected beam. Emittances and beam widths were obtained from measured profiles for comparisons with maximum emittance predictions of a free-energy model and maximum halo-amplitude predictions of a particle-core model. The experiment supports both models and the present theoretical picture of halo formation.

## INTRODUCTION

Control of beam-halo and associated beam losses is a fundamental requirement for high beam availability in high-power proton linacs. More than a decade ago, computer simulation studies [1] identified beam mismatch as the major source of the halo and emittance growth observed in simulations. The emittance growth can be related to the conversion of free energy from mismatch oscillations into thermal energy of the beam. For a given mismatch strength, the free-energy model determines the maximum emittance growth, which results from complete transfer of free energy into emittance [2].

A physical model of halo formation is expected to include both nonlinear and time-dependent forces that drive halo particles to larger amplitudes. Such a mechanism is provided by the particle-core model [3, 4, 5], in which beam mismatch produces an imbalance between focusing, space charge, and emittance, exciting a symmetric or breathing ($x_{rms}$ and $y_{rms}$ in-phase) mode oscillation of the core. The space-charge field of the oscillating core modulates the net focusing force acting on individual particles and drives particles in a nonlinear parametric resonance when $f_{particle}=f_{mode}/2$, where $f_{particle}$ is the betatron frequency of the particle and $f_{mode}$ is the mode-oscillation frequency [4]. The model predicts a maximum resonant-particle amplitude as a function of the mismatch strength [5]. Neither the free-energy, nor the particle-core model predict the growth rates for the halo amplitude and beam emittance, for which numerical simulations are required.

## BEAM-HALO EXPERIMENT

To test the two models, we installed a 52-quadrupole periodic-focusing beam-transport channel at the end of the Los Alamos Low-Energy Demonstration Accelerator (LEDA)[6]. LEDA delivers a 6.7-MeV proton beam from a 350-MHz radiofrequency-quadrupole (RFQ) linac. The beam was pulsed at a 1-Hz rate with a 30-µs pulse length. The channel length of 11 m was sufficient for the development of about 10 mismatch oscillations, enough to observe at least the initial stages of emittance growth and halo formation caused by mismatch. In this paper we present results for a 75-mA proton beam current.

The most important beam-diagnostic elements were the transverse beam-profile scanners [7] that measured the horizontal and vertical distributions. These were installed at nine stations (Fig. 1), each located midway between pairs of quadrupoles. The scanners were labeled with numbers corresponding to the preceding quadrupole-magnet number. The beam was matched, using a least-squares fitting procedure that adjusted the first four quadrupoles to produce equal rms sizes at the last eight scanner locations. For a mismatched beam one must consider not only the breathing mode, but also the antisymmetric or quadrupole mode. The beam was mismatched in nominally pure symmetric or antisymmetric modes by proper settings of the same four matching quadrupoles. The mismatch strength was measured by a mismatch parameter µ, which equals the ratio of the rms size of the initial beam to that of the matched beam. For a matched beam µ=1.

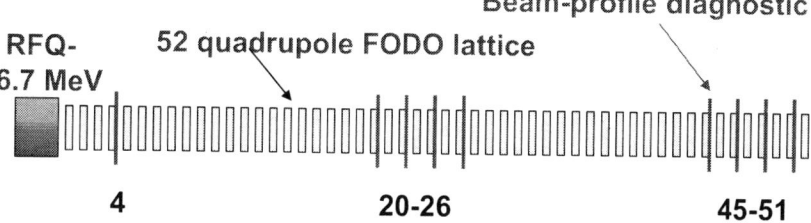

**FIGURE 1.** Block diagram of the 52-quadrupole-magnet lattice showing the nine locations of beam-profile scanners.

Figure 2 shows the matched and mismatched 75-mA beam profiles at scanner 51. The matched beam has a Gaussian-like central profile with an rms beam size of 1.1 mm. For the matched beam a low-density halo is observed to extend as far as 9 rms. This matched-beam halo is observed at all scanners and is most easily explained as a halo that has formed in the injector/RFQ system prior to the periodic quadrupole channel. Direct measurement of the beam-energy distribution with a resolution of about 200 keV, using a dispersive section of the transport line at the end of the periodic quadrupole channel, showed no evidence for low-energy tails that might contribute to this halo. Although collimation can remove this halo, collimation was not implemented in our experiment. Halo caused by mismatch was our main interest, because this mismatch mechanism is expected to involve more particles, and can form halo even at high energy where collimation is more difficult. A breathing-mode-mismatch beam profile for $\mu=1.5$, seen in Fig. 2, shows the growth of shoulders indicating substantial formation of halo.

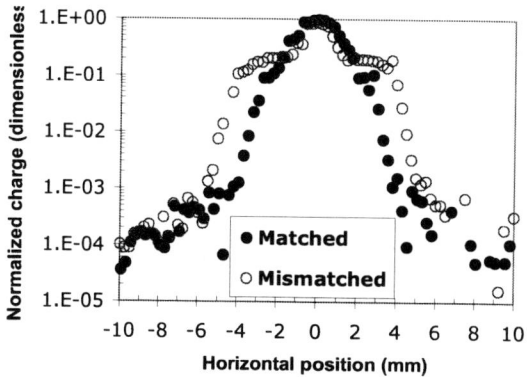

**FIGURE 2.** Horizontal beam profiles at scanner 51 for a 75-mA, $\mu=1$ matched beam (blue solid circles), and breathing-mode $\mu=1.5$ mismatched beam (red open circles).

The rms-size measurements were used to calculate the rms emittances at scanners 20 and 45 [6]. Assuming zero emittance growth in the channel for the matched beam, the tune depression from space-charge was 0.82 immediately after the matching quadrupoles at scanner 4, and was constant at 0.95 after the beam had debunched at approximately quadrupole 16, about 3.5 m from the beginning of the channel. Although the beam was not in a space-charge-dominated regime, significant space-charge effects in mismatched beams were still expected.

The free-energy model can be tested by comparing the measured emittance growths at scanners 20 and 45 with the emittance-growth upper limits from that model. The emittance-growth measurements for mismatched beams show some significant anisotropies (x-y differences). Franchetti, Hofmann, and Jeon [8] report simulation studies of anisotropic beams in uniform focusing channels, in which large (40%) x-y emittance-growth differences are observed that are sensitive to initial x-y tune differences as small as 1%. The sensitivity is not the result of chaotic behavior, but is caused by the parametric resonance discussed earlier, which is sensitive to x-y parameter differences. In our case, anisotropies could be driven by percent-level input x-y emittance differences that are not resolved experimentally. Although the free-energy model was derived for an axisymmetric beam, these authors find that the model can be extended to a 2D anisotropic case if the emittance growth is averaged over x and y.

Figure 3 shows the x-y averaged rms-emittance-growth results (points with error bars) versus $\mu$ at scanner 20 for a 75-mA breathing-mode mismatch. The maximum emittance-growth curves from the free-energy model are shown for the two tune-depression values that bracket the values for the debunching beam, and it can be seen that the theoretical maximum is insensitive to the tune depression over this range. The breathing-mode data in Fig. 3 are consistent at all $\mu$ values with the maximum emittance growth predicted by the model. The breathing mode results at scanner 45 (not shown) show no significant additional emittance growth, consistent with the upper limits from the model and with complete transfer of free energy within only four mismatch oscillations. Quadrupole mismatch data at 75 mA are not available at scanner 20, but are available at scanner 45 (see Fig. 4). These results are also consistent at all measured μ values with the maximum growth of the model. Although an axisymmetric beam is assumed in the model, applicability to the quadruple mode is physically reasonable for a given free energy if equal energy sharing is assumed in x and y. Overall, the data for both mismatch modes indicate a rapid growth mechanism with nearly complete transfer of free energy occurring in less than ten mismatch oscillations.

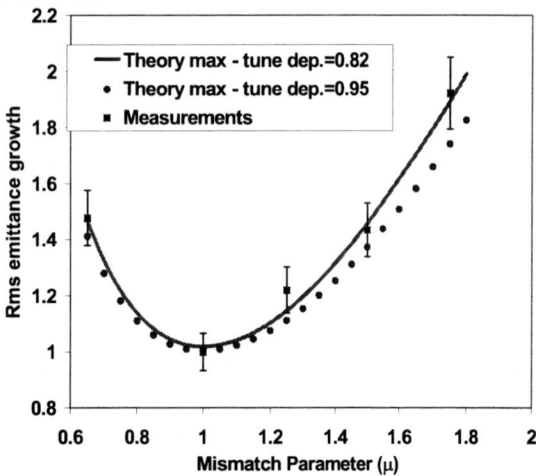

**FIGURE 3**. Points with error bars show measured rms-emittance growth averaged over x and y for 75 mA at scanner 20 for a breathing-mode mismatch. The curve and points without error bars bracket the maximum growth from the free-energy model.

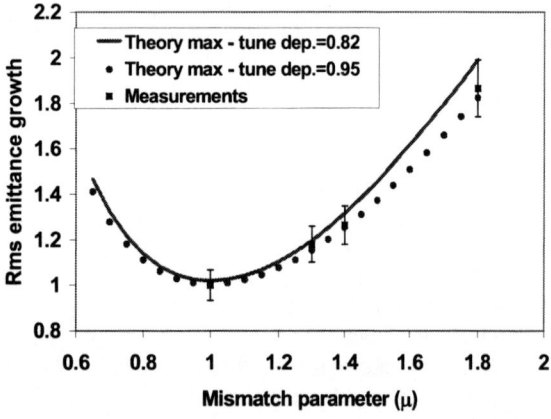

**FIGURE 4**. Points with error bars show measured rms-emittance growth averaged over x and y for 75 mA at scanner 45 for a quadrupole-mode mismatch. The curve and points without error bars bracket the maximum growth from the free-energy model.

The particle-core model predicts the maximum resonant-particle amplitude as a function of mismatch parameter μ [5]. We were unable to determine an experimental maximum amplitude for direct comparison because of background. Instead, we compare the measured amplitudes (x-y averaged half widths of the beam) at three different fractional beam-profile intensity levels (10%, 1%, and 0.1% of the peak) for a breathing-mode mismatch with the maximum amplitude predicted by the particle-core model. A comparison is shown in Figure 5 for scanner 20 at 75 mA. The shapes of all three measured half-width curves are consistent with the shape of the maximum amplitude curve from the particle-core model, and all three measured curves lie below the maximum amplitude curve from the model. Similar results are observed at scanner 51. Although the particle-core model based on a single mismatch mode is a relatively simple description of the beam dynamics, the agreement with the model for the curve shapes and for the consistency of the magnitudes, supports the conclusion that the model incorporates the main physical mechanism responsible for the halo growth.

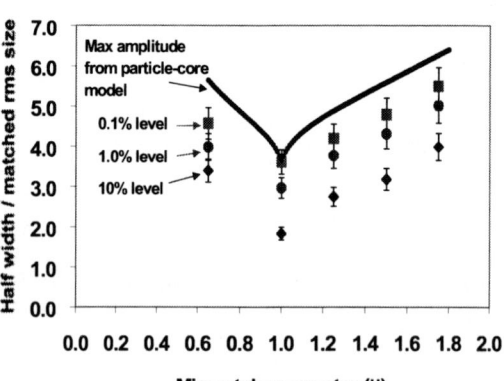

**FIGURE 5.** Measured beam half widths at scanner 20 (75mA and a breathing mode mismatch) at different fractional intensity levels versus mismatch strength μ for comparison with the maximum resonant amplitude (curve) from the particle-core model.

## MULTIPARTICLE SIMULATIONS

Self-consistent multiparticle simulations including space-charge forces were carried out using the macroparticle simulation code IMPACT[9]. The lack of detailed knowledge of the initial beam distribution in phase space is an important issue for the simulations. Our first approach has been to generate three different initial distributions at the entrance of the beam-transport channel, all with the same Courant-Snyder ellipse parameters and emittances; the latter were deduced from the measurements. The three input distributions are: 1) 6D Waterbag, 2) 6D Gaussian, and 3) a distribution called LEBT/RFQ, generated from a simulation through the LEBT and RFQ, starting at the plasma surface at the exit of the ion source. The particle coordinates of the LEBT/RFQ distribution were scaled to produce the correct initial Courant-Snyder parameters and emittances. The transverse phase-space plots of these distributions are shown in Fig. 6.

Figure 6 shows qualitatively an increasing input beam halo as we progress from the Waterbag to the Gaussian to the LEBT/RFQ distribution. For each of these initial distributions we have simulated the beam dynamics through the matched beam channel [10, 11]. For each simulation, using about 2.8 million macroparticles with a computation grid of 65 X 65 X 129, Poisson's equation was solved in cylindrical coordinates with transverse perfect-conducting-wall boundary conditions and a periodic boundary condition, longitudinally.

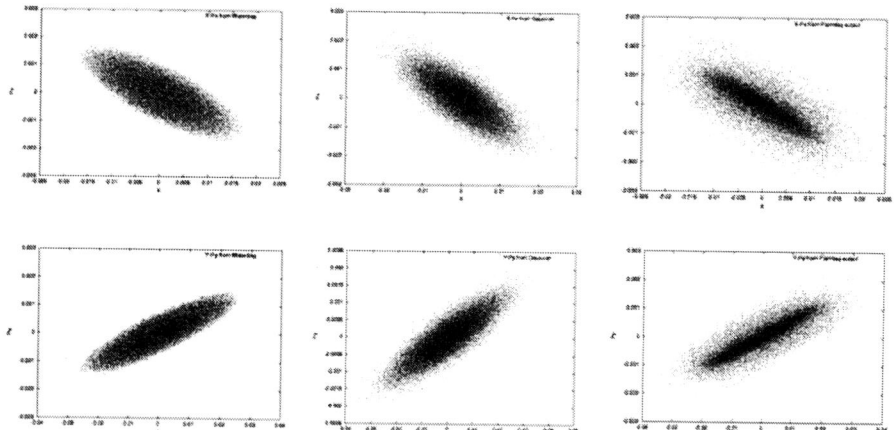

**FIGURE 6.** Transverse phase-space projections for three initial simulation distributions at the entrance of the transport channel for the matched beam: 6D Waterbag (left), 6D Gaussian (middle), and RFQ/LEBT (right). The upper plots show x-$p_x$ phase space, and the lower plots show y-$p_y$ phase space.

All three initial distributions predict a nearly matched transverse rms beam size, in good agreement with the matched-beam measurements. In addition to the rms sizes, we also measured the projected density distributions, i.e. beam profiles in $x$ and $y$ at nine locations along the transport channel. The density profiles from the simulations of the matched beam are compared in Fig. 7 with measurements at the final detector. In general, the LEBT/RFQ simulation agrees best with the measured profiles, especially in the core region. However, none of the distributions reproduce well the tails observed in the measured profiles.

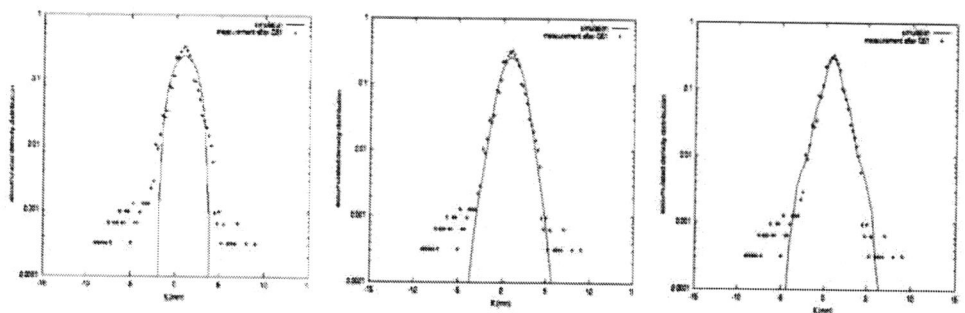

**FIGURE 7.** Horizontal profiles from measurements (points) and simulations (curves) at the final profile detector for 75-mA matched beam. The initial distributions for the simulations are: 6D Waterbag (left), 6D Gaussian (center), and LEBT/RFQ (right).

Figure 8 shows the horizontal-profile comparison at the final detector, for the LEBT/RFQ simulation and for a breathing-mode mismatch with an initial rms beam size that is 50% larger than the matched case. Although the agreement at some of the scanners is better than that seen in Fig.8, simulations for all three initial distributions fail to reproduce the broad shoulders seen in Fig. 8, which are induced in the measured beam profiles by the mismatches. The broader shoulders for the real beam are evidence of a more rapid halo growth rate in the real mismatched beam than in the simulations.

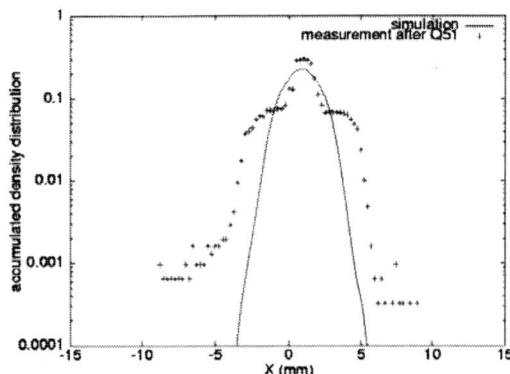

**FIGURE 8.** Horizontal profile from measurements (points) and the LEBT/RFQ simulation (curve) at the final profile detector for a 75-mA breathing-mode mismatched (by 50%) beam.

Figure 9 compares the rms emittance growth at the end of the channel calculated from the measurements at 75 mA for the breathing-mode mismatched beam (initial rms size 50% larger than the matched size) with those from the three simulations. We find that the emittance-growth rates from simulations increase as we progress from the 6D Waterbag to 6D Gaussian to the LEBT/RFQ distribution, which means that the distributions with greater initial beam-halo have more emittance growth. We interpret this as a result that would be expected from the particle-core model [5], because the resonant particles that form the halo lie outside the beam core. The emittance-growth rate calculated from measurements is larger than those from any of the three simulations. Our interpretation is that the initial distributions assumed for the simulations do not adequately populate the tails, which include the resonant particles that are main source of the halo and emittance growth.

## CONCLUSIONS

Our experimental results support both the free-energy model and the particle-core model of halo formation in mismatched beams. This conclusion is important because these models predict upper limits to emittance and halo-amplitude growth in high-current transport channels and linacs, and allow estimation of focusing strength and aperture requirements in new designs. We also conclude that using only the known Courant-Snyder parameters and the emittances as input parameters is not sufficient information for reliable simulations of beam halo formed in mismatched beams. Our interpretation of the simulation results is that the higher emittance-growth rate for the real beam is caused by a higher particle density in the initial beam tails, and consequently, a greater population of the region of phase space that leads to resonant halo growth. We conclude that knowledge of the initial particle distribution, especially the density in the tails, is important for accurate predictions of the beam-halo from simulations.

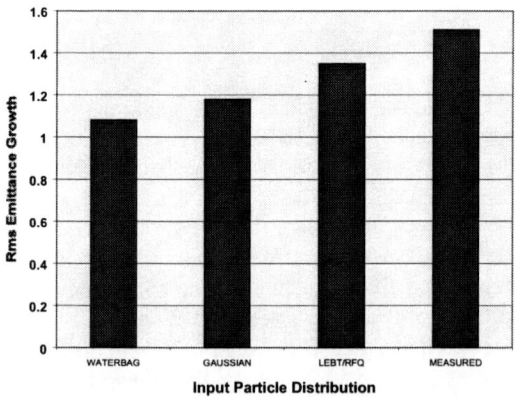

**FIGURE 9.** Emittance growth from three simulations and from the experiment, for a 50% breathing-mode mismatch at 75 mA.

## ACKNOWLEDGMENTS

The experiment was carried out by a scientific team that in addition to the author included C. K. Allen, K. C. D. Chan, P. Colestock, K. R. Crandall, R. W. Garnett, J. D. Gilpatrick, W. Lysenko, J. Qiang, J. D. Schneider, M. E. Schulze, R. Sheffield, and H. V. Smith. We thank the dedicated LEDA personnel who made the experiment possible. This work was supported by the U.S. Department of Energy. Simulations in this research were performed in part using resources of the National Energy Research Scientific Computing Center, which is supported by the Office of Science of the U.S. Department of Energy.

## REFERENCES

[1] A. Cucchetti *et al.*, Proc. of IEEE 1991 Part. Accel. Conf., ed. by Lizama and Chew (IEEE, New York, 1991), p.251.

[2] M. Reiser, *Theory and Design of Charged Particle Beams* (John Wiley & Sons, Inc., New York, 1994), p.477; M. Reiser, J. Applied Phys. **70**, 1919 (1991).

[3] J.S.O'Connell, T.P.Wangler, R.S.Mills, and K.R Crandall, Proc. of 1993 Part. Accel. Conf., IEEE Catalog No. CH3279-7, 3657-3659.

[4] R.L. Gluckstern, Phys. Rev. Lett. **73**, 1247 (1994).

[5] T.P.Wangler, K.R.Crandall, R. Ryne, and T.S.Wang, Phys.Rev. ST-AB **1** (084201) 1998.

[6] C.K.Allen et al., "Beam Halo Measurements in High-Current Proton Beams," Phys. Rev. Lett. **89**, No.21, (214802) (2002).

[7] J.D.Gilpatrick *et al.*, Proc. 2001 Part. Accel. Conf., IEEE Catalog No. 01CH37268, 525-527.

[8] G. Franchetti, I. Hofmann, and D.Jeon, "Anisotropic Free Energy Limit in High Intensity Accelerators," Phys. Rev. Lett. **88**, (254802) (2002).

[9] J. Qiang, R.D.Ryne, S. Habib, and V. Decyk, J. Comput. Phys. 163, 434 (2000).

[10] Ji Qiang, et al., "Macroparticle Simulation Studies of a Proton Beam Halo Experiment, "Phys.Rev.ST Accel. Beams **5**, (124201) (2002).

[11] T.P.Wangler and Ji Qiang, "Los Alamos Beam Halo Experiment: Comparing Theory, Simulation, and Experiment, " to be published in Proc. of Advanced Accelerator Concepts Workshop, June 23-28, 2002, Oxnard, CA.

# Observation of Emittance Growth at KEK PS

S. Igarashi, T. Miura, E. Nakamura, Y. Shimosaki, M. Shirakata, K. Takayama and T. Toyama

*KEK 1-1 Oho, Tsukuba, Ibaraki 305-0801, Japan*

**Abstract.** Emittance growth has been observed in the transverse direction at the injection period of the 12 GeV main ring of the KEK proton synchrotron. Measurement of the beam profiles using flying wires has revealed a characteristic temporal change of the beam profile within a few milliseconds after injection. Horizontal emittance growth was observed when the horizontal tune was close to the integer. The effect was more enhanced for higher beam intensity and could not be explained with the injection mismatch. Resonance created by the space charge field was the cause of the emittance growth. A multiparticle tracking simulation program, ACCSIM, taking account of space charge effects has qualitatively reproduced the beam profiles.

## INTRODUCTION

The beam intensity of the KEK PS 12 GeV main ring has significantly increased since the K2K neutrino oscillation experiment started. Efforts to minimize beam loss have been continuously made. One of the issues is to reduce the loss during the injection period. Nine bunches of protons with the kinetic energy of 500 MeV are injected with the interval of 50 ms. About 30% of protons are lost during the injection period of 510 ms. The highest operating intensity of the main ring is $1.4 \times 10^{12}$ protons per bunch at injection. The nominal operational tune has been optimized to be 7.15 and 5.25 for the horizontal and vertical tune respectively. The main ring has a circumference of 340 m and four-fold symmetry. A super period consists of seven FODO cells.

The incoherent tune shift is estimated to be 0.5 for the highest operating intensity without considering the effect of the image field or dispersion. The large value of the space charge tune shift is partly due to a small emittance of the injection beam. Emittance dilution and particle loss would occur under the circumstances. It is empirically known that emittance dilution observed after injection to the main ring depends on the beam intensity and tune.

## PROFILE MEASUREMENTS

Flying wire transverse beam profile monitors have been operated at the main ring. An analysis procedure has recently been established to reconstruct the beam profile that quickly changes with a time scale of 1 ms or less [1].

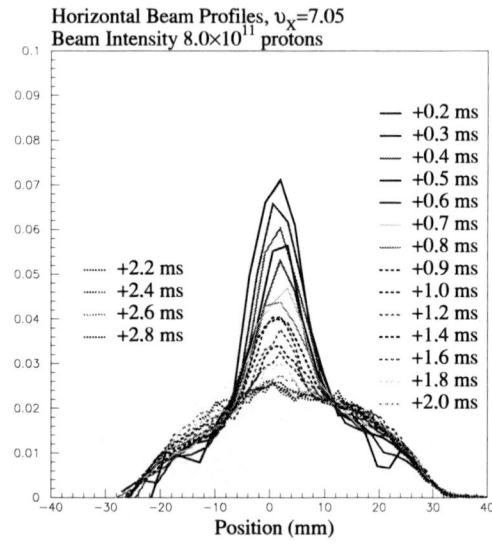

**FIGURE 1.** Horizontal beam profiles 0.2 ~ 2.8 ms after injection when the horizontal tune was 7.05 and the injection beam intensity was $8.0 \times 10^{11}$ protons.

When the injection beam intensity was set to $8.0 \times 10^{11}$ protons, the beam profile 0.2 ms to 2.8 ms after injection were reconstructed as in figure 1. The horizontal tune in this case was 7.05 which was not the nominal operational value. The vertical tune was 5.22. A significant beam loss was observed within 1 ms after injection under the condition. The reconstructed profile shows a notable change of the distribution. The profile at 0.2 ms after injection consists of a narrow peak and a broad distribution. The narrow peak diminishes in 2 ms and only the broad dis-

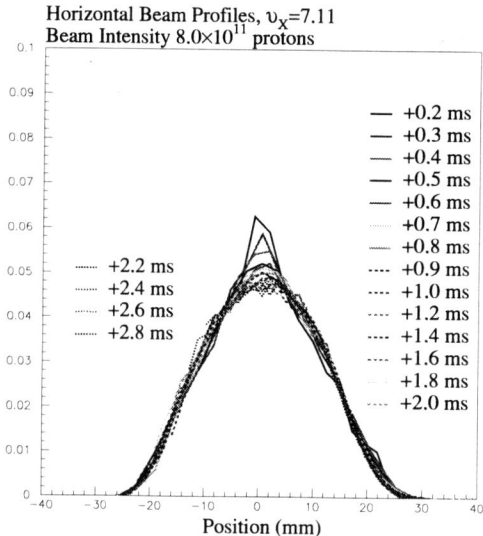

**FIGURE 2.** Horizontal beam profiles 0.2 ~ 2.8 ms after injection when the horizontal tune was 7.11 and the injection beam intensity was $8.0 \times 10^{11}$ protons.

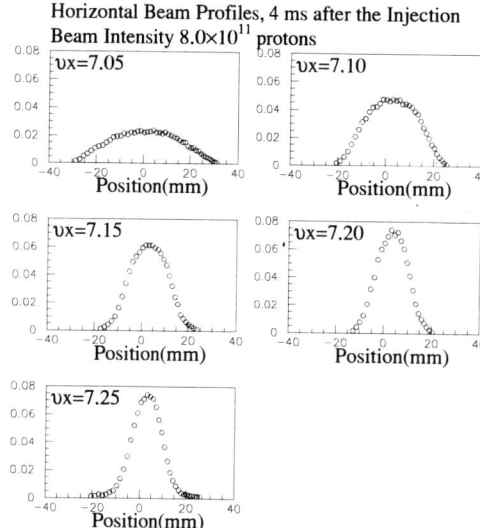

**FIGURE 3.** Horizontal beam profiles 4 ms after injection when the injection beam intensity was $8.0 \times 10^{11}$ protons.

tribution remains.

The same procedure was applied for the horizontal tune of 7.11 and the vertical tune of 5.21 which was near the nominal operational value. The reconstructed profiles are shown in figure 2. The profile at 0.2 ms after injection still consists of a narrow peak and a broad distribution. The narrow peak diminishes in 1 ms, and only the broad distribution remains as in the case of the tune of 7.05. The narrow peak of this case is, however, less significant than that of the previous tune, and the beam loss is not either significant in this case.

Horizontal beam profiles after injection were measured for the injection beam intensity of 2.2, 3.9 and $8.0 \times 10^{11}$ protons. The measurements were performed for a range of the horizontal tune from 7.05 to 7.26. The vertical tune was maintained to be between 5.23 and 5.32. The trigger was set to initiate the wire scanning to take beam profiles of about 4 ms after injection when the rapid change of the profile was settled. The profiles for the intensity of $8.0 \times 10^{11}$ protons are shown in figure 3. It was observed to be wide when the horizontal tune was 7.05, and became narrower as the tune was away from the integer.

The injection beam $\sigma$ emittance was measured at the beam transfer line to be 3.4 and 2.7 $\pi$mmmrad for the horizontal and vertical direction respectively. The emittance did not depend on the beam intensity. The measurement, however, has uncertainty from short understandings of transfer line lattice parameters.

Injection steering error, betatron function mismatch and dispersion function mismatch were measured and the effects to the injected beam to main ring were considered. Figure 4 shows the reconstructed $x - x'$ for the first 80 turns of injected beam from measurement of two beam position monitors for the tune of 7.05 to 7.26. The measurements were performed immediately after the flying wire profile measurements. The mismatch parameters were observed to show little dependence on the horizontal tune between 7.05 and 7.26 as shown. Beam emittance with the mismatch effects as a function of the horizontal tune was estimated. Little dependence on the horizontal tune was observed as in figure 5.

Each beam profile was fitted with either a Gaussian or a parabolic function and the emittance that includes the 87% fraction of the density distribution was estimated and plotted in figure 5. Emittance growth was observed when the tune is close to the integer for all the measured intensity settings. The tune dependence can not be explained only with the mismatch effects. A tune range where the emittance growth occurs depends on the intensity. It is inferred that the emittance growth is due to the space charge field.

## ACCSIM SIMULATIONS

A multiparticle tracking simulation program, ACCSIM [2], taking account of space charge effects has been performed to understand the observed phenomena. Transverse space charge forces have been calculated for 10000 macro particles with a hybrid fast-multipole technique and grids of 1 mm $\times$ 1 mm every 0.76 m step. Thin lens kicks have been applied to simulate sextupole and oc-

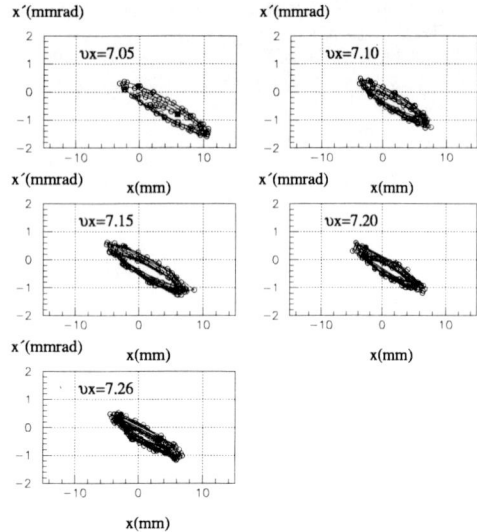

**FIGURE 4.** Reconstructed $x - x'$ for the first 80 turns of injected beam from measurement of two beam position monitors for the tune of 7.05 to 7.26.

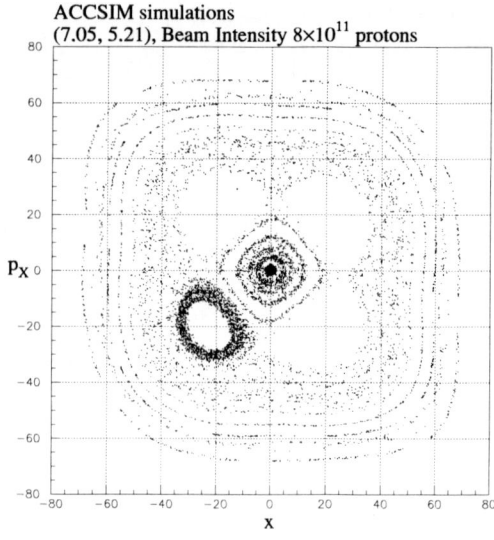

**FIGURE 6.** ACCSIM simulation of the $x - p_x$ phase space plots of 20 test particles when the horizontal tune was 7.05 and the injection beam intensity was $8 \times 10^{11}$ protons.

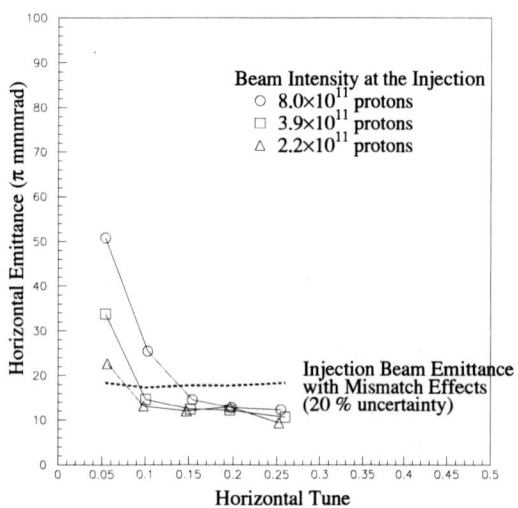

**FIGURE 5.** Horizontal 87% emittance as a function of the horizontal tune for the beam intensity of 2.2, 3.9 and $8.0 \times 10^{11}$ protons.

tupole magnets. A fringing field from an injection septum magnet was suspected as one source of closed orbit distortion and included in simulations for some cases. Parameters for the injection beam emittance were based on the transfer line profile measurements. Another tracking simulation code, PATRASH [3], has also been applied and the results agreed with the ACCSIM results.

Figure 6 is the $x - p_x$ phase space plot of 20 test particles for 400 turns when the horizontal tune is 7.05. It shows patterns of fourth order resonance that was created by the space charge force. Particles having the tune of 7 by the incoherent tune shift make a resonant condition with the space charge field [4]. Octupole type space charge field creates the resonance. The resonant tune is 7/4 for a super period, because the main ring has a four-fold symmetry.

ACCSIM results of the horizontal beam profiles 0.6 ms after injection is shown in figure 7 for the beam intensity of $8 \times 10^{11}$ protons. Profiles were fitted with a Gaussian plus a parabolic function for the tune of 7.05 and 7.10, and a parabolic function for the tune of 7.15 and 7.20. Only the parabolic distribution was assumed to remain in each profile after 4 ms and the emittance was evaluated to include the 87% fraction of the density distribution to compare with the measured emittance.

Figure 7 shows the comparison between the ACCSIM results and flying wire measurements for 87% emittance. The ACCSIM results reproduced the tune dependence qualitatively. The agreement is good for the intensity of $4 \times 10^{11}$ protons. The ACCSIM results, however, are about 1.4 times of the measured emittance for the intensity of $8 \times 10^{11}$ protons. The discrepancy may be due to uncertainty in the measurement of the injection beam emittance. Mechanism to modify the resonance width, otherwise, has to be considered such as effects of the betatron function modulation.

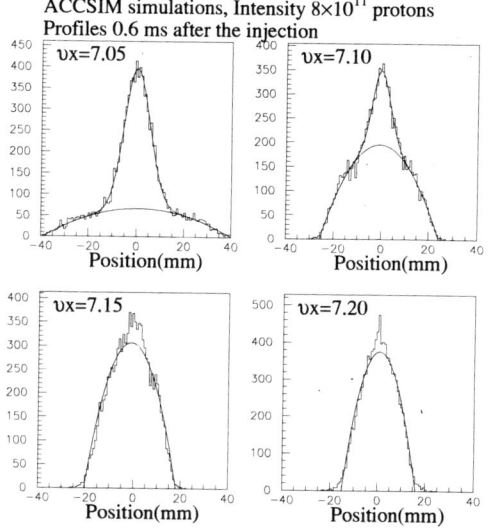

**FIGURE 7.** ACCSIM simulation of the horizontal beam profiles 0.6 ms after injection when the injection beam intensity was $8 \times 10^{11}$ protons and the horizontal tune was 7.05, 7.10, 7.15 or 7.20.

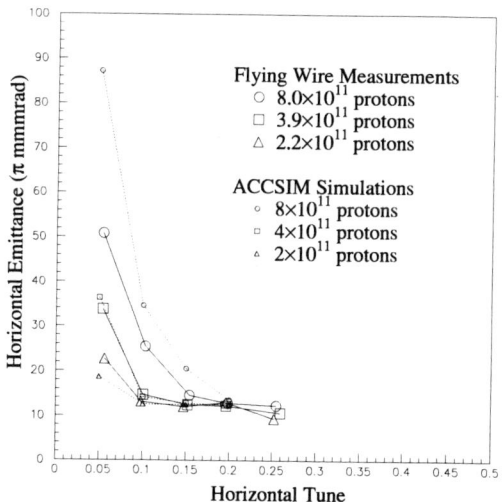

**FIGURE 8.** ACCSIM results of the horizontal 87% emittance as a function of the horizontal tune are shown in small symbols and dotted lines. Flyingwire measurement results are shown in large symbols and solid lines.

## CONCLUSIONS

Measurement of the transverse beam profiles using flying wires has revealed a characteristic temporal change of the beam profile within a few milliseconds after injection. Horizontal emittance growth was observed when the horizontal tune was close to the integer. The effect was more enhanced for higher beam intensity. Resonance created by the space charge field was the cause of the emittance growth. A multiparticle tracking simulation program, ACCSIM, taking account of space charge effects has qualitatively reproduced the beam profiles.

## ACKNOWLEDGMENTS

We thank F. Jones for valuable advices and installation of ACCSIM in our computer. We also thank H. Sato and K. Sato for useful comments.

## REFERENCES

1. Igarashi, S., Arakawa, D., Koba, K., Sato, H., Toyama, T., and Yoshii, M., *Nuclear Instruments and Methods in Physics Research A*, **482/1-2**, 32–41 (2002).
2. Jones, F., *Accsim Reference Guide Version 3.5s*, pp. 1–61 (1999).
3. Shimosaki, Y., and Takayama, K., *KEK Preprint*, **2001-111** (2001).
4. Machida, S., *Nuclear Instruments and Methods in Physics Research A*, **309**, 43–59 (1991).

# Tune-Based Halo Diagnostics

## Peter Cameron

*Collider-Accelerator Department*
*Brookhaven National Laboratory*
*Upton, NY 11973*

**Abstract.** Tune-based halo diagnostics can be divided into two categories - diagnostics for halo prevention, and diagnostics for halo measurement. Diagnostics for halo prevention are standard fare in accumulators, synchrotrons, and storage rings, and again can be divided into two categories - diagnostics to measure the tune distribution (primarily to avoid resonances), and diagnostics to identify instabilities (which will not be discussed here). These diagnostic systems include kicked (coherent) tune measurement, phase-locked loop (PLL) tune measurement, Schottky tune measurement, beam transfer function (BTF) measurements, and measurement of transverse quadrupole mode envelope oscillations. We refer briefly to tune diagnostics used at RHIC and intended for the SNS, and then present experimental results. Tune-based diagnostics for halo measurement (as opposed to prevention) are considerably more difficult. We present one brief example of tune-based halo measurement.

## INTRODUCTION

The incoherent tune distribution is modified by a variety of conditions. These include space charge, linear and non-linear chromaticity, the beam-beam interaction, magnet non-linearities, electron cloud effects, and coupling. This modified tune distribution can be modulated by power supply ripple, mechanical magnet vibrations, and (in the case of uncogged beams) by the beam-beam interaction. We begin with a very brief review of tune measurement methods, followed by the application of these methods to the prevention of halo formation. The discussion is heavily biased towards our experience with SNS and RHIC diagnostics.

## TUNE MEASUREMENT

Three basic tune measurement systems are employed in RHIC; kicked beam, Schottky, and PLL. These systems are described in detail in recent workshop[1] and conference[2] proceedings. Similarly, both kicked beam and Schottky systems are planned for SNS[3]. A PLL system is precluded in the SNS due to the pulsed nature of the machine, but BTF measurements will be accomplished in the SNS with a system very similar to the RHIC PLL.

## TUNE BROADENING

Mechanisms that modify the tune distribution are of interest because of the role they play in halo formation. The primary effect is that they cause some portion of the distribution to overlap resonances, resulting in emittance growth and halo formation.

### Space Charge

Space charge tune spread is not a significant consideration in RHIC, but it is a primary consideration in the SNS. Figure 1 shows calculated space charge tune depression in the SNS ring at three times during the accumulation cycle[4]. The blue dot represents the coherent tune. Even early in the cycle, the space charge effect is large due to the small emittance early in the painting process. As indicated in the figure, halo particles will have minimum space charge tune depression. Measurement of the incoherent tune distribution in the SNS ring will be accomplished with a combination of Schottky and beam transfer function techniques[3]. The resonant pickup frequency is selected so that the frequency spreads due to chromaticity and slip factor/harmonic number will approximately cancel.

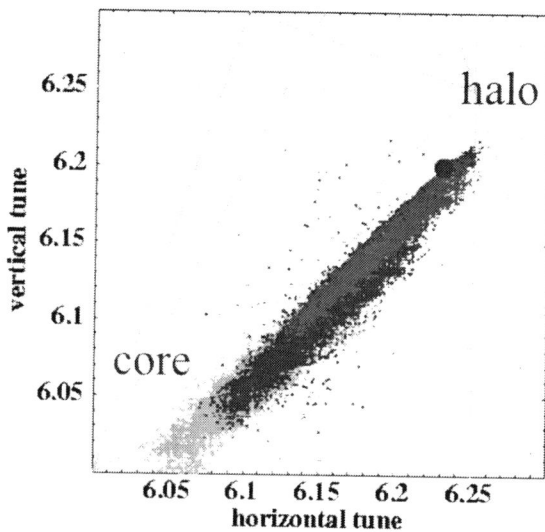

**FIGURE 1.** SNS Accumulator Ring tune footprints after 263, 526 and 1060 turns, $10^{14}$ beam

## Chromaticity

Good chromaticity control is essential to minimize halo formation. Wrong sign chromaticity renders the beam vulnerable to head-tail instability. The effect of too-large chromaticity is similar to space charge, broadening the incoherent tune and potentially causing some particles to cross resonances. Figure 2 shows continuous chromaticity measurements in the horizontal (black) and vertical (blue) planes during a RHIC acceleration cycle with deuteron beam. Duration of the ramp was 4 minutes, with transition crossing at about 1 minute. The measurement was accomplished by modulating the radius with 200μ amplitude at a frequency of 1Hz, and measuring the resulting tune modulation with the PLL. Data in the vertical plane before transition was probably not valid. The change of sign in chromaticity at transition can be seen. Correction of the negative vertical chromaticity near the end of the ramp resulted in improved beam survival during the next ramp.

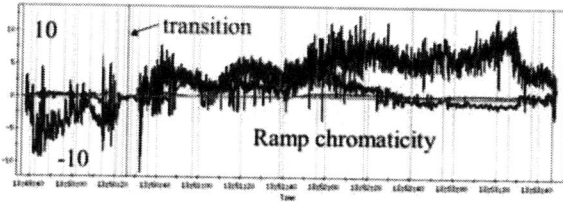

**FIGURE 2.** Chromaticity measurement during a RHIC acceleration ramp

Measurements with both Schottky and PLL show that non-linearity chromaticity can contribute to tune broadening. Figure 3 shows chromaticity measured by the Schottky system during an 8hr polarized proton store in RHIC, with only 28MHz acceleration cavities on for most of the store. Measured horizontal chromaticity (the blue and yellow traces) appears to increase linearly in both rings through the store, while vertical appears to decrease slightly. Blue horizontal was decreased near the end of the store, and then the 200MHz storage RF was turned on, which resulted in large changes in the measured blue horizontal and yellow vertical chromaticities.

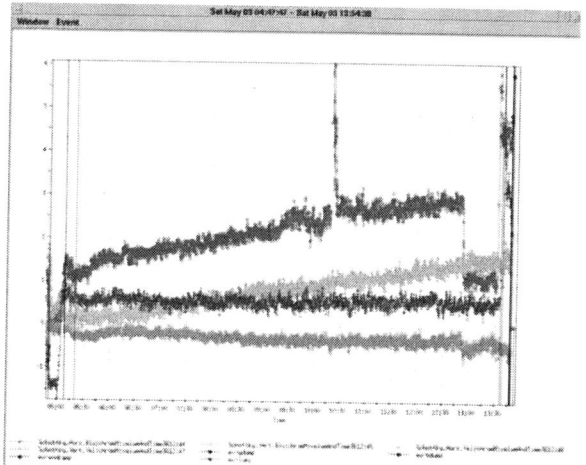

**FIGURE 3.** chromaticity measurement at store

The upper panel of figure 4 shows the effect on tune (as measured by the PLL) of a 400μ radial modulation at 0.4Hz. The FFT of the PLL tune in the lower panel shows a 30dB peak at 0.8Hz (twice the modulation frequency), due to non-linear chromaticity.

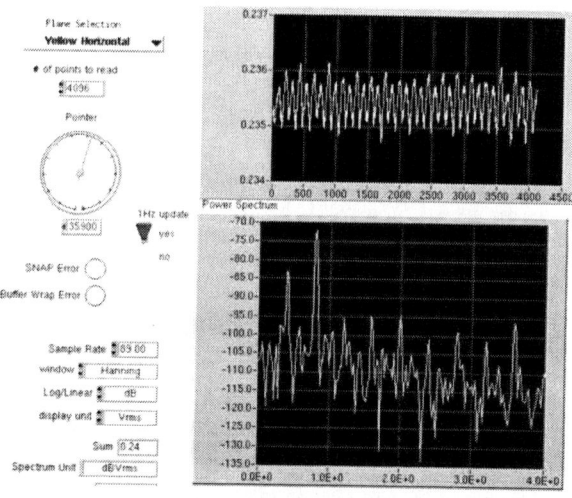

**FIGURE 4.** non-linear chromaticity measurement

## Non-linearities

Figure 5 shows variation of tune (as measured by the PLL) during a position bump in a RHIC triplet, with the upper and lower panels showing data before and after corrections were implemented[5].

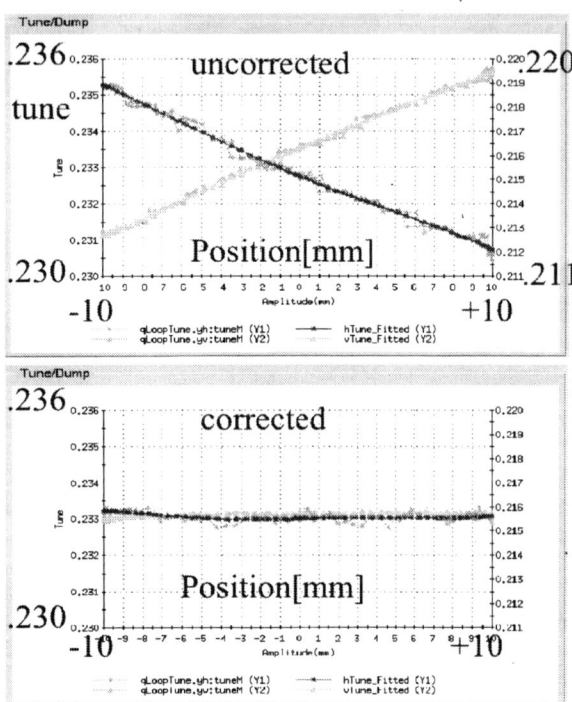

**FIGURE 5.** non-linearity correction in RHIC triplet

## Electron Cloud

Unlike space charge, electron cloud affects both the coherent and incoherent tune. Figure 6 shows coherent tune variation from the front to the back of the bunch train during beam studies of electron cloud in RHIC[6]. The coherent tune is elevated by about .001 at the end of the bunch train. Incoherent tune as measured by the PLL was lowered by a similar amount during these studies. This result is not understood.

## Coupling

Minimizing coupling serves to reduce the effect of coupling resonances, increasing the area available in the tune plane. The traditional 'closest tune approach' method of coupling measurement is not easily applied during an acceleration ramp. The 'N-turn map' method has recently been implemented at RHIC[7], and shows promise for measuring coupling during ramping. Yet another method[8] looks at the 2nd harmonic generated by coupling in the presence of skew quad modulation. The upper panel of figure 7 shows variation of $\sim 10^{-4}$ in the horizontal tune in the presence of 0.6A of 2Hz skew quad modulation in two families, 180 degrees out of phase. The lower panel shows the FFT of the difference between horizontal and vertical tunes. The 35dB line at 4Hz is a measure of coupling.

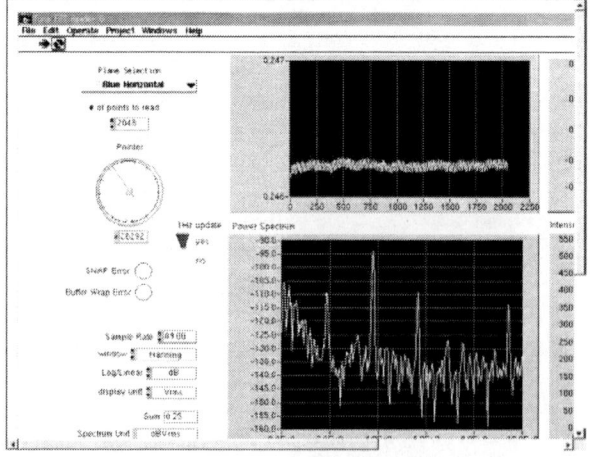

**FIGURE 7.** FFT of PLL tune measurement

## TUNE MODULATION

Tune modulation drives halo formation by causing the affected particles to cross resonances. A partial list of tune modulation mechanisms would include beam-beam effects, magnet power supply ripple, and mechanical vibrations of gradient magnets. Tune modulation due to power supply ripple has not yet been found to be a problem in RHIC.

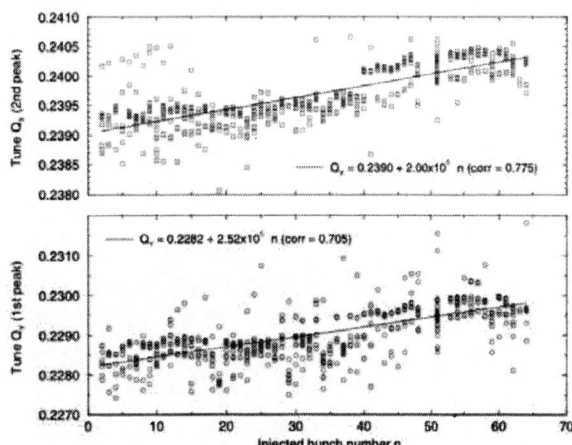

**FIGURE 6.** Effect of electron cloud on coherent tune in RHIC

## Cryostat Vibration

Tune modulation thought to be due to cryostat vibration has been observed in RHIC. Figure 8 shows a dominant line at ~11Hz and harmonics at 22Hz and 33Hz. The absolute and relative amplitudes of these lines seem to be affected by beam steering, suggesting the excitation is accomplished by fields of order higher than quadrupole. The 11Hz line seems to be correlated with cryostat vibrations measured with an accelerometer[9]. The measured Q of the 11Hz line is greater than 1000.

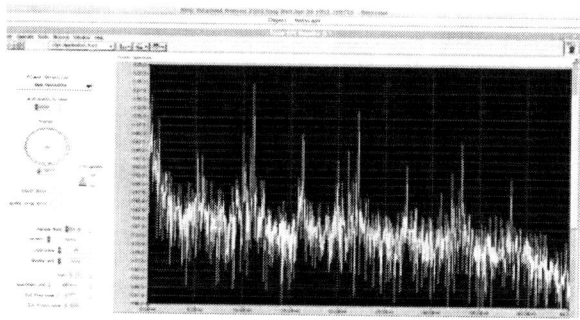

**FIGURE 8.** FFT showing lines thought to be due to cryostat vibration

## Beam-Beam Tune Shift

Figure 9 shows PLL tune near the end of a RHIC ramp. The counter-rotating beams were not 'cogged', so the revolution frequencies were slightly different. The spikes in tune at the extreme left of the figure are when beams from the two rings were not simultaneously present in the interaction regions. Shortly after flat-top the bunches were cogged and precessed to properly align the abort gaps, then uncogged. The resulting incoherent tune shifts are clearly visible in the figure[10].

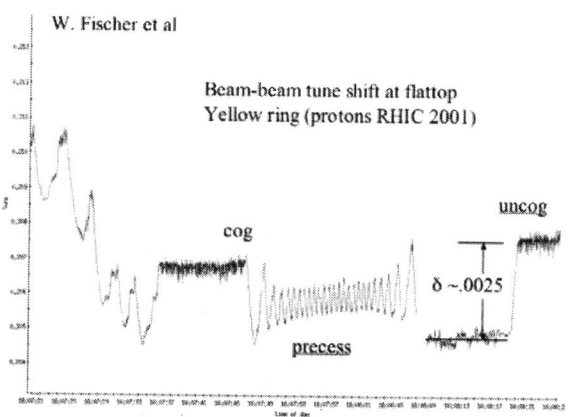

**FIGURE 9.** Tune modulation due to beam-beam

## HALO MEASUREMENT

Figure 10 shows PLL data taken with no kicker excitation, with the horizontal plane locked to the 2/9 resonance. The FFT shows a surprisingly large peak at 33Hz. If the PLL is locked on the large amplitude oscillations of beam in islands, this demonstrates oscillation amplitude dependence of both the strength and frequency of beam excitation. This data is one of the few cases in which tune measurement is used as a halo *measurement* diagnostic, rather than a halo *prevention* diagnostic. One can imagine using this diagnostic to minimize the resonance strength via resonance compensation[11].

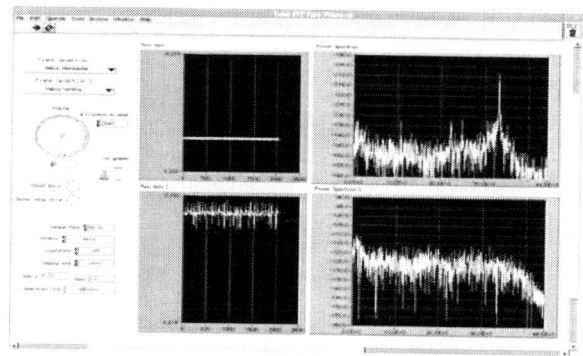

**FIGURE 10.** FFT of PLL tune measurement with no beam excitation, showing lock in both planes.

## REFERENCES

1. P. Cameron et al., "Tune Measurement in RHIC", BIW02, Brookhaven.
2. P. Cameron et al., "RHIC Third Generation PLL Tune Measurement System", PAC03, Portland.
3. P. Cameron et al., "Tune Measurement in the SNS Ring", BIW02, Brookhaven.
4. A. Fedotov et al., PAC01, Chicago, p.2848.
5. F. Pilat et al., "Nonlinear Effects in the RHIC IRs: Modeling, Measurement, Correction", ibid.
6. S. Zhang et al., "RHIC Pressure Rise and Electron Cloud", ibid.
7. W. Fischer, "Linear Coupling Correction with N-turn Maps", ibid.
8. T. Roser, private communication (2002).
9. C. Montag et al., "Vibration Studies on a RHIC IR Quadrupole Triplet", PAC03, Portland.
10. W. Fischer et al., "Observation of Strong-strong and Other Beam-Beam Effects in RHIC", ibid.
11. V. Ptitsyn et al., "Corrections of Non-linear Resonances at RHIC", ibid.

# Wide Dynamic-Range Beam-Profile Instrumentation For A Beam-Halo Measurement: Description And Operation[*]

J. Douglas Gilpatrick[*],

[*] *Los Alamos National Laboratory, MS H808, LANL, Los Alamos, NM, 87545*

**Abstract.** Within the halo experiment conducted at the Low Energy Demonstration Accelerator (LEDA) at LANL, specific beam instruments that acquire horizontally and vertically projected particle-density distributions out to > 100000:1 dynamic range are located throughout the 52-magnet halo lattice. We measured the core of the distributions using traditional wire scanners, and the tails of the distributions using water-cooled graphite scraping devices. The wire scanner and halo scrapers are mounted on the same moving frame whose location is controlled with stepper motors. A sequence within the Experimental Physics and Industrial Control System (EPICS) software communicates with a National Instruments LabVIEW virtual instrument to control the motion and location of the scanner/scraper assembly. Secondary electrons from the wire scanner 0.033-mm carbon wire and protons impinging on the scraper are both detected with a lossy-integrator electronic circuit. Algorithms implemented within EPICS and in Research System's Interactive Data Language subroutines analyze and plot the acquired distributions. This paper describes the beam instrument and our experience with its operation.

## INTRODUCTION

At LEDA, a 100-mA, 6.7-MeV beam is injected into a 52-quadrupole-magnet lattice (see Fig. 1). Within this 11-m FODO lattice, there are nine wire scanner/halo scraper (WS/HS) stations, five pairs of steering magnets and beam position monitors, five loss monitors, and three pulsed-beam current monitors [1]. The WS/HS instrument's purpose is to measure the beam's transverse projected distribution. These measured distributions must have sufficient detail to understand beam halo resulting from upstream lattice mismatches [2,3]. The first WS/HS station, located after the fourth quadrupole magnet, verifies the beam's transverse characteristics after the RFQ exit. A cluster of four WS/HS located after magnets #20, #22, #24, and #26 provides phase space information after the beam has debunched. After magnets #45, #47, #49, and #51 reside the final four WS/HS stations. These four WS/HS acquire projected beam distributions under both matched and mismatched conditions. These conditions are generated by adjusting the first-four quadrupole magnetic fields so that the RFQ output beam is matched or mismatched in a known fashion to the rest of the lattice. Because the halo takes many lattice periods to fully develop, this final cluster of WS/HS are positioned to be most sensitive to halo generation.

**FIGURE 1.** The 11-m, 52-magnet FODO lattice includes nine WS/HS stations that measure the beam's transverse projected distributions.

As the RFQ output beam is mismatched to the lattice, the WS/HS actually observe a variety of distortions to a properly matched Gaussian-like

---

[*] Work supported by the US Department of Energy.

distribution [2,3]. These distortions appear as distribution tails or backgrounds. It is the size, shape, and extent of these tails that predict specific types of halo. However, not every lattice WS/HS observes the halo generated in phase space because the resultant distribution tails may be hidden from the projection's view. Therefore, multiple WS/HS are used to observe the various distribution tails.

## WS/HS DESCRIPTION

Each station consists of a horizontal and vertical actuator assembly (see Fig. 2) that can move a 33-μm-carbon monofilament and two graphite/copper scraper sub-assemblies. The carbon wire and scrapers are connected to the same movable frame. Attached to this movable frame is a linear encoder that provides the wire and scraper edges' relative position to within a typical rms error of 5 μm, and an additional linear potentiometer provides an absolute approximate position for LEDA's run-permit systems. A stepper motor coupled to a ball lead screw is used to drive the moveable frame. A motor-brake and micro-switches limit the frame's movement.

**FIGURE 2.** The WS/HS assembly contains a movable frame on which a 0.033-mm carbon wire resides between two water-cooled graphite scrapers.

The carbon wire, which senses the beam's core, is cooled by thermal radiation. If the beam macropulse is too long, the wire temperature continues above 1800 K resulting in the onset of thermionic emission. Thermionic emission causes an inaccurate appearance to the distribution by exaggerating the core's current density. To eliminate these effects for the halo experiment, the maximum pulse length and repetition rate is limited to approximately 30 μs and 1 Hz, respectively.

The halo scrapers are composed of a 1.5-mm thick graphite plate brazed to a water-cooled 1.5-mm thick copper plate. Since 6.7-MeV protons average range in carbon is approximate 0.3 mm, the beam is completely stopped within the graphite plate. Cooling via conduction lowers the average temperature of the scraper sub-assembly and allows the scraper to be cooled more rapidly than the wire. The lower average temperature and faster cooling allows the scraper to be driven in as far as 2 rms widths from the beam distribution peak without the peak temperature increasing above 1800 K.

The movement and positioning of each wire and scraper pair is controlled by a motion control system that contains a stepper motor, stepper motor controller, a linear encoder, and an electronic driver amplifier. The controller's digital PID loop controls the speed and accuracy at which the assembly is moved and placed.

The target position, as defined by the WS/HS operator, is relayed from the EPICS control screen via a database process variable to a National Instruments LabVIEW Virtual Instrument (VI). The VI also calibrates the relative position of the linear encoders based on the measured position of the limit switches, and provides some error feedback information. The total error between the target wire position and the actual wire position attained is within a total 4% range of a typical 1-mm rms-width beam.

As the wire is moved through the beam, it senses the projected beam core distribution. A small portion of the beam's energy is imparted to the wire causing secondary electron emission to occur. The secondary electrons leaving the wire are replaced by negative charge flowing from the electronics. This current flow for both axes is connected through a bias battery to an electronic lossy integrator circuit and followed by an amplification stage.

The integrator capacitance and amplifier gain are set to allow a very wide range of values of accumulated charge. Data are acquired by digitizing the accumulated charge through the lossy integrator at two different times within the beam pulse. This charge difference, acquired by subtracting the two values of charge, provides a low noise method of relative beam charge acquisition. The wire and scraper accumulated charge signals are digitized using 12- and 14-bit digitizers, respectively. The analog noise floor has been measured to be 0.03 pC, a noise level slightly lower than the scraper digital LSB noise level of 0.15 pC using the highest gain settings within the detection electronics.

The front-end electronic circuitry, mounted on a daughter printed circuit board, is connected to a motherboard that has all of the necessary interface electronics to communicate with EPICS via a controller module within the same electronics crate. A software state machine sequence was written within

EPICS to control and operate WS/HS instrumentation. The state machine instructs the VI to move the wire and scraper to a specific location, acquire synchronous distribution data from either the wire or scraper, trigger the IDL routine to normalize the acquired charge with a nearby toroidal current measurement, graph the normalized data, and write the distribution to a file. The sequence also instructs IDL to calculate the first through fourth moments, fit a Gaussian distribution to the wire scanner data, and calculate the point at which the beam distribution disappears into the background noise.

To plot the complete beam distribution for each axis, the wire scanner and two scraper data sets must be joined. To accomplish this joining, several analysis tasks are performed on the wire and scraper data including, scraper data are spatially differentiated and averaged, wire and scraper data are acquired with sufficient spatial overlap, and differentiated scraper data are normalized to the wire beam core data.

The scraper data need only be normalized in the relative charge axis since the distances between each wire and scraper edge are known to within 0.25-mm. In addition, the first four moments and the point at which the beam distribution disappears into the noise are also calculated for the combined distribution data.

## ACQUIRED DISTRIBUTIONS

Figs. 3 and 4 show typical data from the vertical and horizontal profiles from the WS/HS #51 under a matched and a mismatched condition. The projected distributions displayed in these two figures have had the axis offset displacement subtracted and have been normalized so that their integrals are equal. The matched condition was acquired by comparing all of the final eight WS/HS acquired profile root-mean-square (rms) beam widths and adjusting the first four quadrupole lattice magnets so that all eight beam widths in a single axis are equal. The initial-four quadrupole magnets fields were then adjusted to change the match so that the beam's mismatch parameter was increased by 50%.

Under specific beam mismatch conditions, additional "shoulders" develop in both horizontal and vertical profiles at this particular lattice location. While showing an increase, the rms beam width and full-width-half-maximum (FWHM) do not provide sufficinet descriptive information of the beam profile changes. Since these shoulders are most pronounced between 1- to 10-% of the peak of the projected distribution, further describing the distribution by comparing distribution values of full-width-10%-maximum (FW10%M) and full-width-1%- maximum (FW1%M) values was also very instructive. A more complete description of the matching process and the related beam physics can be obtained from references written by Wangler and Colestock [2,3].

The typical profiles acquired by the WS/HS show distributions with a dynamic range of ~$10^5$:1 and provide distribution information to >5X to 7X times typical rms-beam widths. Table 1 summarizes figs. 3 and 4 distributions' statistical beam-width information. Note how the FW10%M and FW1%M show the added distribution "shoulder" widths resulting from beam mismatched condition whereas the FWHM actually shows a slight decrease in beam width.

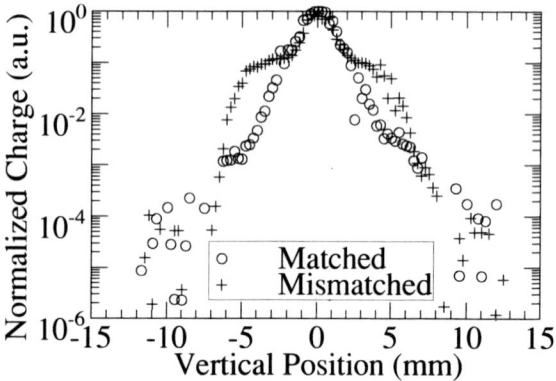

**FIGURE 3.** Acquired vertical beam distributions from WS/HS #51 show "shoulders" resulting from a beam mismatch condition.

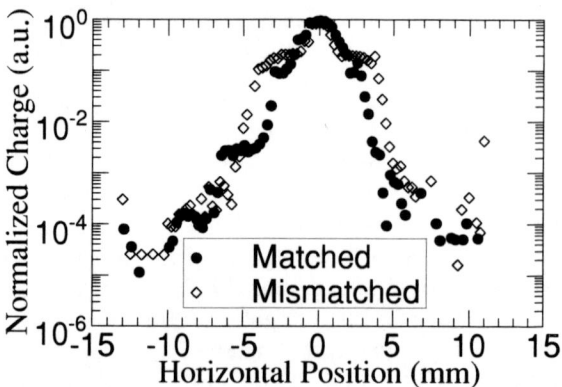

**FIGURE 4.** Acquired horizontal beam distributions from WS/HS #51 show "shoulders" resulting from a beam mismatch condition.

Table 1. Distribution width statistics, such as FW10%M, can further describe "shoulders" in the mismatched profile beam data shown in the Figs. 3 and 4 data.

| Width (mm) | Vertical Mtchd. | Vertical Msmtchd. | Horizontal Mtchd. | Horizontal MsMtchd. |
|---|---|---|---|---|
| R.m.s. Width | 1.15 | 1.89 | 1.10 | 1.82 |
| FWHM | 2.29 | 1.92 | 2.13 | 1.56 |
| FW10%M | 4.46 | 7.21 | 4.41 | 8.06 |
| FW1%M | 7.48 | 11.92 | 6.70 | 9.48 |

## WIRE AND SCRAPER PHYSICS

The WS wire is biased negative to optimize secondary emission (S.E.) yield, where the yield is defined as the ratio of the emitted secondary electron current and the proton beam current intercepted by the wire. All of the wires in the halo lattice WS are configured with a 33-μm, carbon monofilament. The HEBT WS is configured with a 100-μm SiC wire. The choice of bias potential was determined by measuring the wire and scraper currents as a function of bias potential. The resulting data showed that the wire is optimally biased at -6 to -12 V and the scraper is optimally biased at +20 to +30 V [4].

As the wire bias is positively increased from 0 V to > +100 V, the wire secondary electron emission is inhibited and the net wire current reduces to very near zero. As expected, a large positive bias reduces the wire detection signal. Furthermore, it appears that the wire collects positive ions with < -25 V bias potentials well after the beam pulse. This ion collection additionally limits the amount of negative bias that is applied to the wire for proper secondary emission operation.

The scraper detection goal is to inhibit secondary emission and detect only 6.7-MeV protons. With approximately +25 V bias applied to the scraper, the secondary emission is almost entirely inhibited and the net current reduces to the nominal proton current.

## SUMMARY

A wire scanner and halo scraper have been integrated into a single beam profile instrument capable of $10^5$:1 dynamic range. This WS/HS combination was used extensively to acquire wide dynamic range data in order to understand beam halo generation. The WS/HS beam data were analyzed under matched and mismatched beam conditions. Statistics, such as FW10%M, further described the irregular-shaped beam distributions resulting from mismatch beam conditions. The scanner and scraper V-I curves showed that the wire and scraper are optimally biased at –12 V and +25 V, respectively.

## REFERENCES

1. J. D. Gilpatrick, et al., "Experience with the Low Energy Demonstration Accelerator (LEDA) Halo Experiment Beam Instrumentation," *Proceedings of the 2001 Particle Accelerator Conference*, June 18-22, 2001, pp.2311-2313.

2. T. Wangler, "Physics Results from the Los Alamos Beam-Halo Experiment," this workshop.

3. P. L. Colestock, et al., "Measurement of a Beam Halo Generation in an Intense Proton Beam," *20th ICFA Advanced Beam Dynamics Workshop on High Intensity and High Brightness Hadron Beams*, Fermilab National Laboratory, April, 8-12, 2002.

4. J. D. Gilpatrick, et al., "Biasing Wire Scanners and Halo Scrapers for Measuring 6.7-MeV Proton-Beam Halo," *Proceedings of the 2002 Beam Instrumentation Workshop* held at Brookhaven National Laboratory, on May 6-9, 2002.

# Scintillator Telescope in the AGS Extracted Beamline*

D. Gassner, K. A. Brown, I. H. Chiang

*Brookhaven National Laboratory, Upton, NY 11973*

**Abstract.** An instrument for probing beam halo and obtaining beam profiles is discussed. The device described here is a prototype version, to obtain data and prepare for a more permanent device. The goals of the permanent device are to allow slow extracted beam emittances to be more routinely measured and to have a diagnostic for probing the wings of the beam distribution. The device works on secondary emission from thin targets, as well as scattering into two scintillator telescopes. The targets are movable over the entire aperture at the device. Data will be presented, as well as a description of the design of the system.

## INTRODUCTION

The purpose of this prototype instrument [1] is for probing beam halo and obtaining beam profiles of the resonant extracted beam at the AGS. The goals of the possible subsequent next generation permanent device are to allow emittances of low current, but high intensity slowly extracted beams (SEB) to be accurately measured, and to have a diagnostic for probing the wings of the beam distribution. The device works on secondary emission (SE) from thin targets as well as scattering into two scintillator telescopes. The targets are movable over the entire 4-inch aperture at the device. We were motivated to build a new device by the very high intensity beams now routinely being extracted from the AGS. We typically run at intensities that are as high as $6 \times 10^{13}$ protons per AGS pulse. The AGS Switchyard was originally designed to operate at $1 \times 10^{13}$. The central core emittance of the beam does not change too greatly with beam intensity.[2] With the increased AGS injection energy, that came with the AGS Booster, it has been found that to reach these high intensities, the full acceptance of the AGS was being used at injection.

This implies that the normalized emittance is increased. Previous measurements show that the emittance of the resonant extracted beam is more than twice as big as it was in the pre-Booster era. What is more significant is the twiss parameters were significantly changed. In effect the orientation of the phase space had not changed, but we now were extracting a beam that was fatter.[3] Modeling simulations agreed with the measured results[4] as shown in Table 1.

Table 1: Summary of emittance measurement results (from [1]).

(note: $\beta$ and $\alpha$ are referred to start of SEB line)

|  | $\epsilon_x^{95\%,N}$ | $\beta_x$ (m) | $\alpha_x$ | $\epsilon_y^{95\%,N}$ | $\beta_y$ (m) | $\alpha_y$ |
|---|---|---|---|---|---|---|
| FY82 | 31.9 | 57.6 | -6.6 | 38.8 | 3.25 | 0.87 |
| FY96 | 64.4±9.60 | 8.8±1.4 | -0.9±0.2 | 54.7±5.0 | 4.2±0.4 | 1.0±0.09 |

The performance of this new device has exceeded our expectations. We were very concerned about singles rates in the area, since the telescopes were located inside the beam enclosure and had effectively no shielding. The singles rates were not insignificant, as high as several MHz, but the triple coincidence circuitry had no problems contending with these rates.

---

* Work performed under the auspices for the U.S. Department of Energy.

# DISCUSSION

The device consists of two 2.5mm diameter tungsten targets, one that scans across the beam horizontally, and the other vertically. It is located in the upstream AGS Switchyard, before the electrostatic splitters. The vacuum at this location is in the range of $10^{-7}$ torr, making it very good for looking at SE. For the SE to be seen we apply a voltage to the wires, to repel any stray electrons that may wish to collect back onto the targets. Good signals were obtained at voltages down to about 20 volts. Above that we saw little change in the signal. Since the majority of the electrons knocked out of the target have energies in the range of less than 10eV, it isn't surprising that we didn't need very much voltage. The SE electronics included a gated integrator with a 1000pF capacitor, V/F converter, and a scalar.

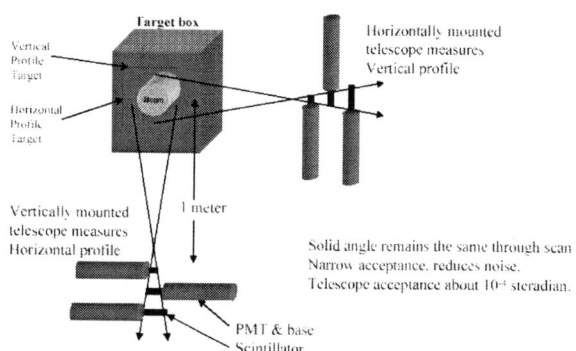

**FIGURE 1.** Telescope diagram.

The scintillator telescopes consisted of three EMI 9813KB photomultiplier tubes (PMT), covered with mu-metal shielding and steel pipe shielding. The stray magnetic fields at the PMT's were estimated to be in the range of a few gauss to at most 10 gauss. There are two telescopes, one in the vertical plane and the other in the horizontal plane, each at $90°$ in the lab frame. The first detector is located 1m from the target and the detectors are separated by 10cm. The solid angle acceptance of each telescope is about $10^{-4}$ steradian. We were initially concerned about temperature problems with the targets, since they were electrically isolated, and relatively massive. Initial calculations, a small amount of simulation, and tests made with an electron-beam welder (in a $10^{-3}$ torr vacuum), all showed that the targets and holder assembly would be very stable and temperatures would not reach any significant levels. One unique concern was the significant change in solid angle seen by the telescopes due to the movement of the targets. In order to cope with this we designed the sizes of the scintillators such that they accepted the same solid angle and could accept a source changing in angle relative to the alignment of the scintillators. The area of the scintillators increases much more than just the linear change in distance from the target. This allowed the horizontally moving target to be viewed with the vertically mounted telescope, and give very little change in observed solid angle over the range of movement of the target.

# RESULTS

Presented data is with beam currents of about 2.5e12 protons/spill, 10 times below saturation levels. The beam at this time had a definite momentum tail on it, which is most easily seen in Figure 3. Figure 2 shows that for a normal plot the two telescopes basically give the same curve. But in Figure 3 it is seen that the horizontal telescope shows a wider beam, more clearly, than the vertical.

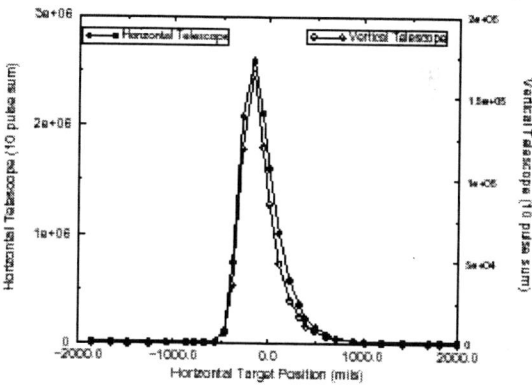

**FIGURE 2.** Horizontal and vertical triple coincidences for horizontal scan.

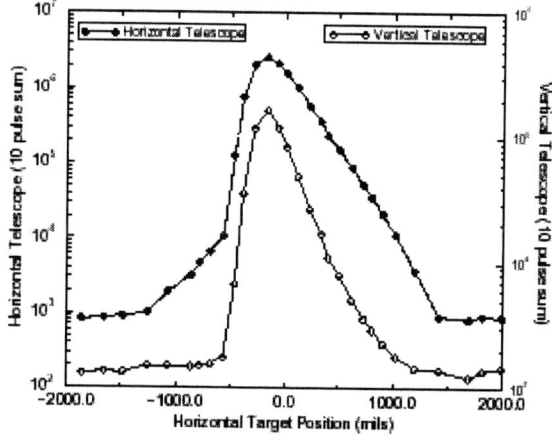

**FIGURE 3.** Horizontal and vertical triple coincidences for horizontal scan.

This is the effect of the solid angle changing more for the horizontal telescope than for the vertical, which has much less change in solid angle over the same range. Figure 4 shows the curves for the secondary emission from the target, on linear and on logarithmic scales. The secondary emission curve closely follows the vertical telescope.

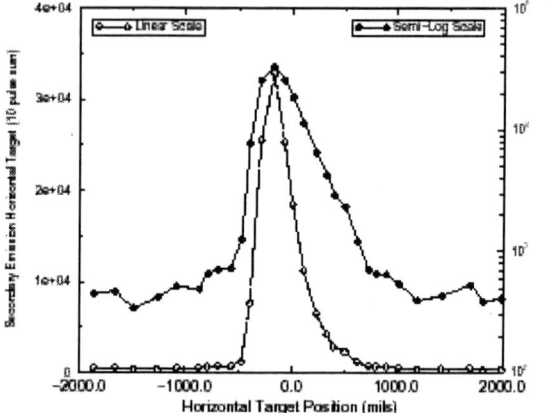

**FIGURE 4.** Secondary emission from target for a horizontal scan.

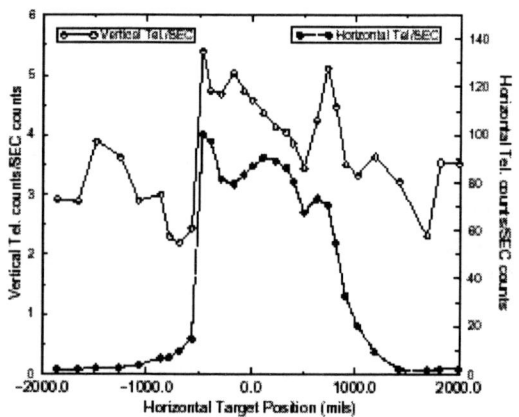

**FIGURE 5.** Ratio of telescope counts to secondary emission.

Figure 5 demonstrates the ratio of telescope counts to secondary emission counts for each. Again the vertical telescope has much less variation than the horizontal.

## CONCLUSIONS

The performance of the device has exceeded our expectations. We see a clean dynamic range between 4 to 5 orders of magnitude from the telescopes, giving significant resolution of the wings of the beam distribution. The SE dynamic range yielded 2 to 3 orders of magnitude. Unfortunately the device does not perform well at the higher beam intensities. Background singles rates are larger than we anticipated and at high intensities become a serious problem. But we are actually encouraged, since this is at least a parameter we have control over. We could reduce the mass of our targets without affecting the performance, and reduce the solid angle acceptance of the telescope without greatly affecting the dynamic range. The effort that went into considering the solid angle effects for the telescope produced a fairly flat response for the vertical telescope when moving the horizontal target. The same compensations done for the horizontal telescope yields a similarly flat response when targeting on the vertical target.

## ACKNOWLEDGMENTS

The AGS instrumentation group, supervised by T. Curcio, along with the excellent mechanical skills of the AGS Beam Components group, supervised by D. Lehn, did a tremendous job putting together this entire system in less than 6 months.

## REFERENCES

1. K.A. Brown, I.H. Chiang, D. Gassner, et al. "A Scanning Target Profile Monitor for the Slow Extracted Beam at the AGS", 1997 IEEE Particle Accelerator Conference, Vancouver, B. C., Canada, p.2149

2. K. A. Brown and R. Thern, "Beam Size Versus Intensity for Resonant Extracted Beam at the Brookhaven AGS", 1995 IEEE Particle Accelerator Conference, Dallas, TX, p.3212

3. AGS/AD/Tech. Note 445; "FY96 SEB Emittance Measurements", K. Brown, M. Blaskiewicz, H. Brown, K. Reece, Oct. 1996

4. AGS/AD/Tech Note 447; Mad Simulation: HEP/SEB Extraction (FY1996), M. (Sanki) Tanaka, Oct. 1996

# Beam Tail Measurements using Wire Scanners at DESY

Suren Arutunian[1], Matthias Werner[2], Kay Wittenburg[2]

*[1]Yerevan Physics Institute, [2]DESY*

**Abstract.** Wire scanners are used usually to measure the profile of the core of the beam. Especially at high beam currents the wire has to flip very fast through the beam, otherwise it will burn This is not true for the tails of the beam where the wire can be moved very slowly (or even stay stationary). Two effects are described which can be used to determine the tail distribution of the beam with the help of a slow wire scan. First: Counting of the scattered beam (tail) particles by scintillation counters. The rate versus the wire position will give directly the particle distribution in the tail. Second: The temperature of the wire depends on the amount of beam particle interactions with the wire. A temperature increase leads to an elongation of the wire and a dramatic change of its tension. This can be measured by the change of the natural frequency of the wire. Both methods are very sensitive and have a large dynamic range. The devices will be described and first results will be presented.

## INTRODUCTION

Wire scanners are used in many laboratories to measure very precisely the beam profile. Different readout philosophies are applied, depending on the specific beam parameters and geometries: detection of - secondary emission current, - bremsstrahlung, - scattered beam particles, - shower particles. Most of the applications are limited in their signal to noise ratio and therefore in their dynamic range. However, carefully designed systems may reach a dynamic range of $10^5 - 10^6$ with a single scanner. Other authors used a combination of wire scanner and scraper to achieve a high dynamic range [1].

Two different readout systems for wire scanners with a dynamic range of up to $10^8$ will be discussed in this report. One method uses the readout of shower particles: During a fast scan in the beam core the signal of an adjacent scintillator is analyzed by an ADC, during a very slow scan in the beam tails the signals from the scintillator are counted. The second method is based on a very sensitive measurement of the temperature of the wire, which is heated by the crossing beam particles. In the following sections the two ideas are described and first measurements are presented.

## TAIL SCANS USING COUNTING MODE

The standard wire scanner used in nearly all accelerators at DESY is based on a design of the wire scanner developed at LEP and is described elsewhere [Ref. 2, 3]. At DESY either 7 μm Carbon wires (proton beams) or 15 μm SiC wires (electron beams) are used. The driver and readout electronic developed at DESY [4] makes it possible to scan with a maximum speed of 1 m/s through the beam core or to make very slow scans, or even to stop the wire at a certain position. Both modes have a position resolution of 1 μm. The signal from the adjacent fast scintillator/photomultiplier (NE104/XP2243B) is sampled bunch by bunch with a sampling rate of 10.5 MHz with a resolution of 14 bit. Additionally the electronics provide a counting option, which simply counts the number of applied fast TTL signals over a certain time period.

For the sensitive tail studies at the HERA proton accelerator a fast constant fraction discriminator was connected additionally to the photomultiplier. A fast NIM-to-TTL converter (Type: LeCroy 688AL) converted the NIM signals from the discriminator and generated the TTL signals with a length of about 50 ns.

The threshold of the discriminator can be varied and was set to avoid early saturation (the maximum count rate is the bunch repetition rate of 10.4 MHz at HERAp). The threshold defined the overall sensitivity of the counting system. The following picture 2 shows the spectrum of the deposited energy in the scintillator for $E_{dep} > 1$ keV calculated by the Monte Carlo program GEANT 4. The number of primary protons hitting the 8 µm Carbon wire was $3.91 * 10^7$ while 496 make an energy deposition $E_{dep} > 1$ keV. The efficiency is then $496/3.91 * 10^7 = 1.27 \times 10^{-5}$; however, the efficiency can simply be reduced by increasing the threshold to higher values of $E_{dep}$. Figure 3 shows the expected count rates at different positions of the wire at an efficiency of $10^{-7}$. Wire positions between 2 σ and 6 σ can expect to provide sufficient count rates for tail measurements.

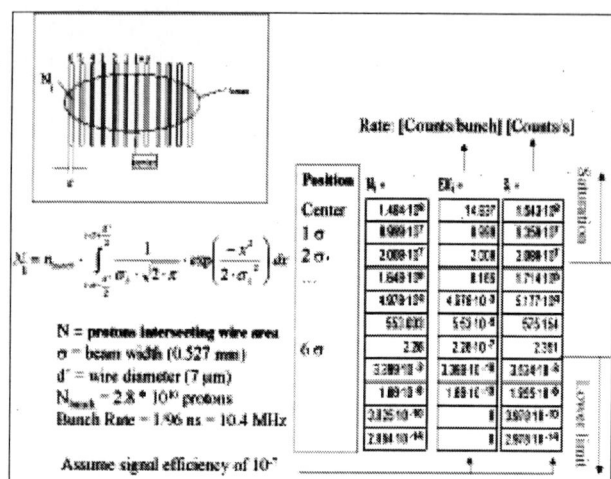

**FIGURE 3.** Expected count rate assuming an efficiency of $10^{-7}$ and a bunch rate of 10.4 MHz at different wire positions.

## Measurements

To achieve a large dynamic range, first a normal scan was made and the profile of the beam was stored and fitted by a gaussian. After that, a slow scan starting from the outside was made, in discrete steps. At each position the number of counts within 9 s was recorded. During this period the wire was kept at the same position. After the recording the wire was moved to the next position and the counts were recorded again. Since the fast and slow scan was made with exactly the same wire, the position information is identical. The overlap of the data from the fast and slow scan at around 2 σ (non saturated data) was used to normalize the counting data with the ADC data. In figures 4a, b, c the profile from the fast scan together with the gaussian fit (4a) and two tail scans with different sensitivities together with the same fit are shown. These results show well the extended dynamic range as a result of the counting technique; it may reach up to $10^8$! Driving one scraper to about 6 σ from the beam core initiates the creation of the larger tails while the smaller tails were measured without any scraper.

### More measurements

During the last years the HERA B experiment had take a lot of data. The internal wire targets of the experiment were brought to certain positions in the beam tails to achieve an adequate interaction rate. The experiment detected the interaction rate with an efficiency of about 50%. A lot of tail studies were

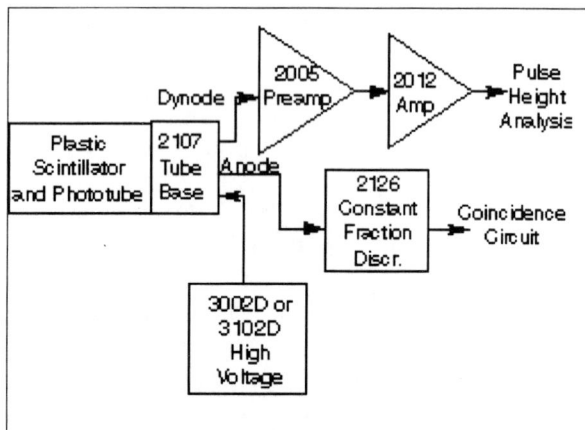

**FIGURE 1.** Block diagram of the readout

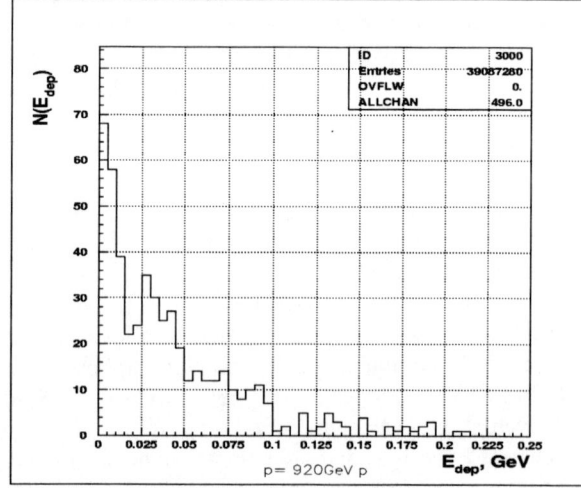

**FIGURE 2.** Spectrum of the deposited energy in the scintillator calculated by GEANT. The real geometry of the wire scanner setup in HERAp was used.

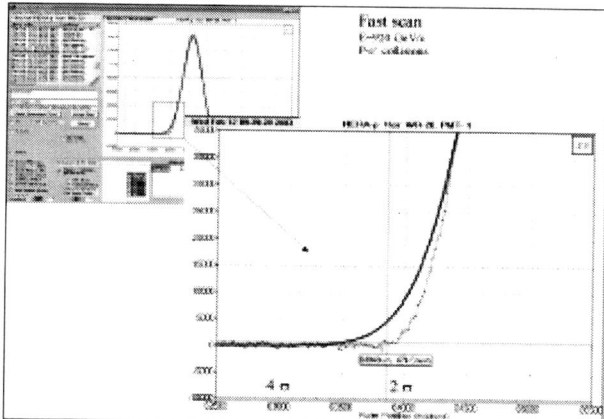

**FIGURE 4A.** Horizontal beam profile and a zoom on the tail distribution around 2 σ (fast scan). Note that this profile is the sum of 100 single bunch profiles. The black line shows a gaussian fit to the data.

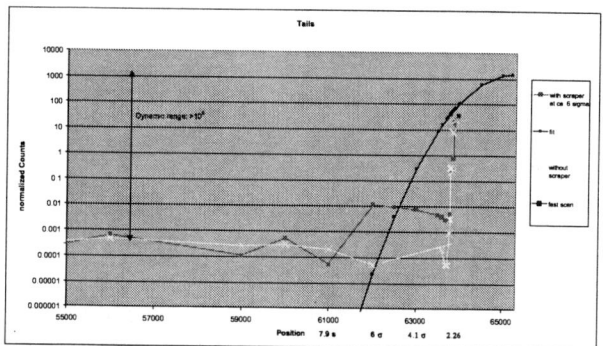

**FIGURE 4B.** Tail distributions around 6σ measured by the counting method and normalized to the fast scan data of a single bunch.

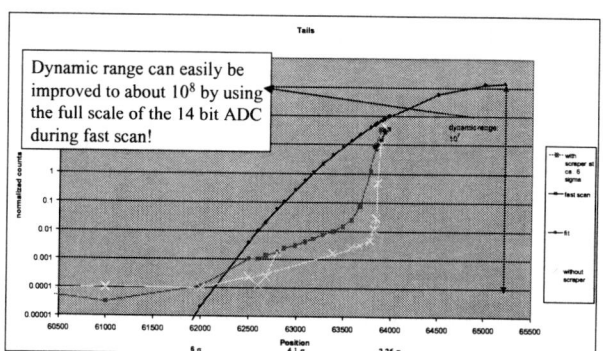

**FIGURE 4C.** Tail distributions around 6σ measured by the counting method and normalized to the fast scan data of a single bunch. This measurement was done with a reduced threshold and therefore with an increased sensitivity.

done using the infrastructure of the experiment. Beside a much higher efficiency, the experiment provides also a high-resolution timing system. By gating the counts within successive time slots of 24 ns or 0.1 ns, coasting beam between bunches and even the occupation of neighbor buckets (500 MHz) were observable, respectively. These measurements are described in detail in [10]. However, this technique can be applied easily to the one-channel-readout of a wire scanner. Commercial available TDC or similar techniques have resolutions of much better than 1 ns. Also photomultipliers with rise times of below 1 ns are available.

### Improvements

Some improvements are planned to increase the resolution and the dynamic range: i) At high count rates the measurement should be linearised by applying Poisson statistic. ii) Applying a more reasonable fit to the tails of the fast measurement will improve the overlap conditions. iii) The lower count rate was limited by real beam losses to about 3 Hz at the higher threshold and about 40 Hz at the lower threshold. Applying a telescope technique with a triple coincidence can reduce this background. An example is given in [5]. vi) Applying of a time resolving counting technique, like the HERA-B technique.

## SCANNING EXPERIMENTS USING VIBRATING WIRE

The principle of operation of the vibrating wire scanner is based on the dependence of the wire natural frequency $f_0$ on the beam intensity at the given location. The energy deposition of the beam particles in the wire causes heating of the wire. Hence the stretched wire temperature can be obtained by measuring its natural oscillations frequency by an autogenerator electronic circuit with a positive feedback loop. Initial experiments on profiling were done using laser beams [6, 7]. The effective temperature precision was estimated to be about $10^{-4}$ degrees C (without noise). Later experiments [8, 9] used the low intensity electron beam of the injector of the Yerevan synchrotron (bunches with RF of 2797.3 MHz with pulse duration of 2 μs and a repetition rate of 50 Hz). A beryllium-bronze wire of 90 μm diameter was used to scan the beam profile. Fig. 5 shows the result of the beam profile for the first scanning together with a normal distribution with σ = 1.48 mm and a beam central position at 30.87 mm. The overall current of the beam was set to $I_0$ ~ 10 nA. Only half of the beam could be scanned because of the short throw of the scanner.

### Calculations of "tail sensitivity"

Predictions about the sensitivity according to the measurements were done in [9]. With a sensitivity of 0.01 K (assumed thermal sensitivity in the presence of noise), the estimations show that for a 90 μm beryllium-bronze wire and 100 mA proton beam with $\sigma_x = 0.6$ cm (PETRA conditions) the value $T_m$ of about 0.01 K (vibrating wire thermal sensitivity at presence of electromagnetic noises) is achieved at $x = 5.7\ \sigma$. The following estimation agrees well with this calculation: At $2\ \sigma$ the wire was hit by $1.8 \cdot 10^8$ e/s (assuming 10 nA beam current), showing the first increase of the temperature. For a 100 mA beam with a width of $\sigma = 0.6$ cm the same amount of hitting particles (and therefore the same increase if the temperature) will be reached at about $6.1\ \sigma$.

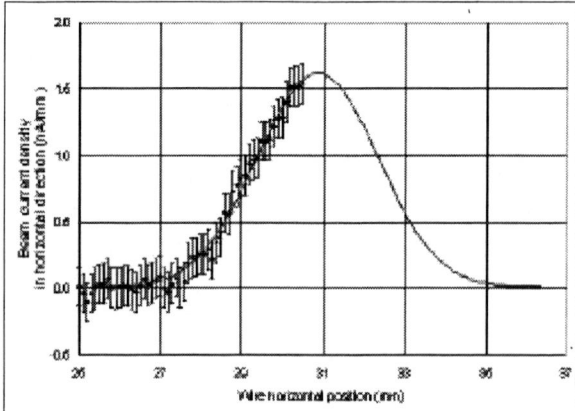

**FIGURE 5.** Reconstructed horizontal profile of an electron beam with a current of about 10 nA. The reconstruction was done after a detailed noise analysis; see [8, 9].

*Further studies*

One vibrating wire scanner is already installed in the PETRA proton ring at DESY. Further test are planned to determine: i) the influence of mechanical vibrations of the device on the measurement, ii) how well the theoretical predictions compare with measurements. The theory for the heating of the wire was developed mainly for DC beams. The authors expect, that it can be applied in the same way to bunched beams. Detailed studies with different bunch repetition rates are foreseen. iii) Higher order modes (HOM) might depose energy in the vibrating wire. At the SPS at CERN this effect was strong enough to damage the conducting Carbon wire [11]. The induced heating of the wire will disturb the vibration measurements. Test will to be done to damp the resonant modes using ferrites.

## ACKNOWLEDGMENTS

A. Batalov (IHEP, Moscow) provided the Monte Carlo calculations. His results and discussions with him helped a lot for understanding the results.

## REFERENCES

1. R. Macek, "Halo measurement diagnostic for the extracted beam in PSR", this workshop; D. Gilpatrick, "Wide Dynamic-Range Beam-Profile Instrumentation for a Beam-Halo Measurement: Description and Operation"; this workshop

2. B. Bouchet, et al., "Wire Scanners at LEP.", in CERN-SL-91-20-DI. *IEEE PAC San Francisco*, CA, 6 - 9 May 1991, p. 1186-1188.

3. C. Bovet, A. Burns, F. Ferioli, Q. King, J. Koopman, J. Mann, H. Michel, R. Schmidt, L. Vos, "Experience with LEP Wire Scanners" in *CERN-LEP-COMMISSIONING-NOTE-24*, Apr 1990. 6pp.

4. M. Werner, K. Wittenburg, "A New Wire Scanner Controll Unit" in 5th European Workshop on Diagnostics and Beam Instrumentation, 13th -15th May, 2001, ESRF, Grenoble.

5. D. Gassner, "A Scanning Target Profile Monitor in the Slow Extracted Beam at the AGS" this workshop

6. Arutunian S.G., et al., "Vibrating Wire for Beam Profile Scanning" *In:* Phys. Rev. Spec. Top. Accel. Beams 2 (1999), pp.122801.

7. Arutunian S.G., et al., "Nonselective receiver of laser radiation on the basis of vibrating wire." - *Proc. Conference Laser 2000* (November 2000, Ashtarak, Armenia)

8. Arutunian S.G. et al, "Vibrating wire scanner: first experimental result on the injector beam of Yerevan Synchrotron" *Phys. Rev. ST Accel. Beams* 6, pp. 042801 (2003)

9. Arutunian et al, "First Experimental Results and Improvements on Profile Measurements with the Vibrating Wire Scanner", in *6th European Workshop on Diagnostics and Beam Instrumentation* 5 - 7 May, 2003, Mainz, Germany

10. S. Pratte, DESY, „Bestimmung der Wechselwirkungsrate des HERA-B Targets und Untersuchungen des Coasting Beam am HERA Protonen Ring". Thesis, University Dortmund, June 2000.

11. F. Caspers, et al., "Cavity Mode Related Breaking of the SPS Wire Scanners and Loss Measurements of Wire Materials", in *6th European Workshop on Diagnostics and Beam Instrumentation* 5 - 7 May, 2003, Mainz, Germany

# Collimation Experience at RHIC[1]

K.A. Drees, R. Fliller, D. Trbojevic* and V. Kain[†]

*Brookhaven National Laboratory, Upton, NY 11973
[†]CERN, Geneva, Switzerland

**Abstract.** In the Relativistic Heavy Ion Collider (RHIC) the abort kicker magnets are the limiting aperture. Continuous losses at this location could deteriorate the kicker performance. In addition, losses especially in the triplet area cause backgrounds in the experimental detectors. The RHIC one-stage collimation system was used to reduce these backgrounds as well as losses at the abort kickers. Collimation performance and results from various runs with even and uneven species (Au-Au, pp and d-Au) are presented and compared. Upgrades of the system for the upcoming high luminosity runs are outlined.

## INTRODUCTION

Beam halo, large beam profiles and beam losses induce high experimental backgrounds throughout the stores as well as contribute to the reduction of the lifetime of accelerator components. In superconducting machines quenches due to uncontrolled beam losses during beam steering, the acceleration ramp or fault conditions are likely. Collimators used as the limiting aperture can help prevent damage. Figure 1 sketches the geometry of RHIC with the collimators and the five RHIC experiments.

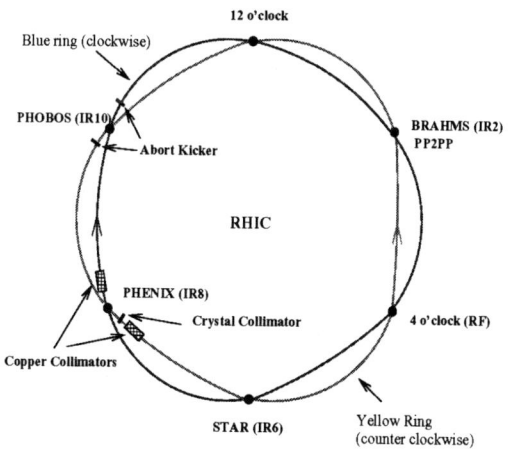

**FIGURE 1.** Location of the collimators and experiments in the RHIC rings.

Table 1 lists all RHIC runs to date. The operation period in the year 2001/2002 consisted of a heavy ion run followed by 8 weeks of a polarized proton run. Both runs had different needs for collimation. During the Au-run several ramps were introduced implementing $\beta^*$-squeezes from 10m to 5m, from 10m to 2m and finally from 10m to 1m (at PHENIX only). There were no squeezes during the proton run, instead $\beta^* = 3$ m was used for injection as well as storage for all IRs. In the year 2003, the pp run was preceded by a run with uneven species (d and Au). In both runs $\beta^*$ was squeezed to several different values at the IRs. The many configurations introduced an increased demand for collimation in the FY03 runs.

**TABLE 1.** List of RHIC runs 1999-2003.

| year | run | Energy | $\beta^*$ |
|---|---|---|---|
| 1999 | pilot au | 10 GeV | 10 m all |
| 2000 | rhic_au_00 | 70 GeV | 3 m, 8 m |
| 2001 | rhic_au_01 | 100 GeV | 2 m, 1 m |
| 2002 | rhic_pp_02 | 100 GeV | 3 m all |
| 2003 | rhic_dau_03 | 100 GeV | 2 m, 3 m, 4m |
|  | rhic_pp_03 | 100 GeV | 1 m, 2 m, 10 m |

## LAYOUT

The RHIC collimation system [1] layout is shown in figure 2.

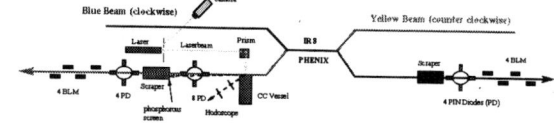

**FIGURE 2.** The RHIC collimation system

---

[1] Work performed under Contract Number DE-AC02-98CH10886 with the auspices of the US Department of Energy.

It consists of two 450 mm long L-shaped copper scrapers placed downstream of the PHENIX detector in each ring. Each collimator is moved by three stepper motors, which control the horizontal and vertical positions and rotate the collimator about the vertical axis. The step size is approximately 0.5 $\mu$m. Fully retracted the vertical jaws are about 56 mm and the horizontal jaws about 52 mm from the center of the beam pipe. In addition the yellow (counter-clockwise) ring has a 5 mm long, O-shaped silicon crystal. The crystal is bent 0.44 mrad. More details can be found in [1]. Because of the negligible dispersion at the location of the collimators [2], the scrapers cannot be used for momentum collimation. Since RHIC so far lacks any other collimators, a combination of fast kickers and the scrapers was used to excite and remove off-momentum beam particles during the last run [3]. A dedicated gap cleaning system including additional kickers is currently being installed during the FY03 shutdown.

**TABLE 2.** List of run configurations during the rhic_dau_03 run. a = IR 6 & 8, b = IR2, c = IR10. "chg." refers to 'changing', i.e. the cavity status changed rapidly during this period.

| date | fill | $\beta^*$ (m) | | | n.o.b. | RF | |
|---|---|---|---|---|---|---|---|
| | | a | b | c | | B | Y |
| bef. 01/11 | - | 2 | 2 | 2 | 55 | n.a. | n.a. |
| 01/11 | 2715 | 2 | 2 | 2 | 110 | n.a. | n.a. |
| 01/17 | 2758 | 2 | 2 | 2 | 110 | on | on |
| 01/21-23 | 2799 | 2 | 2 | 2 | 110 | chg. | chg. |
| 01/24 | 2812 | 2 | 2 | 2 | 110 | on | off |
| 01/30 | 2843 | 2 | 2 | 2 | 110 | off | off |
| 02/02 | 2883 | 2 | 2 | 2 | 110 | on | on |
| 02/09 | 2952 | 2 | 2 | 4 | 110 | on | on |
| 02/24 | 3056 | 2 | 2 | 4 | 110 | off | on |
| 02/27 | 3070 | 2 | 2 | 4 | 55 | off | on |
| 03/05 | 3108 | 2 | 3 | 3 | 55 | off | on |
| 03/11 | 3154 | 2 | 3 | 3 | 55 | on | on |
| 03/18 | 3221 | 2 | 3 | 3 | 55 | off | on |
| 03/19 | 3235 | 2 | 3 | 3 | 55 | on | on |

## HEAVY ION OPERATION

During the 2001 au_01 and the later dau_03 run the ions were accelerated up to $\gamma \approx 107$. During the acceleration ramp various processes, such as orbit variations and radial shifts, are potentially leading to beam losses. It turned out that the abort system kickers acted as limiting aperture (see [2] for more details). In the run au_01 the collimators were moved in to a predefined position during all ramps squeezing to $\beta^* \leq 2$ m, starting Oct. 26, 01. However, they were not used routinely for the ramps in the 2003 dau_03 run.

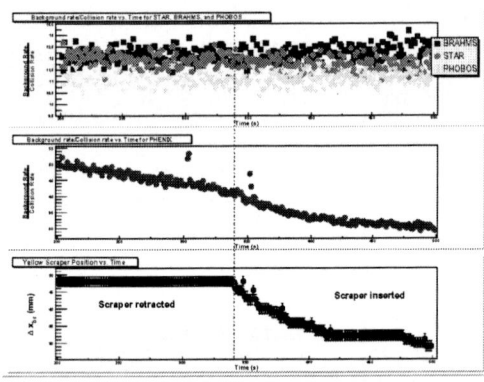

**FIGURE 3.** Background as a function of scraper position during Au operation. $\beta^* = 2m$ at all IRs.

The scrapers were also used in an attempt to prevent experimental background during storage. Figure 3 shows the yellow scraper position and the experimental background rates based on ZDC [4] signals during fill 1759. The yellow collimator was moved in horizontally after orienting it such that it was parallel to the direction of the beam. There is no obvious effect on the background signal except a small drop at PHENIX. Indeed, the scrapers had no significant effect on the background rates at any time during the au_01 run leading to the conclusion that background was mainly due to local causes such as beam-gas interactions or colliding beams. Attempts to use the crystal for further experimental background reduction during the au_01 and dau_03 run were unsuccessful.

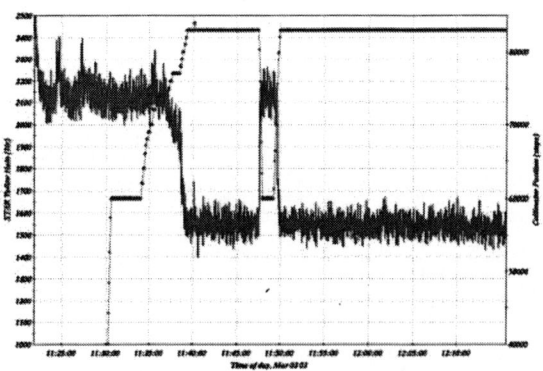

**FIGURE 4.** STAR ZDC background signal (line) and yellow vertical collimator position (stars) during the dAu store 3094.

Table 2 gives an overview of the different running configurations during the dau_03 run. The usage of the storage cavities (RF B and Y) would mainly cause higher debunching rates, increasing the necessity of gap cleaning. The different $\beta^*$ values as well as the number of bunches (n.o.b.) likely lead to losses and background problems especially during the ramp development phase. In general, background conditions relaxed when $\beta^*$ was increased at IR10 and later at IR2. However, experimental background conditions did require collimation for the most part of the run. Figure 4 shows a background signal from the STAR detector associated with background caused by the yellow (i.e. Au) beam. When the yellow vertical scraper is moved in (i.e. towards higher number of steps), the background signal drops by some 30 %. Collimation clearly helped to reduce the background. Therefore effects local to the individual IR can be excluded as a sole cause for high background conditions. Collimators were used for background reduction more or less during every fill but the collimation efficiency did not appear to be sufficient.

## POLARIZED PROTON OPERATION

During both polarized proton runs beam was ramped to $\gamma \approx 107$. pp_02 used $\beta^* = 3$ m at all IRs while pp_03 used several settings. They are listed in table 3 below. In FY02 there was no need to use the scrapers during the ramp and neither scraper nor crystal were used routinely at any time during the pp_02 run. Figure 5 shows the effect of both, crystal and scraper, on experimental backgrounds in STAR (top) and BRAHMS (center). The experimental background rates are normalized to the collision rate. While the crystal has no obvious effect on the background signal in either IR, the retracted scraper increases the signal at IR6 (STAR) by 6% while decreasing it slightly at IR2 (BRAHMS). Background signals were derived from the experimental luminosity monitors (ZDC), which are situated close to the beam pipe. However, when looking at other signals from STAR [5]

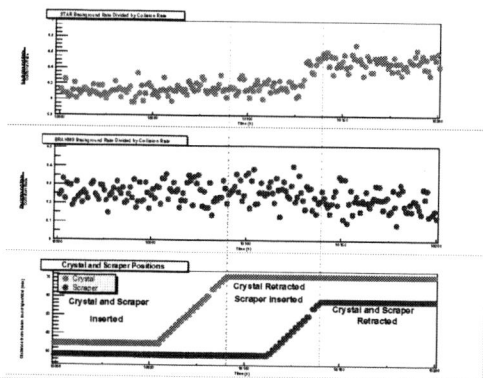

**FIGURE 5.** Background signals as a function of collimator position during fill 2185 in run pp_02.

coming from a detector some 2m away from the beam pipe, background is increased when the scraper is moved in. The results from the crystal are inconclusive since it increases as well as decreases the background compared to the scraper being in alone. In either case, the rates are higher than with both devices out by several 10%. The crystal experiment was repeated in the following year in the dau_03 run and is reported in [6]. It confirmed that a

**TABLE 3.** List of configurations during the pp_03 run with 55 bunches. "-" refers to "off" or "no collisions". A snake value of 1.88 refers to a full snake in one ring plus 88 % in the other. a = IR6 & 8, b = IR2 & 10, (*) IR2 only

| date | fill | $\beta^*$ (m) a | b | snk. | rotrs. (IR) | RF B | Y |
|---|---|---|---|---|---|---|---|
| 04/08 | 3427 | 2 | 2 | full | - | - | - |
| 04/12 | 3459 | 1 | 3 | full | - | - | - |
| 04/14-15 | 3480 | 1 | 3 | full | - | on | on |
| 04/16 | 3496 | 1 | 3 | - | - | - | - |
| 04/19 | 3502 | 1 | 3 | 1.88 | - | - | - |
| 04/25 | 3541 | 1 | 3 | 1.88 | 8 | - | - |
| 04/30 | 3591 | 1 | 3 | 1.88 | 8 | on | on |
| 05/15 | 3720 | 1 | 3 | 1.88 | 8 & 6 | on | on |
| 05/18 | 3735 | 1 | 3 | 1.88 | 8 & 6 | on | - |
| 05/19-21 | 3744 | - | 10* | 1.88 | - | on | - |
| 05/21 | 3757 | 1 | 3 | 1.88 | 8 & 6 | on | - |
| 05/23 | 3767 | 1 | - | 1.88 | 8 & 6 | on | on |

collimation system including the crystal as primary collimator at its current location cannot reduce experimental backgrounds. During a special experiment in 2002, several detector signals from PHENIX [7] were monitored during a dedicated end-of-fill background study. When the scrapers are moved in aggressively, the MUID (PHENIX muon identification chambers, extending to meters away from the pipe) rate drops by a factor of 8 while others, BBCLL1 and NTC (both collision signals from detector components close to the beam pipe) remain constant. More details can be found in [2].

The FY03 polarized proton run again consisted of many ramp commissioning periods. The various schemes are listed in table 3. The include changes in $\beta^*$, as well as various states of the snake magnets (needed to maintain transverse polarization during the RHIC ramp) and rotator magnets around PHENIX and STAR (needed to locally change transverse polarization into longitudinal polarization). They sum up to a total of eight commissioning periods in eight weeks! In addition, the $\beta$-squeeze approach was changed. Instead of squeezing during the energy ramp, a dedicated squeeze-ramp was performed at flattop energy. In short, the conditions during run pp_03 were quite different from pp_02. Collimation was essential during most parts of the run. Background conditions were bad enough to prevent experiments from turning on some of their detectors.

Figure 6 shows background signals from PHENIX during the period when beams collided in IR6 and IR8 only. The scintillators were installed on either side close to the Q3 magnets, some 40 meters upstream of the IR. It turned out that their signal represented the background conditions for the MUID detector fairly well, indicating that background was caused by beam scraping in the triplet area. The signals are shown in percent of the value before collimation starts. When the horizontal jaw is

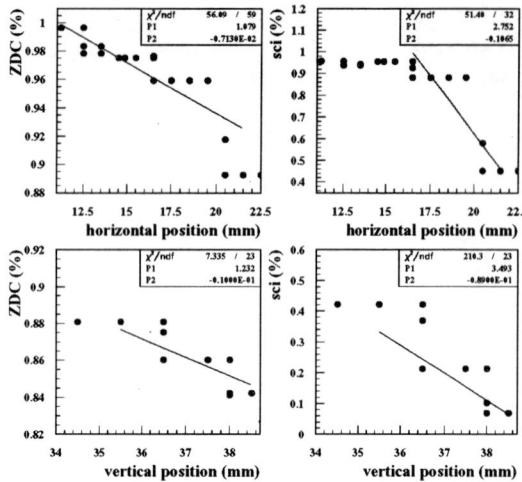

**FIGURE 6.** *PHENIX scintillator and ZDC signals as a function of the collimator jaw position during fill 3780.*

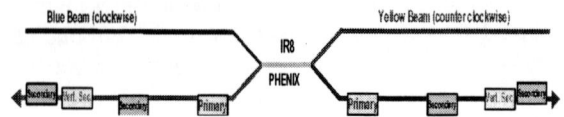

**FIGURE 7.** *Sketch of the planned conventional 2 stage collimation system.*

moved (top two figures), the slope of the linear fits differs by approximately a factor of 100 between the ZDC and scintillator signal. This difference is reduced to a factor 10 when the vertical jaw is moved closer to the beam. The horizontal jaw was moved first. Also, in this plane the drop of the scintillator signals begins later (by about 4 mm) than the decrease in the ZDC signal. However, while the the ZDC signal is reduced by about 16% only, the other signal drops to about 5% of its original value. At this point the MUID detectors could be turned on. This difference is due to the low sensitivity of the ZDC detector to beam background in general.

## UPGRADE PLANS

Anticipating higher beam currents, higher luminosity and therefore higher backgrounds, the crystal will be removed and a conventional 2 stage system will be installed in its place [8]. The efficiency of the crystal collimation strongly depends on the local lattice function and was shown to have insufficient background removal efficiencies [6]. The 2 stage system includes two secondary horizontal collimators and one secondary vertical collimator. The primary collimator consists of both, a horizontal and vertical jaw. All collimators will be, for this upcoming run, positioned around PHENIX, in the warm straight section between the triplet (Q3) and Q4. Thus there will be three independently movable horizontal collimators and two vertical collimators per ring. The vertical secondary collimator will not be operational at the beginning of the next run. Fig. 7 shows a sketch of the collimator system as being installed. Each collimator will be furnished with four PIN diode loss monitors allowing a feedback system based on the loss rate at the individual jaw.

In addition to the secondary collimators, stainless steel shielding [9] is being installed in the tunnel around the PHENIX IR. Thus, a significant reduction of the background rates is expected in particular in the PHENIX Muon Identification chambers (MUID) for the next run.

## CONCLUSION

Background conditions were particularly challenging during the past two runs, dau_03 and pp_03, most likely due to the horizontal crossing angles (dAu), the ramping and squeezing scheme (pp) and in general the high number of running configurations during both runs. Although background could be reduced at various locations with the primary collimators alone it remained difficult, if not impossible, to reduce background rates to a tolerable level at all experiments at the same time for the most part of the last two runs. The existing crystal collimator at its location could not enhance the overall collimation efficiency. Therefore, additional secondary collimators are being installed while the crystal collimator is dismounted. The new, mainly horizontal system, will be operational for the next run.

## REFERENCES

1. R. P. Fliller III et. al. "The Two Stage Crystal Collimator for RHIC", Proceedings of the PAC 01, Chicago (2001).
2. A. Drees et al.,"RHIC Collimator Performance",Proceedings of the EPAC 02, Paris, June 2002.
3. A. Drees, L. Ahrens et al. "Abort Gap Cleaning in RHIC", Proceedings of the EPAC 02, Paris, June 2002.
4. C. Adler, et al., nucl-ex/0008005.
5. W. Christie, "STAR", RHIC Retreat 2003, http://www.rhichome.bnl.gov/AP/RHIC2003/Retreat/.
6. R. Fliller et. al. "Crystal Collimation at RHIC", these Proceedings.
7. M. Perdekamp, "PHENIX", RHIC Retreat 2003, http://www.rhichome.bnl.gov/AP/RHIC2003/Retreat/.
8. R. Fliller, A. Drees, "Collimation", RHIC Retreat 2003, http://www.rhichome.bnl.gov/AP/RHIC2003/Retreat/.
9. Kin Yip, "Shielding",RHIC Retreat 2003, http://www.rhichome.bnl.gov/AP/RHIC2003/Retreat/.

# SNS Longitudinal and Transverse Halo Measurement

D. Gassner, P. Cameron, R. Witkover*

*Brookhaven National Laboratory, Upton, NY 11973*
*\*TechSource, Inc., Santa Fe, NM*

**Abstract.** Stringent particle loss constraints for the SNS accumulator ring require that the beam gap must be kept clean, and beam in the halo be lost in a controlled manner. A fixed amplitude resonant strip line kicker system capable of reversing polarity turn by turn is being designed for gap cleaning. Detectors will be installed in the Ring to observe the radiation from the interaction of the kicked gap beam on a movable scraper using a gated photomultiplier with a scintillator. This detector will also be used to measure the transverse halo of the circulating beam.

## INTRODUCTION

The Spallation Neutron Source (SNS) [1] being built at Oak Ridge National Laboratory (ORNL) by a collaboration of 6 laboratories, each responsible for specific sections of the accelerator, is designed to deliver $1.5 \times 10^{14}$ protons at 1.0 GeV in one bunch at 60 Hz to a liquid mercury target. To achieve this without excessive activation, an uncontrolled loss criterion of 1 part $10^4$ (~1 W/m) has been specified. BNL has responsibility for the design of the longitudinal and transverse halo measurement systems. The High Energy Beam Transport carries the beam to the Ring where it is stripped to protons and accumulated in a single 695 nsec bunch over the 1 msec injection pulse. The design pulse beam current in the Linac and HEBT is 38 mA. In the Ring it will increase to more than 40 Amps at the end of the pulse, for a total of $1.5 \times 10^{14}$ protons. This high average beam power makes it crucial that uncontrolled losses be minimized through careful design with beam dumps and collimators handling losses in a controlled manner.

---

SNS is managed by UT-Battelle, LLC, under contract DE-AC05-00OR22725 for the U.S. Department of Energy. SNS is a partnership of six national laboratories: Argonne, Brookhaven, Jefferson, Lawrence Berkeley, Los Alamos, and Oak Ridge.

It is the job of the halo measurement system to provide data to help minimize uncontrolled losses due to longitudinal halo (beam-in-gap), and diagnose transverse halo growth mechanisms. The possible transverse and longitudinal ring halo measurement solutions described here are handled separately.

## TRANSVERSE HALO

In the effort to reduce beam loss in the ring during the accumulation cycle, it is essential to understand transverse halo growth characteristics. This requires profile measurements of beam tails 4 orders of magnitude below the peak beam intensities, which range from 40mA to 40 Amps, yielding tails in the range of 4uA to 4mA. We propose a method of direct halo measurement using instruments in the ring to measure the intercepted beam.

Wire scanners (built by LANL) [2] and Ionization Profile Monitors (by BNL) will provide beam profile measurements. Due to the elevated levels resulting from controlled losses near the ring collimators, these instruments are located far away, on the other side of the arc downstream of the extraction region as shown in Figure 1. The wire scanners will measure profiles of the accumulated beam by detecting the hadron shower created when the 32-micron carbon wire intercepts the beam while it is slowly scanned across the 8-inch aperture. Measurements will be taken using the Beam Loss Monitor (BLM) System [3] ion chamber detectors, and photomultiplier based Fast Beam Loss Monitors (FBLM). Secondary emission signals plan to be acquired from the wire as a backup method. Unfortunately, these systems are expected to yield poor resolution in the profile tails.

Beam scrapers will be installed near the wire scanners. They will be similar in design to the HEBT momentum scrapers [4], except they will have limited motion range due to heating issues. The 20mil thick carbon-carbon jaw will insert far enough so a cross calibration can be done with the data from the other profile measurement devices.

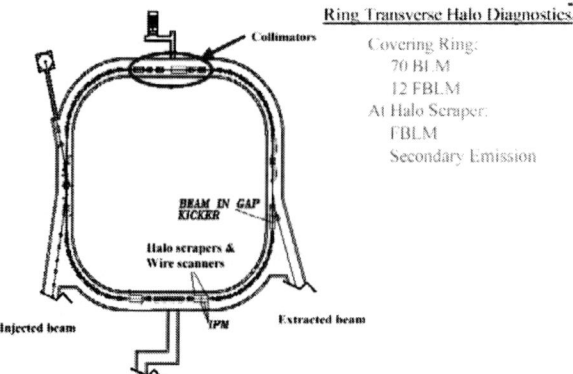

**FIGURE 1.** SNS layout, and locations of profile and halo measurement components.

The ion chamber detectors and the FBLM's used in the global BLM system, will be positioned to measure the hadron shower resulting from the interaction with the scrapers and scanning wires. Particle tracking simulations are needed to determine the best locations to install these instruments.

## LONGITUDINAL HALO

In theory there should be no beam in the gap from Linac, the chopper should be 100% efficient. But we expect longitudinal halo due to nuclear scattering, foil losses, rf noise, collimation inefficiency, etc. Since the uncontrolled loss budget is $10^{-4}$, care has to taken to see that it is not used up due to subsequent losses of gap beam.

### Beam In Gap Kicker

Using a resonant gap cleaning technique [5], the gap beam will be vertically kicked onto the collimators during normal operations for the last 60 to 100 turns of the accumulation cycle by three 1.5m long 50 Ohm stripline kickers during normal operations. This process can be done anytime in the cycle to study the development of the beam in gap.

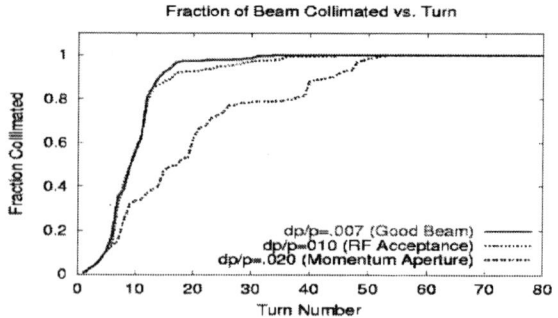

**FIGURE 2.** Data from simulation of SNS beam-in-gap kicker. S. Cousineau, et al, EPAC 2002.

The typical bunch length is 675ns, and a gap length 275ns. The design calls for a rise time of 10-20ns for 10-90%. The kickers will be powered by 7kV pulsers driving opposing striplines with opposite polarity capable of reversing polarity turn-by-turn. The kick strength and number was determined utilizing particle-in-cell tracking code, see Figure 2.

### Beam In Gap Monitor

To avoid saturation from controlled losses while the bunch passes by, it will be necessary to use a gated detector. Several types of gated high gain, high speed detectors are being considered, these include a Burle 85104 MCP/PMT, and a Hamamatsu gateable microchannel plate photomultiplier tube R5916U-50 which has a $10^8$ switching ratio, $10^5$ gain, 0.2ns rise time, and variable gate widths. Analog signals will be buffered and amplified by a VME based wideband amplifiers from CAEN, then digitized by the Acqiris DP235 PCI digitizer card. This card provides 8-bits, dual channels, 500MHz bandwidth, 1Gsample/sec sampling rate, and 4MB memory option.

Loss patterns will be observed at three locations; the collimators, which is the destination of the kicked gap beam, at the halo scraper, which will become a temporary limiting aperture when inserted during studies, and at the extraction region where the majority of the uncleaned gap beam will be lost at the septum.

*Ring Beam In Gap Measurement*

The beam cleaned in the gap will be lost in the collimators, 75% primary, and 25% secondary as shown in Figure 3. These collimators will be in place all the time, which allows measurements during normal operations. Unfortunately, this region will eventually have elevated background levels that may make fine measurements difficult. The beam scrapers located near the wire scanners can be inserted to obtain a higher resolution measurement. They will become the new limiting aperture, conveniently located at a lower background region. The scrapers cannot be used continually for this type of measurement due to the potential for high losses.

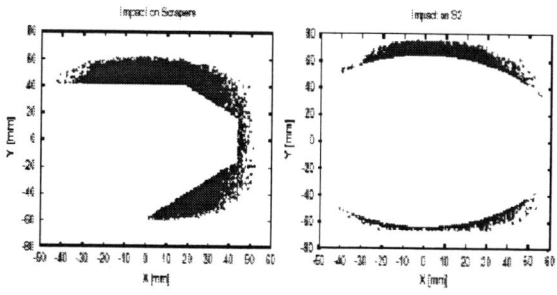

**FIGURE 3.** Data from simulation of SNS beam-in-gap kicker showing losses on the primary and secondary collimators. S. Cousineau et al, EPAC 2002.

*Extraction Beam In Gap Measurement*

An alternative method of determining the amount of gap beam is to extract beam normally to the target at various times in the cycle and observe losses with a gated detector. The immediate benefit will be a factor of 15-30 improvements in instantaneous loss over the resonant cleaning scheme. In addition, the losses are easier to measure since the extraction septum is not buried in shielding, as is the collimator. An estimate of resonant gap cleaning efficiency can be determined by measuring losses at the extraction septum during the gap with the resonant gap cleaner off, and on.

## CALIBRATION TECHNIQUES

Two calibration methods are described, one for when using the resonant gap cleaner kicker, the other for when using the extraction kicker. The machine must be in a dedicated studies mode cycle for these measurements.

*Beam in Gap Kicker*

Inject 1 turn (0.1% of full beam, 15mA). Execute resonant gap cleaning during the beam pulse. Measure turn-by-turn change in total charge with Beam Current Monitor. Measure losses at the collimator or halo scraper turn-by-turn over the resonant kick duration using a gated MCP/PMT. Also measure losses with the FBLM's and BLM's.

*Extraction Kicker*

Inject 1 turn; adjust the timing of the extraction kicker trigger so kicker current begins to rise while the leading edge of the bunch is passing the kicker. Calibrate losses as measured at the extraction region and at any other significant loss locations.

## SUMMARY

Although measurement and calibration solutions are presented here, several tasks lay ahead. They include the analysis, simulation of beam interaction with the beam scrapers, choice of a gated MCP/PMT and design and test of its support electronics. Also, further simulations and analysis to better understand loss patterns at the extraction region due to the beam in gap cleaner, and while extracting on one turn for calibration.

## REFERENCES

1. J. Alonso, "The Spallation Neutron Source Project," Proceedings of the 1999 PAC, NY, p. 574. For current status see http://www.sns.gov

2. M. Plum, "HEBT, Ring, & RTBT Wire Scanner System Update" SNS Diagnostics Review Proceedings, BNL, March 26, 2003.

3. D. Gassner, P. Cameron, C. Mi, R. Witkover, "Spallation Neutron Source Beam Loss Monitor System", IEEE Particle Accelerator Conference 2003, Portland, OR.

4. S. Cousineau, D. Davino, N. Catalan-Lasheras, J. Holmes, "SNS Beam-In-Gap Cleaning and Collimation" Proceedings from EPAC 2002, Paris, France.

5. N. Simos, H. Ludewig, et al, "Collimator Design for the HEBT Line of the SNS" Proceedings from EPAC 2002, Paris, France, p. 1055.

# The IPM as a Halo Measurement and Prevention Diagnostic*

R. Connolly, M. Grau, R. Michnoff, and S. Tepikian

*Brookhaven National Lab, Upton, New York, U.S.A.*
*e-mail: connolly@bnl.gov*

**Abstract.** Four ionization beam profile monitors (IPM's) are in RHIC to measure vertical and horizontal profiles in the two rings. Each IPM collects and measures the distribution of electrons in the beamline resulting from residual gas ionization during bunch passage. The ionized electron signal provides an accurate beam profile and the detectors are capable of measuring individual gold bunches. However the detectors are extremely sensitive and their performance has been limited by backgrounds from radiation spray, rf coupling to the beam, and secondary electrons. In 2002 two IPM's were rebuilt using design changes which greatly reduced the backgrounds. During the summer 2003 shutdown another rebuild will increase the sweep electric field, make the electric field more uniform, and add a calibration system. The improvements in the electric field will increase the sensitivity to beam without increasing backgrounds and the calibration system will allow channel-channel gain variations to be removed. These improvements should increase the detectors sensitivity to a level where the IPM can be considered as a beam halo monitor.

## INTRODUCTION

Beam emittances in RHIC [1] are measured with ionization profile monitors (IPMs) [2,3,4,5]. An IPM measures the transverse beam profile by collecting and measuring the distribution of electrons generated by beam ionization of residual gas in the beamline.

An electric field across the beam pipe sweeps the electrons into an 8cm x 10cm microchannel plate (MCP) [6] which amplifies the electron flux by $10^4$-$10^6$. Electrons exiting the MCP are collected on a circuit board with 64 parallel strip anodes spaced 0.6mm apart. The transducer head is placed in the gap of a 0.14T dipole magnet to force the electrons to travel perpendicular to the collection board.

The first IPM detectors that were installed in RHIC gave single-bunch profiles but also were sensitive to noise and backgrounds. Noise was from rf coupling to the bunched beam and backgrounds were from radiation spray from upstream beam loss and from background electrons. Two IPMs were rebuilt during the 2002 shutdown. The new design greatly reduces noise and backgrounds so the spurious signals on the outer channels average about 2-3% of the center-channel signal.

The dramatic noise and background reductions in the 2002 IPMs made evident the need for three more design improvements. These three improvements are being implemented on all RHIC IPMs during the 2003 shutdown.

Until the 2003 run the premise was that the trajectories of the electrons were completely determined by the magnetic field and so no effort was made to shape the electric field. However this year we observed that the electric field had a small effect on the electron collection. For this reason the high voltage plate which generates the transverse electric field is being rebuilt to give nearly parallel field lines.

It was also discovered that the electric potential drop between the center of the beampipe and the entrance to the MCP was too small. The electrons generated near the MCP were not being detected. To remedy this the voltage on the sweep electrode is being raised.

Finally the IPMs are sensitive enough that the small variations in individual channel gains need to be removed. To do this we are placing an Electrogen™ [7] electron generator array in a cutout in the high voltage sweep-field electrode. This will flood the MCP with a uniform electron flux for calibration.

Increasing the signal and reducing the channel-to-channel gain variations will increase the sensitivity of the detectors to a level where it may be possible to see beam halo effects. This paper compares data from the 2003 beam run to earlier data and describes the changes being implemented in the 2003 rebuild.

---

*Work performed under the auspices of the U.S. Department of Energy.

# THE 2002 DESIGN

## Vacuum Chamber and Detector Head

Figure 1 is a photograph of the original IPM detector head which was installed in RHIC. The transducer assembly is mounted on a 10" conflat flange and it is inserted through a side port into a 100cm x 150cm rectangular beam pipe. The electron collector board and MCP is on the bottom of the picture and the electrodes which create the electron sweep field are on the top. This IPM was extremely susceptible to beam coupling and a phosphor-bronze shield was added which completely enclosed the electron-collection electronics. This shield has an opening covered with copper mesh over the MCP and it made contact via rf spring-finger stock with the flange face.

Since the IPM assembly was inside the beam line, it created both an aperture restriction and an impedance bump. One goal of the rebuild was to move the detector components outside of the beam image-current path. To do this the vacuum chambers were rebuilt as shown in fig. 2.

A large detector chamber was installed in the rectangular beam pipe. A section of the beam pipe with rectangular cutouts on two sides is inserted into this chamber and makes rf contact with the beamline on both ends, fig. 3. One of the cutouts allows electrons to pass to the MCP and collector board and the other is for a high-voltage sweep electrode. Both of the openings are covered with an hexagonal aluminum mesh from Laird [8] which has 95% open area and attenuates rf by 100dB. This places the electronics out of the path of the image current and presents a continuous beamline cross section to the beam. Also the MCP is in an alcove which shields it from radiation spray from upstream beam loss.

**FIGURE 1.** Photograph of original IPM transducer head installed in RHIC.

**FIGURE 3.** The new IPM detector being inserted into the beam line

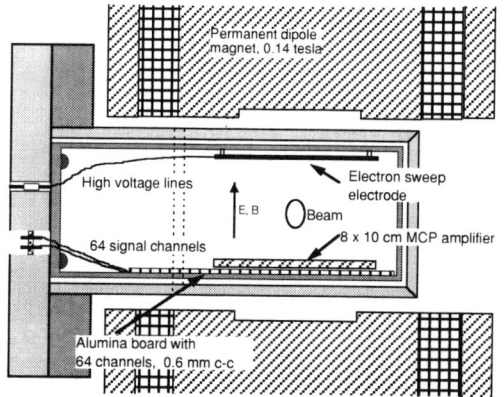
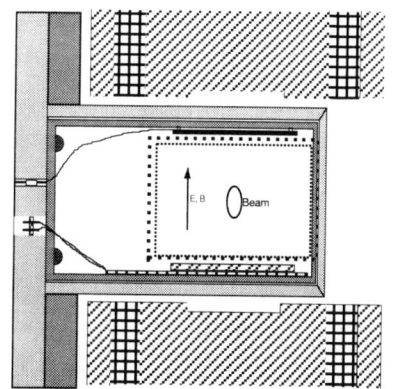

**FIGURE 2.** A) Original IPM transducer in beam pipe. B) New design will have all components "outside" of beam pipe (dotted rectangle).

As the RHIC beam intensity has increased the detector is proving to be sensitive to background electrons created in the beamline. To reduce this background the electric sweep plates are extended 3cm beyond the detector to remove electrons before they can drift into the MCP.

## PERFORMANCE OF 2002 DETECTOR

The new IPMs are extremely immune to rf coupling to the beam. In past runs the beam intensity has increased to a level where the amplifiers were saturated by pick up and the IPMs became useless. In the 2003 beam run the new IPMs delivered clean profiles at the highest beam intensities.

A beam test was done to show the radiation immunity of the new design. The beam was intentionally steered into the beam pipe upstream of the two IPMs in the Blue ring. The beam-loss monitors showed high radiation levels at the IPMs. The results are shown in fig. 4. The horizontal detector is the new design and the vertical detector is the old design with the electronics inside the beampipe. The large "error" bars are artifacts caused by the electron counting rate being far smaller than the sampling rate.

Figure 5 shows gold-beam profiles from the old detector in the 2002 run, left, and the new detector in the 2003 run. Gold beams generate large numbers of background electrons. The reduction in sensitivity to background electrons in the new design is evident in the far lower baseline in the 2003 profile. In the 2003 profile the baseline is about 2-3% of the peak channel.

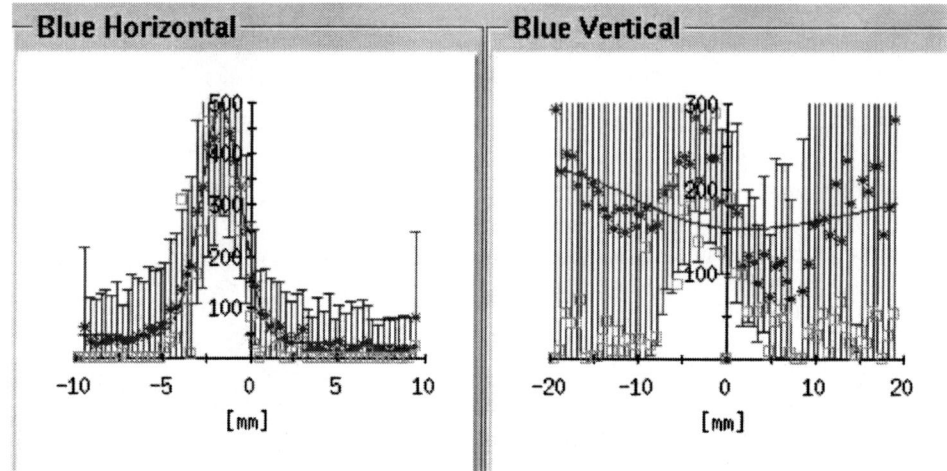

**FIGURE 4.** Beam was intentionally steered into the beampipe upstream of the IPMs which showered the detectors with radiation. The horizontal profile is from the new detector and the vertical profile is from the old detector.

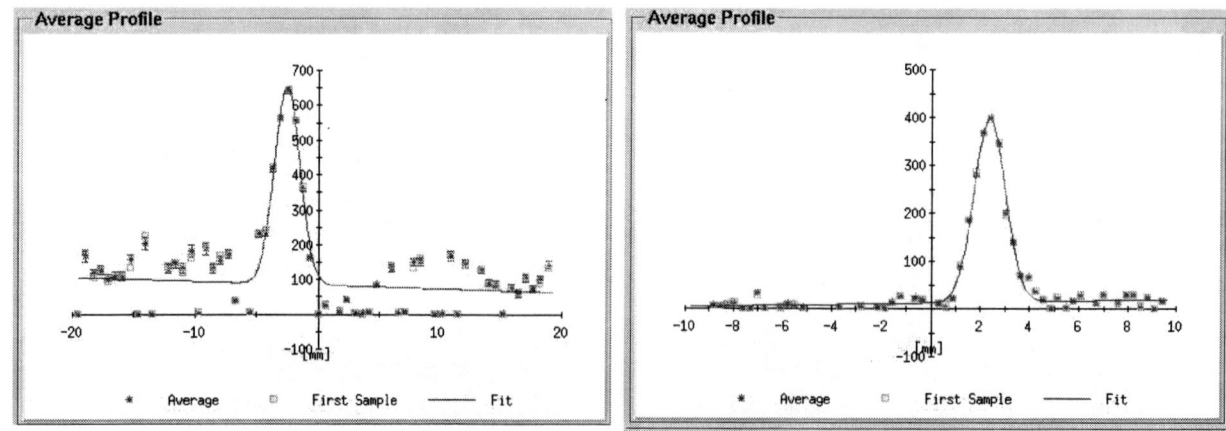

**FIGURE 5.** Profiles of gold beams with large electron backgrounds. The left profile was taken in 2001 with the old detector and the right profile was taken in 2003 with the new detector. The electron background is almost gone in the new detector.

## THE 2003 DESIGN

An IPM has 64 channels each with its own amplifier, transmission line and digitizer. The original detectors had two systems for pulsing simultaneously all of the channels to check that they were functional. However these systems were not sufficiently uniform for individual channel calibrations. The rebuild in 2003 adds an Electrogen™ [7] Electron Generator Array to each detector head. The EGA will flood the MCP with a uniform electron flux. The signal will be measured and the gains of the channels adjusted in software to give a uniform response. This system will allow the slight channel-channel gain variations to be removed and the nonuniform aging of the MCP to be compensated.

Data from the 2003 beam run show the measured beam width and position varied slightly with the transverse sweep-field voltage. For this reason the high voltage plate which generates the transverse electric field is being rebuilt so it wraps halfway around the beampipe insert. This simple change in the elctrode structure will create a far more parallel electric field. Figure 6 shows the calculated equipotential lines.

It was also discovered that the electric potential drop between the center of the beampipe and the entrance to the MCP was too small. The electrons generated near the MCP were not being detected. To remedy this the voltage on the sweep electrode is being raised from 5kV to 10kV. This increase in sweep voltage will also increase the background electron suppression.

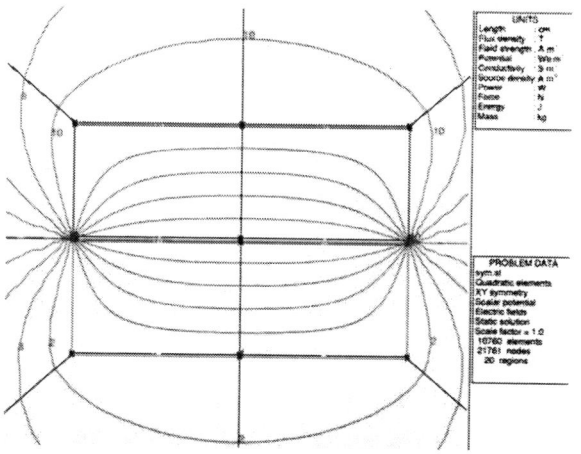

**FIGURE 6.** Plot of electric equipotential lines generated by new electrode geometry.

## DISCUSSION

The best profiles measured during the 2003 beam run had backgrounds of about 2% of the center channel counts. Our new detectors should give significantly larger beam signals from the increased electric field without any increase in noise or background. In some cases over half of the signal electrons were not detected. The increase in the electric sweep field will also further reduce the detector sensitivity to background electrons.

Beam steering caused by the poorly shaped electric field in the earlier detectors caused the measured profile to wander around on the detector. This together with the lack of accurate individual channel gain calibration created measurement artifacts which mimicked beam halo. The improved field quality and the accurate calibration system of the 2003 IPMs will reduce these effects.

The improvements to the IPMs being implemented in 2003 should make them sensitive to beam halo at the level of a few percent of the central beam density.

## ACKNOWLEDGMENTS

The 2002 and 2003 detectors were assembled by Chuck Trabocchi and Al Ravenhall. Testing, magnet modifications, and beamline installation were done by Robert Sikora and John Cupolo.

## REFERENCES

[1] http://www.rhic.bnl.gov/
[2] Connolly, R., Cameron, P., Ryan, W., Shea, T.J., Sikora, R., and Tsoupas, N., *A Prototype Ionization Profile Monitor for RHIC*, Proc. 1997 PAC, Vancouver, BC.
[3] Connolly, R., Cameron, P., Michnoff, R., Radeka, V., Ryan, W., Shea, T., Sikora, R., Stephani, D., Tapikian, S. and Tsoupas, N., *The RHIC Ionization Beam Profile Monitor*, Proc. 1999 PAC., New York.
[4] Connolly, R., Michnoff, R., Moore, T., Shea, T., and Tepikian, S., Nucl. Instr. and Meth. A 443 (2000) 215-222.
[5] Connolly, R., Cameron, P., Michnoff, R., and Tepikian, S., *Performance of the RHIC IPM*, Proc. 2001 PAC, Chicago.
[6] J. Wiza, *Microchannel Plate Detectors*, Nucl. Instr. Meth. 162 (1979) 587-601.
[7] Burle Electro-Optics Corp., Sturbridge, MA 01566.
[8] Laird Technologies, www.lairdtech.com

# HERA Beam Tail Shaping by Tune Modulation

Christoph Montag

*BNL, Upton, NY 11973*

**Abstract.** To study CP violation, the HERA-B experiment uses an internal wire target in the transverse halo of the stored HERA proton beam. Operational experience shows that the resulting interaction rates are extremely sensitive to tiny orbit jitter amplitudes. Various methods have been studied to stabilize these interaction rates by increasing diffusion in the transverse proton beam tails without affecting the luminosity at the electron-proton collider experiments ZEUS and H1. Tune modulation was found to be a promising method for this task. Experiments performed in recent years will be reported.

## INTRODUCTION

The electron-proton collider HERA consists of a superconducting 920 GeV proton storage ring and a 27.5 GeV electron storage ring. Beams collide in two of the four interaction regions (IRs), which are equipped with the detectors ZEUS (IR South) and H1 (IR North). In the interaction region East the experiment HERMES with its polarized internal gas target is installed in the electron ring to measure the spin structure of protons and neutrons using the stored polarized HERA electron beam. The HERA-B detector is installed in the straight section West of the HERA proton ring to study CP violation in the B-meson decay $B^0 \to J/\psi K_s^0$, using an internal wire target in the beam halo. To achieve the design interaction rate of $\dot{N}_{\text{interaction}} = 40\,\text{MHz}$, each bunch crossing has to provide about four interactions. To enable reconstruction of these interactions, a sufficient spatial distribution is required. This is achieved by arranging four target wires around the beam, operated simultaneously. Each of these wires is equipped with a stepping motor to ensure equal contribution of all four wires to the overall interaction rate. These motors allow for accurate target movements in nominal steps of 50 nm. The interaction rate is kept constant by a 10 Hz wire position feedback which counteracts slow beam orbit variations by target position adjustment.

Observations demonstrate that this slow wire position feedback is not sufficient to ensure the required interaction rate stability. At high interaction rates above some 20 MHz large fluctuations were observed, with relative rms deviations

$$\sigma_{\text{rel}} = \frac{\sqrt{\langle(\dot{N}_{\text{interaction}} - \langle\dot{N}_{\text{interaction}}\rangle)^2\rangle}}{\langle\dot{N}_{\text{interaction}}\rangle} \quad (1)$$

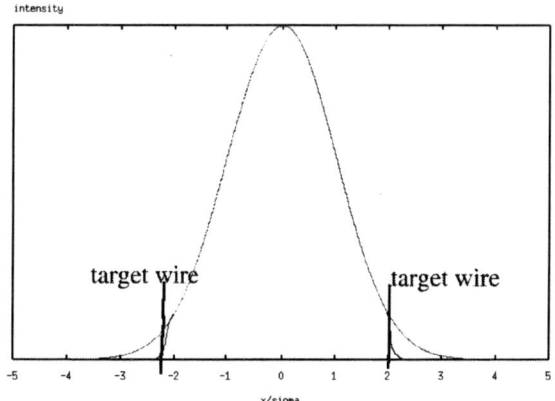

**FIGURE 1.** Schematic view of the horizontal HERA-B wire target in the halo of the stored proton beam. Due to insufficient diffusion the wires have to move ever closer to the beam core, thus scraping away the halo. The resulting "hard edge" of the modified distribution leads to large interaction rate fluctuations when the beam orbit varies due to effects such as ground motion.

above 50%. This effect can be explained by lack of diffusion into the transverse beam tails, requiring the eventual positioning of the target wires extremely close to the beam center, typically at 3 to 4 transverse rms beam sizes. As Figure 1 shows, this tight target aperture leads to sharp "edges" of the initially Gaussian distribution, resulting in a high sensitivity of the interaction rate to disturbances such as beam orbit jitter.

To overcome this difficulty, a re-population of the transverse proton beam tails by increased diffusion is required. However, since HERA has to simultaneously provide high luminosity electron-proton collisions at the colliding-beam experiments ZEUS and H1, this "tail shaping" must be performed without any detrimental effects like luminosity degradation or increased background rates. A simple slow transverse blow-up of the beam is therefore not tolerable, and a more sophisticated method has to be chosen.

Tune modulation in connection with a strong nonlinearity, in particular the beam-beam interaction, is well known to result in increased diffusion in the transverse beam tails. The beam core, where the linear part of the beam-beam force dominates, is practically unaffected [1, 2, 3, 4]. This is therefore an ideal approach for "tail shaping".

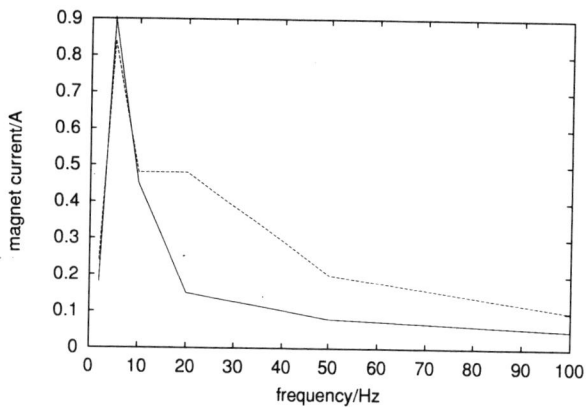

**FIGURE 2.** Measured AC peak magnet current as function of frequency for an input voltage of 1 V and for two DC magnet currents of 30 A (lower) and 60 A (upper), respectively. The lines are drawn to guide the eye.

## EXPERIMENTAL SET-UP

The main quadrupoles of the 920 GeV superconducting HERA proton storage ring are connected in series with the main dipole magnets. For fine-adjustment and control of the tunes two families of superconducting quadrupole magnets are installed in each HERA quadrant. The chopper power supplies of these quadrupoles in the quadrant West are modified to provide an additional control input, allowing for tune modulation with frequencies up to about 1.5 kHz.

The tune modulation signal is generated by a PC with DATEL PC-420 wave generator board [5], remotely controllable from the HERA control room. This board generates an arbitrary, continuous, periodic signal that is used as the input signal for the quadrupole chopper power supplies.

The large inductance of the superconducting quadrupole magnets leads to an effective suppression of high frequency modulation signals, while low frequencies of a few Hertz are partially counteracted by the power supplies' internal feedback loops. Figure 2 shows the measured transfer function of this assembly.

## EXPERIMENTS

The effect of tune modulation on the beam halo was studied in two steps. First a single target wire was moved into the beam halo until the interaction rate reached 10 MHz. Then the wire was retracted by 100 $\mu$m and the target postion feedback was turned off. After an initial decay time the interaction rate reached a stable value of about 0.3 MHz. When the tune modulation was turned on, the interaction rate increased until a new equilibrium was reached. This equilibrium value was recorded for different tune modulation parameters, as shown in Figure 3, with

$$U(t) = \sum_i U_i \cdot \sin(2\pi f_i t) \qquad (2)$$

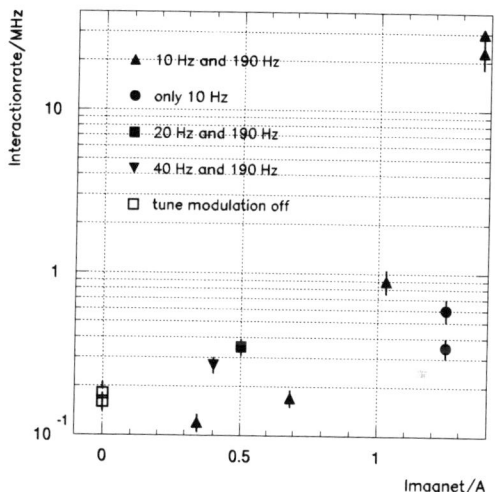

**FIGURE 3.** Interaction rates with stationary wire for different tune modulation parameters. For each setting, the input signal voltages for each frequency component are identical, but their sum might be varied. Due to the strong frequency dependency of the magnet power supply output current, the resulting total tune modulation amplitude is different for different frequency parameter sets.

being the power supply AC input voltage. The amplitude of all frequency components $U_i$ of the input signal was equal for all measurements, but due to the frequency dependency of the magnet currents the resulting tune modulation depths were not equal.

The maximum interaction rate was achieved by simultaneous application of two modulation frequencies:

$f_2 = 190\,\text{Hz}$ and $f_1 = 10$, 20, or 40 Hz, see Figure 3. At an AC magnet current of 1.37 A, an interaction rate of 23 resp. 30 MHz could be achieved using $f_1 = 10\,\text{Hz}$ and $f_2 = 190\,\text{Hz}$. When the 190 Hz component was turned off, the resulting interaction rate dropped significantly to 0.35 resp. 0.6 Hz, though the relative reduction in the total AC peak magnet current to 1.25 A was much smaller. With the 190 Hz component turned on again and reduced amplitudes of both components, the resulting 0.9 MHz interaction rate was significantly larger than with 10 Hz only though the total resulting tune modulation depth was reduced. Replacing the 10 Hz frequency component by a 20 or 40 Hz component while keeping the 190 Hz component constant shows a clear correlation between total tune modulation amplitude and resulting HERA-B interaction rate, see Figure 3. But since these frequencies are already strongly suppressed by the large magnet inductance the maximum achievable interaction rate is still much smaller than in the case with the 10 Hz component.

In a later dedicated HERA run the effect of tune modulation on HERA-B interaction rate stability was studied. For this study the wire position feedback was switched on and a single target wire was inserted into the beam halo until a stable interaction rate of 20 MHz was reached. To avoid interference effects with the 10 Hz wire position feedback, the lower frequency $f_1$ was set to 8 Hz resp. 12 Hz during these experiments.

With the lower modulation frequency set to $f_1 = 12\,\text{Hz}$, the effect of the second frequency component $f_2$ was studied in steps of 100 Hz for input modulation signal amplitudes of 0.5 V and 1.0 V, respectively. As an example, Figure 4 shows the effect on the interaction rate stability for one configuration. Figure 5 depicts the resulting relative rms interaction rate width $\sigma_{\text{rel}}$ as a function of $f_2$ for different tune modulation depths.

In contrast to the previous measurement at a fixed wire position, the resulting interaction rate stability does not correlate with the second frequency $f_2$. The rate stability exhibits a strong dependency on the amplitude of the input signal, and thus on the resulting modulation depth. This behavior was also observed when the 12 Hz component was replaced by an 8 Hz signal.

With the tune modulation switched on, the interaction rate per single proton bunch is clearly correlated to the intensity of the corresponding electron bunch, see Figure 6. This indicates that the observed stabilization effect is indeed dominated by tune modulation in combination with the beam-beam interaction, rather than a simple orbit effect due to orbit offsets in the modulated quadrupoles.

During these experiments the specific luminosity at the two colliding beam experiments ZEUS and H1 was recorded. As shown in Figure 7, these tune modulation

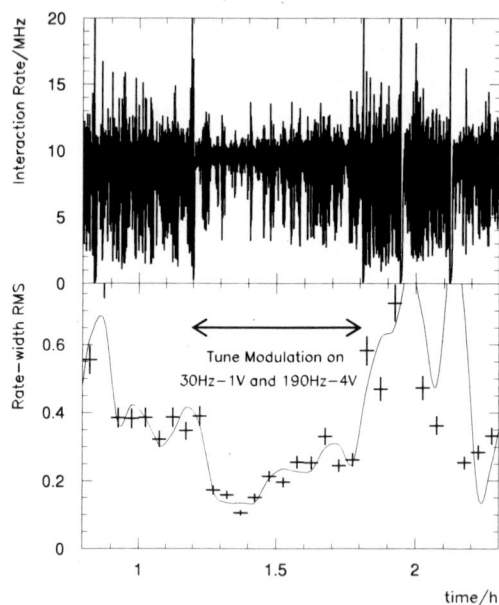

**FIGURE 4.** Example of the effect of tune modulation at $f_1 = 30\,\text{Hz}$, $f_2 = 190\,\text{Hz}$ on rate stability. When the modulation is turned on, the relative width of the interaction rate drops from some 50% to roughly 25%, and increases immediately to the previous level after the tune modulation turned off.

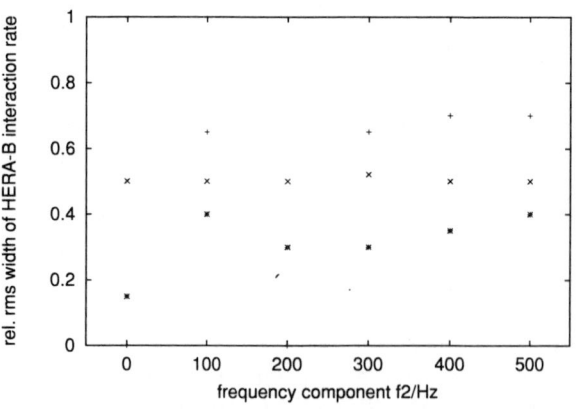

**FIGURE 5.** Interaction rate stability for different tune modulation frequencies $f_2$ and three different amplitudes ($+$ : 0.5 V, $\times$ : 1.0 V, $*$ : 2.0 V). The case without a second frequency component, i.e. 12 Hz only, is indicated as $f_2 = 0\,\text{Hz}$.

studies did not result in any decrease of specific luminosity. Furthermore, no significant experimental background increase was observed.

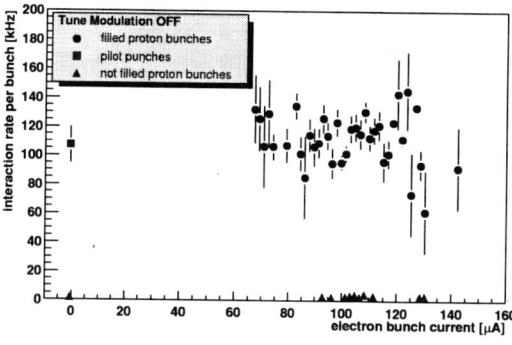

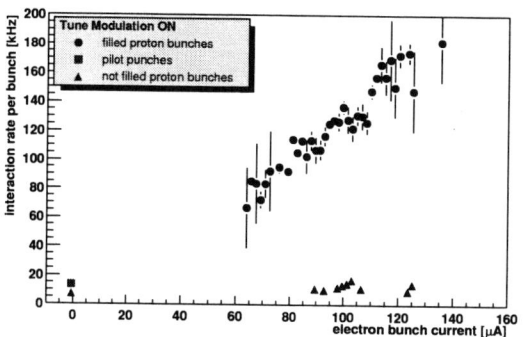

**FIGURE 6.** HERA-B interaction rate per bunch vs. bunch current of the corresponding electron bunch. Note that the last two proton bunches ("pilot bunches") in each 60 bunch train have no counterpart in the electron beam.

## CONCLUSION

Interaction rate stabilization at the internal halo wire target of the HERA-B experiment by means of tune modulation has been successfully demonstrated. Since this method does not result in any detrimental effects such as luminosity degradation or increased background rates at the colliding beam experiments ZEUS and H1, it has become a standard procedure in regular HERA runs.

## REFERENCES

1. S. Peggs, Nonlinear Dynamics Experiments, in: A. W. Chao, M. Tigner (eds.), Handbook of Accelerator Physics and Engineering, World Scientific, Singapore, 1999
2. L. Evans, The Beam-Beam Interaction, Proc. CERN Accelerator School, CERN 84-15
3. O. S. Bruening, Emittance Growth in Proton Storage Rings due to the Combined Effect of Non-Linear Fields and Modulation Effects with more than one Frequency, DESY HERA 93-10
4. O. S. Bruening, An Analysis of the Long-Term Stability of the Particle Dynamics in Hadron Storage Rings, DESY 94-085
5. DATEL PC-420 User Manual
6. I. Ludwig, private communication

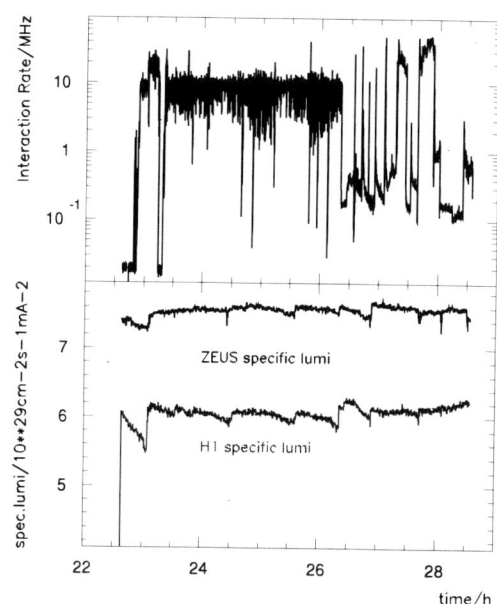

**FIGURE 7.** HERA-B interaction rate (top) and specific luminosity (bottom) during a dedicated tune modulation study run. The specific luminosity tends to deteriorate due to slow orbit drifts, but recovers completely to the initial value after orbit correction is applied.

# Working Group III:
# Halo Collimation

# Collimation Working Group Summary Report

A. Drees* and N.V. Mokhov†

*Brookhaven National Laboratory, Upton, NY 11973*
†*Fermi National Accelerator Laboratory, P.O. Box 500, Batavia, IL 60510*

## INTRODUCTION

The creation of beam halo and corresponding beam loss is unavoidable at any accelerator. The consequences to the machine and detector components can range from minor to severe. An accidental beam loss can cause catastrophic damage to the complex equipment. Only with a very efficient beam collimation system can one reduce uncontrolled beam losses in the machine to an allowable level. This Working Group overviewed and discussed the status of the collimation system developments, design and performance in three machine categories: high-power proton machines, hadron colliders and $e^+e^-$ linear colliders. Status, performance and outstanding issues of beam collimation in each category are described in the first paragraphs of the sections below, with additional observations by the group at the section ends.

## HIGH-POWER PROTON MACHINES

In his plenary talk, J. Wei (BNL) overviewed the collimation issues at high-power proton accelerators. He demonstrated the vital necessity of collimation for a Megawatt beam machines to minimize uncontrolled beam loss. Using the Spallation Neutron Source as an example, he described guiding principles, beam cleaning strategies and collimator design and handling issues. Based on the experience at the existing machines, he concluded that two-stage collimation in both transverse and longitudinal directions is essential in achieving a high efficiency. Cleaning of electron cloud and suppression of electron generation are important aspects in the design of new rings.

C. Warsop (RAL) described an approach to beam loss control at the ISIS synchrotron. Collimation is done by betatron and momentum collectors (graphite-copper-graphite devices 5 to 30 cm long) to shave off 2 kW halo out of a 240 kW 800-MeV beam. The system is instrumented with 40 BLMs. A halo growth rate is 10 $\mu$m per turn. Simulations and measurements on beam loss are in a good agreement. Further upgrades plans exist.

E. Prebys (FNAL) presented a well-designed two-stage collimation system for the Fermilab 8-GeV Booster to be installed in the tunnel this fall. It consists of two scattering foils followed by three integrated collimator-shielding modules about $1 \times 1 \times 1$ m$^3$ each, well controlled and instrumented. The system will be intercepting up to 2 kW halo out of a 64-kW beam, providing $\leq$ 0.1-1 W/m beam loss rates in the rest of the ring. Prompt radiation, sump water activation, residual dose and dose to sensitive components are within the stringent regulatory limits.

N. Nakao (KEK) described results of the MARS14 collimation and shielding studies for the 3-GeV ring of the J-PARC project. He impressed the audience by the MARS model built for the entire ring with all machine and tunnel components, shielding and soil, almost $10^5$ elements total. A 4-kW collimation system and sophisticated shielding designed with this model, will provide prompt and residual radiation levels within the regulatory limits.

The SNS collimation system design, shielding and handling were reviewed in great detail by N. Simos (BNL), H. Ludewig (BNL) and G. Murdoch (ORNL). The system will be intercepting 2 kW on each of the seven collimators, providing less than 1 W/m beam loss rate in the ring. Impressive ANSYS analysis and experiment were performed on thermo-mechanical response of the critical components to beam accidents. Estimates of dose and residual activity in the collimation region were done via a MCNPX-CINDER90-ORIGEN-MCNP chain of the computational codes. A thorough consideration was given to mechanical engineering design of collimators and their windows, collimator removal procedure, fast disconnects, remote vacuum clamps etc. A very complex cooling system raised some questions.

In all the machines considered, collimators take about 2 kW of beam power, resulting in a very similar scope and level of radiation, cooling, thermal, mechanical, engineering and handling problems. It was recognized that dry runs are very useful, such as collimator removal procedure etc. Although the codes used for the collimator and shielding design are quite reliable, benchmarking of residual dose predictions would be useful (contribution of low-energy neutrons, borated water in the SNS collimators etc.).

## HADRON COLLIDERS

In his plenary talk, N. Mokhov (FNAL) described principles and realization of a reliable beam collimation system required to sustain favorable background conditions in the collider detectors, provide quench stability of superconducting magnets, minimize irradiation of hadron collider equipment, maintain operational reliability over the life of the machine, and reduce the impact of radiation on personnel and the environment. He discussed sources, beam loss and scraping rates at hadron colliders and an approach to design an optimal two-stage collimation system. In addition he presented the Tevatron and HERA collimator performance, needs for a multi-component approach to collimation at LHC, and possible ways to improve collimation efficiency such as use of a bent crystal.

D. Still (FNAL) presented results on the Tevatron Collider Run II collimation system performance. The proton and anti-proton halo removal is done by 12 L-shape collimators catching about 0.4% of intensity for each beam. The flexible system is programmable, well-controlled and instrumented with motion based on BLM feedback. It is efficient and handled automatically at the beginning of every store.

R. Schmidt and J.B. Jeanneret (CERN) described Beam loss scenarios and strategies for machine protection and on-going work on collimation at the LHC. A BPM system will be used as a protection system for collimators. BPMs will catch beam motion (both planes) and give a warning to dump the beam before damage can happen. BLMs are not reliable enough at low halo level. Two serious issues were discovered recently in their studies: impedance created by collimators and baseline materials won't withstand errant beams. Studies are underway to solve these problems. I. Rakhno (FNAL) presented a sophisticated movable collimator system designed on a basis of thorough MARS calculations to protect the LHC machine and collider detectors at an unsynchronized beam abort. D. Kaltchev (TRIUMPH) showed results on the LHC collimator performance as calculated with three different codes.

A. Drees (BNL) reported on collimation experience at RHIC to date. Background conditions were more challenging in the last two runs than before (due to horizontal crossing angles (dAu), the ramping and squeezing scheme (pp)). Experimental background could be reduced with the primary collimators alone but not to a tolerable level at all experiments at the same time. Additional secondary collimators will be installed for the next run.

R. Fliller (BNL) described first results on crystal collimation of relativistic heavy ion beam. A fraction of the beam deflected by a bent crystal ended up at a secondary collimator. The result of studies was that at the present location, ion and proton beam channeling could be observed but crystal contribution to collimation efficiency is negligible. The crystal will be removed and replaced by a standard primary collimator.

M. Kostin (FNAL) discussed simulation aspects of beam collimation and their remedies in the MARS14 code.

The need for efficient collimation systems is apparent in all considered hadron colliders. Collimation and understanding of halo production mechanisms is vital for many different reasons, including lifetime of equipment, protection of personnel and accelerator components as well as experimental backgrounds. Bent crystals as a technique to potentially improve collimation efficiency are still being studied. The performance of the collimation system in use is directly linked to integrated luminosity and thus overall machine performance (Tevatron, RHIC). With the next generation of high energy and high intensity hadron colliders (LHC) the protection of the collimators themselves plays an important role in the collimation system design.

## $E^+E^-$ LINEAR COLLIDERS

In his plenary talk, T. Raubenheimer (SLAC) discussed $e^+e^-$ linear collider halo and collimation issues with some focus on the NLC. One main requirement is to prevent large amplitude particles radiating photons into the IR determining the collimation depth to be $\pm 10\sigma_x$ and $\pm 31\sigma_y$ respectively. Machine protection and the collimation of tails without generating muon flux at the IP are additional requirements. Calculations predict $< 10^{-5}$ beam in the tails but the NLC design is ready for (hopefully conservative) $10^{-3}$. In general, linac beams with sizes of $10 \times 1\mu m$ and high train currents of $1\text{-}2\ 10^{12}$ are difficult to handle and have the potential of damaging material. Collimation concepts for LC are based on thin spoilers and thick absorbers. Collimator wakefields are an important limit and define the spoiler taper angles for compensation. In addition, the LC powerful beams demand the design of collimators allowing a certain amount of damage (consumable/renewable collimators) as well as the adoption of novel techniques such as octupole doublets for tail folding.

W. Kozanecki (Saclay) presented detector background requirements at $e^+e^-$ linear colliders. He analyzed contributions of different source mechanisms to the background rates in major detector components (silicon vertex detector, calorimeter and muon system) and discussed possible shielding strategies.

A. Seryi (SLAC) described a novel approach to collimation via nonlinear optics with focusing using an octupole doublet folding in position tails. This provides no beam outside a certain sigma (beam size). Its real effectiveness still needs to be demonstrated.

A. Seryi (SLAC) presented results of detailed analysis of collimation system performance for NLC, TESLA and CLIC under the same 0.1% beam loss scenario. It was shown that the current collimation schemes in the beam delivery systems perform well at the NLC and CLIC while there is some problem at TESLA. Work is in progress to fix this problem.

F. Zimmerman (CERN) described collimation for CLIC. CLIC has crossing angles, conflicting requirements and only one-stage momentum collimators. The collimation depth is $\pm 10\sigma_x$, $\pm 80\sigma_y$ and $\pm 1.5\%$ in momentum. Laser wire signals are dominated by particle losses. He discussed possible blow-up of a vertical beam size for collimator survival and a non-linear collimation system.

T. Markiewicz (SLAC) presented results of the GEANT3 simulation of the NLC collimation, featuring rescattering, energy deposition, synchrotron radiation and energy cutoffs, a house-made interface between MAD and GEANT. He showed difficulties and possible solutions in dealing with Bethe-Heitler muons. The MUCARLO code was used for tracking and transport of 50-GeV muons. Their fluxes at the detector are reduced by betatron and momentum collimation, and additionally by a 9-m steel shielding wall magnetized with "donuts" (tunnel magnetized spoilers). He also highlighted novel technologies with renewable and consumable collimators. Consumable collimators: rotating wheel concept or tape collimators, wheel prototype not as good as requirements (yet) but encouraging, Renewable spoilers: liquid metal, Tin and Molybdenum.

H. Schlarb (SLAC) described design and performance of the TESLA Test Facility (TTF) collimation system. The purpose is the undulator protection (damage of Nd-FeB permanent magnet), with the additional complication of high magnitude dark currents from RF. The collimation scheme uses two spoilers and three absorbers.

P. Tennenbaum (SLAC) gave an excellent overview of the collimator wakefield problem. Two effects contribute: resistive wall (the non-zero resistivity of the the wall adds a kick) and geometric wall (the change in cross section adds a kick). Results from a wakefield test box include: (i) measured/predicted (MAFIA) agreed quite well but disagreed with calculations; (ii) resistivity (copper vs. graphite) agrees with calculations.

Linear collider collimation challenges include small beam size, high repetition rate and high beam power. Detector specifications are vital to finalize collimation system design goals. There are two major components to deal with: muons and synchrotron radiation. A careful layout of beam delivery systems is necessary. Two novel technologies look promising: octupole halo folding and renewable/consumable collimators.

# Beam Loss Control on the ISIS Synchrotron: Simulations, Measurements, Upgrades

## C M Warsop

*Rutherford Appleton Laboratory, Oxfordshire, UK*

**Abstract.** The ISIS 800 MeV proton synchrotron presently provides $2.5 \times 10^{13}$ protons per pulse at 50 Hz, corresponding to a mean power of 160 kW. A dual harmonic RF system upgrade, now being installed, is expected to increase the intensity and power to about $3.75 \times 10^{13}$ ppp and 240 kW respectively. This paper describes work presently underway to understand and optimise beam loss control, which is a dominant factor determining operational performance. The main features of the collimation system are described, and Monte Carlo simulations of the loss control process are used to understand variations of efficiency with beam loss mode (growth rate, plane). Results of simulations are compared with measurements and operational data. Improvements to measurements are also outlined.

## INTRODUCTION

The ISIS synchrotron provides beam powers of up to 160 kW for neutron and muon production, with upgrades to 240 kW under way. Beam loss limits the running intensity and, although collimation systems generally work well, some problems with machine damage and planned operation at higher power have motivated a renewed study of their performance.

### Ring Parameters and Upgrades

The ISIS ring [1] has a mean radius of 26 m, and presently accumulates $2.8 \times 10^{13}$ protons per pulse (ppp), over the 130 turn, charge-exchange injection process. The injected beam is painted over both the transverse acceptances ($\sim 400\,\pi$ mm mr), and is effectively unbunched. The machine cycles at 50 Hz, accelerating protons from 70 to 800 MeV in 10 ms, using 6 ferrite tuned RF cavities which provide up to 140 kV per turn. The RF system is h=2 and the frequency ranges from 1.3 to 3.1 MHz. The ring has nominal tunes of $Q_h=4.31$, $Q_v=3.83$, which are varied through the machine cycle with AC trim quadrupoles. AC dipoles provide similar time dependent orbit correction. The machine consists of 10 super-periods, with most loss localised near the collimator/extraction straight. Fast extraction uses kickers with a rise time of ~200 ns to deflect beam into the extraction septum, before transport to the target. Upgrades, using four additional h=4 RF cavities to increase longitudinal acceptance and bunching factor [2], should allow acceleration of about $3.75 \times 10^{13}$ ppp, with similar loss levels. This increased current is required in time for the ISIS second target station, which is due to be operational by 2007.

### Ring Losses

The dominant losses on ISIS are during the non-adiabatic trapping process, when the unbunched beam is captured longitudinally within the first 1 ms of the 10 ms magnet ramp. About 8% of the beam is lost over this time, below energies of ~100 MeV. As the cycle progresses, losses rapidly tail off to 0.01%.

At the highest intensities, space charge enhances longitudinal and transverse loss during trapping, with betatron tune shifts peaking at -0.4. Although the largest contribution is expected to be momentum loss of untrapped particles, high intensity effects lead to a combination of loss mechanisms, with contributions from all three planes at a wide range of possible growth rates. The distribution and control of loss is seen to be very sensitive to many aspects of machine set up, presumably as the beam loss characteristics change. Beam dynamics and loss mechanisms vary substantially through the cycle, leading to a complex time dependent optimisation.

# THE LOSS CONTROL SYSTEMS

## Basic Systems and Upgrades

Loss control is achieved with a vertical betatron system and a combined horizontal betatron-momentum system [3]. The latter intercepts most beam and is designed principally for the expected, dominant longitudinal loss. Untrapped particles spiral radially inwards as the magnet field ramps and hit the primary jaw, located near the dispersion peak. The horizontal system operates mainly in a single stage mode, with secondary jaws included for the protection of downstream components. Secondary jaws, at optimal betatron phases, have recently been added to intercept out-scatter that may occur at high energies. The vertical system consists essentially of a primary jaw and a betatron secondary.

All collectors are located in one straight section, which covers only about 50° in betatron phase in both planes, and is immediately followed by a main lattice dipole. Out-scattered loss is seen to spread over ~180° following the primary collimators, into the downstream super-periods. Presently, due to practical constraints, secondary betatron collectors are only placed at $\theta=18°$ and not $180°-\theta=162°$, as standard practice would suggest. Collimation is at about 80% of full aperture in both planes.

Primary jaws consist of a ~40 mm upstream graphite section, followed by a shorter copper section which sits 0.1 mm closer to the beam. Particles impacting deeply interact with the graphite, with reduced activation, whilst particles with shallow impacts interact with the copper with enhanced scattering [3]. For the horizontal jaw the copper section is 15 mm, removing most protons ≤ 100 MeV in a turn. The vertical copper lip is 0.1 mm long and acts as a pre-deflector. Recently, 250 mm graphite sections have added to the primaries to enhance high energy performance. All secondary jaws are now 300 mm long (graphite) for the same reason. Extra vertical jaws have been added to allow halo scraping should it be needed, and horizontal jaws on the outside radius have been added for enhanced protection. All jaws are angled in the longitudinal direction to follow the beam envelope. Systems to measure deposited power have been added, which use flow rate and temperature change of the cooling water.

## Important Features and Problems

Good loss control is generally achieved, but still significant loss escapes into the downstream dipole and super-period. Loss is also very sensitive to the precise machine settings, and has on occasion damaged RF shields in the nearby dipole. Although redesigns of the dipole should prevent further damage, there is a need to understand loss control. During the machine design, it was expected that most particles would be lost longitudinally, at low energy and with fast radial growth rates, and thus be easily captured in a single stage system [3]. Collimation efficiencies suggested by these assumptions have not been routinely achieved at high intensity, and understanding why is the purpose of this study. Possible reasons are the appearance of loss modes at high intensity, in planes, growth rates and energies not originally expected in significant numbers. Confining betatron losses to regions of phase shift <50° of the primary jaw is difficult, particularly at slow growth rates. Collector geometry, materials, and alignment with respect to the beam are also important, and their detailed effects require study.

# LOSS MEASUREMENT

## Determination of Loss Distributions

There are comprehensive protection systems on ISIS, which monitor beam loss levels and distributions on every pulse, and trip the beam off if necessary. There are 40 beam loss monitors (BLM's) covering the entire inner circumference of the ring, 4 per super-period. These are 3 m long, argon filled coaxial ionisation chambers, that respond to the isotropically emitted evaporation neutrons generated when the beam hits machine components [4]. These give an indication of the spatial and time variation of loss - but with a strongly energy dependent sensitivity. Beam toroids, measuring the number of circulating particles through the machine cycle, can be used to calculate lost power during given intervals. The combined use of BLM's and toroids allows estimates of lost power distributions around the machine, as a function of time through the cycle.

A more detailed analysis is now being used to (a) provide a more precise knowledge of total loss with time, and (b) estimate the spatial distribution as a function of time. The former is important as it indicates when in the cycle beam loss is occurring, and the energy of lost particles. The latter is essential to understand and check the loss control process, which will vary with particle energy and potentially as a function of the time dependant machine optimisation. Analysis is taken over short 0.5 ms intervals, so that BLM sensitivity variations do not distort the distributions.

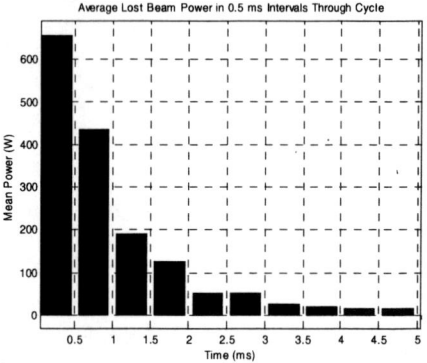

**FIGURE 1.** Measured mean beam loss power (toroid, errors about ± 50 W), in 0.5 ms intervals, for first 5 ms, normal operation.

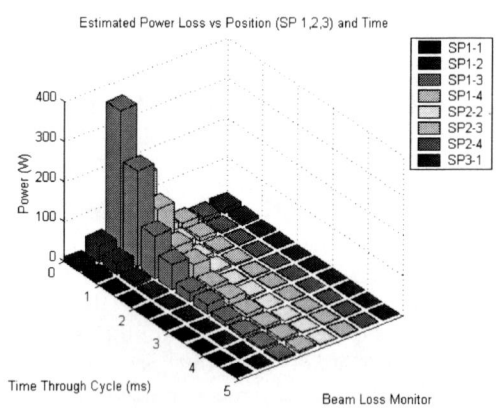

**FIGURE 2.** Measured loss distribution in space and time, for first 5 ms (toroids and BLM's), normal operation.

## Typical Results and Key Parameters

Measured loss distributions for normal operation are shown in Figures 1 and 2. Figure 1 shows the mean lost power over the first 5 ms of the cycle: most loss occurs before 2 ms (≤ 100 MeV). Large amounts of loss at higher energies would significantly affect loss control. Figure 2 shows the loss distribution at and near the collimation straight, as a function of time. Concentration of loss at the upstream end of the collector straight (SP1-3), through the cycle indicates consistent loss control.

The analysis of spatial loss distributions with time, during normal running, shows that relative distributions for 0.5 ms intervals over the first 2 ms are very similar. Averaged relative values are used for the representative measurement of loss distribution given in Table 1. BLM's are presently the best measurement of loss distribution available, and so provide essential data. However, there are some uncertainties (presently estimated at ±10%), caused by overlapping response, dependence on local geometry and materials. Deposited power measurements on primary jaws indicate relative levels of vertical and horizontal/momentum loss to be about 30% and 70% respectively.

## SIMULATIONS

### Monte Carlo Code and Approach

A Monte Carlo code, already used in studies of loss control on the ESS rings [5], has been used to model ISIS. This simulates all important proton interactions, and has been tested carefully against experimental data. A 3D representation of the collimator jaws and detailed lattice model is included. The intention here is to model, understand and control proton loss. The complex high intensity processes leading to loss are not yet fully understood, and so simplified 'loss modes' are used. This identifies the essential properties of lost beam (e.g. plane, growth rate, energy), and then studies how they affect collector performance. It is therefore the *collimation process* that is studied. Understanding how high intensity settings affect the appearance of loss modes is another essential, but separate, part of the puzzle.

### First Simulation Results For ISIS

The modes of loss on ISIS are not known in detail, therefore a set of simulations, covering a range of growth rates (10-100 μm/turn) for all three planes were carried out. Simulations were at 100 MeV, roughly where most loss occurs. This gives an indication of the dependence of collimation efficiency on the loss mode, as well as data for comparison with measurements. Simulations include a detailed model of the apertures, with beam to jaw distances similar to those measured on the machine: the effects of orbit errors, etc. have not yet been studied.

**TABLE 1.** Percentage Loss in Locations Covered by Beam Loss Monitors: Simulations and Measurement

| Super Period - Loss Monitor Plane | Growth Rate | 1-1 | 1-2 | 1-3 | 1-4 | 2-1 | 2-2 | 2-3 | 2-4 | 3-1 | rest of ring |
|---|---|---|---|---|---|---|---|---|---|---|---|
| momentum | 100 μm/turn | 0 | 0 | 85 | 11 | 3 | 0 | 0 | 0 | 0 | 0 |
| momentum | 10 μm/turn | 0 | 0 | 56 | 28 | 11 | 3 | 0 | 1 | 1 | 1 |
| horizontal | 100 μm/turn | 0 | 0 | 78 | 17 | 4 | 1 | 0 | 0 | 0 | 1 |
| horizontal | 10 μm/turn | 0 | 0 | 49 | 30 | 14 | 3 | 1 | 1 | 1 | 1 |
| vertical | 100 μm/turn | 0 | 0 | 44 | 34 | 13 | 3 | 2 | 2 | 1 | 1 |
| vertical | 10 μm/turn | 0 | 0 | 52 | 29 | 13 | 2 | 1 | 0 | 1 | 1 |
| measurement - *normal operation* | | 2 | 6 | 54 | 30 | * | 4 | 3 | 1 | 2 | 0 |

Statistical uncertainty in simulations ±1%. Uncertainty in measurements ~ ±10%, see text. [* monitor unavailable]

An example of the loss distribution as predicted by the code, for vertical loss at 10 μm/turn, is given in Figure 1: this shows loss distribution in the first three super-periods, at ~0.5 m intervals. Full results are summarised in Table 1: for comparison with measurements, losses are binned in regions corresponding to lengths of BLM's. The sensitivity of loss distribution to growth rate is important, indicating that loss control will be very dependant on details of the beam loss process.

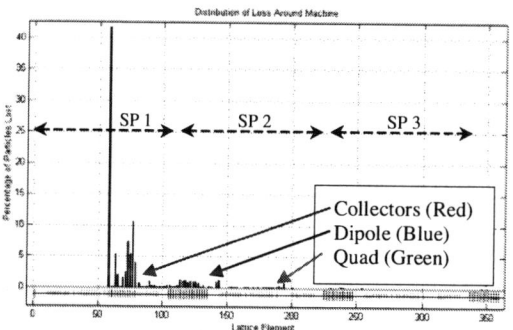

**FIGURE 3.** Loss distribution predicted by simulation in the collimator straight and neighbouring super-periods.

## Comparison with Measurements

The results of these simulations are compared with measurements for typical running in Table 1. Note that the measurements represent a sum over many possible, and essentially unknown, loss modes. There is some agreement between measured distributions and simulated results, which suggests there are no great discrepancies. Simulations at slow growth rates seem to follow measurements particularly closely. However, more needs to be known about the relationship between the measured loss distribution and the underlying proton loss. It is likely that overlap in the BLM response smears the measured distribution to some extent - further study is planned.

## SUMMARY AND PLANS

New simulations and improved measurements of beam loss on the ISIS ring are giving a better understanding of the key factors affecting loss control. Further work is required in developing and understanding beam loss measurements, possibly with new devices. More detailed experiments and simulations are now planned to quantify and understand the effects of different loss modes. The information provided should then allow the optimal collector configuration to be identified. The dependence of loss control on the high intensity beam behaviour means that more detailed control and understanding of the beam is highly desirable.

## ACKNOWLEDGMENTS

The work reported here owes much to many members of the ISIS Accelerator Division.

## REFERENCES

1. B Boardman, Spallation Neutron Source: Description of Accelerator and Target, RAL Report RL-82-006

2. C R Prior, Studies of Dual Harmonic Acceleration on ISIS, ICANS XII, 1993, RAL Report RAL 94 025

3. G H Rees et al, SNS 70-100 MeV Horizontal Beam Loss Collection, ISIS Internal Report, SNS/φ/N2/83.

4. M A Clarke-Gayther, Global Beam Loss Monitoring Using Long Ionisation Chambers at ISIS, Proc. EPAC94

5. C M Warsop, Beam Loss Control on the ESS Accumulator Rings, AIP Conf Proc CP642, 2002.

# MARS14 Collimation and Shielding Studies for the 3 GeV Ring of J-PARC Project

Noriaki Nakao*, Nikolai Mokhov†, Kazami Yamamoto**, Yoshiro Irie* and Alexander Drozhdin†

*High Energy Accelerator Research Organization (KEK), Tsukuba, Ibaraki 305-0801 JAPAN
†Fermi National Accelerator Laboratory (FNAL), Batavia, IL 60510-0500, USA
**Japan Atomic Energy Research Institute (JAERI), Tokai-mura, Ibaraki 319-1195, JAPAN

**Abstract.** MARS14 Monte Carlo simulations were performed for collimation and shielding studies of the J-PARC 3 GeV ring. A 400 MeV proton beam loss distribution, calculated with the STRUCT code, was used as a source term. The module locations in the ring and the curved tunnel sections were described by the MAD-MARS beam line builder and a deep penetration calculation with good statistics was carried out using a 3-dimensional multi-layer technique. Prompt dose-rate distributions were calculated inside and outside the concrete and soil shield, and an effective shielding design was made. The residual dose rates for various beam line materials were also calculated to estimate the external-exposures during maintenance. In this paper, the calculation results are exemplified for the region from the injection through the collimator.

## INTRODUCTION

A high energy, intense proton accelerator project (J-PARC) exists in Japan. In the facility with a few MW beam-power machine, neutron penetration in a shield, activation of materials near beam line due to secondary particles produced by a beam loss are serious problems.

In this work, simulations by the MARS14 Monte Carlo code [1] for collimation and shielding studies of the J-PARC 3 GeV ring were performed, and effective designs of the tunnel and local shields were determined based on design standards of the dose rate at J-PARC. Besides, the residual dose rates at the surfaces of machine mudules were calculated in order to estimate the external-exposures to workers during maintenance.

## J-PARC 3 GEV RING

The J-PARC facility consists of a 400-MeV linac, a 3-GeV ring, a 50-GeV ring and some experimental halls. Fig. 1 shows a horizontal cross section of the MARS14 calculation geometry for the whole beam line tunnel of the 3-GeV ring, that is the triangle shape with 3 arc regions. The circumference of the ring is 348.3 m and the position of the charge-exchange foil is set to be 0.0 m, as shown in the figure. 400 MeV protons from the linac are injected into the septum, followed by the collimator region in the first straight section. Kickers and an extraction septum are located at the second straight

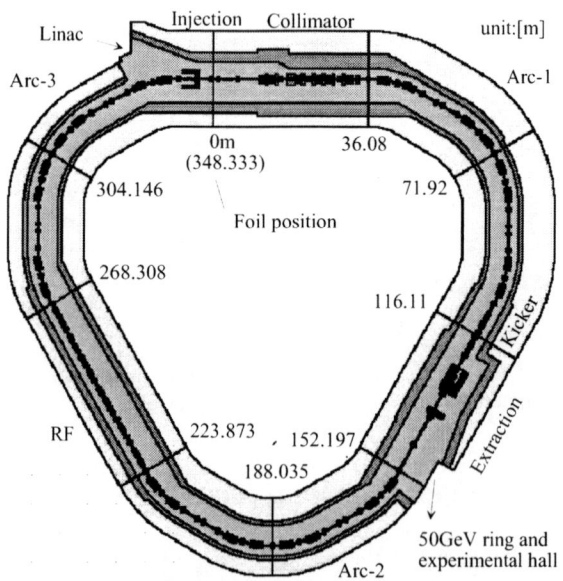

**FIGURE 1.** Overview of the MARS14 calculation geometry for the J-PARC 3 GeV ring.

section, and 3 GeV protons are transported to a 50 GeV ring or a neutron-muon experimental area. RF cavities are located at the third straight section, and 8 dipole magnets are located at each arc. The maximum beam current is scheduled to be 0.333 mA, which is ~133 kW at 400 MeV and ~1 MW at 3 GeV.

# BEAM LOSS DISTRIBUTION

Using a multi-turn tracking Monte Carlo code, STRUCT [2], 400 MeV protons of the beam halo were traced along the whole ring, and the beam-loss distribution was estimated. The beam pipe and the collimator aperture and the magnet fields were taken into account in the calculation. If the particles go outside the beam pipe boundaries, or lose more than 30% of the primary kinetic energy, the traces are terminated and particle information is stored for the MARS14 calculation. Almost all of the beam loss occurs in the collimator region based on this calculation. The loss in the collimator region was normalized to 4kW, which is 3% of the total injected beam power predicted based on the space-charge calculation with Simpsons code [3].

Possible location of the beam loss at the injection area would be at the injection septum magnet, where the incoming and the circulating beams come very close with each other. From view point of shielding design of the synchrotron vault, the loss at the septum magnet was assumed to be 1kW, which is less than 1% of the total injected beam power.

# MARS14 CALCULATION

The machine modules, such as the magnets, collimators, scrapers and peam pipes, were taken into account as calculation geometries. The locations of these modules and the curved tunnel structures in Fig. 1 were described by the help of the MAD-MARS beam line builder. The 2D-magnetic fields at the dipole and quadruple magnets, calculated with the POISSON code [4], were also taken into account.

Fig. 2 shows horizontal and vertical cross sections of the calculation geometry at the injection and collimator region along the beam axis; Fig. 3 shows the vertical cross section, which is perpendicular to the beam axis. The beam line is 11.8 m below the ground level and 1.2 m above the main-tunnel floor. A sub-tunnel is arranged below the main tunnel for the electric cables and cooling water pipes. The thicknesses of the concrete shield walls of the tunnel were determined based on the final calculated prompt dose rate at the concrete-soil boundary and the ground level. Local shields, which consist of iron and concrete, were equipped at the high beam loss areas, such as the injection septum and collimators. Massive local shields are required, especially for the upward direction, to reduce the prompt dose rate at the ground level.

Flux detectors for simulation, which are described as rectangular cells, as shown in the Fig. 3, were defined at various positions of the concrete wall and soil region to obtain the particle fluxes and prompt dose rates at four directions from the beam line. The fluxes are determined using the particle track lengths across the detector volume. The residual dose rates were also estimated at parts of the machine module surfaces and of the inner surface of the concrete tunnel wall.

In order to obtain good statistics of particles up to the ground level through 3.5-m concrete and 5.1-m soil shields with a reasonable computing time in the Monte-Carlo simulation, a three-dimensional multi-layer technique [5] was used in the MARS14 calculation. The shielding geometry was three-dimensionally divided into several layers of about 1-m thickness, and a step-by-step calculation was carried out to multiply the number of penetrated particles at the boundaries between the layers.

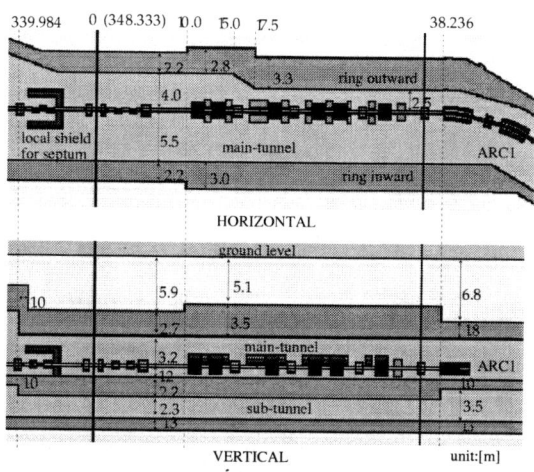

**FIGURE 2.** Horizontal and vertical cross sections of tunnel at the region from injection through the collimator.

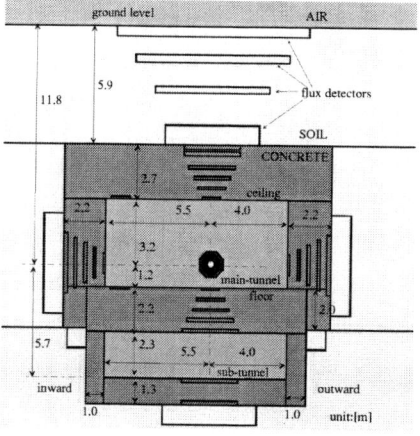

**FIGURE 3.** Vertical cross section of the beam line tunnel perpendicular to the beam axis.

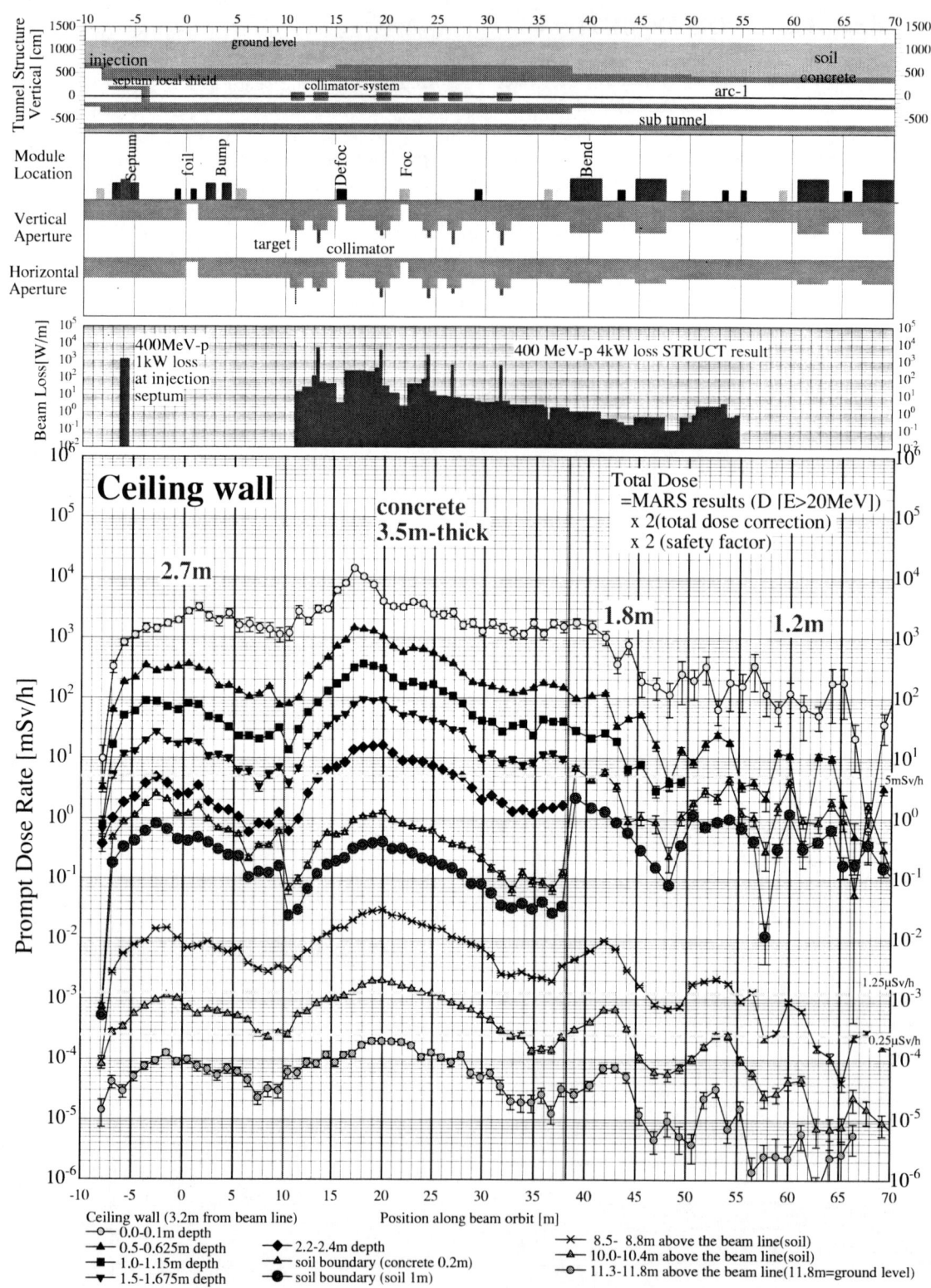

**FIGURE 4.** Vertical shield structure, machine module location, aperture structure and beam-loss distribution in the region from injection through the collimator. The calculated prompt dose rate distributions inside the shield in the ceiling wall direction (upward) are also shown in the same region.

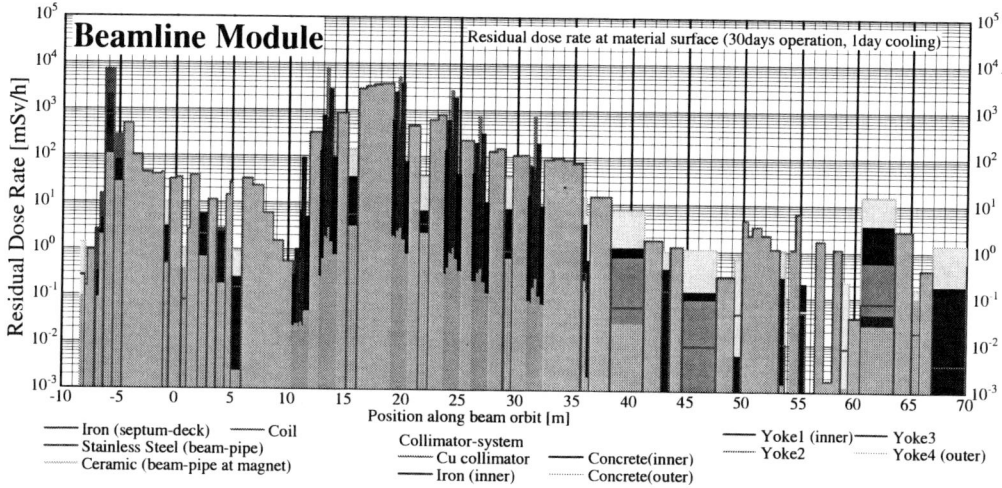

**FIGURE 5.** Calculated residual dose rate distribution at those parts of the machine module surfaces in the region from injection through collimator.

## RESULTS AND DISCUSSIONS

A 4 kW beam-loss distribution at the collimatior region, obtained by a STRUCT simulation, and a 1 kW loss at the injection, which were used as the source term in a MARS14 calculation. This distribution is shown in Fig.4, together with the vertical tunnel structure, the machine module location and the aperture structure. The calculated prompt dose rate distributions inside the concrete and soil shield in the ceiling wall direction are also shown in the figure. Although the beam losses at the positions of the septum, target and 5 collimators are dominant, the local shields at these areas attenuated prompt dose rates at the inner surface of the ceiling wall. From the dose distribution, the MARS14 results satisfy the J-PARC standard of 5 mSv/h and 0.25 $\mu$Sv/h in the soil regions at the concrete-soil boundary and the ground level, respectively.

The calculated residual dose rate distribution at the surfaces of the machine modules are shown in Fig.5 under an assumption of 30-day operation and 1-day cooling. Several parts of the modules, such as the septum coil, copper collimators, iron shields and beam pipe near the collimator, exceed 1 Sv/h, which is too high for maintenance. Therefore, a low beam-loss design [6] and/or an active handling system would be introduced.

The calculation for extraction region and the same study as well as injection reigon were also carried out and will be published elsewhere [7].

## SUMMARY

MARS14 Monte-Carlo simulations were performed for collimation and shielding studies of the J-PARC 3 GeV ring. An effective shielding design was made using the calculated prompt dose-rate distributions inside and outside the shield. The residual dose rates at the machine module surfaces were also calculated for the estimation of the external exposures.

## ACKNOWLEDGMENTS

The authors wish to thank the members at the Beams Division of the Fermi National Accelerator Laboratory (FNAL) for their great support to this work.

## REFERENCES

1. Mokhov, N. V., "The Mars Code System User's Guide", Fermilab-FN-628 (1995), Mokhov, N. V. and Krivosheev, O. E., "MARS Code Status," Fermilab-Conf-00/181 (2000), http://www-ap.fnal.gov/MARS.
2. Baishev, I., Drozhdin, A., and Mokhov, N. V., "STRUCT Program User's Manual", SSCL-MAN-0034 (1994).
3. Machida, S., *Nucl. Instr. Meth. A*, **309**, 43–59 (1991).
4. Billen, J. H., and Young, L. M., "POISSON SUPERFISH", LA-UR-96-1834 (1996).
5. Nunomiya, T., Nakao, N., Iwase, H., and Nakamura, T., "Deep-Penetration Calculation for the ISIS Target Station Shielding using the MARS Monte Carlo Code", KEK Report 2002-12 (2003).
6. Sakai, I. et al., "H$^-$ Painting Injection System for the JKJ 3-GeV high-intensity proton synchrotron", Proc. EPAC2002, Paris, France, 1040-1042 (2002).
7. Nakao, N. et al., "MARS14 Shieding calculations for the J-PARC 3 GeV RCS", KEK Report (2003), to be submitted.

# SNS Collimating System Design – Performance and Integration

N. Simos[1], H. Ludewig[1], D. Raparia[1], N. Catalan-Lasheras[3], J. Brodowski[1], G. Murdoch[2]

[1]*Brookhaven National Laboratory, Upton, New York, USA 11973*
[2]*Oak Ridge National Laboratory, Oak Ridge, TN, USA*
[3]*CERN, 1211 Geneva 23, Switzerland.*

**Abstract.** The collimating system in the accumulator ring and transfer lines of the Spallation Neutron Source (SNS) project is responsible for stopping 0.1% of the 2 MW beam of 1.0 GeV protons that are in the beam halo. The collimating structures are a combination of movable beam scrapers and stationary absorbers. Specifically, pairs of charge-exchange foils or scrapers moving in-and-out of the beam in the vertical and horizontal directions help guide the halo protons into respective absorbers which consist of an intricate design of a double wall beam tube, a water-cooled particle bed and radial shielding. Off-momentum protons, with the help of respective charge exchange foils and a dipole magnet, are directed to a momentum dump consisting of a cooled particle bed downstream of a double-walled window separating it from the vacuum space. Addressed in this paper is the thermo-mechanical response and survivability of key components of the collimating system (such as the collimating beam tube in the absorbers, the beam windows and the primary element of the bean scraper structure) in the event of intercept of the full beam under accident conditions. While the potential for the full beam to be intercepted by these components is remote, still special attention will be paid in assessing the amount of full beam (or number of pulses) they can tolerate.

## INTRODUCTION

Under normal SNS operation the collimator absorber array in the transfer lines and accumulator ring is expected to intercept 0.1% of the 2 MW, 1 GeV proton beam. The final design of the collimating scheme was aided by beam optics and energy deposition calculations performed by a synergy of particle interaction codes and thermo-mechanical analyses. Specifically, the final system consists of movable and "stationary" collimating elements integrated at key locations with charge exchange foils. The stationary collimators/absorbers consist of an intricate design of a double wall beam tube, a water-cooled particle bed responsible for stopping the incoming protons, and heavy radial shielding surrounding the particle bed. The survivability of the double beam tube intercepting halo protons over a relatively small footprint under normal operating conditions and potentially the full beam under accident conditions is of primary concern. In both beam transfer lines and in the accumulation ring there is always the potential of the full beam, due to system failure upstream (such as kicker failure) going through the double-walled Inconel-718 beam tube over the transition region (geometric) in a fashion similar to going through a beam window. It is postulated that in the event of an accident, during which more beam than anticipated is lost during charge exchange, the accelerator will be tripped after two pulses.

Under similar conditions the momentum absorber widows, which intercepts off momentum protons over two tight spots and potentially the full beam in an accident scenario, has been carefully evaluated and designed. Specifically, two decoupled thin windows, one on the vacuum side and one on the absorber side, see two distinct beam spots under normal operating conditions and a single spot under accident conditions. The latter case is a more serious one since it implies that the entire proton bunch is directed to the momentum beam stop. For both critical systems (beam tube and momentum absorber windows) special attention was paid in the material selection. Thermo-mechanical considerations and irradiation data availability led to the selection of Inconel-718 as the material of choice. The movable collimator of the accumulator ring consisting of a high-Z material that is guided into the beam halo can potentially intercept the fully developed beam pulse. Attention has been paid in evaluating the severity of the shock load that will be induced by the interception of protons during an accident scenario.

## SYSTEM INTEGRATION

Different beam collimating schemes are integrated in the transfer lines and the accumulator ring.

The design of the HEBT clean-up system is a combination of charge exchange foils and "stationary" absorbers. Figure 1 depicts the layout of the HEBT line where collimation is taking place. The actual collimating scheme consists of two pairs of charge-exchange foils moving in-and-out of the beam in the vertical and horizontal directions that help guide the halo protons into the respective absorbers. Two stationary collimators form the intercept scheme of the halo protons. The collimating elements are integrated, as shown in Fig.1, with a series of quadropole magnets that, in conjunction with the charge exchange foils, clean-up the halo protons while generating the footprint on the collimator front beam tube transition shown in Figure 5. The guided halo protons penetrate the double wall structure and are stopped by the water-cooled particle bed approximately at the middle of the absorber. A secondary shielding envelope surrounds both HEBT stationary collimators. The off-momentum protons are directed to a momentum dump with the help of respective charge exchange foils and a dipole magnet. Figure 7 depicts the off-momentum charge exchange process. Shown are the two distinct tails of halo protons that are directed into the dump by the dipole action. The momentum dump consists of a cooled particle bed and is surrounded by shielding. It interfaces with a double wall window separating the vacuum space from the rest of the dump.

In the accumulator ring the collimating system is placed along the straight section after beam injection from HEBT into the ring and consists of a primary stationary collimator integrated with an upstream set of movable collimators or scrappers whose role is to induce enough scattering angle on the halo protons such that they are diverted into the envelope of the stationary absorber. Figure 10 depicts the integration of the movable collimator and the primary absorber as well as the surrounding shielding enveloping both systems. Detailed discussion on the design of the systems and their performance will be given in a later section. The ring collimating system is enhanced with two additional stationary collimators (secondary and tertiary) placed along the same straight section. These two absorbers are expected to intercept halo protons that are missed by the primary collimator as the proton bunches accumulate in the ring. These two absorbers are also enclosed in a secondary shielding envelope.

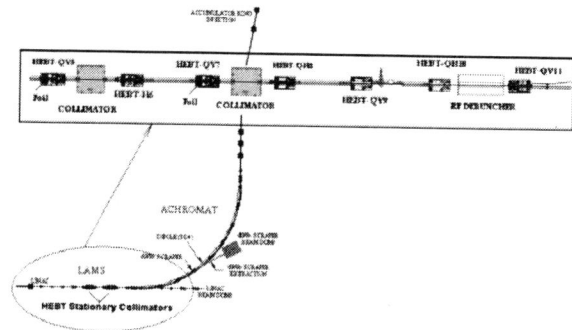

**FIGURE 1.** Collimation system integration in HEBT.

In the RTBT line which transfers the full beam to the target the clean-up system consists of two stationary absorbers that are designed to intercept protons in a failure mode. During normal operating conditions the proton pulse to the target has been shaped and cleaned in the accumulator ring. Failure of upstream optical elements (kicker magnets, etc.) will cause the shifting of the full proton pulses. The role of the RTBT absorbers is to intercept the protons that lie outside the nominal beam aperture. Safety provisions will only allow two full pulses under a system failure mode before the beam is tripped at the source.

## SYSTEM PERFORMANCE UNDER NORMAL & ACCIDENT CONDITIONS

The design of the SNS beam clean-up system has been primarily guided by anticipated accident scenarios and system failure modes. While there is great structural similarity between the stationary absorbers in transfer lines and the accumulator ring, except for the aperture and shape of the vacuum tube that traverses each absorber, the anticipated accident scenarios and the impact on the integrity of the components are different due to the structure of the proton bunches and their intensity. Figure 2 depicts the pulse structure in the HEBT and the accumulator ring. Specifically, the micro-pulses in the HEBT that form the 1ms - 60 Hz pulse structure are injected into the ring where they accumulate over many turns around the ring and are extracted from the ring as a single 1_s bunch. The RTBT sees the full 1_s bunch with 60 Hz frequency.

## Stationary Absorbers

The stationary absorber design, shown in Figure 3, consists of a vacuum tube that has its own special design according to the beam optics, a granular water-cooled structure that surrounds the beam tube and radial shielding to arrest the generated secondary particles. The integrity of the overall systems relies solely on the ability of the beam tube to maintain its structural integrity during long exposure to radiation and under accident conditions when potentially the full proton beam will be intercepted creating high temperatures and stresses within a small area. It is thus critical that a material exhibiting good mechanical strength properties and has good resistance to radiation damage must be selected. The requirements are further enhanced because of the coolant (light water) flowing on the back face of the beam tube and through the particle bed. Inconel-718 possesses both the required strength and radiation resistance that is supported by experience data. The result of all requirements was a double Inconel-718 wall separated by a gap filled with pressurized helium. The purpose of the helium is twofold. On one hand it provides a heat transfer path to the outer wall and on the other it can be detected in the beam tube vacuum space if a leak in the inner wall occurs. The double wall concept also allows for the separation of the critical inner wall from the light cooling water. In order to help the heat transfer path from the inner wall to the flowing coolant, copper wire is introduced within the helium gap as shown in the detail of Figure 4.

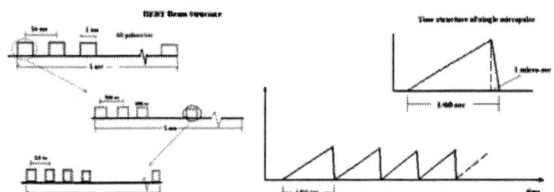

**FIGURE 2.** Proton pulse structure in HEBT and ring.

As seen in Figure 3 a very narrow cooling annulus has been introduced in the front transition section to help increase the velocity of the cooling water and subsequently the heat transfer capacity of the system around this critical location. Shown in Figure 5 is the distinct "painting" of halo protons on the transition section of the HEBT absorber beam tube. Different beam footprints will be experienced by the ring collimators. Specifically, in the primary ring collimator (whose vacuum tube follows the dynamic evolution of the proton bunch) is expected to intercept protons even over the downstream transition. In the upstream transition the interception of halo protons is governed by the scrapper position into the beam. In the RTBT line the transitions are expected to intercept protons of the "good" beam during a failure mode.

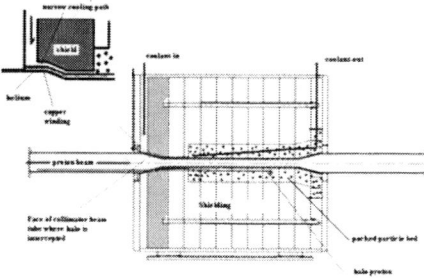

**FIGURE 3.** SNS stationary absorber cross-section (HEBT).

Thermal calculations revealed that the section shown in the details of Figure 3 and 4 is the one most severely heated and it requires higher thermal heat transfer capacity. A coolant flow of 5gpm is required in order for temperatures to remain low. Under an accident condition (conservatively assumed as two full pulses) the generated thermal stresses are quite small and pose no concern of failure in the double Inconel-718 wall. Figure 6 depicts the temperature and stress conditions at the beam tube transition region after intercepting a single pulse with full intensity.

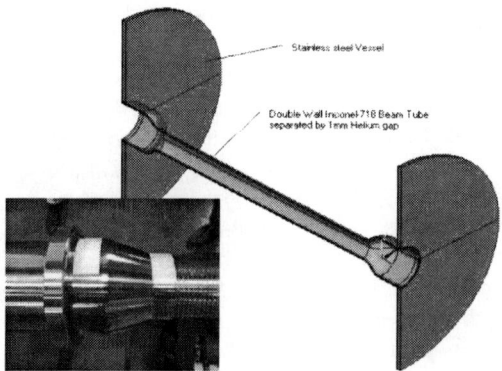

**FIGURE 4.** 3-D view of the double-walled collimator beam tube and details of the fabricated transition section.

## HEBT Momentum Beam Absorber

Figure 7 depicts the process by which the off-momentum protons in the HEBT line are diverted to an absorber situated off the HEBT achromat. This takes place with the help of the shown momentum scrapers which turn $H^-$ into $H^+$ and the action of the dipole that separates the $H^-$ from the two $H^+$ tails. The $H^+$ scraped beam is absorbed into the body of the momentum dump after going through two Inconel-718 windows that separate the vacuum space from the cooled space of the absorber. Under normal operating conditions two distinct beam spots will be seen by the two beam windows while generated heat in the vacuum window will be removed by natural convection in the space between the two windows and radiation heat exchange. The window on the absorber side is being cooled by the flowing water coolant. A postulated accident scenario calls for the interception of the entire H$^-$ HEBT beam by the momentum scraper (beam wobbling or excess movement of scraper into the vacuum space) that sends the entire H$^+$ beam to the momentum dump and over a single off-centered beam spot. Detailed finite element analyses of the two windows revealed that the structure can safely withstand two consecutive full intensity pulses.

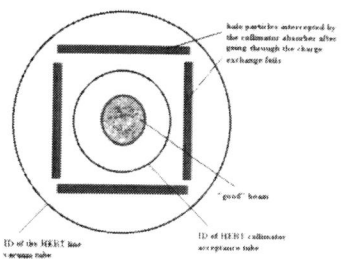

**FIGURE 5.** Footprint of intercepted halo protons in HEBT.

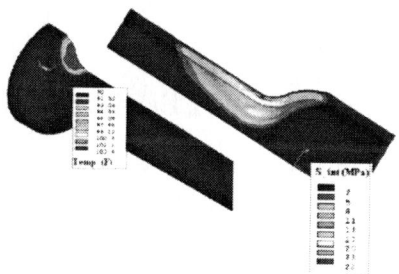

**FIGURE 6.** Thermal stress conditions in the collimator vacuum tube under accident conditions (single full pulse).

## Accumulator Ring Scraper

To assist in the interception of halo protons in the ring the concept of beam scrapers has been introduced. The idea behind it is that the protons in the halo experience small energy loss as well as large multiple Coulomb scattering deflection angle. At 1 GeV, the energy lost in a thin scraper, when compared with the total energy of the beam, is not negligible. A 1 GeV proton traversing ~1cm of heavy material may loose more than 2% of its initial energy. Potentially the scraped protons may end up outside the RF bucket or the longitudinal acceptance of the ring and be lost in high dispersion regions. Also, the scraper needs to induce enough of a kick the halo particles to drive them into the secondary and tertiary collimators. This angular kick is primarily produced by multiple Coulomb scattering (mCs), which in turn is a function of the scraper material. Figure 9 depicts the mean scattering angle for different materials while the thickness of the scrapper has been adjusted so that the average momentum loss for the 1 GeV protons is 1%. Studies have been focused on platinum and tantalum. Finally, tantalum has been selected as the scraping material for the fabricated system.

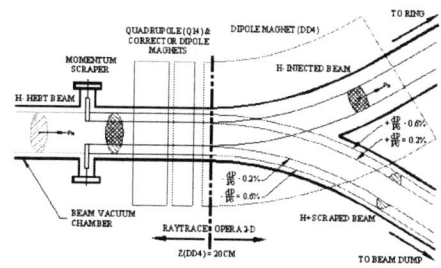

**FIGURE 7.** Off-momentum proton clean-up in HEBT.

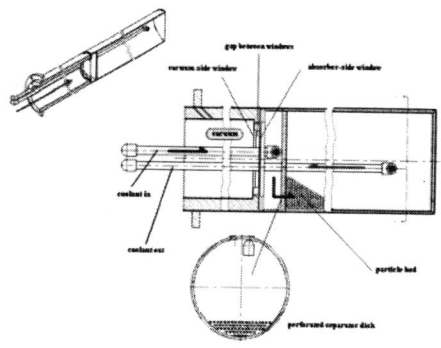

**FIGURE 8.** Design of HEBT momentum absorber.

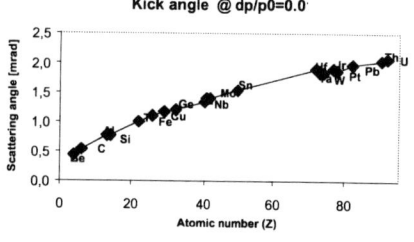

**FIGURE 9.** Scattering angles for various scraper materials.

Shown in Figure 10 is the assembly of the four (4) ring scrapers integrated with the primary ring collimator and the outer shielding structure. As seen in the details of Figure 10, the scraper material is supported by a block of copper that is being cooled continuously. Studies both for normal and accident conditions have been performed to ensure the integrity of the scraper structure. Some of the results are shown in Figure 11.

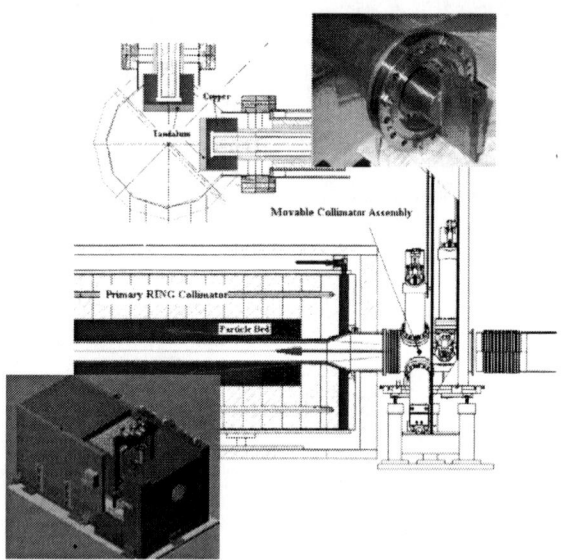

**FIGURE 10.** SNS scraper/primary collimator integration.

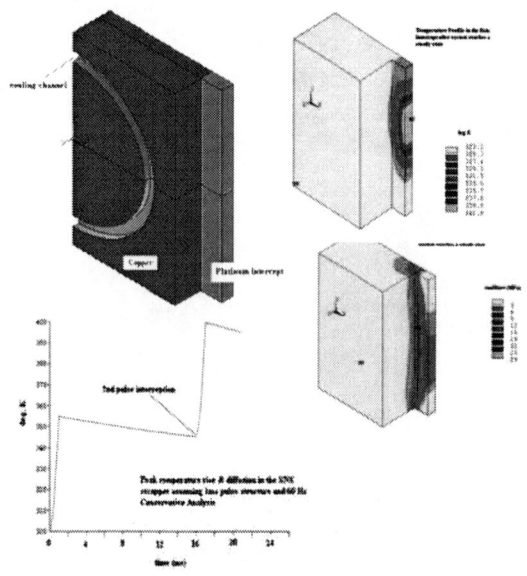

**FIGURE 11.** Scraper thermo-mechanical analysis results.

### *Particle Bed Assessment*

In the course of the conceptual design of the various elements of the collimating system, extensive use of the particle bed concept has been made and eventually introduced in the final design and fabrication. Specifically, the particle bed concept has been integrated into the stationary absorbers and the momentum dump. Given that the body of these absorbers will at times act as a target (when the full beam accidentally is directed into it), the granular form of the proton-stopping material will help it survive potential shattering and help attenuate the thermo-mechanical shock. Further, and under normal operating conditions, it is expected to enhance the heat transfer capacity of the system. Figure 12 below depicts a series of snapshots of the propagation of stress waves in a set of particles that make up the water-cooled bed. It is observed that the amplitude of stress away from the heated zone reduces significantly due to energy losses over the contact surfaces.

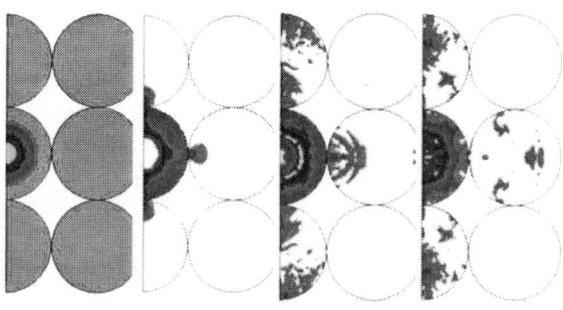

**FIGURE 12.** Stress waves in a particle bed.

## ACKNOWLEDGMENTS

Work performed under the auspices of the US DOE

## REFERENCES

1. N. Simos, *et al.*, "Thermal Shock Analysis of Windows Interacting with Energetic, Focused Beam of the BNL Muon Target Experiment", PAC2001, TPAH085, 2001
2. N. Simos, *et al.*, "Double Wall Collimator Design of the SNS Project", PAC2001, TPAH084, 2001
3. Catalan-Lasheras, N., et al., "Optimization of the Collimator System for the SNS Accumulator Ring," *Physical Review Special Topics–Accelerators and Beams*, Vol.3, 2000
4. Chapman, A.J., "Heat Transfer," 3$^{rd}$ Edition, McMillan, 1974
5. MCNPX Users Manual – Version 2.1.5, LANL, 1999
6. ANSYS Eng. Analysis of Systems, 1999

# Estimate of Dose and Residual Activity in the SNS Ring Collimation Straight

H. Ludewig, N. Simos, D. Davino*, S. Cousineau+, N. Catalan-Lasheras**, J. Brodowski, J. Tuozzolo, C. Longo, B. Mullany, and D. Raparia

*Brookhaven National Laboratory, Upton, New York, USA*
*\*Univesita' del Sannio, Benevento, Italy*
*+Oak Ridge National Laboratory, Oak Ridge, Tennessee, USA*
*\*\* CERN, 1211 Geneva 23, Switzerland*

**Abstract.** The collimation system in the SNS ring includes a two-stage collimator consisting of a halo scraper and an appropriate fixed aperture collimator. This unit is placed between the first quadru-pole and the first doublet in the collimation straight section of the ring. The entire structure is surrounded by an outer shield structure. The downstream dose to the doublet and the attached corrector magnet will be estimated for normal operating conditions. In addition, the activities of cooling water, tunnel air, and dose to cables will be estimated. The dose at the flange locations will be estimated following machine shutdown. Finally, the implied dose to surroundings during the removal of an exposed collimator will be made.

## INTRODUCTION

A two-stage halo cleaning system, consisting of a beam scraper and a fixed aperture collimator is placed between the first quadru-pole and the first doublet in the collimation straight. Halo particles that are intercepted by the scrapers undergo Coulomb scattering that either deflects them directly into the primary collimator or one of the other collimators in the collimation straight. The scrapers will be placed at approximately 140 πmm-mrad, and the fixed aperture of the collimator will be at approximately 300 πmm-mrad. The beam tube of the collimator will be elliptical with y/2=48 mm, and x/2=79 mm, respectively.

The fixed aperture collimator consists of a double walled tube manufactured of Inconel-718 [4], surrounded by a water-cooled bed of stainless steel spheres, which in turn is surrounded by a stainless steel shield. This entire configuration is contained in a pressure vessel. The space between the inner and outer Inconel-718 tubes if filled with pressurized helium and a copper wire wrap, which enhances the heat transfer from the inner tube to the outer tube. Collimators are enclosed in an outer iron shield that reduces dose to the tunnel environment during operation, and in particular reduces the dose to the scraper drive motors. The primary collimator, scraper mechanism, and the common mounting plate are shown in Fig. 1. The remaining fixed aperture collimators used in the HEBT, Ring, and the RTBT are similar to the primary collimator described above. The primary difference is that they have circular apertures (40 mm in the HEBT, 62 cm in the Ring, and 86 cm in the RTBT). The use of a particle bed makes it feasible to absorb the thermo-mechanically enhanced thermal stresses resulting from a full power beam loss on the collimators. Detailed descriptions can be found elsewhere [5-8].

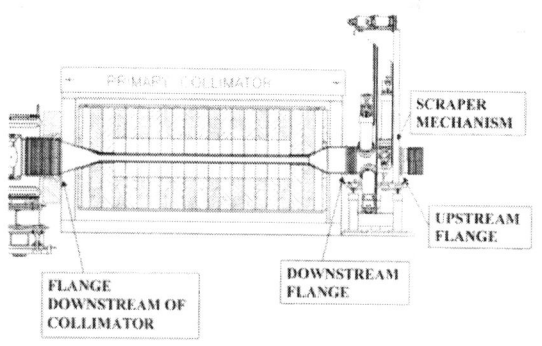

**FIGURE 1:** Beam Scraper and Fixed Aperture Primary Collimator

All fluxes, and heat depositions were determined by the Monte Carlo code MCNPX [1]. The activation estimates were carried out using fluxes determined by MCNPX, nuclear data from the CINDER90 library [2], and transmutation and gamma-ray source determinations by a suitably modified version of the ORIGEN code [3]. The activation following machine shutdown was determined by using the gamma-ray source in each volume of interest as a source input in a photon transport calculation using MCNPX.

The loss distribution of scattered particles along the inner surface of the primary collimator tube is required to determine the resulting doses to surrounding components. The distribution assumed in this study, as a percentage of the particles lost on the tube surface, is shown on Fig. 2. This loss profile is used as input to the Monte Carlo code MCNPX, which determines the fluxes of secondary particles.

## RESULTS

In these sections we will discuss the results of doses and residual activation in the vicinity of flange areas on

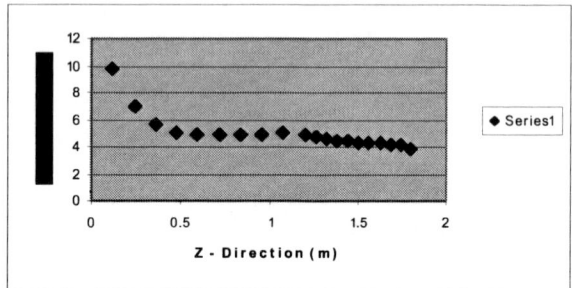

**FIGURE 2:** Primary Collimator Loss Profile

either side of the primary collimator and scrapers, to the scraper drive motors, the downstream magnets, cables, and tunnel air.

### Doses to Components Around the Primary Collimator and Scraper

During normal operation of the machine the dose for one year is given below at the flanges (both upstream and downstream of the collimator and scraper), and the downstream magnet.

**TABLE 1.** Dose to Flanges while Machine is Operating (rad/yr)

| Location Description | Collimator Dose | Scraper Dose | Total Dose |
|---|---|---|---|
| Flange upstream of scraper | 1.20(5)* | 1.71(6) | 1.83(6) |
| Flange downstream of scraper | 4.87(5) | 7.54(6) | 8.03(6) |
| Flange downstream of collimator | 2.47(7) | -- | 2.47(7) |
| Magnet downstream of collimator | 4.83(7) | -- | 4.83(7) |

*$1.20(5) = 1.20 \times 10^5$

The above dose to the flanges requires that they be radiation hardened against neutron, proton and gamma-ray radiation. The magnet placed downstream of the collimator requires that the insulation be as radiation hardened as possible. Assuming that Kapton will be used as the magnet insulating material, and assuming that its life is approximately 2(9) rads, then the magnet life under normal conditions should be ~ 40 years. Higher losses or a lower radiation damage resistance limit for the Kapton insulation, would both reduce the magnet life in proportion to the change.

Following machine shutdown the level of radiation at the flanges is of interest, since they may have to be accessed for maintenance purposes. An estimate was made of the potential dose at the flange locations following a 180 day period of operation at 2 MW average power, and with a loss fraction of 35 % of the total ring loss on the scraper-collimator assembly. The estimated dose at the four locations described above, immediately following shutdown, are given below.

**TABLE 2.** Dose to Flanges Following Machine Shutdown

| Location Description | Dose (mrad/hr) |
|---|---|
| Flange upstream of scraper | 84 |
| Flange downstream of scraper | 5025 |
| Flange downsream of collimator | 490 |
| Magnet downstream of collimator | 790 |

An estimate of the reduction in the above values following a decay period can be obtained by determining the reduction in activation for both the tantalum scrapers and the stainless steel collimator components.

**TABLE 3.** Normalized Decay of Activation of Tantalum and Stainless Steel Following Machine Shutdown

| Material | Time following machine shutdown | | | | |
|---|---|---|---|---|---|
| | 0.0 | 4 hrs. | 1 day | 7 days | 30 days |
| Stainless Steel | 1.0 | 0.776 | 0.677 | 0.537 | 0.328 |
| Tantalum | 1.0 | 0.462 | 0.284 | 0.125 | 0.061 |

These estimates indicate that the collimator will decay at a much slower rate than the scrapers. However, after a decay period of approximately one day the dose should be reduced by ~ 50 %. Finally, if the losses in the ring are concentrated at the primary collimator these doses would increase by a factor of three.

The dose to the scraper drive motors while the machine is operating is of concern, since eventually a high enough dose will compromise the motor insulation. Estimates of dose at the motor location were made, assuming that 5 % of the loss occurs on the scrapers, and an iron shield was inserted between the scraper mechanism and the motors. Two values for the shield thickness were assumed, and the corresponding doses estimated. Results are given below, and indicate that an iron shield of nominal thickness implies an acceptable motor life, assuming the same radiation damage resistance as that given above for the magnets.

**TABLE 4.** Dose to Scraper Drive Motors

| Iron Shield Thickness (cm) | Dose (rad/yr) |
|---|---|
| 15 | 1.1(5) |
| 30 | 5.5(4) |

## Collimator Cooling Water Residual Activity

An estimate was made of the residual activity in the collimator and scraper cooling water immediately after shutdown and following 4 hours of decay time. These estimates assume that the ring loss is confined to the primary collimator. Immediately following shutdown, the activity is dominated by the short half-lived isotopes $^{15}O$, $^{16}N$, $^{11}C$, and $^{12}B$. After a 4 hour decay period the dominant contributors to the activity are $^{7}Be$, and $^{14}C$. The activity of the collimator water following a 4 hour decay period is given in Table 5.

**TABLE 5.** Residual Activity of Collimator Cooling Water 4 Hours after Machine Shutdown (Curies)

| Isotope | Activity |
|---|---|
| Tritium ($^{3}H$) | 2.6(-8) |
| Beryllium ($^{7}Be$) | 6.1(-2) |
| Carbon ($^{14}C$) | 2.2(-5) |

Scraper cooling water activity is a fraction of the above collimator cooling water activity. It is to be expected that the $^{7}Be$ activity will be reduced by at least an order of magnitude and the tritium and $^{14}C$ activity will be reduced by approximately a factor of six. The primary cooling circuit, and the intermediate heat exchanger should be embedded in an appropriate shield, since the presence of $^{7}Be$ implies a source of approximately 0.5 MeV gamma-rays.

## Dose to Cables in Collimation Straight

A cable tray was assumed to be located along the wall of the ring, at an elevation of 1 m above the centerline of the beam. The maximum dose to cables was found to be opposite the primary collimator/scraper location. It was assumed that the entire loss fraction at the primary collimator is ~ 35 % of the ring total ring loss. Under these conditions it was found that the maximum dose to cable insulation will be ~ 250 rads/hour during machine operation. Assuming a dose limit of approximately $1 \times 10^8$ rads for cable material implies a life of ~ 80 years. This value will vary depending on loss fraction, radiation hardness of the cable insulation, and final position of the cable tray.

## Tunnel Air Activation

An estimate was made of the tunnel air activation in the vicinity of the primary collimator-scraper assembly. The atmosphere included oxygen, nitrogen, hydrogen, and argon. Immediately following shutdown of the machine the activation was dominated by $^{13}N$, $^{37}Ar$, $^{39}Cl$, and $^{16}N$. Following four hours of decay time the dominant activity is due to $^{14}C$. At shutdown the activity is ~ 0.06445 Curies, following the four hour decay time the activity drops to 0.00426 Curies, and then decreases gradually as the $^{14}C$ decays. These estimates were made assuming no atmospheric motion. In a real situation, the air would likely be circulated and vented following an appropriate holdup period to ensure that the short-lived isotopes have decayed. In this manner the buildup of activity can be controlled.

## Dose Between Primary and Secondary Collimators

An estimate of dose between the primary collimator, the down-stream doublet magnet, and the secondary collimator was made following machine shutdown and a decay period of four hours. The machine was assumed to operate at 2 MW for 180 days before the shutdown. In these calculations, unlike those reported on above, the tunnel walls were also included. In addition, the moveable lead shielding is assumed to be placed axially along the vacuum chamber. The results obtained for the dose estimates are given in Table 6.

**TABLE 6.** Shielded Doses Between the Primary and Secondary Collimators Following Machine Shutdown Behind Moveable Shield (mrads/hr)

| Location | Dose |
|---|---|
| Next to primary collimator (r~0.5m) | 300 |
| Next to primary collimator (r~1.0m) | 180 |
| Opposite first quadru-pole (r~0.5m) | 80 |

| Location | Dose |
|---|---|
| Opposite first quadru-pole (r~1.0m) | 60 |
| Opposite second quadru-pole (r~0.5m) | 60 |
| Opposite second quadru-pole (r~1.0m) | 80 |
| Next to secondary collimator (r~0.5m) | 400 |
| Next to secondary collimator (r~1.0m) | 250 |

These values are consistent with those reported in table 2, despite the fact that different primary beam loss profiles were assumed in the two calculations. The dose closest to the collimator faces is due primarily to the residual activity of the outer collimator shielding. The dose furthest from the collimators, and behind the moveable shield is not entirely zero, primarily due to reflections off the tunnel walls.

## Effect of Magnetic Field on Dose to Doublets

An estimate of the dose to the quadru-pole magnets, including the effect of the doublet magnetic field on the secondary particles created by the halo, will be made. The dose estimates were made using the MCNPX code, with an appropriate modification that allows tracking of charged particles through magnetic fields [9,10]. The magnetic field profiles were generated by the COSY INFINITY code [11]. The two quadru-pole magnets making up the doublet are 0.7 m and 0.55 m in length, and have magnetic fields of 0.5 T and 0.55 T respectively.

Dose estimates were carried out assuming the machine operates at 1.5 MW, with a loss fraction of 0.001 on the scrapers and the front face of the collimator. The results (Table 7) are for two faces of the quadru-pole magnet windings. Face number 1 is at 90° to face number 2. The remaining two faces are symmetrical to these two faces.

**TABLE 7.** Dose with and without magnetic field of the doublet included (mrad/s)

| Magnet/Face Number | Without Magnetic Field | With Magntic Field |
|---|---|---|
| First Quadru-pole/ face 1 | 6.0(6) | 6.7(6) |
| First Quadru-pole/ face 2 | 4.5(6) | 1.2(7) |
| Second Quadru-pole/ face 1 | 3.5(7) | 1.5(7) |
| Second Quadru-pole/ face 2 | 5.1(7) | 3.5(7) |

It is seen that the inclusion of the magnetic field increases the dose to the first magnet but decreases it to the second magnet. The overall conclusion regarding expected magnet insulation life still applies, but the importance of including either magnetic or electric fields in the trajectories of secondary charged particles and their effect on the dose, or heating, is evident.

## Dose in the Vicinity of a Partially or Unshielded Collimator

In the event that a collimator should be moved, following machine operation at 2 MW for 180 days, and shutdown for four hours, the doses in the vicinity of the collimator will vary with position and the assumed loss pattern. In this study it was assumed that the loss takes place primarily at the entrance of the collimator tube, and that the loss fraction is 0.001 of the full beam intensity.

The expected dose pattern around a collimator indicates that it is "flash-light" like in structure. At a distance beyond ~ 2 m axially downstream the dose is approximately constant radially with a flat peak along the centerline. This pattern also applies to the other end of the collimator as well. However, due to the assumed loss pattern, the values are lower by approximately two orders of magnitude. At the outer radius if the collimator inner shield the dose is ~ 0.01 rads/hr. At the axial distance of ~ 2m in front of the collimator the dose has dropped to ~ 5 rads/hr compared to a value of ~ 10,000 rads/hr inside the collimator.

## Dose Around HEBT Momentum Dump

The shielding requirements, and thus the implied doses at the HEBT momentum "dump" is of interest. The actual dump consists of a rectangular shaped water cooled stainless steel particle bed 30 cm x 30 cm x 100 cm. It has a water-cooled window at one end, and wheels for installation at the other end. The dump is surrounded by an iron inner shield 58 cm thick on all sides. The outer shield consists of concrete, which is 80 cm thick at the inlet, and 120 cm thick at the back end, and of variable thickness on the lateral sides. Values of the dose while the machine is operating at the front, back and top of the dump are given in Table 8.

**TABLE 8.** Dose at various positions around the HEBT momentum dump (rad/y)

| Lateral Concrete Thickness | Front | Top | Back |
|---|---|---|---|
| 50 cm | 2.4(5) | 1.3(6) | 1.0(5) |
| 75 cm | 2.4(5) | 4.2(5) | 1.0(5) |
| 100 cm | 2.4(5) | 1.6(5) | 1.0(5) |

It would thus be prudent to increase the concrete thickness to approximately 100 cm, in order to ensure no radiation damage induced problems with the cables following years of machine operation.

## CONCLUSIONS

The following conclusions can be drawn from this study:

(1) Dose estimates based on the current loss profile and fraction will require the exchange of the downstream magnets at least once during the life of the machine,

(2) Dose to the scraper drive motors is likely to be low enough to ensure that they will operate for the machine life,

(3) The intermediate heat exchanger for the collimator and the scraper cooling water will require a shielded pipe to reduce the intensity of the gamma-ray from the Be-7 decay.

(4) Maintenance work on the flanges will best be carried out following a decay period of at least one day to one week, in order to reduce the dose to the workers.

(5) Doses to cables in the collimation straight section and in the vicinity of the HEBT momentum dump should be low enough to allow operation for the life of the machine.

(6) Long-term tunnel air activation is dominated by $^{14}$C, which can be mitigated by a ventilation system.

(7) In the event that a collimator needs to be removed following a long period of operation, care should be taken to shield the gamma-ray shine from each end of the opened vacuum chamber.

(8) The presence of a magnetic field interacts with the secondary particles created by the halo particles to increase the dose to magnet components. Future dose estimates of operating machines should include both magnetic and electric fields.

## ACKNOWLEDGEMENTS

SNS is managed by UT-Battelle, LLC, under contract DE-AC05-00OR22725 for the U.S. Department of Energy. SNS is a partnership of six national laboratories: Argonne, Brookhaven, Jefferson, Lawrence Berkeley, Los Alamos and Oak Ridge.

## REFERENCES

[1] *MCNPX Users Manual-Version 2.1.5*, L.S. Waters, ed., Los Alamos National Laboratory, Los Alamos, NM. TPO-E83-G-UG-X-00001. (1999).

[2] W.B. Wilson, "Accelerator Transmutation Studies at Los Alamos with LAHET, MCNP, and CINDER-90", Los Alamos National Laboratory, Los Alamos, NM. LA-UR-93-3080 (1993).

[3] A.G. Croff, ""ORIGEN2 – A revised and updated version of the Oak Ridge isotope generation and depletion code", Oak Ridge National Laboratory, Oak Ridge, TN. (1977).

[4] W. Sommer, R. Werbeck, S. Maloy, M. Borden, and R. Brown' "Materials selection and qualification processes at a high-power spallation neutron source", Materials Characterization, 43, 97 – 123. (1999).

[5] H. Ludewig, N. Simos, J. Walker, P. Thieberger, A. Aronson, J. Wei, and M Todosow,"Collimation system for the SNS Ring",PAC, New York City, [1999].

[6] H. Ludewig, N. Simos, N. Catalan-Lasheras,"Preliminary estimates of dose and residual activity of selected components in the ring collimation straight of the SNS",EPAC, Vienna, [2000].

[7] N. Simos, H. Ludewig, N. Catalan-Lasheras, J. Brodowski, and J. Wei," Thermo-mechanical response of the halo intercepts interacting with the SNS proton beam",PAC, Chicago, [2001].

[8] N. Simos, H. Ludewig, N. Catalan-Lasheras, and S. Crivello, "Double-walled collimator design of the SNS project",PAC, Chicago, [2001].

[9] H. Ludewig, A. Mallen, N. Catalan-Lasheras, N. Simos, J. Wei, and M. Todosow,"Effect of magnetic fields on the dose estimates due to beam halo loss in the Ring collimation straight of the SNS",PAC, Chicago, [2001].

[10] J. Favorite et al.,Trans. Am. Nuc. Soc., Vol. 81, 1999.

[11] M. Berz, "COSY INFINITY Users manual", 1998.

# Handling High Activity Components on the SNS*
(Collimators and Linac Passive Dump Window)

G. Murdoch**, A. Decarlo**, K. Potter**, T. Roseberry**, J. Schubert**,
J. Brodowski+, H. Ludewig+, J. Tuozzolo+, N. Simos+, J Hirst++

**Oak Ridge National Laboratory, Oak Ridge, TN 37830, USA,
+Brookhaven National Laboratory, Upton, New York, 11973,
++Rutherford Appleton Laboratory, Didcot, Oxon, UK

**Abstract.** The Spallation Neutron Source accelerator will provide a 1 GeV, 1.44 MW proton beam to a liquid mercury target for neutron production. The expected highest doses to components are in the collimation regions. This paper presents the mechanical engineering design of a typical collimator highlighting the features incorporated to assist with collimator removal once it is activated. These features include modular shielding, integrated crane mounting, remote water fittings and vacuum clamps. Also presented is the design work in progress at present to validate the remote vacuum clamp design. This includes a test rig that mimics an active handling scenario where vacuum bellows can be compressed and clamps removed/replaced from a safe distance.

## INTRODUCTION

Handling high activity components can be extremely difficult if the various technical areas associated with removal of a radioactive piece of equipment are not addressed at the design stage. The collimators in the SNS accelerator by definition will receive high doses from the circulating proton beam and therefore require to be designed with a view to handling in the active state. Also, the Linear Accelerator (linac) passive dump window that separates the linac beam dump flight tube medium from the accelerator vacuum will have an H- beam impinging on it during tuning periods, removal /replacement may be required consequently active handling must be addressed.

## Collimator Design & Layout

Collimators are distributed around the ring lattice with two in the high-energy beam transport (HEBT) line, three in the ring collimation straight and two in the ring to target beam transport (RTBT) line. The ring collimation straight [1] incorporates a beam scraper and primary collimator with two tertiary collimators. Halo particles that are intercepted by the scrapers will be either deflected directly to the primary collimator or into one of the other two collimators immediately downstream in the collimation straight. The basic design of the collimators [2] consists of a double walled co-axial Inconel-718 vacuum vessel surrounded by a water-cooled bed of stainless steel balls. The space between the co-axial tubes is maintained by the addition of a copper strip wound in a helix around the outer diameter of the inner tube, this space is also filled with pressurized helium to enhance the heat transfer. This configuration limits the possibility of water to vacuum leaks; if the inner vessel wall is breached then helium will be detected in the vacuum system. The collimators will be cooled from a dedicated closed loop cooling skid incorporating features such as double contained pipe-work, shielded resin beds, pressurized back flush etc. The cooling water will be fed to the collimators via stainless steel flexible hoses attached to remote water fittings these fittings are discussed later.

Figure 1 shows the layout of the HEBT collimators and the ring collimation straight. The HEBT collimators are shown with outer shielding in place, the total assembly weight including shielding is ~50Ton. The ring collimators are shown without outer shielding.

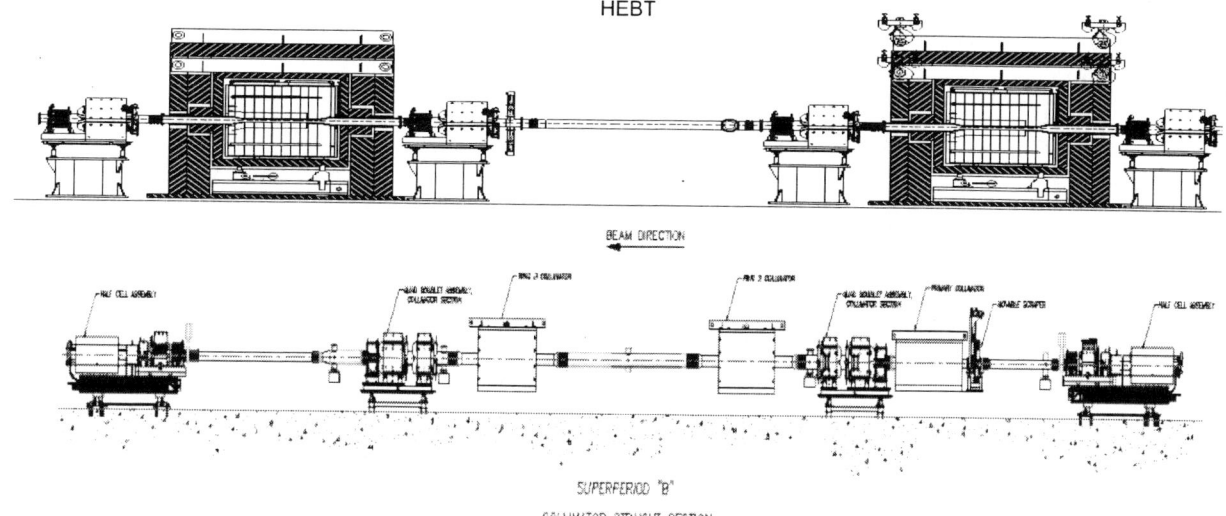

**Figure 1.**

## Collimator Active Handling

Several design features are incorporated into the collimator to minimize dose to technicians during removal when active. Three areas have been addressed; water fittings, vacuum clamps and lifting features. A remote water fitting will be designed based on an existing proven fitting [3] and it will be modified to suit the needs of the collimator arrangement. The fitting utilizes a bayonet system that incorporates a copper seal with multiple seal faces on the body of the fitting. The bayonet can be removed or engaged easily by means of long handled tongs. Once the bayonet is engaged in the fitting body a bail is swung over and a seal is made between both the bayonet and body by applying torque to the bail screw. Minimum torque is required to attain a reliable seal. Figure 2 & 3 show a cross section of the fitting that will be modified to suit the collimator geometry.

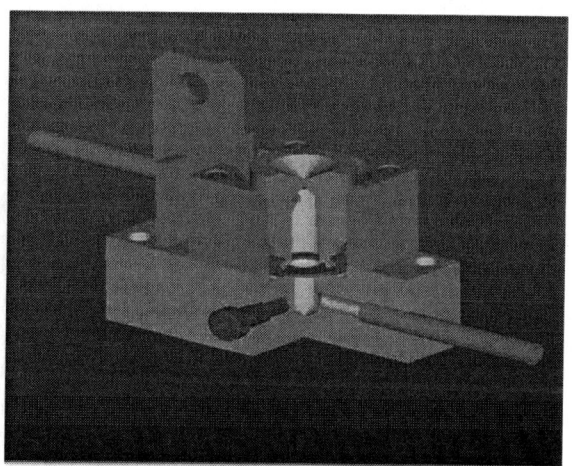

**Figure 3.**

Two remote vacuum clamp concepts have been designed and built, primarily these are to support the linac passive dump window design but one of these concepts will be chosen to upgrade to a larger clamp for use on the collimators. The first concept is based on a standard EVAC type chain clamp, the drive shoes are modified to accept a left/right hand thread that allows the clamp to be opened and closed by means of two drive shafts. The clamp arrangement is attached to a support that allows it to be positioned relative to the vacuum flange it will enclose. Dowel pins are added to each shoe to limit the sag when opened. Figure 4 shows the 3-D model of the development clamp. The second concept works on the same drive principal. The shoes are attached either by links or a thin steel band and the drive pins are offset to allow vertical removal of the flanges with minimal compression of any adjacent vacuum bellows.

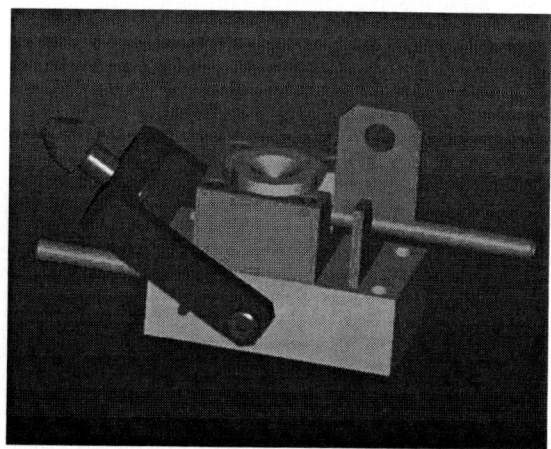

**Figure 2.**

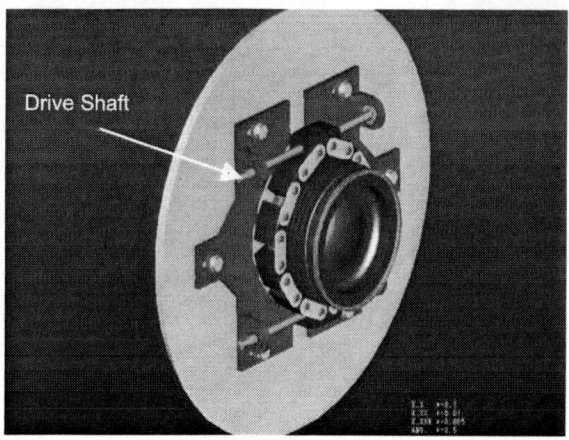

**Figure 4.**

The drive shafts are spring loaded within the support frame allowing the clamp to open and close like a clamshell around the flanges as it engages. Figure 5 shows the 3-D model of the second concept.

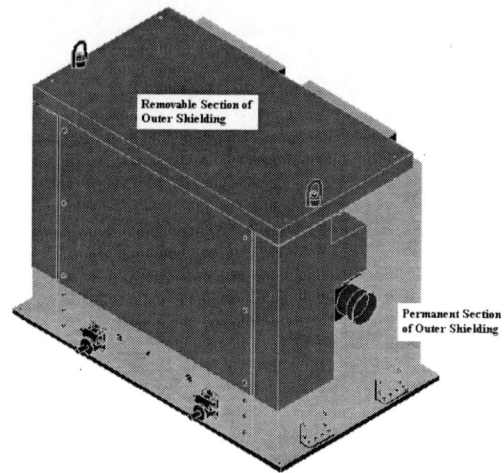

**Figure 6.**

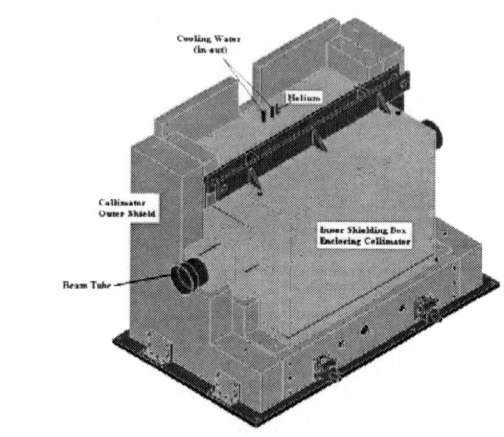

**Figure 7.**

**Figure 5.**

The collimator is designed to be modular with integrated shielding and lifting fixtures as the main handling features. The outer shielding assembly can be picked off in one operation and set aside as a freestanding unit. Once this is removed the collimator can then be accessed. An integral lifting fixture on top of the collimator allows easy crane attachment and removal from the permanent shielding. Removal of the collimator reveals three mounting jacks, these can be adjusted from the front face of the collimator. Figures 6 & 7 show the collimator with and without shielding.

## Linac Passive Dump Window

The linac passive dump window [4] separates the accelerator vacuum from the dump flight tube medium. The domed window is manufactured from Inconel-718 and has been designed to last the lifetime of the machine but in the event that replacement is required tooling has been designed to facilitate this. The window, shown in Figure 8, is incorporated into a bellows/flange assembly that will be removed as a unit. To successfully remove an active window assembly remote vacuum clamps are required as well as a bellows compression mechanism. One of the vacuum clamp designs discussed earlier will be used along with the bellows cradle, shown in Figure 9. Manipulating the drive pins on the cradle rotates the cam plates and consequently the drive plates extend or retract. The spring assisted drive plates pick up on the machined wings of the bellows assembly and either compress or extend the bellows.

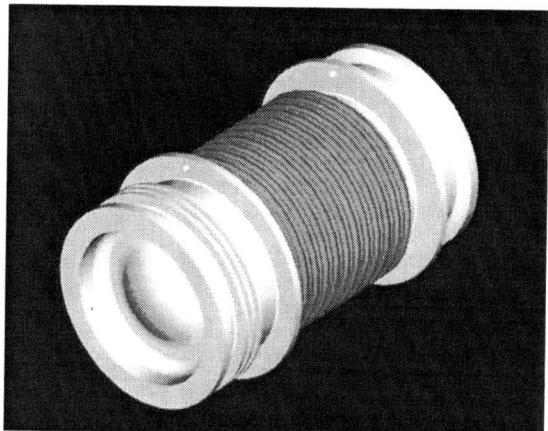

**Figure 8.**

Once the cradle has compressed the bellows the whole unit can be lifted out with the bellows assembly captive to the cradle.

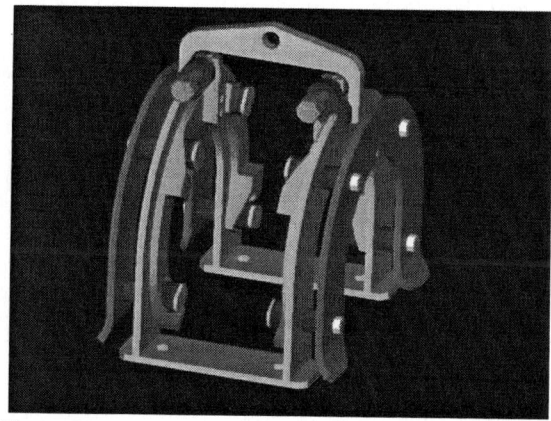

**Figure 9.**

A test rig has also been built to test and validate the remote vacuum clamps and bellows compression cradle. The test rig is designed to mimic a window change scenario. This will allow technicians to practice making and breaking vacuum seals from a distance as well as going through the procedure for removing and replacing a window bellows assembly. The 3-D model of the test rig is shown in Figure 10.

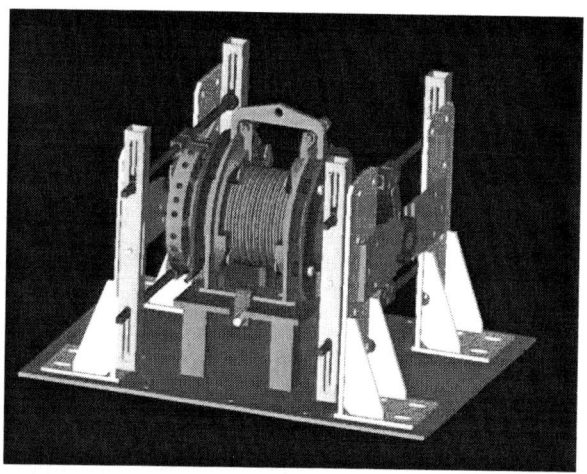

**Figure 10.**

# REFERENCES

1. Ludewig H. et al, "Integration of the Beam Scraper and Primary Collimator in the SNS Ring", These Proceedings

2. Simos N. et al, "SNS Collimating System Design – Performance and Integration", These Proceedings.

3. Hirst J., Private Communication, Rutherford Appleton Laboratory, UK.

4. Murdoch G. et al, "Beam Dump Window Design for the Spallation Neutron Source", PAC'03'.

\*SNS is a collaboration of six US National Laboratories: Argonne National Laboratory (ANL), Brookhaven National Laboratory (BNL), Thomas Jefferson National Accelerator Facility (TJNAF), Los Alamos National Laboratory (LANL), Lawrence Berkeley National Laboratory (LBNL), and Oak Ridge National Laboratory (ORNL). SNS is managed by UT-Battelle, LLC, under contract DE-AC05-00OR22725 for the U.S. Department of Energy.

# The Tevatron Collider Run II Halo Removal System

D. Still, J. Annala, M. Church, B. Hendricks, B. Kramper, A. Legan,
Fermi National Accelerator Laboratory*, Batavia, IL, 60510 USA

**Abstract.** The Fermilab Collider Run I (1994-1996) experienced limitations in the Tevatron halo removal system that motivated upgrades for the halo removal system for the Collider Run II. The upgrade provided a new 2 stage collimator design, new designs for collimators and collimator motion control incorporating loss monitor and beam intensity feedback. A central process is used to coordinate the 12 collimator microprocessors that utilize local feedback to produce an automated halo removal system. The halo removal system and experiences for the Tevatron Collider Run II will be described.

## INTRODUCTION

During the Collider Run I (1994-1996) the Tevatron halo removal system experienced limitations that prompted a complete design of a new system for the Collider Run II. The new design specified that the entire halo removal process needed to be conducted in approximately 5 min [1]. This implied that the halo removal process would have to move toward automation. The new design would incorporate a multiple 2 stage collimator design including 8 newly built 1.5m collimators and 4 targets. New motion control hardware capable of fast processing of beam loss monitor and beam intensity feedback control with motor speeds that would allow the 2 inch full travel of the collimator to take 15sec were also specified. A central control software system was also developed to coordinate the global sequence of motion for all 12 collimators while incorporating the halo removal system into the Tevatron Collider sequencer software. The Collider Run II halo removal system was installed, commissioned and operational since June 2001.

## 2 STAGE COLLIMATION SYSTEM

The Collider II Halo removal system was designed to be a 2 stage collimation system [2] which employs an "L" shaped primary collimator consisting of a 5-mm thick tungsten wing that is used to scatter particles which are then intercepted with 2 secondary collimators placed a appropriate phase advance. The primary collimator is placed a $5\sigma$ and the secondary collimators are placed at $6\sigma$ from the beam. Secondary collimators are 1.5m in length and made of stainless steel also "L" shaped. Figure 1 is an example of the placement of the collimators in the 2 stage collimator system.

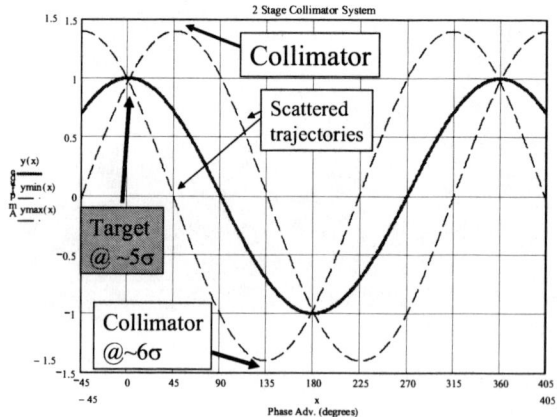

**FIGURE 1.** Placement of the target and secondary collimators to produce a 2-stage collimator system.

The Tevatron Collider II halo removal system requires 12 collimators of which there are 4 primary collimators or targets and 8 secondary collimators. The collimators are arranged in 4 sets: 2 proton and 2 antiproton sets and are installed around the Tevatron ring as shown in Figure 2. Placement of collimators in the Tevatron is limited to a few locations since there is limited warm space and the proton and antiproton beams are on helical orbits.

* Operated by Universities Research Association Inc., under contract with the U.S. Department of Energy.

CP693, *Beam Halo Dynamics, Diagnostics, and Collimation*, edited by J. Wei, W. Fischer, and P. Manning
© 2003 American Institute of Physics 0-7354-0166-7/03/$20.00

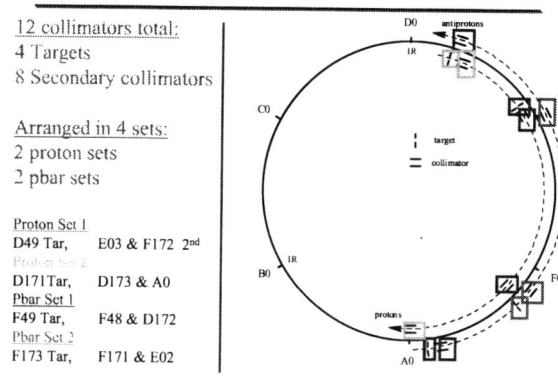

**FIGURE 2.** Tevatron Collider Run II Halo Removal Collimator Layout. CDF and D0 detectors are located at B0 and D0 respectively.

## HALO REMOVAL CONTROLS

### Hardware Controls

The hardware consists of a Motorola VME 162 processor and Advanced Controls System Corp. Step/Pac stepping motor drivers that interface to the VME processor [3]. LVDT's (linear voltage differential transformers) are used to read collimator positions. Figure 3 is the block diagram for the hardware controls for a single collimator.

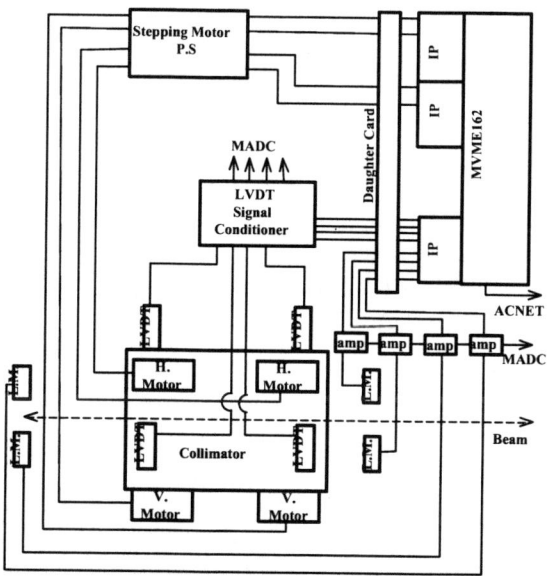

**FIGURE 3.** Block diagram of the hardware controls for a single collimator.

The Collider II collimator halo removal system was designed with the capability of incorporating feedback into the motion of a collimator. The system uses two sources for feedback. The first source is feedback from a local beam loss monitor. 4 standard Tevatron beam loss monitors and amplifiers are interfaced to the VME processor to provide loss monitor feedback. 2 of these loss monitors are used to detect losses in the proton direction and 2 in the antiproton direction. 2 loss monitors for each particle are used to provide redundant loss monitor signals in case of failure during collimator movement. Figure 4 denotes the tunnel loss monitor layout for the system. The second source of feedback comes from a beam intensity signal. A fast bunch integrator [4] is used to provide beam intensities signals for both proton and antiproton beams at a 360Hz update rate. Feedback is accomplished by encoding proton and antiproton intensity signals on to the global machine data link (MDAT). The MDAT signal is decoded by each of the VME processors at a 720Hz rate.

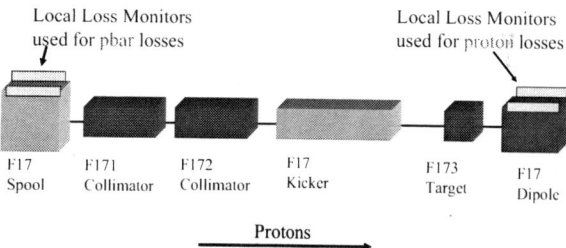

**FIGURE 4.** Tunnel layout of warm straight section that houses F17 collimators and location of beam loss monitors for feedback.

Processing the feedback internal to the VME is accomplished by sampling the loss monitor and/or beam intensity signal periodically while the collimator is moving. The smallest step the collimator can make is 1 mil. This minimum step takes 20msec to complete. A wait step occurs after the move step to provide more flexibility to timing movements. During this step, loss monitor signals and/or beam intensity signals are sampled every 4msec and are compared to a loss limit value or beam intensity percentage to remove value to decide if the collimator is to be halted for the next step. Each collimator VME front end has 17 parameters that the user can change to specify details about feedback processing.

### Software Controls

The halo removal system also utilizes software that allows global coordination of all 12 local collimator VME front ends. This global coordination software is called an open access client (OAC). An OAC is a central process that runs on a VAX and has controls hooks into the main Tevatron sequencer software [5]. The OAC employs a finite state machine that is configurable by the user to preprogram one or many collimators to complete a task on a transition of a state. For example, on the state "Goto injection positions" all collimator front ends are preprogrammed with local parameters that define their out of beam positions. The OAC owns a configurable matrix of states verse collimators and the user specifies which collimators are to move when the state is transitioned. Once the state is transitioned, all collimators will be moved back to their injection positions. There are currently 11 defined collimator states with names like: Goto injection Positions, Begin Halo Removal Scraping and Retract Proton Collimators. There is one special collimator state which is "Global Collimator Abort". A transition of this state will stop all 12 collimators immediately. Figure 5 is a block diagram of the OAC.

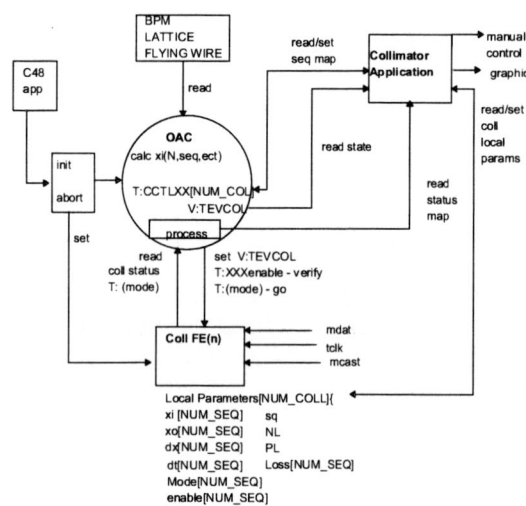

**FIGURE 5.** Block diagram of the OAC process which coordinates global movement of all 12 collimators.

# COLLIDER RUN II HALO REMOVAL EXPERIENCE

The halo removal process is conducted in the Tevatron at the flattop energy of 980gev after and the proton and antiproton beams have been brought into collisions. This process is initiated by the Tevatron sequencer software. There are 4 sub-sequence operations that are necessary in order to complete halo removal. 1) *Move Collimators to Initial Positions*: This sub-sequence moves all the collimators at 50mils/sec into the beam to the "half way" point to the beam. The motivation of this sub-sequence is to speed up the process. 2) *Intermediate Halo Removal*: Here each set (proton and antiproton) of collimators and targets are moved together under beam loss monitor feedback until a small loss is detected and all collimator in the set stop. This sub-sequence is also preformed in order to reduce the total amount of time the halo removal process takes. 3) *Perform Halo Removal*: Each secondary collimator and target is moved serially into the beam. Secondary collimators are moved under loss monitor feedback with a step size of 1 mil or .025mm until they reach the edge of the beam to shadow the losses by the primary collimator. After all secondary collimators are placed next to the beam, each target is moved under loss monitor and beam intensity feedback until 0.4% of each beam (proton and antiproton) is removed. 4) *Retract Collimators For Store*: After targets and secondary collimators have reached their final assignment, they are retracted approximately 1mm. This is the position they remain at for the duration of the store. This roughly leaves the targets and secondary collimators at the 5 and 6 $\sigma$ points discussed in the 2-stage collimation system chapter.

## Merit of Halo Removal Efficiency

The merit of halo removal efficiency is to simply record the proton and antiproton halo losses at CDF and D0 IP's before halo removal divided by the same losses recorded at the completion of halo removal. For stores 1220 through 1340, the average reduction in loss roughly due to halo removal is shown in Table 1. The fact that there is no reduction in the D0 proton halo loss is attributed to the fact that for the proton direction the CDF IP acts as an addition collimator to reduce proton halo losses at D0.

**TABLE 1. Merit of Halo Removal Efficiency**

| Halo Loss Counter at CDF or D0 IP | Factor of reduction of halo losses after halo removal |
|---|---|
| CDF proton halo loss | 9 |
| CDF anti proton halo loss | 28 |
| D0 proton halo loss | 1 |
| D0 antiproton halo loss | 100 |

# Tuning Halo Removal System for Maximum Efficiency

During the Collider Run II at the beginning and through out stores, the Tevatron would experience periods that recorded more proton and antiproton halo losses at the IP's. In attempts to minimize these losses, studies with the collimation system were conducted in attempts to understand the necessary parameters for minimizing halo losses at the IP's. Two procedures were developed from these studies that are used routinely to verify proper operation for the collimator system. The first procedure is to verify that target and corresponding secondary collimators roughly have the correct 5 and 6 $\sigma$ retraction from the beam. Targets are left into the beam while each secondary collimator is retracted to the out of beam position. Halo losses vs. collimator losses are plotted and a local minimum will denote the optimal location to place the secondary collimator. This is repeated for all secondary collimators. Figure 6 is an example of a scan using the E03 horizontal secondary collimator with the corresponding target at D49.

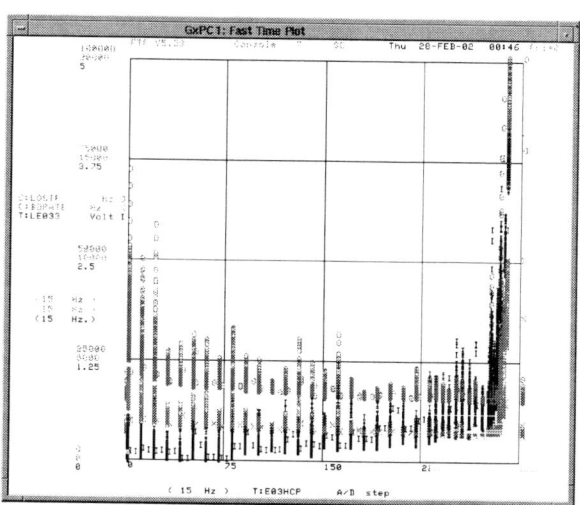

**FIGURE 6.** Retraction scan of E03 horizontal secondary collimator to verify correct position with respect to D49 target. Plot is proton halo losses vs. E03 collimator position.

Collimators are originally retracted 1 mm from the beam after halo removal is complete and this roughly satisfies this relationship. The second procedure is to verify that the collimator to beam is parallel. A collimator that is grossly misaligned with respect to the beam can cause a factor of 1.5 to 2 times larger halo losses. By moving each end of the collimator independently as a function of the local beam loss monitor, alignment can be accomplished. It has been found that collimators with angles less that $\pm$ 166 $\mu$,rad are considered in good alignment and angles over $\pm$ 500 $\mu$,rad require realigning of the collimator.

## CONCLUSIONS

The Collider Run II halo removal system overall has been a successful redesign from Run I. The system provides adequate reduction of proton and antiproton halo losses at the CDF and D0 detectors reducing background losses. Experience during Run II has provided procedures for best maximizing the halo removal efficiency by conducting collimator scans to verify proper 2-stage operation and also performing collimator to beam alignments to ensure that the beam is parallel to the collimator. The halo removal system is a necessary and integral part of Tevatron Collider operations. The halo removal system is completely automated and benefits operations with ease of use. The entire process takes 12 min. to complete and has been completed in as fast as 7 min.

## REFERENCES

[1] J. Annala, R. Joshel, "Collider Shot Setup for Run II Observations and Suggestions", FERMILAB-TM-1961, Mar 1996. 11pp.

[2] N. Mokhov et.al, "Tevatron Run-II Beam Collimation System", Proc. PAC 1999, Fermilab-Conf-99/059.

[3] Church M., An Update on the Tevatron Collimator System for Run II", in 7th ICFA Mini-Workshop on High Intensity High Brightness Hadron Beams, Lake Como, Wisconsin, ed N. Mokhov, W. Chou (1999).

[4] G. Vogel, B. Fellenz and J. Utterback, "A Multi-Batch Fast Bunch Integrator for the Fermilab Main Ring," BIW98.

[5] "Tevatron Sequencer Software Help and Explanation" http://wwwbd.fnal.gov/webhelp_edit/pa1028/pa10281_1.html

# Beam Loss and Collimation at LHC

J.B. Jeanneret*, O. Aberle*, I.L. Ajguirei[†], R. Assmann*, I. Baishev[†], J-P. Bojon*, L. Bruno*, E. Carlier*, E. Chapochnikova*, E. Chiaveri*, B. Dehning*, S. Fartoukh*, A. Ferrari*, B. Goddard*, J.M. Jimenez*, D. Kaltchev**, V. Kain*, I. Kourotchkine[†], H. Preis*, F. Ruggiero*, R. Schmidt*, P. Sievers*, J. Uythoven*, V. Vlachoudis*, L. Vos* and E. Vossenberg*

*CERN, Geneva, Switzerland
[†]IHEP, Protvino, Russia
**TRIUMF, Vancouver, Canada

**Abstract.** After a short review of past collimation work at LHC, the conception and baseline of the LHC collimation system are described. Abort-gap cleaning and beam loss monitoring are also discussed.

## 1. INTRODUCTION

A bit more than ten years ago, collimation in proton colliders was still in a kind of prehistory. Collimation was used mainly to limit the background to the experiments. This changed with the SSC and LHC projects, where stored beam energies opened a new scale for loss control and machine protection. Beam induced quenches became an issue which deserved a quantitative approach. The emphasis was then focused on the optics of a two-stage collimation system [9, 10]. For LHC, a long straight section was early on dedicated to the betatron collimation system [1, 2]. Two codes were developed, STRUCT for SSC and K2 for LHC, in order to quantify precisely the residual losses associated to a collimation system [3, 4]. These codes combined scattering at the edge of a jaw and tracking around the ring. They allowed to compute beam loss densities along the ring in the presence of collimators. Using the cascade codes MARS and FLUKA [5, 6], peak energy and power densities were simulated and compared to quench limit calculations [7]. Low-Z materials were obvious good choices for collimator jaws. With their large radiation length they accept substantially higher rates than high-Z ones. The best candidates were (and still are) graphite and beryllium. The first one was rejected for poor vacuum properties and a potential for dust release. Beryllium was discarded for toxicity reason, even if the use of massive blocks in vacuum is not really problematic. Eventhough as little as 5% of a proton bunch can already damage a piece of copper at 7 TeV in LHC in case of frontal impact, it was initially decided to use aluminum and copper, for primary and secondary collimators respectively. The dogma was to rely on safe and clever operation. It was considered that modern control systems would allow to detect early enough any drift away from safe predefined conditions. More

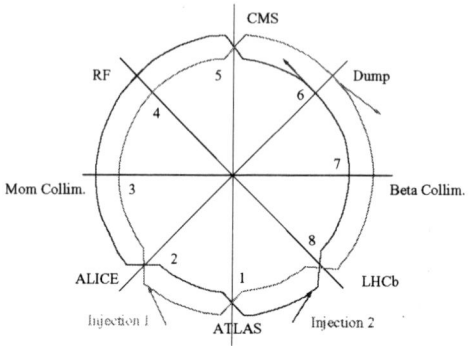

**FIGURE 1.** A schematic layout of the LHC ring

recently, after a review of the failure rate of the dump kicker error, which induce severe beam losses, the probabilities of such events was raised substantially. Then, operational and machine integrity issues were considered in a more balanced way. A substantial revision of the collimation system was therefore initiated recently [11, 12], together with the design of a structured machine protection scheme [13]. Finally, the use of additional single turn beam absorbers associated with the injection and dump systems is presently studied coherently with multi-turn collimation [14].

## 2. BASIC CONCEPTS

In order to avoid the early dump of a beam during injection, ramping or steady collisions, it is mandatory to avoid a quench resulting from bad but not rare conditions. In Table 2, transient and steady losses are compared to quench limits [7]. The efficiency of collimation must be $\eta < 10^{-4}\,\mathrm{m}^{-1}$. The quantity $\eta$ is the rate of losses per meter of ring at an aperture limitation divided by the primary loss rate. This high efficiency requires the use of a two-stage collimation system [8]. The op-

**TABLE 1.** Some LHC beam parameters.

| | | | |
|---|---|---|---|
| Luminosity | $10^{34}$ | $cm^{-2}s^{-1}$ | |
| $\sigma^*$ at crossing | 16 | $\mu m$ | $\beta^* = 0.5$ m |
| Nominal bunch | $1.05\ 10^{11}$ | protons | |
| Stored beam | $3\ 10^{14}$ | protons | 2800 bunches |
| Beam energy | 450 | Gev | Injection |
| Beam energy | 7000 | Gev | Collision |
| Injected energy | $2\ 10^6$ | J | $24 \times 4$ kg Cu * |
| Stored energy | $340\ 10^6$ | J | $2 \times 800$ kg Cu |
| Loss rate | $3\ 10^9$ p/s | 3 kW | $\tau_{beam} = 30$ hr |
| Peak loss rate | $10^{11}$ p/s | 100 kW | $\tau_{beam} \sim 1$ hr |

* Melted copper

**TABLE 2.** Expected regular transient (top part of the table) and steady (bottom part) losses compared to quench limits. At injection we consider the loss of 5% of an injected batch, associated with an error of the damping of the injection oscillation. At ramping, 10% of the beam may be off-bucket because of RF phase errors. At collision, degraded conditions, which require some time to be corrected, may lower the beam-lifetime down to $\tau = 1$ hour. All these values are indicative.

| Case | $\Delta N$ [p] | $\Delta N_q$ [p/m] | $\eta = \Delta N_q / \Delta N$ [1/m] |
|---|---|---|---|
| Injection | $1.25\ 10^{12}$ | $10^9$ | $8\ 10^{-4}$ |
| Ramping | $3\ 10^{13}$ | $2.5\ 10^{10}$ | $8\ 10^{-4}$ |
| | $\dot N$ [p/s] | $\dot N$ [p/m/s] | $\eta = \dot N_q / \dot N$ |
| 7 TeV | $8\ 10^{10}$ | $8\ 10^6$ | $10^{-4}$ |

**TABLE 3.** Correlated phase advances between primary and secondary jaws $\mu_x$ and $\mu_y$ and $X - Y$ jaw orientations $\alpha_{Jaw}$ for three primary jaw orientations $\alpha$ and four scattering angles $\phi$ with $\mu_o = \cos^{-1}(n_1/n_2)$.

| $\alpha$ | $\phi$ | $\mu_x$ | $\mu_y$ | $\alpha_{Jaw}$ |
|---|---|---|---|---|
| 0 | 0 | $\mu_o$ | - | 0 * |
| 0 | $\pi$ | $\pi - \mu_o$ | - | 0 * |
| 0 | $\pi/2$ | $\pi$ | $3\pi/2$ | $\mu_o$ * |
| 0 | $-\pi/2$ | $\pi$ | $3\pi/2$ | $-\mu_o$ * |
| $\pi/4$ | $\pi/4$ | $\mu_o$ | $\mu_o$ | $\pi/4$ |
| $\pi/4$ | $5\pi/4$ | $\pi - \mu_o$ | $\pi - \mu_o$ | $\pi/4$ |
| $\pi/4$ | $3\pi/4$ | $\pi - \mu_o$ | $\pi + \mu_o$ | $\pi/4$ |
| $\pi/4$ | $-\pi/4$ | $\pi + \mu_o$ | $\pi - \mu_o$ | $\pi/4$ |
| $\pi/2$ | $\pi/2$ | - | $\mu_o$ | $\pi/2$ |
| $\pi/2$ | $-\pi/2$ | - | $\pi - \mu_o$ | $\pi/2$ |
| $\pi/2$ | $\pi$ | $\pi/2$ | $\pi$ | $\pi/2 - \mu_o$ |
| $\pi/2$ | 0 | $\pi/2$ | $\pi$ | $\pi/2 + \mu_o$ |

* Also used for momentum collimation

## 3. JAW MATERIALS

A dump kicker error at top energy is the worst case for damage of a jaw. For quite a time, it was believed that the probability of such an event was larger than 20 years$^{-1}$. It was therefore considered that the small risk of destruction of a collimator was acceptable. This probability was revised two years ago to $\sim 1$ year$^{-1}$ and the collimator jaws had to be redesigned to survive to this kind of events. The choice of materials discussed in Section 1 was therefore revised. Two kinds of dump kicker errors may occur. An external spurious trigger may fire the whole set of 15 kicker modules, or one module may auto-trigger [18]. In both cases the beam is spread quite uniformly between the ring beam axis and the dump channel axis. In the latter case, the other modules must be re-triggered rapidly, in order to avoid that most of the circulating beam is kicked inside the aperture of the ring. The former re-triggering time was $\tau_{re-trig} = 1.3 \mu s$, with 20 bunches impacting a jaw. Even the best species of graphite would suffer some damage. The internal dump timing system was revisited and now offers $\tau_{re-trig} = 0.7 \mu s$, with the impact of 8 bunches only, see table 4. FLUKA [6] and ANSYS simulations indicate that good species of graphite can stand this impact. The second best material is beryllium. Because of its large Young modulus and in spite of its smaller atomic number, it is already eight times worse than graphite, in terms of ultimate tensile strength, see table 4. Beryllium can be envisaged only for those jaws which cannot be touched by dump kicked beam (approximately a half of them) while some injection failure modes remain to be fully explored. All other materials must be discarded, see Table 5. An absorber (TCDQ) must be installed behind

tics of a two-stage two-dimensional collimation system is designed by considering that protons scattered out of a collimator occupy the whole (x',y') space, even if large values are unlikely. This 'stochastic coupling' imposes the use of several secondary jaws per primary collimator (in LHC four of them), in order minimize the size of the secondary halo. In a true 2D-2-stage collimation system, an optimum is obtained with well defined correlated transverse betatron phase advances between primary and secondary collimators [10]. If the primary aperture defined by the primary jaws is $n_1 \sigma_\beta$ and the secondary one defined by the secondary jaws is $n_2 \sigma_\beta$, all the phases and the normalized skew angles of the jaws are expressed with either the angle $\mu_o = \cos^{-1}(n_1/n_2)$ or a rational fraction of $\pi$, see Table 3. At ramping, momentum collimation must be used [9, 10]. Conflicting optics requirements imply to use separate insertions for betatron and momentum collimation even if their optics are similar [10, 17]. Figure 1 shows the location of the cleaning insertions in the LHC ring. In an insertion of finite length, the best correlation of the phase advances can never be reached. It can at best be optimized. The location and the transverse tilt of the jaws are calculated numerically [15, 16]. The optics of the two insertions of LHC are discussed in [17]. They are presently revised, in order to satisfy new requirements related to impedance considerations, see Section 6.

**TABLE 4.** Number of bunches impacting the collimator jaws in case of dump errors. The quantity $\tau_{re-trig}$ is discussed in the text. SDF stands for 'survival deficit factor', a value which is related to the ultimate tensile strength. A good case is SDF < 1.

| Case | Nb. bunches on jaw | SDF Graphite | Beryllium |
|---|---|---|---|
| Auto-trigger | | | |
| $\tau_{re-trig} = 1.3\ \mu s$ | 20 | 1.6 | 12 |
| $\tau_{re-trig} = 0.7\ \mu s$ | 8 | 0.7 | 5 |
| External trigger | 5 | 0.4 | 3 |

**TABLE 5.** Radiation length and temperature increase after impact of 8 nominal bunches spread between 5 and 10 $\sigma_\beta$ on a jaw for some materials.

| Material | Z | $X_0$ [cm] | $\Delta T_{max}$ [°C] |
|---|---|---|---|
| Beryllium | 4 | 35.3 | 310 |
| Graphite | 6 | 18.8 | 800 |
| Aluminum | 13 | 8.9 | 2700 |
| Titanium | 22 | 3.6 | ≥ 5000 |
| Copper | 29 | 1.4 | ≥ 5000 |

the dump kickers at a phase advance of $\mu = \pi/2$ and at a depth of $\sim 10\sigma_\beta$, in order to protect the low-$\beta$ insertions which are located in between the dump area and the collimation insertion [19], see Fig. 1. The TCDQ will be thick enough to avoid damages of the nearby machine elements. In case of kicker error, similar beam densities will impact the TCDQ and the collimators. The former one too will thus be made of graphite. The vacuum problems expected with graphite have been studied recently. A dump device made of graphite was installed in the SPS ring, and revealed no dust release. New treatments of graphite were worked-out [20, 21]. Initial outgassing is dominated by long hydrocarbon chains, which are residues of materials used during the compaction process of the graphite block. In order to improve its vacuum properties, a heat treatment at 1000 °C burns and expels most of these chains. Later regular in-situ bakeout at 300°C exhibits the usual low-atomic mass spectra. Out-gassing rates are quite similar to usual metallic vacuum components.

## 4. ABORT GAP CLEANING

Apart from the case of ramping, where uncaptured protons induce a flash of losses soon after the beginning of the ramp [22, 23], momentum collimation is needed to capture off-bucket protons which loose momentum by synchrotron radiation at top energy [24]. Long storage time of particles with large momentum offset must be avoided. Their detuning with momentum can be quite large (the momentum aperture of the ring is $\approx 6 \times 10^{-4}$)

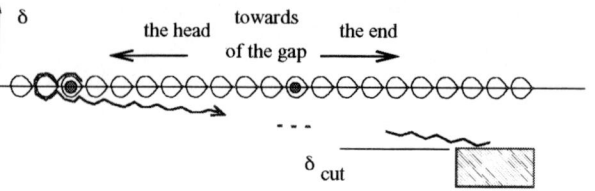

**FIGURE 2.** The longitudinal motion of a proton which left the bucket. It looses momentum by synchrotron radiation and is finally captured by the primary momentum collimator.

and thus the effective aperture may differ from the nominal one for these. In addition, these particles creep along the bunch structure and invade the abort gap. If their density is too large there, a quench will occur in the magnets downstream the dump system even during regular dump actions. The phenomenon is similar to the dump error dicussed in Section 3. A detailed description of this effect is in preparation [25] and is illustrated in Figure 2. The peak density in the abort gap is given here by the very simplified expression

$$\hat{\rho}_0 \simeq 0.7 \frac{N_0}{\tau_{long} L_{ring}} \frac{\delta_{cut}}{\dot{\delta}} = 2.2\ 10^7\ \text{p/m}, \quad (1)$$

with $N_0$ the number of stored protons (see Table 1), $\tau_{long} = 10$ h a somewhat low longitudinal beam lifetime, $L_{ring} = 26660$ m, $\dot{\delta}_{cut} \simeq 10^{-3}$ the momentum cut made by the momentum collimation system at top energy and $\dot{\delta} = U_0 f_r / E_{beam} = 10^{-5}$ the momentum loss per second by synchrotron radiation with $U_0 = 7$ keV/turn, $E_{beam} = 7$ TeV and $f_r = 1.1\ 10^4$ Hz the rotation frequency. The coefficient 0.7 is obtained by the integration of the synchrotron motion between $\delta_{bucket}$ and $\delta_{cut}$ and by summing over all occupied buckets. This value is case specific and shall be used only indicatively, see [25] for a complete formalism. The peak density is reached at the rear side of the abort gap, because particles with negative $\delta_p$ creep forward. In our case, the density at the head is $\rho_{head} \approx \hat{\rho}_0/2 = 1.2 \times 10^7$ p/m. This value is larger than the critical $\rho_{tol} \approx 0.4 \times 10^7$ p/m, above which a quench is induced behind the dump system [19]. It is intended to make use of the transverse damper, used in an excitation mode, in order to grow the betatronic amplitude of the particles which are present in the abort gap, and thus accelerate their capture. It is very fortunate that the dangerous part of the abort gap is located at its head (this is where the dump kicker starts to rise and sprays the beam at low amplitude). The creeping protons must traverse the entire gap before reaching the head. This allows to let the damper work mostly in the central part of the gap, leaving enough time for turning on and off the excitation mode.

## 5. BEAM LOSS MONITORING

In addition to detectors which will be installed all along the ring and in every critical location, in particular near experiments, beam loss monitors will be installed near every collimator [26]. The monitors will be connected to the dump trigger system and used to check on-line if the collimation setting is correct. They may also be used, to some extent, to diagnose damage of the collimator jaws. The data recorded near collimators will not be easy to use and interpret. At high energy, the cascade developed in a jaw and in the surrounding material will induce a signal in all the monitors which are installed nearby and downstream of it. In order to understand how to use the signals, we made a preparatory simulation with MARS, which develops cascades into the entire momentum cleaning section (7 collimators and monitors), including vacuum chambers, magnets with their field, tunnel, ground, etc [27]. A primary impact map was generated with K2. The partial fluences as issued from every collimator were recorded at each monitor, allowing to build a matrix which allows to compute the normalized rate $s_i$ at every monitor as a function of the primary rate $r_i$ at each collimator. For the nominal working condition at injection energy, for $\vec{s} = \mathbf{M}\vec{r}$, $\mathbf{M}$ is equal to

| .0178 | .0    | .0     | .0    | .0    | .0    | .0    |
|-------|-------|--------|-------|-------|-------|-------|
| .4662 | 1.19  | .0     | .0    | .0    | .0    | .0    |
| .0268 | .0291 | 1.081  | .0004 | .0    | .0    | .0    |
| .0432 | .0389 | 1.085  | 1.044 | .0    | .0    | .0    |
| .0079 | .0036 | .138   | .3245 | .9891 | .0    | .0    |
| .0036 | .0017 | .03858 | .1187 | .513  | .9848 | .0    |
| .0012 | .0007 | .0099  | .0349 | .1642 | .5093 | .9445 |

Further work will include a variation of the jaw depth $n_i$ one by one, in order to map $\mathbf{M}$ as a function of $\vec{n}$. With beam, $\mathbf{M}$ may be constructed by sending a pilot bunch on every jaw one after one. With the high value of many non-diagonal terms in $\mathbf{M}$, we are not yet sure that unambiguous calculations of the loss rate on every collimator can be deduced with this approach. It may be necessary to add more counters, in order to be overdeterministic and to remove ambiguities.

## 6. BASELINE

A new baseline design was defined recently. It deserves further detailed studies and discussion. It is just outlined here. In phase 1, during the early period of operation with relaxed beam parameters (half stored current, in particular), the jaws will be made of graphite. They will not only survive dump errors, but allow for quite degraded lifetime conditions. At 7 TeV, the resistive $Z_\perp$ will be too high with all the secondary jaws at $n_2 = 7$. With $n_2 = 10$, and $n_1 = [6-8]$, yielding a secondary halo size of $A_{\text{sec}} \approx 13\sigma_\beta$, $Z_\perp \sim 330$ MΩ/m. This value is compatible with the damping strength of the Landau octupoles. The aperture of the experimental triplet must be $A_{\text{triplet}} \approx A_{\text{sec}}$, and therefore $\beta^* \approx 0.85$ m. The luminosity will thus be slightly reduced during this phase. The penetration of the EM fields in the graphite will dissipate $P \sim 400$ W in the 1m-long secondary jaws. A water cooling of the jaw is therefore mandatory, even in the absence of direct beam power deposition. In phase 2, in order to to reach the nominal performance, $n_2$ will be reduced by replacing some graphite jaws (graphite with thin film coating, graphite with higher conductivity, beryllium or other good conductors). Operational experience will tell us which jaws are not subject to impact following kicker errors and in the meantime, these alternative will be explored and tested.

## REFERENCES

1. Design study of the Large Hadron Collider, Chapter 9, CERN/91-03, 1991.
2. J.B. Jeanneret, CERN SL/EA/Note 90-01, 1990.
3. I. Baichev, et al., SSCL-MAN-0034, Dallas, 1994.
4. T. Trenkler and J.B. Jeanneret, CERN SL/Note 94-105 (AP), 1994.
5. I.L.Azhgirey, I.A.Kurochkin and V.V.Talanov, in: Ann. of XV Conf. on Charg. Part. Accel., p.74, Protvino, 22-24.10.96 (in russian).
6. A. Fasso et al., Proc. Monte-Carlo 2000 Conf., Lisbon, Springer-Verlag Berlin, p. 955, 2000.
7. J.B. Jeanneret et al., CERN LPR 44, 1996.
8. The Large Hadron Collider, CERN/AC/95-05(LHC), Chap.4,1995.
9. T. Trenkler and J.B. Jeanneret, Part.Acc., **50**, (1995) 287, January 1995.
10. J.B. Jeanneret, Phys.Rev. ST-AB, **1**, 081001, Dec. 1998.
11. R. Assmann et al., CERN LPR 599 and EPAC02, 2002.
12. R. Assmann et al., CERN LPR 640 and PAC03,2003.
13. R. Schmidt, these proceedings.
14. H. Burkhardt, B. Goddard and V. Mertens, CERN LPR 641 and PAC03,2003.
15. D.I. Kaltchev et al., CERN LPR 134 and PAC1997,1997.
16. D.I. Kaltchev et al., CERN LPR 194 and EPAC98,1998.
17. D.I. Kaltchev et al., CERN LPR 305 and PAC1999,1999.
18. R. Assmann, B. Goddard, E. Vossenberg and E. Weisse, CERN LPN 293,2002.
19. A.I. Drozhdin et al., Fermilab Project Note FN-0724,2002.
20. J-P. Bojon et al., CERN AT Note, to be issued,2003.
21. P. Chiggiato, Private communication, 2002.
22. J.B. Jeanneret, SL/Note 92-56 (EA), 1992.
23. I. Baishev et al., CERN LPR 309,1999.
24. J.B.Jeanneret, CERN/SL/92-44(EA).
25. E. Chapochnikova,S. Fartoukh and J.B. Jeanneret, CERN LPN, to be issued 2003.
26. J.B.Jeanneret et al., CERN LHC-BLM-ES-0001.00 rev 1.1, EDMS doc 328146, 2003.
27. I. Ajguirei, I.S. Baichev, J.B. Jeanneret and I.A. Kourotchkine, CERN LPN, to be issued, 2003.

# Beam loss scenarios and strategies for machine protection at the LHC

R.Schmidt, R.Assmann, H.Burkhardt, E.Carlier, B.Dehning, B.Goddard, J.B.Jeanneret, V.Kain, B.Puccio, J.Wenninger

*CERN, Switzerland, Geneva*

**Abstract.** At the Large Hadron Collider (LHC) with nominal parameters at 7 TeV, each proton beam has an energy of more than 330 MJ threatening to damage accelerator equipment in case of uncontrolled beam loss. To prevent such damage, kickers are fired in case of failure deflecting the beams into dump blocks. The dump blocks are the only elements that can safely absorb the beams without damage. The time constant for particle losses depends on the specific failure and ranges from microseconds to several seconds. Starting with some typical failure scenarios, the strategy for the protection during LHC beam operation is illustrated. The systems designed to ensure safe operation, such as beam dump, beam instruments, collimators / absorbers and interlocks are discussed.

## INTRODUCTION

To deliver proton-proton collisions at the centre of mass energy of 14 TeV with a nominal luminosity of $10^{34}$ cm$^{-2}$s$^{-1}$, the LHC will operate with high-field dipole magnets using NbTi superconductors cooled below the λ-point of helium. The most important parameters for the LHC as proton collider are given in Table 1. One of the main challenges is the safe operation with beam parameters pushed to the extreme in order to achieve design luminosity. Whereas the proton energy is a factor of seven above accelerators such as SPS, Tevatron and HERA, the energy stored in the beams is more than a factor of 100 higher. For damage of equipment the transverse energy density is the relevant parameter, a factor of 1000 higher.

The beams must be handled in an environment with superconducting magnets that could quench in case of fast losses of $10^{-8}$-$10^{-7}$ of the nominal number of protons (at 7 TeV). This value is orders of magnitude lower than for any other accelerator with superconducting magnets. Any uncontrolled release of the beam energy could cause serious damage to equipment. As an example, a fast loss of about 5 % of a single nominal 7 TeV bunch in one spot could already damage material with high Z, such as copper.

| TABLE 1: LHC Parameters | | |
|---|---|---|
| Energy at collision | 7 | TeV |
| Dipole field for 7 TeV | 8.33 | T |
| Luminosity | $10^{34}$ | cm$^{-2}$s$^{-1}$ |
| Protons per bunch | $1.1 \cdot 10^{11}$ | |
| Number of bunches / beam | 2808 | |
| Nominal bunch spacing | 25 | ns |
| Normalised emittance | 3.75 | µm |
| Beam size at IP for 7 TeV | 15.9 | µm |
| Typical beam size in arcs (rms) for 7 TeV | 300 | µm |
| Arc magnet coil inner diameter | 56 | mm |

Other accelerators operate with beams that have much less stored beam energy but where protection of equipment from beam losses is still a concern. For example, the SPS beam drilled holes through the vacuum chamber when the protection systems did not work correctly. Both at SPS and TEVATRON radiation damage of silicon detectors of the experiments occurred after beam losses.

## BEAM LIFETIME

Under optimum condition without collision the beam lifetime could exceed, say, 100 h. This would be very comfortable since the beam deposited power into the equipment is only about 1 kW. If the lifetime decreases to 10 h, the LHC would rely on collimators to capture the losses in two cleaning insertions with normal conducting magnets [1]. The collimation system is designed to accept a lifetime of about 0.2 h for a 10 s long transient, e.g. when changing the betatron tune. This corresponds to a power deposition of 500 kW. If the lifetime becomes even smaller, in particular after equipment failure, the beams will have to be dumped immediately. Depending on the type of failure, dumping the beams must be very fast (see Table 2). The dump blocks are the only elements that can absorb the energy stored in the LHC beam without being damaged.

| Table 2: Lifetime of the LHC beams | | |
|---|---|---|
| Beam lifetime | Beam power into environment (1 beam) | Comments |
| 100 h | 1 kW | Healthy operation |
| 10 h | 10 kW | Operation acceptable, collimation must absorb large fraction of beam energy |
| 12 min | 500 kW | Operation only possibly for short time, collimators must be very efficient |
| 1 s | 330 MW | Failure of equipment - beam must be dumped rapidly |
| 15 turns | Several 100 GW | Failure of D1 normal conducting dipole magnet - detect beam losses, beam dump as fast as possible |
| 1 turn | ~ TW | Failure at injection or during beam dump, potential damage of equipment, passive protection relies on collimators |

## COLLIMATORS AND BEAM ABSORBERS

Three of the eight insertions are dedicated to machine protection. One insertion is for the beam dump systems, one with collimators capturing protons with large betatron amplitudes, and a third with collimators in locations with non-zero dispersion catching protons with large momentum deviations. Additional collimators are installed in most other insertions.

In order to limit the beam losses in superconducting magnets, the LHC will be the first machine requiring collimators to define the mechanical aperture through the entire cycle. For efficient beam cleaning, the collimators are adjusted between 5-9 σ. For operating at 7 TeV, the opening between two collimators jaws is about 3-4 mm. More than 99.9% of the protons in the beam halo should be captured in the cleaning insertions. The elements that are closest to the beam are the collimator jaws. In case of failures causing fast movement of the beam, the jaws intercept particles and prevent damage of other equipment until the beams are safely extracted.

## FAILURE SCENARIOS AND PROTECTION SYSTEMS

Most likely are failures in the magnet and powering system, with about 8000 magnets powered in 1700 electrical circuits. Other failures are due to aperture restrictions, with two beams circulating through 53 km of beam vacuum pipe with the beam screen inside, helium feedthroughs, interconnects, RF shielding etc. Many vacuum valves and more than 100 collimator jaws could also obstruct the beam passage.

Beam losses due to a failure can be in a single turn, or during many turns:

- One turn failures: ultra-fast losses
- Multiturn failures: very fast losses in less that 5 ms, and fast losses in more than 5 ms
- Steady losses (one second or more), discussed in the next chapter

### One-turn failures (ultra-fast losses)

Beam can be lost in less than one turn due to a failure at injection or due to a failure when extracting the beam into the dump blocks.

<u>Injection</u>: The beam is accelerated in the SPS to 450 GeV and then sent through ~2.8 km long transfer lines to the LHC [2]. During extraction from the SPS, beam transfer and injection into the LHC, the beam could damage septum magnets or kickers.

There could also be a failure by septum magnet or injection kicker that steers the beam onto a wrong trajectory. In order to make the beam travel correctly through the first turn, the parameters of all magnets in the LHC should be set to the correct values. If one of the magnets had a wrong current value, or in case of an aperture restriction (for example due to a closed vacuum valve), the beam would be lost. Assuming the worst case that a corrector magnet close to one of the experiments is set to maximum current, the beam would go straight into the detector.

In order to avoid damage during injection it is proposed to ensure correct beam parameters in the SPS before extraction ("beam quality check"). If the parameters were not acceptable, the beam would not be extracted. Collimators in the transfer line will capture bunches that are outside the acceptable trajectory range, to avoid damage of elements in the injection region and the following sectors of the LHC [3]. Additional collimators downstream of the kicker magnet will prevent damage in the LHC in case of a mis-kicked beam.

With no beam circulating in the LHC, only injection of beam with limited intensity (=non destructive) is allowed. Beam exceeding this intensity can only be injected when there is already beam circulating [4]. The damage level at 450 GeV is about 10-20 bunches with nominal intensity.

Extraction: For a clean extraction of the beam into the dump blocks several conditions have to be met [5]:
- The beam dump kicker must by synchronised with the 3 µs long particle free abort gap.
- Since the aperture of the dump channel is tight, closed orbit errors in the dump insertion must be limited to about 4 mm.
- Energy tracking between the main dipole magnets and the elements in the beam dump system (kickers and septa magnets) must be ensured by special hardware with ultrahigh reliability.

A likely failure scenario is the pre-firing of one beam dump kicker module. The other 14 kicker modules would be immediately triggered after such failure, but about 94 bunches deflected up to 4 µs after the pre-firing would not be extracted correctly. The bunches having received the smallest kick would travel through the machine, come back after one turn and then be deflected by the beam dump kicker in the second turn. These bunches could have a large offset at the second deflection, and absorbers must ensure that no equipment is damaged. About 10 bunches would receive a kick such that they reach the cleaning insertion and hit the collimators. Bunches that are deflected with a larger angle would hit an absorber in the dump insertion (TCDQ). Some 40 bunches would hit an absorber in front of the septum magnet (TCDS).

A failure in the synchronisation between RF and abort gap has slightly less severe consequences, since the number of bunches that do not travel correctly through the beam dump channel is smaller.

The design of collimators and absorbers is a major challenge. The objects should not be damaged in case of an impact of about 10-20 bunches [1]. To stand such an event, the material for the collimators, originally Al and Cu, is being reconsidered.

During extraction, the 3 µs long gap must be free of particles. Debunching caused by RF noise, intrabeam scattering, etc. populates the beam abort gap [1]. At an energy of 7 TeV, non-captured protons will lose energy by synchrotron radiation and therefore be captured in the momentum cleaning insertion.

Whenever a failure during extraction occurs, bunches will oscillate with large amplitudes around the closed orbit. Orbit excursions exceeding, say, 4 mm limiting the aperture could lead to damage in case of unclean beam dump. Hence, the closed orbit around the machine must be well controlled [6].

## Multiturn failures (fast / very fast losses)

Failures that could drive the beam unstable are mainly quenches of superconducting magnets, of a single magnet or several magnets. The current decay in the quenching magnet is approximately Gaussian, from the maximum (corresponding to the initial current) down to zero current. Other hardware failures require fast discharge of magnets with an exponential current decay. There could also be a failure in the power converter control - for example the power converter ramps current with maximum voltage, or a wrong reference value for the current. An electric short in the coil of a normal conducting magnet is also considered. After a failure in the RF system, the beam is dumped immediately, since the beam would debunch and the abort gap would be filled with particles. A clean dump would no longer be possible. There are operational failures, and combined failures (for example after Mains disturbances).

A failure of a normal conducting magnet in two insertions with physics experiments (ATLAS and CMS) is most critical. The deflection by the D1 magnet installed at a location with a β-function of more than 4000 m leads to a fast change of the closed orbit. In [7], several failures were considered:

- Powering failure of the normal conducting dipole magnets D1 for beam separation at β = 4000 m
- Quench of the superconducting dipole magnets for beam separation at β = 2000 m
- Quench of the very strong low-β triplet quadrupole magnets at β = 4000 m with orbit offset due to the crossing angle. This will also lead to a change of tunes and betatron functions
- Quenches of superconducting arc dipole magnets

Since the collimators are very close to the beam, protons in the tails of the distribution would touch the jaws already after several turns. A failure of the normal conducting D1 magnets leads to losses within the shortest time. The loss would exceed more than $10^9$ protons after about 15 turns, and possibly damage the collimators after 30 turns (assuming that the collimators can withstand a beam loss of about $10^{12}$ protons). In Fig.1, the change of closed orbit after a failure of D1 at the position of three different collimators is given (lower part of the figure). In the upper part the number of protons touching the collimators is shown for a Gaussian particle

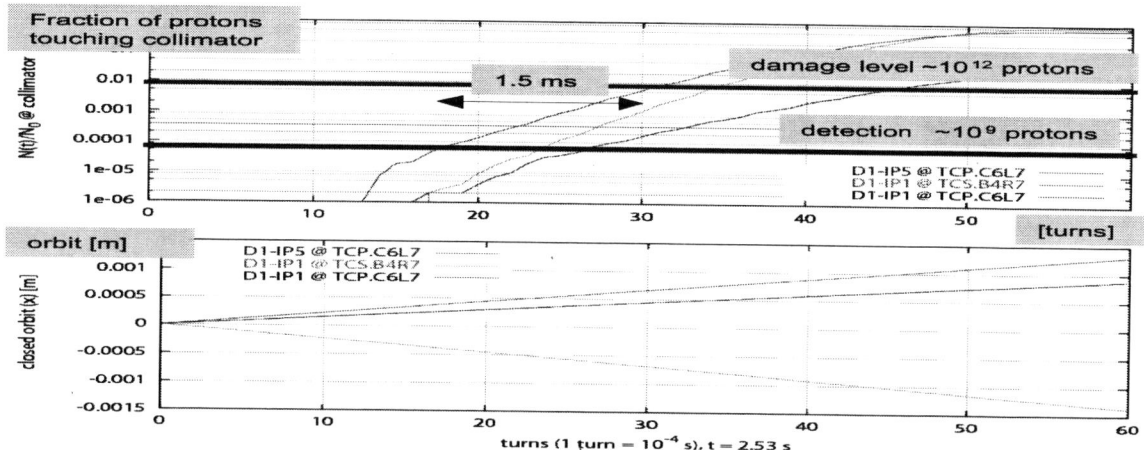

Fig.1: Beam losses (upper curve) and orbit movement (lower curve) after a failure of the D1 magnets as function of the number of turns. Three collimators that are affected most by a failure have been considered.

distribution. If the particle distribution is not Gaussian, the time between detection of the loss and damage of the collimators could be even shorter.

## STRATEGIES FOR PROTECTION

<u>Ultrafast losses</u>: The hardware of injection and extraction system should be as reliable as possible. However, it is known from other accelerators that such failures cannot be completely excluded. Protection relies on collimators and beam absorber that need to be correctly positioned with respect to the beam to capture the particles.

<u>Very fast losses</u>: For failures that lead to very fast beam movements, the beam losses close to aperture limitations (collimators, beam absorbers, low-beta quadrupoles, etc.) would be detected. The beam is dumped if the threshold is exceeded. It is also proposed to detect rapid beam position changes. If beam orbit movements exceed a predefined value, say, about 0.2 mm / ms, the beams are dumped. Such a system could detect failures earlier than beam loss monitors and is a redundant system for protection.

<u>Fast Losses</u>: Beam loss monitors around the machine and signals from equipment in case of hardware failure for many systems will be used to generate beam dump requests and complement fast beam loss and beam position monitors.

<u>Steady losses</u>: the beam losses and heat load at collimators will be monitored. A beam dump in case of unacceptable lifetime is also being considered.

<u>Other options</u>: During the most critical part of operation at 7 TeV with squeezed optics to $\beta^* = 0.5$ m, the triplet magnets focusing the beams into the interaction point have a very tight aperture. Absorbers in front of the triplet magnets would protect the superconducting quadrupoles in case of failure.

Critical elements for the protection of the LHC are the TCDQ collimators in the dump insertion. The TCDQ must be adjusted to about 10 σ from the beam. For protection against dump failures a jaw on one side would be sufficient. However, it is suggested to install absorbers from two sides in order to avoid a large orbit offset that would compromise the TCDQ functionality.

As it has been shown, a failure of the D1 magnets would lead to very fast orbit changes. Increasing the inductance in the electrical circuit with D1 magnets by about a factor of five, possibly with a superconducting solenoid in series with the magnets, would increase the time constant for orbit changes and relax the parameters for the protection system.

## REFERENCES

1. Jeanneret, J.B. et al., "Collimation at LHC", these proceeding and Assmann, R. et al, "Designing and Building a Collimation System for the High-Intensity LHC Beam", PAC 2003, Portland, USA, May 2003

2. Hilaire, A. et al., "Beam Transfer to and injection onto the LHC", EPAC '98, Stockholm, June 1998

3. Burkhardt, H., "Do we need collimation in the transfer lines?" in *Chamonix XI*, March 3-8, 2003, CERN-AB-2003-008 ADM, April 2003

4. Schmidt, R. and Wenninger, J., "LHC Injection Scenarios", LHC Project Note 287, CERN, March-2002

5. Goddard, B., "Apertures During Beam Abort", in *Chamonix XI*, March 3-8, 2003, CERN-AB-2003-008 ADM, April 2003

6. Wenninger, J., "Orbit Control for Machine Operation and Protection", in Chamonix XI, March 3-8, 2003, CERN-AB-2003-008 ADM, April 2003

7. Kain, V., "Studies of equipment failures and beam losses in the LHC", Diploma thesis, Wien, October 2002

# Collider and Detector Protection at Beam Accidents[1]

I. L. Rakhno,[2] N. V. Mokhov, A. I. Drozhdin

*Fermilab, P.O. Box 500, Batavia, IL 60510, USA*

**Abstract.** Dealing with beam loss due to abort kicker prefire is considered for hadron colliders. The prefires occured at Tevatron (Fermilab) during Run I and Run II are analyzed and a protection system implemented is described. The effect of accidental beam loss in the Large Hadron Collider (LHC) at CERN on machine and detector components is studied via realistic Monte Carlo calculations. The simulations show that beam loss at an unsynchronized beam abort would result in severe heating of conventional and superconducting magnets and possible damage to the collider detector elements. A proposed set of collimators would reduce energy deposition effects to acceptable levels. Special attention is paid to reducing peak temperature rise within the septum magnet and minimizing quench region length downstream of the LHC beam abort straight section.

## INTRODUCTION

An accidental beam loss caused by an unsynchronized abort launched at abort system malfunction, can cause severe damage to a collider equipment. Such a malfunction can be initiated, *e.g.*, by a spontaneous high voltage discharge in a kicker generator module or high energy cosmic particle crossing a sensitive element of the abort system trigger. Statistical data accumulated at Tevatron for Run I and Run II indicate that such prefires happened, on the average, a few times a year.

In the LHC a single prefired kicker module induces coherent beam oscillations with an amplitude up to $21\sigma$ of the beam at collisions. Simulations show that if this happens at the top energy, starting from 70-80% of the kicker strength, the misbehaved beam ends up in the IP5 inner triplet causing destruction of its components and damage to the CMS detector near-beam elements [1]. To avoid this, the other kicker modules are fired immediately after the prefired one (thus producing a full, unsynchronized, abort), but this does not prevent beam loss completely. A set of stationary collimators for the IP5 has been proposed in [1] to protect its inner triplet against irreversible consequences of a fast beam loss. Alternatively, a movable collimator TCDQ, in IP6 as close to the cause as possible, has been proposed in [2] to protect the entire LHC machine. There were two major unresolved energy deposition problems associated with the effect of an unsynchronized beam abort in the LHC IP6 [3]: (i) a peak temperature rise at the upstream end of the septum magnet MSD exceeds the limit of 100°C; (ii) a peak energy deposition in superconducting (SC) coils downstream of the Q5 quadrupole remains above the quench limit in more than 50% of the magnet string resulting in an unacceptably severe quench. The solutions to these problems has been recently proposed based on the updated LHC beam optics (version 6.4) and detailed MARS [4] calculations.

## TEVATRON

For Tevatron Run I (from December 1993 till February 1996) there were 10 abort kicker prefires (AKP) and "fails-to-fire", while for the initial stage of Run II (from March 2001 till November 2002) there were 7 documented AKP. The average rate of one AKP per about three months is quite high taking into account recovery of the machine after severe quench of the SC magnets and possible damage to sensitive electronics. For example, three recent AKP in Tevatron caused quenches of several SC dipoles in the A-sector and BØ region and 2-Gy instantaneous dose in the CDF central detector that gave rise to damage of silicon ladders.

To protect the CDF detector and Tevatron components in an event of AKP, an existing collimator at the A11 straight section will be used and a new collimator is to be installed at the A48 location during the summer 2003 shutdown [5, 6]. The A11 collimator will protect the Tevatron dipoles. Detailed calculations with the MARS [4] and STRUCT [7] codes have shown that a 0.5-m long steel collimator at A48 will intercept one of

---

[1] This work was supported by the Universities Research Association, Inc., under contract DE-AC02-76CH03000 with the U.S. Department of Energy.
[2] rakhno@fnal.gov

36 proton bunches when such an incident occurs, providing reliable protection of the CDF main detector at an AKP. It will also mitigate the backgrounds induced by elastic beam-gas interactions upstream of BØ. Although the Roman Pot detectors downstream of the A48 collimator will see an increased background, the amount of radiation they will receive either resulting from beam halo interactions in the collimator or during an AKP will not damage their sensitive parts.

A bunch lost at A48 during an AKP represents more than $2 \times 10^{11}$ protons "instantaneously" interacting with the collimator material. Although the A11 and A48 collimators protect the downstream SC dipoles against a damage in such an event, secondaries generated in them create a significant radiation load on the dipoles which will likely result in a quench of the first one. Optimization of the A48 collimator based on detailed MARS simulations [6] allows to reduce the peak energy deposition density in the SC coil from 17 to 3.5 mJ/g (see Figs. 1 and 2). Due to the space constraints at this location, this is the maximum achievable reduction. The radiation load is reduced, but the first dipole is still subject to quench because the above values exceed the quench limit of 0.5 mJ/g per pulse at fast beam loss [8].

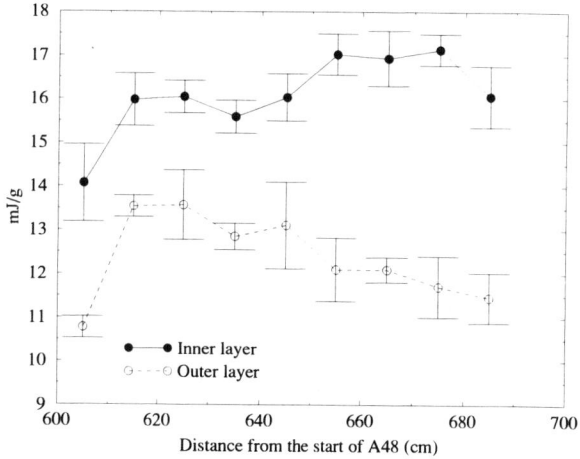

**FIGURE 1.** Energy deposition in the inner and outer SC coils of the Tevatron A48 first dipole for the baseline 0.5-m stainless steel collimator.

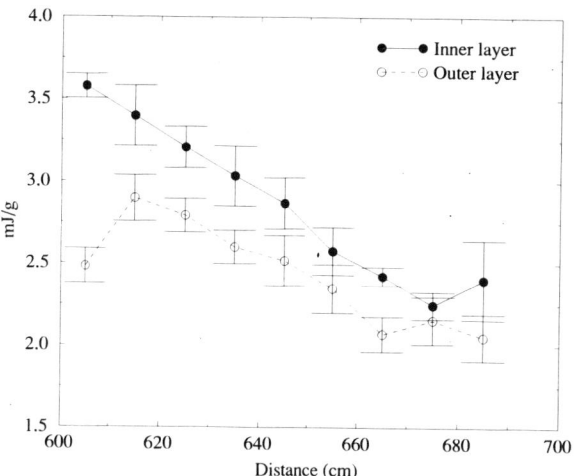

**FIGURE 2.** Energy deposition in the horizontal plane of the inner and outer SC coils along the Tevatron A48 first dipole for the final collimator arrangement: a composite A48 collimator (0.3-m tungsten followed by 0.37-m stainless steel) with a 0.34-m tungsten mask with a round 2.5-cm radius aperture placed immediately upstream of the first dipole.

## LHC

### IP6 Model and Beam Parameters

A central part of the current LHC IP6 calculation model, optimized over years in thorough MARS calculations, is shown in Fig. 3. A stationary 5-m long rectangular graphite collimator TCDS is placed at 0.1335 mrad with respect to the circulating beam axis. Its width increases gradually from 24.5 to 25.2 mm when going from a non-IP facet to IP one. The facets are placed at 14 and 14.9 mm from the circulating beam axis, respectively. A composite 9.5-m long graphite (8 m) and aluminum (1.5 m) collimator TCDQ is placed at a radial position of 9.1 mm, corresponding to $8\sigma_x$ of the circulating beam at collision energy of 7 TeV, plus orbit deviations. It is movable, *i.e.* the jaws are retracted at injection to accommodate a larger beam size.

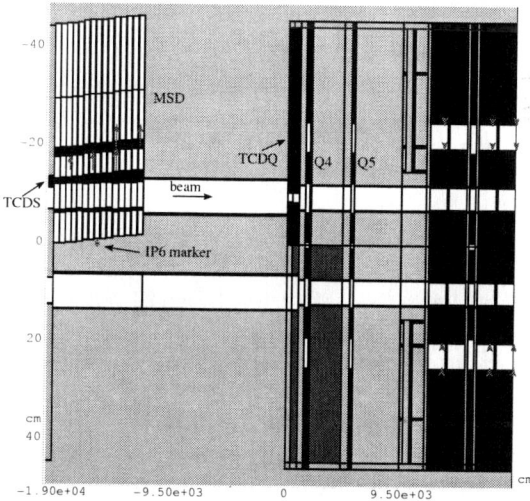

**FIGURE 3.** Central part of the IP6 MARS model.

The beam parameters used are described in Table 1. In our calculations, we assume that bunches are distributed uniformly around the lattice and there is only one 3 $\mu$s abort gap in the circulating beam. To cover the abort kicker rise time of 3 $\mu$s (see Table 1), 280 bunches are considered which correspond to 7 $\mu$s. It enables us to investigate effect of delay time $\tau$, i.e. time elapsed between prefiring the single kicker and firing the other ones. Results presented below were obtained for a kicker strength $B \cdot l$ = 63.05 kG·m, which corresponds to an angle $\alpha$ = 0.27 mrad at 7 TeV.

**TABLE 1.** LHC beam parameters used in the study.

| | |
|---|---|
| Proton energy | 7000 GeV |
| Normalized transverse emittance ($\sigma$) | 3.75 mm·mrad |
| Protons per bunch | $1.05 \cdot 10^{11}$ |
| Number of bunches | 2835 |
| Total intensity | $3 \cdot 10^{14}$ |
| Horizontal crossing angle in the IP5 | 150 $\mu$rad |
| Bunch separation (10 RF buckets) | 24.95 ns |
| Abort gap (127 missing bunches) | 3.17 $\mu$s |
| Number of abort kicker modules | 14 |
| Abort kicker rise time | 3 $\mu$s |

**TABLE 2.** Instantaneous peak temperature rise $\Delta T$ (C) in the collimators and MSDA1 dipole at baseline luminosity.

| Module | Delay time $\tau$ ($\mu$s) | | |
|---|---|---|---|
| | 1.2 | 3.0 | 4.0 |
| TCDS (5 m) | 698 | 680 | 694 |
| MSDA1 (4.5 m) | 522 | 450 | 504 |
| TCDQ1 (4 m) | 456 | 810 | 1170 |
| TCDQ2 (4 m) | 155 | 246 | 348 |
| TCDQ3 (1.5 m) | 5 | 14 | 34 |

## Temperature Rise in IP6 Components

The results of previous MARS [2, 3] calculations are summarized in Table 2. A single prefired kicker is not strong enough to deflect the beam significantly for it to hit the TCDS collimator. Deflected bunches hit the collimator when the other fired kickers attain a given strength. That is why almost no dependence of instantaneous peak temperature rise in the TCDS collimator and MSD magnet on delay time is observed in the results. The peak temperatures in the TCDS and TCDQ graphite are well below the shock wave limit of about 2200°C. The Tevatron beam abort dump operates for more than 20 years with the peak temperature in the graphite core of 1000°C. At the same time, the peak temperature rise in the septum magnet MSDA1, immediately downstream of the 5-m graphite TCDS collimator, is unacceptably high. Heating a magnet up to about 110°C with subsequent cooling down gives rise to irreversible changes in magnetic properties and field homogeneity [9]. The problem can be solved by increasing the graphite length by 1 m and adding at the TCDS downstream end a 0.5-m long iron section of the same profile as the graphite one. The resulting peak temperature distribution along such a composite 6.5-m TCDS collimator and MSDA magnet is shown in Fig. 4. One sees that the maximum instantaneous temperature rise in the MSDA1 at an unsynchronized beam abort is about 12°C which is quite acceptable. This value almost does not depend on the delay time.

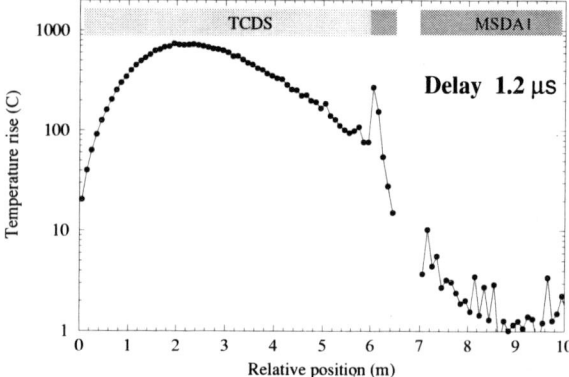

**FIGURE 4.** Peak temperature rise along the TCDS collimator and MSDA1 magnet at baseline luminosity.

## Minimizing Quench Region Length

The SC magnets in the IP6 and farther in the machine are reliably protected against destruction at an unsynchronized beam abort by the TCDQ collimator proposed in Ref. [2]. There is no easy way to avoid quench of the first SC quadrupoles Q4 and Q5, but what matters is the length of the quenched string afterwards. It is required to limit the number of quenched dipoles to less than 50% of the string. The peak energy deposition density $\varepsilon_{max}$ in the SC coils should be compared to the quench limit that can be estimated as 0.5 mJ/g per pulse for the LHC magnets at fast beam loss ($\leq$ 1 ms) [8]. To reduce peak energy deposition $\varepsilon_{max}$ downstream of Q5, two 2-m steel masks were implemented. The first one, with the aperture of $20\sigma_{col}$, is downstream the Q5 quadrupole and second one, with aperture of $21\sigma_{col}$, is in front of the MBA1. The masks enable one to limit the quench region to one ($\tau$ = 1.2 $\mu$sec) or two ($\tau$ = 4 $\mu$sec) dipoles with other magnets downstream remaining in the superconducting state (see Fig. 5). It reduces $\varepsilon_{max}$ in the MBA1 magnet by almost a factor of ten. Unfortunately, the value of $20\sigma_{col}$ is less than $10\sigma_{inj}$ and that requires for the two masks to be movable ones.

Larger apertures of the masks, if they are fixed, do not give such a significant protection effect. Therefore, other options should be considered in further studies to avoid using movable masks, provide $\varepsilon_{max}$ in the Q4, Q5, MBA1 and MBB1 as low as possible and mitigate collimator alignment problem. The possibilities are to split the 9.5-

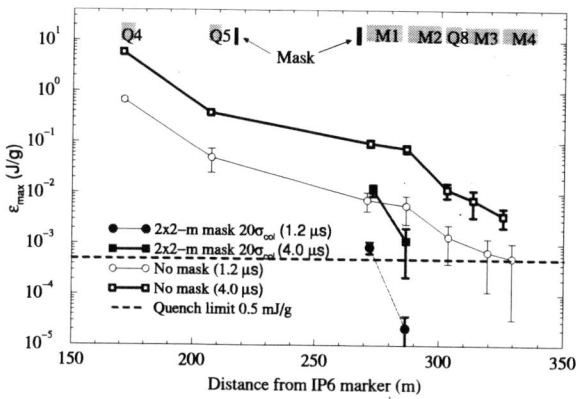

**FIGURE 5.** Peak energy deposition density $\varepsilon'_{max}$ in SC coils *vs.* distance from the IP6 marker, where M1 - M4 denote the first four dipoles (MBA1, MBB1 etc).

m TCDQ collimator into two shorter sections with about 10 m between them, and to use a V-shaped aperture for a stationary TCDQ with separate passes for injected and 7-TeV beams by means of a constant field dogleg.

The radiation loads in the rest of the machine downstream of the M4 dipole are well below the limits, with undetectable beam loss rate at all critical locations.

## Radiation Levels in Tunnel

An unsynchronized beam abort gives rise to instantaneous irradiation in the tunnel. Distributions of equivalent dose can be useful to estimate dose load to monitoring electronics during such an accident and determine the most dangerous regions. The distributions due to a beam abort have been calculated with MARS in the tunnel in the vicinity of the vessel and near the farthest concrete wall. The extracted beam is supposed to be directed to an external beam dump and, therefore, contribution to dose in the tunnel due to the beam was not considered. According to the calculations, the hottest spots in the tunnel are near the TCDS and TCDQ collimators: for a 1.2-$\mu$s delay time the instantaneous dose in the vicinity of the vessel equals to 800 and 100 Sv, respectively. The instantaneous dose around other components in the region is substantially lower: 0.1 to 1 Sv.

## CONCLUSIONS

The performed studies revealed that for both Tevatron and LHC the proposed collimators ensure reliable protection against destruction (melting down) of the SC coils and detector components due to abort kicker prefire, and drastically reduce the overall radiation loads. However, in both the cases the first SC magnets immediately downstream of such collimators will quench. The number of such magnets is one at Tevatron and seven at LHC. Possibilities to reduce the number of the magnets subject to quench are under investigation. The studies revealed that, with the LHC collimators TCDS and TCDQ in the appropriate position, the machine and detector components are reliably protected against any damage at an unsynchronized beam abort with the peak temperature rise in the IP6 components being quite acceptable. A re-evaluation of the peak energy deposition is required to take into account the recently confirmed shorter delay time of 0.7 $\mu$s instead of the previously assumed 1.2 $\mu$s. Slow (continuous) beam loss on TCDQ is of interest in the case it becomes a secondary collimator.

## ACKNOWLEDGMENTS

Authors thank B. Goddard, M. Gyr, L. Nicolas, R. Schmidt and D. Still for useful discussions.

## REFERENCES

1. Drozhdin, A. I., Mokhov, N. V., and Huhtinen, M., "Impact of the LHC Beam Abort Kicker Prefire on High Luminosity Insertion and CMS Detector Performance," in *Particle Accelerator Conference-1999*, IEEE Conference Proceedings 99CH36366, IEEE, New York, 1999, pp. 1231–1233.
2. Mokhov, N. V., Drozhdin, A. I., Rakhno, I. L., Gyr, M., and Weisse, E., "Protecting LHC Components Against Radiation Resulting from an Unsynchronized Beam Abort," in *Particle Accelerator Conference-2001*, IEEE Conference Proceedings 01CH37268, IEEE, Chicago, 2001, pp. 3168–3170.
3. Rakhno, I. L., Drozhdin, A. I., Mokhov, N. V., Goddard, B., Gyr, M., Sans, M., and Weisse, E., Further Studies on Protecting LHC Components against Radiation resulting from an Unsynchronized Beam Abort, Tech. rep., Fermilab-FN-724 (2002).
4. Mokhov, N. V., Status of MARS Code, Tech. rep., Fermilab-Conf-03/053 (2003).
5. Church, M. D., Drozhdin, A. I., Moore, R. S., and Still, D. A., Tevatron Abort Kicker Prefire Simulations, Tech. rep., Fermilab Beams-doc-648 (2002).
6. Nicolas, L. Y., and Mokhov, N. V., Impact of the A48 Collimator on the Tevatron B0 Dipoles, Tech. rep., Fermilab-TM-2214 (2003).
7. Baishev, I. S., Drozhdin, A. I., and Mokhov, N. V., STRUCT Program User's Reference Manual, Tech. rep., SSCL-MAN-0034 (1994).
8. Fermilab Superconducting Accelerator Design Report, Tech. rep., Fermilab (1979).
9. Rijk, G. D., *Private Communication*, **CERN** (2002).

# Crystal Collimation at RHIC

R. P. Fliller III*, A. Drees*, D. Gassner*, L. Hammons*, G. McIntyre*, S. Peggs*, D. Trbojevic*, V. Biryukov[†], Y. Chesnokov[†] and V. Terekhov[†]

*BNL, Upton, NY, 11793
[†]IHEP, Protvino, Moscow Region

**Abstract.** Crystal Channeling occurs when an ion enters a crystal with a small angle with respect to the crystal planes, The electrostatic interaction between the incoming ion and the lattice causes the ion to follow the crystal planes. By mechanically bending a crystal, it is possible to use a crystal to deflect ions. One novel use of a bent crystal is to use it to channel beam halo particles into a collimator downstream. By deflecting the halo particles into a collimator with a crystal it may be possible to improve collimation efficiency as compared to a single collimator.

A bent crystal is installed in the yellow ring of the Relativistic Heavy Ion Collider (RHIC). In this paper we discuss our experience with the crystal collimator, and compare our results to previous data, simulation, and theoretical prediction.

## INTRODUCTION

The usual collimation system for a collider consists of multiple sets of jaws that are used to remove the beam halo. The primary set of jaws is used to define the primary aperture of the machine. Particles with low impact parameters on these jaws have a finite probability of scattering out of the collimator and form a secondary halo [1]. A secondary set of jaws is used to remove this halo and increase the collimator efficiency.

Because the proper placement of the secondary jaws is crucial to the overall performance of the collimation system, it is advantageous to find a way to relax this constraint. By using bent crystal channeling, it is possible to give a well defined angular kick to the particles that enter the crystal. A secondary collimator can then be used to remove these particles. This process should reduce the amount of secondary halo that is generated and improve collimation efficiency. An added advantage is that the secondary collimator can be placed further away from the beam, while still keeping the impact parameter high enough to reduce the probability of particles scattering out of it. This paper discusses our experiences with a bent crystal collimator in the yellow ring of the Relativistic Heavy Ion Collider (RHIC).

---

[1] Work performed under the auspices of the U.S. Department of Energy

## CRYSTAL CHANNELING

If ions enter a crystal with small angles relative to the crystal planes, the ions are channeled by the interplanar potential, even if the crystal is mechanically bent [2]. This makes it possible to give a large angular kick to the channeled ions within a short distance. In order for the ions to be channeled, the energy of their motion perpendicular to the crystal planes must be smaller than the maximum interplanar potential, $U(x_c)$. $x_c$ is the transverse interplanar location where the incident ion enters the electron cloud of the lattice atoms. $U(x_c)$ is approximately $Z_{ion} 16$ eV for silicon, the crystal material. This condition gives the maximum angle for channeling to be

$$\theta_c = \sqrt{\frac{2U(x_c)}{pv}}. \quad (1)$$

where $p$ and $v$ are the momentum and velocity of the ion respectively. At larger angles, the incoming ions will experience a potential similar to an amorphous solid. For RHIC energies, $\theta_c = 37$ μrad at injection and 11 μrad at storage energy.

## LAYOUT

The RHIC crystal collimation system is shown in Fig. 1. It consists of a 5 mm long crystal and a 450 mm long L-shaped copper scraper placed downstream of the PHENIX detector in the yellow (counter-clockwise) ring. The crystal is an O-shaped silicon crystal with the (110) planes placed at a slight angle with respect to the nor-

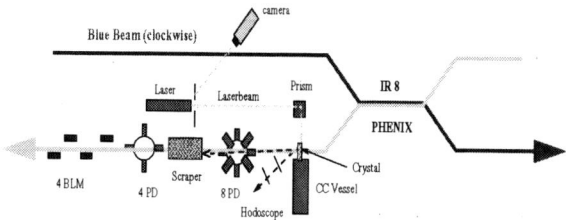

**FIGURE 1.** The RHIC Crystal Collimation system

mal of the input face. The crystal for the 2001-2 RHIC Run and the 2003 Run are of the same design but different bend angles. The former crystal had a bend angle of $\theta_b = 0.37$ mrad and the latter was bent $\theta_b = 0.34$ mrad, as determined from measurements with the beam. There are eight PIN diode loss monitors between the crystal and the scraper (the upstream PIN diodes) to measure particles scattered from the crystal. Four PIN diodes are placed downstream of the scraper to measure particles scattered from the scraper (the downstream PIN diodes). Two scintillators form a hodoscope aligned to the crystal surface to look for particles scattered at large angles. Four ion chamber loss monitors are also located downstream of the scraper [3].

## THEORY

Because the channeling efficiency of the crystal depends on the proper alignment of the crystal to the beam, we have developed a simple model to determine the particle distribution on the crystal face [4]. We assume an initial distribution of

$$f(J,\delta) = \frac{1}{\sqrt{2\pi}\sigma_\delta \varepsilon} \exp\left[-\frac{\delta^2}{2\sigma_\delta^2}\right] \exp\left[-\frac{J}{\varepsilon}\right] \quad (2)$$

where $J$ is the particle action, $\varepsilon$ is the unnormalized rms emittance, $\delta$ is the momentum deviation, $\sigma_\delta$ is the rms momentum spread. By writing $J = J(x, x', \delta)$ and integrating over all momenta, it is possible to transform $f(J,\delta)$ to $f(x,x')$ and obtain the phase space distribution of all particles that will hit the crystal. One can then calculate the average angle, $\theta$, with which the particles will hit the crystal face. Assuming that particles have low impact parameters with respect to the rms beam size, or that the crystal is far from the beam core, one obtains

$$\theta = x_{crystal} \frac{-\alpha \varepsilon + DD'\sigma_\delta^2}{\beta \varepsilon + D^2 \sigma_\delta^2}. \quad (3)$$

The crystal edge is at $x_{crystal}$ from the center of the beam, the dispersion and its slope are given by $D$ and $D'$, and $\alpha$ and $\beta$ are the Twiss parameters at the crystal. For the

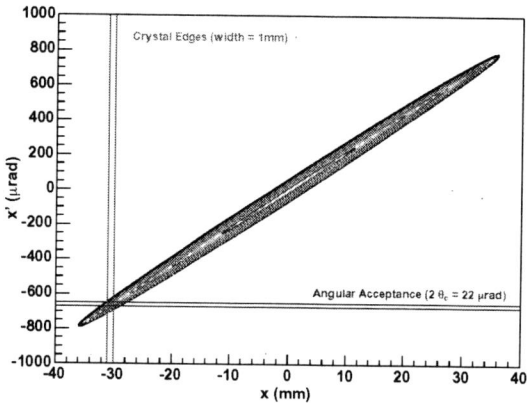

**FIGURE 2.** The Horizontal Phase Space at the Crystal Collimator, The outermost contour is $10\sigma$. The red and blue lines show the crystal size and angular acceptance.

channeling to occur, the crystal planes must be placed at this angle with respect to the beam direction. For typical RHIC parameters, this expression reduces to

$$\theta \approx x_{crystal} \frac{-\alpha}{\beta} \quad (4)$$

and is approximately 700 $\mu$rad.

One can also calculate the rms angular spread of the beam, $\sigma_\theta$, that hits the crystal face. In general, it can be shown that $\sigma_\theta$ is strongly proportional to $\alpha, \beta, D\sigma_\delta$, and the rms impact parameter, and is weakly affected by $\varepsilon$, and $x_{crystal}$ [4]. For the 2003 RHIC parameters, $\sigma_\theta$ is calculated to be 23 $\mu$rad, assuming that particles hit over the entire face of the crystal. Figure 2 shows the horizontal phase space ellipse of the beam at the location of the crystal collimator for a $\beta^*_{PHENIX} = 2$ m lattice.

By knowing the rms angular spread of the beam that hits the crystal face, using Eq. 2.12 from Ref. [5], one can estimate the channeling efficiency.

$$e \approx \frac{2x_c}{d_p} \frac{\pi}{4} \frac{\theta_c}{\sigma_\theta}. \quad (5)$$

where $d_p$ is the distance between the crystal planes. This formula is only valid as long as $\sigma_\theta > \theta_c$. For the 2003 RHIC parameters, Eq.5 predicts $e = 32\%$.

## SIMULATION

To simulate the effect of the crystal in RHIC, the CATCH (Capture And Transport of CHarged particles in a crystal) code was used [6]. A $6 \times 6$ matrix was used to track particles around RHIC so that the effect of multiple turns could be investigated. The particle distribution is chosen identical to that in Eq. 2, using only those particles on horizontal ellipses that will hit the crystal. In the vertical plane the phase space was filled. The design RHIC beam parameters of a $15\pi$ mm-mrad normalized rms horizontal and vertical emittance and a 0.0013 rms momentum spread where used.

Figure 3 shows the effect of the crystal on the particles that strike it. The left plot shows the input particle distribution in the horizontal phase space. The edges of the crystal and the angular acceptance are also marked. The right plot shows the output of the simulation. The band of particles around $-450$ $\mu$rad are the channeled particles. Particles near $-900$ $\mu$rad where not channeled.

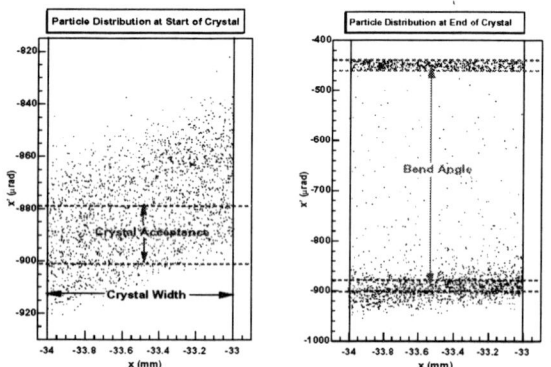

**FIGURE 3.** Simulation of the effect of Crystal Channeling. The left picture is the input particle distribution into the crystal. The right picture shows the output.

Figure 4 shows a simulation of the number of particles scattered from the crystal vs. the crystal angle for varying number of turns in the simulation. The dip in the scattering rate is due to particles channeling. The large portion of the dip occurs when the crystal is aligned to the beam halo. The remaining structure occurs when particles scatter through a fraction of the crystal into a crystal plane and then channel the remaining distance, this is known as volume capture [5]. The depth is dominated by the number of encounters that particles can have with the crystal, and the width is given by the crystal bend angle.

## CRYSTAL CHANNELING

Experiments with the crystal collimator have been performed during the 2001, 2002 and 2003 RHIC runs with fully striped gold and polarized proton beams. The crystal angle was scanned through a range of angles for a variety of different crystal positions, scraper positions, and lattices. Beam losses were recorded by the PIN diodes, hodoscope, and beam loss monitors. Signals from the RHIC experiments were also logged to monitor their background rates. Table 1 lists the available data samples.

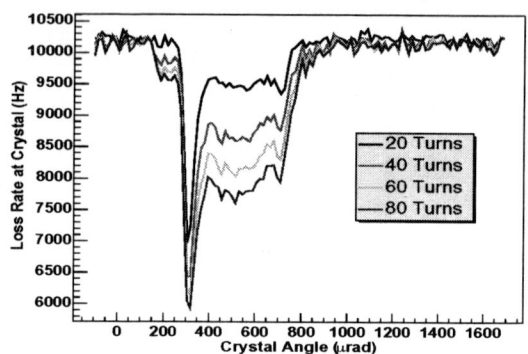

**FIGURE 4.** The Effect of multiple turns of the channeling signal.

**TABLE 1.** Tabulation of Angular Scans

| Species | $\beta^*$ @ IR8 | No. of Scans | $<\sigma_\theta>$ | $<e>$ |
|---------|-----------------|--------------|-------------------|-------|
| Au | 5 m | 27 | 45 $\mu$rad | 20% |
| Au | 2 m | 24 (2001 Run) | 105 $\mu$rad | 28% |
| Au | 2 m | 20 (2003 Run) | 37 $\mu$rad | 26% |
| Au | 1 m | 109 | 69 $\mu$rad | 16% |
| p | 3 m | 119 | 70 $\mu$rad | 26% |

Figure 5 shows the scattering rate seen on an upstream PIN diode during a typical scan from the RHIC 2003 run. The data were averaged over 20 $\mu$rad, the angular resolution of the crystal positioning system. The function for fitting is empirically determined to be two gaussians on a sloping background, with a sloped line connecting them [7]. The efficiency is defined to be the depth of the narrow dip divided by the background. The fitting function generally returns width dip widths and lower efficiencies than the data and simulations show. The values in Table 1 are averages over all of the fits for each RHIC configuration.

## CRYSTAL COLLIMATION

The ultimate measure of the effectiveness of the crystal collimator is its effect on the experimental backgrounds. To measure the collimation performance, the crystal was inserted into the halo and an angular scan was taken. The crystal was then set to the angle which minimized the scattering and thus had the greatest channeling efficiency. Then the copper scraper was moved relative to the crystal until it became the primary aperture. The crystal and scraper where then retracted. The background from the yellow beam was monitored by a detector in the downstream STAR experiment. Figure 6 shows a typical measurement.

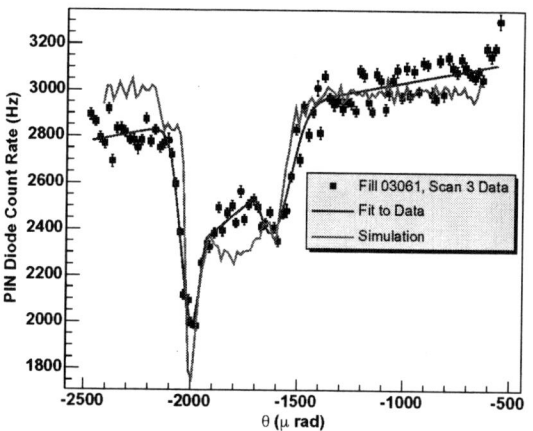

**FIGURE 5.** Data from Fill 03061. The simulation is done with the measured machine optics[8].

As Fig. 6 shows, the background at the STAR detector is correlated with the crystal angle and the scraper position. At the left side of the figure, the crystal is not channeling. Then it is rotated to the channeling angle, and the background increases at the STAR detector indicating that the channeled beam is not caught by the scraper. With the crystal stationary, the scraper is moved, and the background minimized when the scraper becomes the primary aperture. Near the right edge of Fig.6, the crystal is retracted, and the STAR background is lower than with the crystal in the beam, even when the scraper was the primary aperture. The scraper is finally retracted and the background rises.

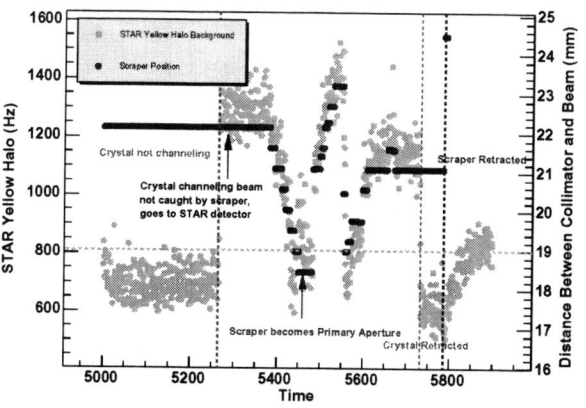

**FIGURE 6.** The Effect of multiple turns of the channeling signal.

We attribute the poor performance of the crystal to the large tilt of the phase ellipse caused by a large $\alpha$ at the crystal location, as seen in Fig. 2. This increases the angular spread of the beam that hits the crystal, and reduces the amount of beam inside of the crystal acceptance. This increases the amount of particles scattered by the crystal. One can see from Fig. 2 that if the crystal was placed in an area with a small $\alpha$, the efficiency would be improved, and collimation using a crystal may be more successful.

## CONCLUSIONS AND FUTURE PLANS

Crystal channeling was achieved in RHIC, and is consistent with simulations and theoretical predictions. Unfortunately, the current setup could not reduce experimental backgrounds. It is hoped that crystal collimation can be used in the future, provided that the lattice position can be chosen such that the angular spread of particles that will the crystal can be reduced.

All of the available warm spaces in RHIC are at locations where $\alpha \approx 10 \rightarrow 20$. To place a crystal at a cold location would be too expensive and time consuming to be done during the summer shutdown. Therefore, it was decided to abandon the crystal collimator in favor of a traditional two stage collimation system which can be installed in the warm section. The current copper scraper will be moved to the current crystal location to serve as the primary collimator. Two horizontal and one vertical secondary collimators will be placed at various locations downstream of it, in the same warm straight. This will be done in both rings.

## REFERENCES

1. T. Trenkler and J.B. Jeanneret., "The Principles of Two Stage Betatron and Momentum Collimation in Circular Accelerators" CERN Note: SL/95-03 (AP), LHC Note 312;
2. J. Lindhard. *K. Dan. Vidensk. Selsk. Mat.-Fys. Medd.* **34**, No. 14 (1965).
3. R. P. Fliller III, et. al. "New Results From Crystal Collimation at RHIC" Proceeding of the 2003 Particle Accelerator Conference, Portland, Oregon (2003).
4. R. P. Fliller III, Ph. D. thesis, State University of New York at Stony Brook (to be published).
5. V. M. Biryukov, Y. A. Chesnokov, and V. I. Kotov. *Crystal Channeling and Its Application at High Energy Accelerators*. Springer–Verlag, Berlin, Heidelberg, 1997.
6. V. Biryukov, "Crystal Channeling Simulation - CATCH 1.4 User's Guide", SL/Note 93-74(AP),CERN 1993.
7. R. P. Fliller III, et. al. "Crystal Collimation at RHIC" Proceeding of the 2002 European Particle Accelerator Conference, Paris, France (2002).
8. T.Satogata,et. al. "Linear Optics Measurement and Correction in the RHIC 2003 Run" Proceeding of the 2003 Particle Accelerator Conference, Portland, Oregon (2003); M. Bai et. al. "Measurement of Betatron Functions and Phase Advances in RHIC with AC Dipoles" Proceeding of the 2003 Particle Accelerator Conference, Portland, Oregon (2003).

# Simulation Aspects of Beam Collimation and Their Remedies in the MARS14 Code [1]

M.A. Kostin*, N.V. Mokhov*, S.I. Striganov* and I.S. Tropin[†]

*Fermi National Accelerator Laboratory P.O. Box 500, Batavia, Illinois 60510*
[†]*Tomsk Polytechnic University, Tomsk, 634034, Russia*

**Abstract.** Simulation aspects of beam collimation are described along with a number of tools and methods developed and used within the MARS14 framework. The tools and methods were implemented in order to relieve the burden of simulations needed for reliable calculations required for design of efficient collimation systems at high-intensity accelerators and colliders.

## 1. INTRODUCTION

Collimators are an essential part of modern accelerators. With an increasing power of machines, the collimators have a major impact on beam, radiation shielding and backgrounds, which needs to be predicted and monitored. The predictions involve simulations with use of several models and requires precise description of physics processes. In this paper some of simulation aspects of beam collimation and related tools and methods designed and developed within the MARS14 framework [1] are addressed.

## 2. COUPLING ACCELERATOR AND RADIATION TRANSPORT CODES

Simulation of beam collimation systems normally consists of two stages. At the first stage, beam halo is transported through the accelerator lattice and lost particles are recorded. At the second stage, the lost particles are propagated through the material of collimators and other elements and their interactions with matter are simulated. There are currently no codes known to effectively perform both the stages within a single framework. Accelerator simulation codes that use a matrix formalism provide a very fast tracking of the particles in lattices but can not simulate interactions in matter. Vice versa, codes that can simulate the interactions implement slow step-wise tracking algorithms. Therefore, at least two codes need to be employed. The variety of models implemented within various frameworks may lead to model inconsistencies, performance penalties and increased time needed for the development.

A MAD-MARS Beam Line Builder (MMBLB) [2] substantially improved recently [3] helps bind accelerator code models with ones implemented in MARS14. The MMBLB places elements of a beam line according to an optics file, created in a MAD style [4]. This format has been chosen given the fact that MAD is an industry standard. So that if an accelerator simulation code uses either a MAD input or optics file to represent a lattice then the model can easily be transfered to MARS14. The examples of recent use of the MMBLB include the Femilab Booster, Tevatron, Proton Driver, NuMI beam line, NLC and JPARC where beam loss distributions and induced radiation effects were studied and beam collimation systems were designed.

An optics file controls the longitudinal positions of elements along a beam line, element orientations and central magnetic field. An element is described with a single line. The first three fields are MAD keyword, type and name, respectively. Those serve for unambiguous identification of an element with a unique structure. The keyword can not be shortened, although this is a normal practice in MAD. This means that the keyword "KICK" can not be used in lieu of "KICKER". There are two methods to identify the elements: 1) using the exact match of the type and name, 2) if this fails, then only the type is used. The next two fields in the optics file are the S-position (path length) of the element end and optical length expressed in meters. The following four parameters are related to the field. They represent the

---

[1] This work was supported by the Universities Research Association, Inc., under contract DE-AC02-76CH03000 with the U.S. Department of Energy.

magnet 'strength' in dipole, quadrupole, sextupole and octupole components, as defined in MAD, multiplied by the length of magnet. At this, the dipole parameter becomes the bend angle of a dipole. The last parameter is the roll angle of an element about the S-axis. The roll angle makes it possible to define a 3-D beam line with practically any orientation of its elements. The position and orientation of the elements are entirely governed by the rules of MAD [4].

There is no restriction on the number of beam lines in the new version of MMBLB. Each beam line is defined with its own optics file. Also, a beam line can be accommodated in an arbitrary place in the MARS global frame. The orientation of the first element of the beam line is specified by the user.

The MMBLB supports the full list of MAD elements except BEAMBEAM, MATRIX and LUMP. The parameters for elements represented with the keywords SOLENOID, HKICKER, VKICKER, KICKER, RFCAVITY, ELSEPARATOR, ECOLLIMATOR and RCOLLIMATOR can not be provided with an optics file. Instead, the parameters must be defined externally in the code. Whereas an optics file controls the longitudinal position of a beam line element, a user must provide several subroutines that describe the geometry, field, materials, volumes and names of sub-zones. In order to bind all those subroutines with a real element from the optics file they must be registered by means of a special subroutine.

The MMBLB offers a possibility to describe a simple beam enclosure. In order to do this, the enclosure must not change its geometry with respect to a reference orbit within some stretch. This method only works for flat 2-D beam lines. The code that describes the geometry of the beam enclosure must be provided by a user in a designated subroutine. Fig. 1 shows a NLC beam delivery section with a corresponding beam enclosure.

## 3. DESCRIPTION OF SCATTERING

A precise description of the scattering processes is exceptionally important for efficient calculations of collimation systems. Four parts are distinguished in a typical scattering spectrum (Fig. 2). With the increase of the momentum transfer $t$, they are respectively: multiple Coulomb scattering (MCS), elastic coherent and incoherent scattering on a nucleus and the scattering at larger angles due to diffractive processes. All four processes are carefully described and treated in the MARS14 code.

Multiple Coulomb scattering is most important for a description of scattering on an edge of collimators. Two methods are widely used: sampling from Moliere [5] and Gaussian [6] distributions. Limits of applicability of these approaches were determined in [7], where the

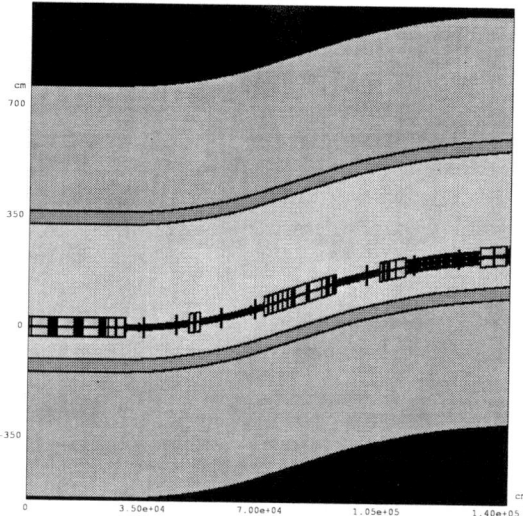

**FIGURE 1.** The NLC beam delivery section modeled with MMBLB.

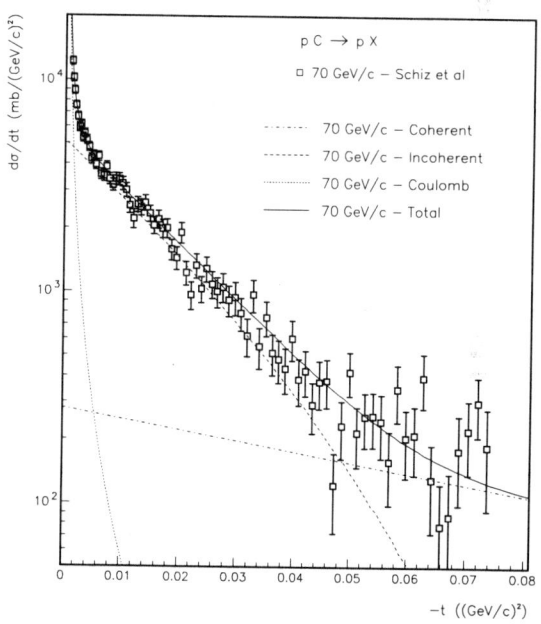

**FIGURE 2.** Proton-carbon scattering spectrum at 70 GeV/c.

Moliere theory of MCS was modified to take into account a nuclear screening. It was shown that the angular distribution obtained in such a way coincided with the Moliere one for the thicknesses of 0.1 - 1 radiation lengths and reached the Gaussian asymptotic for 100 - 1000 radiation lengths (Fig. 3). It was shown [8] that using Moliere Gaussian distributions alone can lead to quite large errors in Monte-Carlo simulations especially at large angles. An efficient method to simulate (MCS) is based on a separate treatment of the "soft" and "hard" interactions. A large number of "soft" collisions is described using a "continuous scattering" approximation; a small number of "hard" collisions is simulated directly. A new analytical expression for a "continuous" angular distribution was recently developed [9]. A boundary angle between "soft" and "hard" collisions is determined as a function of a step-length providing a possibility for fast and precise simulation. Results of simulation using a Gaussian nuclear form-factor agree within 1% with analytical calculations [7], but the new algorithm provides a possibility to include an arbitrary charge distribution of projectile and target nucleus. A corresponding algorithm was implemented in MARS14 [9].

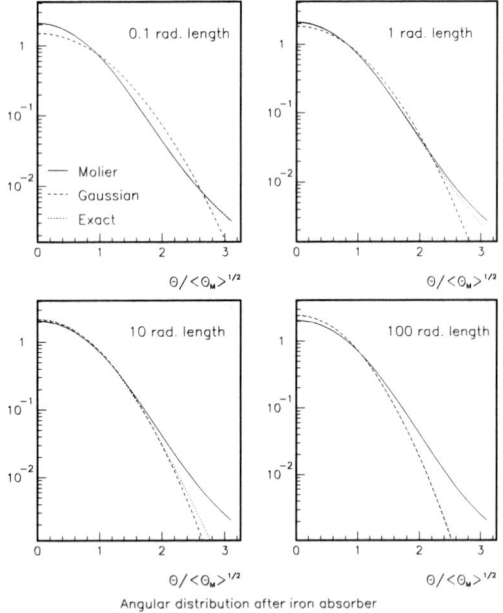

**FIGURE 3.** Angle distributions for particles scattered in iron absorbers of various lengths.

The coherent and incoherent elastic processes have smaller cross-sections compared to MCS but provide larger scattering angles. These processes may dominate for description of collimation systems that intercept beam particles scattered on a residual gas. The density of the residual gas is normally quite low, so that a noticeable angle deviation due to MCS is accumulated only in many turns. Coherent and incoherent elastic scattering are simulated in MARS14 using a fit to experimental data.

Description of diffractive dissociation in MARS14 is based on a triple-Reggeon phenomenology. A leading proton production cross section can be approximated by $exp(B_d t)/(1-x)$, where $x = p'/p$, and $p$ and $p'$ are proton momenta before and after the interaction. The exponential slope $B_d$ is a factor of two larger than the slope of the incoherent elastic scattering. Due to non-marginal energy loss and a larger scattering angle, the diffractive scattering is unimportant for simulation of collimation systems, since in most cases, the particles scattered to large angles are lost in the lattice promptly and typically can not reach the collimators.

## 4. GEOMETRY DESCRIPTION AND VISUALIZATION

The collimation system development includes thorough radiation transport and energy deposition simulations in beam elements, shielding and around to meet regulatory requirements on prompt and residual radiation in and around the tunnel. A quite comprehensive and complex geometry, materials and magnetic field descriptions are needed in order to address all these features. The model complexity can cause errors. Two modules in MARS14 help deal with this problem.

The first one allows a choice of geometry description. Depending on configuration and user's experience, one can use one of the geometry packages available in MARS14: Standard ($r$-$z$-$\phi$), MCNP, Extended and Non-Standard (arbitrary user-defined) [1]. Any part of the system, described within the first three options, can be overwritten by the Non-Standard geometry segments. In the MCNP mode, the code uses input geometry description in the MCNP format [10] (except lattices and universes). The term "Extended" refers to an extension beyond the Standard zones. In the current version, Extended geometry uses a combination of boxes, cylinders, spheres, cones and tetrahedra, similar to the methods used by other Monte Carlo programs, particularly GEANT [11]. These elements can be arbitrary positioned and rotated by a set of transformation matrices, and divided to sub-zones. For example, a box can be sub-divided along all three directions. The elements can overlap.

The second module is a powerful Graphical User Interfaces (GUI), used for visualization and debugging of geometry, materials and magnetic field descriptions, simulated processes and calculated results. This user-friendly MARS-GUI-SLICE tool is absolutely vital in serious studies of accelerator, detector and shielding applications and used with enthusiasm by the MARS14 community worldwide.

A new 3-D extension has recently been developed and added to the MARS14 code system [12]. It is is built on the top of the OPENINVENTOR [13] SGI library. A 3-D view panel is launched from the MARS-GUI-SLICE window (Fig. 4). One can use the OPENINVENTOR 3-D GUI to rotate, zoom in and out and move the scene. Beside those standard OPENINVENTOR functions, other useful features have been added such as changing from a solid to wire-frame rendering, a light editor, ability to save the scene image to a postscript file and modify the background. One can also plot 3-D particle tracks and modify a color and transparency of materials chosen from the MARS-GUI-SLICE window.

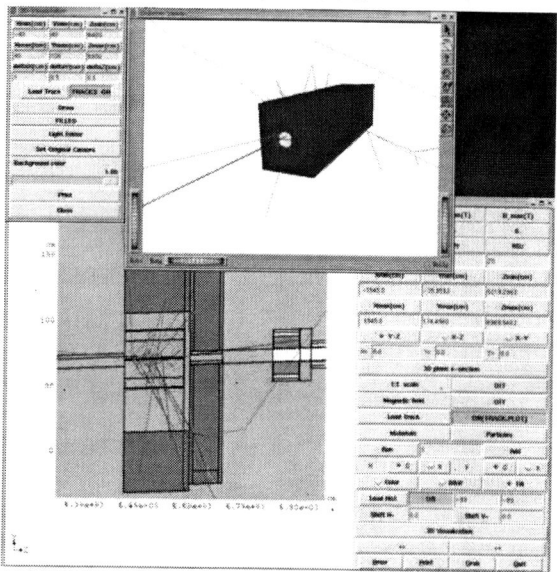

**FIGURE 4.** MARS14-3D-GUI example.

## 5. TRACKING ACCURACY

There are several particle tracking issues vital in collimation applications. All of them are treated as precise and efficient as possible in the MARS14 code. These include:
- *Boundary localization.* It is done exactly in the Extended geometry mode, and controlled by the region-material depended parameter to the accuracy required in the given application. Special options are provided in the code for efficient simulations in a system which includes, for example, tiny objects such as vacuum windows, beam pipes, electrostatic septum wires etc (microns to millimeters), implemented into extended vacuum voids (hundreds meters), followed by the bulk shielding (meters to kilometers).
- *Edge scattering.* This is especially tough problem for charged particle tracking near the vacuum-material boundaries in presence of magnetic field and quasi-continuous processes such as Coulomb scattering.
- *Tracking in magnetic field.* A step-size is chosen according to the field value in the plane perpendicular to the direction of motion, particle momentum and its ratio to the primary beam momentum, with a possible user control on top of that. MARS14 algorithms provide typically a ten micron (microradian) accuracy for the beam line and accelerator systems of several kilometers long.

## 6. CONCLUSIONS

MARS14 is the powerful code that offers a number of tools and methods for simulating collimation systems. The recent advancements, described in this paper, include the MAD-MARS Beam Line Builder, a careful treatment of the scattering processes, several options for geometry description, Graphical User Interface and high tracking accuracy. Two types of simulation that require the most significant computational time are precise tracking in magnetic field of very long beam lines and accelerator lattices and a deep penetration problem in bulk shielding around collimators. A normal approach to speed the calculations up is the use of a number of relatively short computational jobs and then average the results. A paradox arises however: in order to obtain a statistically significant result by averaging the results of the shorter runs, the shorter job outputs have to be statistically significant already. Otherwise, the use of the averaging procedure that assumes the values distributed over Gaussian, would lead to biased both the errors and mean values. A simple way out of this is to use parallel computing. A corresponding work with MARS14 is underway.

## REFERENCES

1. N.V. Mokhov, "The MARS Code System User's Guide", Fermilab-FN-628 (1995); N.V. Mokhov, O.E. Krivosheev, "MARS Code Status", Proc. Monte Carlo 2000 Conf., p. 943, Lisbon, October 23-26, 2000; Fermilab-Conf-00/181 (2000); http://www-ap.fnal.gov/MARS/.
2. O. Krivosheev et al., "A Lex-based MAD parser and its applications," FERMILAB-CONF-01-142-T; Proc. IEEE Particle Accelerator Conference (PAC 2001), Chicago, Illinois, 18-22 Jun 2001.
3. M.A. Kostin, O.E. Krivosheev, N.V. Mokhov, I.S. Tropin, "An Improved MAD-MARS Beam Line Builder: User's Guide", FERMILAB-FN-738, 2003.
4. H. Grote, F.C. Iselin, "The MAD program (Methodical Accelerator Design)", CERN/SL/90-13.
5. G.Z. Molier, Z. Naturforsh 2a, p.133, 1947; Z. Naturforsh 3a, p.78, 1948.
6. B. Rossi and K. Greisen, Rev.Mod.Phys 13, p.240, 1941.
7. I.S. Baishev, N.V. Mokhov and S.I. Striganov, Sov. J. Nucl. Phys. 42, p.745, 1985.
8. S.I. Striganov, Nucl.Phys. B (Proc. Suppl) 51A, p.172, 1996.
9. N.V. Mokhov and S.I. Striganov, LA-UR-03-4262.
10. http://laws.lanl.gov/x5/MCNP
11. http://wwwinfo.cern.ch/asd.
12. J.P. Rzepecki, M.A. Kostin and N.V. Mokhov, "3D Visualization for the MARS14 Code", FERMILAB-TM-2197, Jan. 2003.
13. http://www.openinventor.com.

# Comparison of the TESLA, NLC and CLIC Beam-Collimation System Performance[1]

A. Drozhdin*, G. Blair†, L. Keller**, W. Kozanecki‡, T. Markiewicz**,
T. Maruyama**, N. Mokhov*, O. Napoly‡, T. Raubenheimer**, D. Schulte§,
A. Seryi**, P. Tenenbaum**, N. Walker¶, M. Woodley** and F. Zimmermann§

*FNAL, Batavia, IL, USA
†Royal Holloway College, UK
**SLAC, Stanford, CA, USA
‡DSM/DAPNIA, CEA-Saclay, France
§CERN, Geneva, Switzerland
¶DESY, Hamburg, Germany

**Abstract.** This report briefly describes studies performed in the framework of the Collimation Task Force organized to support the work of the second International Linear Collider Technical Review Committee. The post-linac beam-collimation systems in the TESLA, JLC/NLC and CLIC linear-collider designs are compared using the same computer code under the same assumptions. Their performance is quantified in terms of beam-halo and synchrotron-radiation collimation efficiency. The performance of the current designs varies across projects, and does not always meet the original design goals. But these comparisons suggest that achieving the required performance in a future linear collider is feasible. Further work of the group is briefly described as well.

## INTRODUCTION

We present a summary of comparisons of the collimation-system performance for the three main candidate linear-collider designs: JLC/NLC, CLIC and TESLA. The essence of these results is included in Ref. [1] and more details can be found in Ref. [2].

For the next generation $e^+e^-$ linear colliders (see [1] and Table 1), small fractional beam losses along the transport line, or the presence of particles far from the beam core in the IP region, may strongly affect the background conditions in the detector, as well as cause irradiation and heating of collider components.

All machine designs need to remove this halo to a certain "collimation depth", which is generally set by the synchrotron-radiation fan generated by the halo particles in the last few magnets close to the IP: by definition, all particles within the collimation depth generate photons that should pass cleanly through the IR. Halo particles outside this collimation depth are removed by physically intercepting them with "collimators", which are formed by a thick absorber of many radiation lengths placed in the optical shadow of a thin spoiler, the thickness of which is generally less than one radiation length.

**TABLE 1.** LC parameters for 500 GeV c.m.energy.

| parameter | TESLA | NLC | CLIC |
|---|---|---|---|
| Np, $10^{10}$ | 2 | 0.75 | 0.4 |
| bunch/train | 2820 | 192 | 154 |
| bunch.sep, ns | 337 | 1.4 | 0.67 |
| $F_{rep}$, Hz | 5 | 120 | 200 |
| <I>/beam, $\mu$A | 45.1 | 27.6 | 19.7 |
| <P>/beam, MW | 11.3 | 6.9 | 4.9 |
| $\gamma\varepsilon_{x,y}$, mm·mrad | 10, 0.03 | 3.6, 0.04 | 2.0, 0.01 |
| $\beta^*_{x,y}$, mm | 15.2, 0.41 | 8, 0.11 | 10, 0.05 |
| $\sigma^*_{x,y}$, nm | 553, 5 | 243, 3 | 202, 1.5 |

Analytic estimates predict halo of the order of $10^{-6}$ of the LC beam current. However, given the SLC experience, designers of collimation systems have taken the conservative approach to build a collimation system that would be able to intercept a fractional halo of $10^{-3}$ of the beam – the number we assumed for the present study.

The comparative studies were carried out using the program STRUCT [3]. This package performs particle tracking, taking into account aperture restrictions, interaction of primary beam particles with collimators, beam losses, synchrotron radiation and transport of the photons along the beamline.

---

[1] Work supported in part by US DOE, Contract DE-AC03-76SF00515.

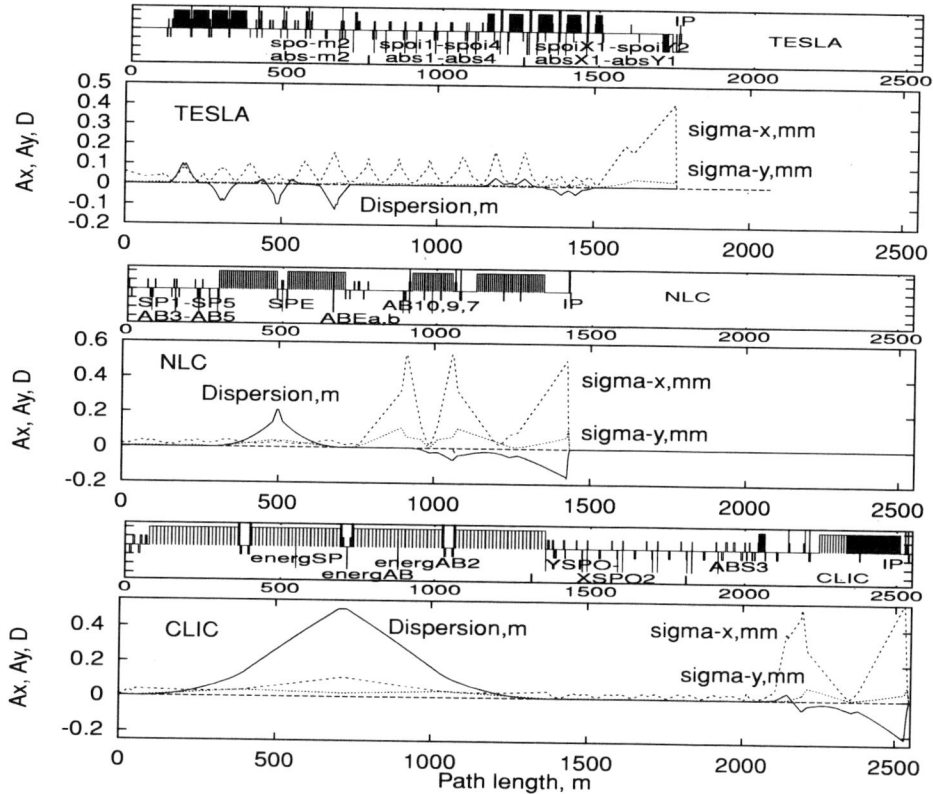

**FIGURE 1.** Optics of the Beam Delivery Systems and collimator locations in TESLA, NLC and CLIC.

# COLLIMATION IN LINEAR COLLIDERS

All designs have a dedicated primary collimation system (betatron and off-energy) located upstream of the final focus system (FFS). Additional secondary or "clean-up" collimators are located in the FFS. The maximum number of halo particles that may be intercepted in this secondary system is limited by the muon flux the detector can tolerate. The primary system — which intercepts most of the halo – should have high enough an "efficiency" to reduce the losses in the secondary system to acceptable levels. At the same time, the combination of primary and secondary collimation must bring the halo population outside the collimation depth in the final doublets within tolerance.

Collimation of the beam requires putting material close to a beam with a high energy density, which in turn creates a risk that a missteered beam might destroy the collimator. In practice, in order to limit the betatron functions in the collimation region, the design relies on thin (0.5-1 radiation length) spoilers which scrape the halo with minimal heating and enlarge the spot size of a missteered beam via multiple Coulomb scattering and energy loss. The enlarged beam is then absorbed in thick (30 radiation lengths) copper absorbers. Absorbers in the primary collimation section should lie in the shadow of their spoiler partner to reduce the probability of being hit directly by a missteered beam.

Table 2 lists the physical properties of the spoilers and absorbers for the three machines. Fig. 1 shows collimator locations, horizontal dispersion and beam sizes in the Beam Delivery Systems reviewed by the Collimation Task Force. The considered BDS have different design. In particular, NLC and CLIC are based on the new compact Final Focus with local chromaticity compensation [4], while TESLA is based on (now) traditional FF design. Both JLC/NLC and CLIC have crossing angle, which allows to separate the issue of spent beam transport and collimation. In contrast, TESLA is using head-on collision that impose additional restriction on the near IP collimation and masking system, since the spent beam needs to share the beamline with the incoming beam for a certain distance. The order of the collimation system is also different in the designs: in TESLA and CLIC the energy collimation precedes the betatron collimation, while this order is reversed in NLC. The order of collimation systems reflects expectations of the probability of certain fault scenarios and this is one of the topics that deserves further careful analysis.

**TABLE 2.** Parameters and achieved performance of the collimation systems. $\sigma_{x,y}$ are the beam sizes at the primary spoiler (including the dispersive contribution); $\sigma_{x,y}^{\beta}$ refer to the betatron contributions alone. The spoiler settings are tighter than the effective collimation depth at the Final Doublet (FD) due to dispersive and higher-order effects.

| units | TESLA | JLC-X/NLC | CLIC |
|---|---|---|---|
| Nominal collimation depth (at spoiler) | | | |
| # $\sigma_{x,y}^{\beta}$ | 12, 74 | 10, 31 | 9, 65 |
| Energy collimator | | | |
| x gap, mm | 3.0 | 6.4 | 3.2 |
| $\sigma_{x,y}$, $\mu$m | 154, 4.5 | 534, 29 | 814, 38 |
| Betatron collimator Final-doublet phase | | | |
| x, y gaps, mm | 3.0, 1.0 | 0.6, 0.4 | 0.68, 0.4 |
| $\sigma_{x,y}$, $\mu$m | 129, 7 | 28, 6.5 | 38, 3 |
| Betatron collimator IP phase | | | |
| x, y gaps, mm | 3.0, 1.0 | 0.6, 0.5 | 0.6, 0.4 |
| $\sigma_{x,y}$, $\mu$m | 128, 7 | 16, 0.8 | 22, 3 |
| Effective collimation depth (at FD) | | | |
| # $\sigma_{x,y}^{\beta}$ | 13, 80 | 15, 31 | 11, 100 |
| Spoiler material and length | | | |
| | Ti | Cu + Be | |
| mm (rad.length) | 35 (1) | 117 ($0.5_{Cu}+0.3_{Be}$) | |
| Absorber length | | | |
| mm (rad.length) | 500 (35) | 429 (30) | |
| Losses in secondary collimation section | | | |
| part./bunch | $2.4 \cdot 10^5$ | 50 | 1000 |
| Achieved primary-collim. efficiency | | | |
| | 0.01 | $< 1 \cdot 10^{-5}$ | $< 3 \cdot 10^{-4}$ |

## METHODOLOGY AND RESULTS

The effectiveness of the collimation system can be quantified in terms of: a) the fraction of initial halo particles that survive (or are rescattered out of) the primary collimation system and hit secondary collimators or other aperture limitations closer to the IP (this is relevant when estimating muon backgrounds); b) the number of halo particles that lie outside the collimation depth when they reach the final doublet (this is relevant when estimating synchrotron-radiation backgrounds).

For simulations of the effectiveness of the three collimation systems and of background conditions at the IP, the beam halo was represented by a large number of rays (typically $5 \times 10^5$) distributed in phase space with $1/r$ amplitude distributions and with a Gaussian momentum distribution of $\sigma(dP/P) = 1\%$.

## Primary-collimation Efficiency

Figure 2 (left) displays, for each machine, the cumulative particle loss, starting at the IP and integrating back to the entrance of the collimation system.

The NLC design achieves a primary-collimation efficiency significantly better than $10^{-5}$, resulting in less than $10^4$ particles per train being lost in the secondary system. The CLIC collimation system achieves a primary-collimation efficiency[2] of about $3 \times 10^{-4}$.

In TESLA, with the primary collimation as currently designed, the loss rate in the secondary system amounts to about 1% of the initial halo population. Because the TESLA bunch spacing is longer than the entire bunch train for the warm machines, TESLA generally quotes background rates per bunch crossing. However the sub-detector most sensitive to muon background, the time projection chamber (TPC), integrates over 150 bunches, so that for the same assumed incident halo fraction of $10^{-3}$, the effective halo population becomes similar to that of NLC and the effective loss in the secondary collimation system amounts to $3 \cdot 10^7$ particles per sensitivity window.

## Halo Photons

The collimation-system performance achieved at the entrance to the final doublet, and the resulting level of halo-induced SR backgrounds, are described in details in [2]. Here we briefly highlight these results.

In NLC, the edge of the collimation depth is sharply defined (Fig. 2 bottom); but for no halo photons to hit the beam pipe near the IP, rather tight collimator settings ($\pm 0.2$–$0.3$ mm) are needed[3]. The halo photon flux hitting the FD SR mask on the incoming-side is low enough; in addition, these photons are rather soft ($\langle E_\gamma \rangle \sim 31$ KeV). The halo hitting the detector masks and the vertex detector is negligible. Photon losses in the outgoing beam line were not calculated for NLC or CLIC because it was assumed that the crossing-angle geometry provides enough flexibility for an ample stay-clear on the spent-beam side.

In TESLA, the boundary of the collimated halo is barely visible. Charged-halo losses on the SR mask amount to about 7400 particles/bunch on the upstream

---

[2] For both NLC and CLIC this efficiency number is a too crude figure of merit as losses vanish sharply after the collimation system. Further studies of muon reaching detector would give a better indication of performance.
[3] TRC studies considered only the more pessimistic case, i.e. without tail folding octupoles, included in NLC BDS, which allow widening the spoiler gaps by a factor of 3 to 4. NLC collimation performance with these octupoles is presented further below and discussed in more details in Ref. [5]

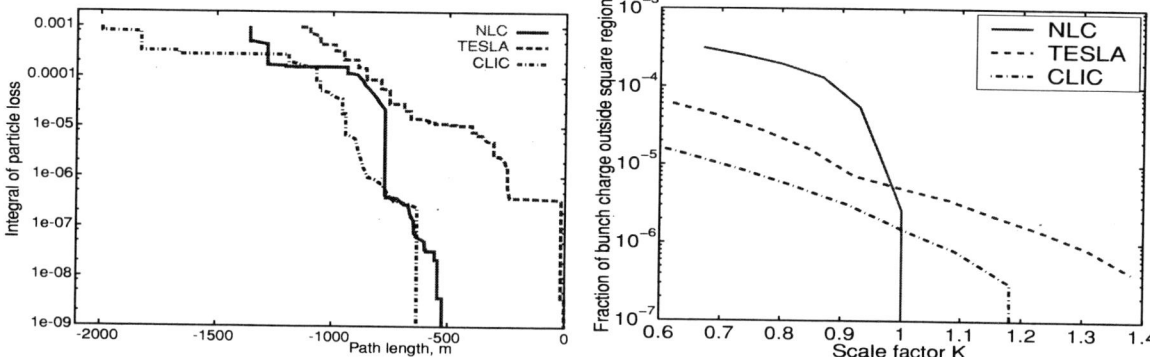

**FIGURE 2.** Collimation-system performance assuming an incident fractional halo of $10^{-3}$. Left: fractional loss of charged-halo particles, integrating back, starting at the IP. The horizontal scale shows the distance from the IP. Right: number of charged-halo particles per bunch, normalized to the nominal bunch charge, in a rectangular $x-y$ window at the entrance to the final doublet, as a function of the collimation depth. The scale factor $K$ defines the window dimension: for $K=1$, the window size corresponds to the effective collimation depth listed in Table 2.

side, and about 250 particles/bunch on the downstream mask. Simulations also indicate that some SR photons from the halo ($> 10^5$ photons/bunch) hit the detector mask located 3 m downstream of IP; their total energy (158 GeV/bunch) is however small compared to that of beam-beam induced pairs. More importantly, one observes a sizeable outgoing photon halo ($\sim 1.2 \times 10^5$ GeV/bunch, corresponding to about $1.2 \times 10^7$ photons) hitting the downstream SR mask 18 m from the IP: the total energy of the halo photons intercepted by this mask is about half of that deposited by outgoing SR photons from the beam core hitting the same mask. Both the mean energy and the number of halo photons per pulse is an order of magnitude larger in TESLA than in NLC, because of significantly stronger bending fields. This remark also applies to SR photons radiated by the core of the incoming $e^{\pm}$ beam.

The halo in CLIC-500 appears reasonably well-behaved, and the number of photons hitting the SR and IR masks is of no concern. This promising performance was however obtained with rather tight collimator settings. Detailed simulations of the 500 GeV CLIC system are only beginning, and its collimator configuration is still in flux.

### Synchrotron Radiation from the Beam Core

A sizeable flux of SR photons produced by the beam core (primarily in the last dipole) hits the SR masks on either side of the IP. In NLC, when integrated over the entire bunch train, the flux of SR photons from the core reaches a level that may deserve attention. In TESLA, about $10^{10}$ core photons/bunch hit the SR mask upstream of the IP, depositing $10^9$ GeV/effective bunch train. While it is plausible that the effectiveness of the TESLA collimation system may be further improved, these results underscore the urgent need for more detailed studies. In CLIC, the flux of intercepted core SR photons is slightly lower than in NLC, presumably due to the fact that the CLIC IR has been optimized for 3 TeV c.m. energy.

## CONCLUSION OF TRC COMPARATIVE STUDIES

Comparative studies of the performance of the post-linac beam-collimation systems in the TESLA, NLC and CLIC designs have shown that the performance of the systems as currently designed is not uniform across projects, and that it does not always meet all the design goals. As of this writing, the CLIC and NLC collimation schemes appear the most promising. Improvements of the TESLA collimation system are expected to result from the ongoing overhaul of their BDS design [6]. Overall, the very existence of an acceptable solution suggests that achieving the required performance in future linear colliders is feasible.

## FURTHER WORK

Further plans include continuing studies of muon backgrounds, evaluation of performance in a non-ideally tuned BDS with both static and dynamic errors, more detailed simulation of SR background, etc. [7].

For TRC study, for the NLC system, only the more pessimistic case without tail folding octupoles was considered. The case with these octupoles has been recently verified with the same codes used for TRC studies and shows good performance. The halo losses along the beamline are very similar to the case without octupoles, as seen in Fig.3, however, because the two octupole doublets squeeze the halo size at the FD, this allowed to open the smallest gaps from ±0.2 mm to ±0.6 mm (i.e. so far we achieved a factor of 3 from the theoretical possible factor of 4; further improvements are expected to be possible). The larger gaps are beneficial, in particular, to reduce wake-fields generated in the collimators. With these octupoles, the SR photon losses are also under control, and photon losses occur only on dedicated masks (two additional photon masks, located several tens of meters from IP, were needed to protect the IP region from SR emitted by the halo, see [5] for more details).

Muon background simulation was not done within TRC. Studies were done before that, and continue now. For example, studies with MUCARLO code (see Lew Keller report in [7]) have shown that assuming (pessimistically) that 0.001 of the beam is collimated, two tunnel-filling spoilers are needed to keep the number of muons per pulse train hitting detector below 10 (which is then considered acceptable). Studies have started at Fermilab with MARS code [8] to cross check these results.

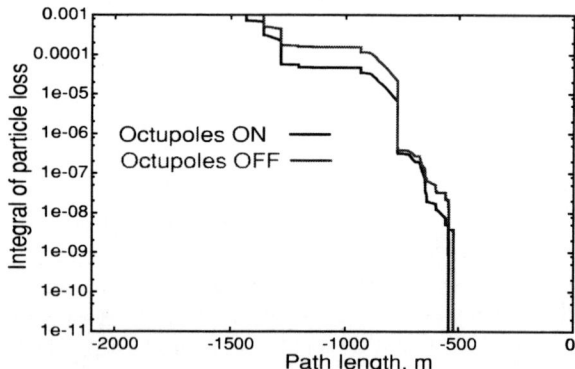

**FIGURE 3.** Fractional loss of charged-halo particles in NLC with and without tail folding octupoles.

## REFERENCES

1. Second ILC-TRC Report, SLAC-R-606, 2003.
2. A. Drozhdin, *et. al.*, FERMILAB-TM-2200, 2003.
3. A. Drozhdin, *et. al.*, 'STRUCT Program User's Reference Manual', http://www-ap.fnal.gov/~drozhdin/
4. P. Raimondi and A. Seryi, Phys. Rev. Lett. **86**, 3779 (2001).
5. A. Drozhdin, *et. al.*, LCC-118, SLAC, to be published.
6. J. Payet, O. Napoly, N. Walker, In Proceedings of PAC2003.
7. Collimation Task Force Workshop, SLAC, Dec. 2002, http://www-project.slac.stanford.edu/lc/wkshp/colltf2002/
8. N. Mokhov, *et. al.*, 'MARS Code System', http://www-ap.fnal.gov/~mokhov/MARS/

# Collimation for CLIC

R. Assmann, H. Burkhardt, S. Fartoukh, J.B. Jeanneret, J. Pancin, S. Redaelli,
T. Risselada, D. Schulte, F. Zimmermann*, A. Faus-Golfe[†], H.-J. Schreiber[**] and
G.A. Blair[‡]

*CERN, AB Division, Switzerland
[†]University of Valencia, Spain
[**]DESY/Zeuthen, Germany
[‡]Royal Holloway, London, UK

**Abstract.** The collimation system of the Compact Linear Collider (CLIC) must fulfil a number of conflicting requirements, namely it should (1) remove beam halo to reduce the detector background, (2) provide a minimum distance between collimators and collision point for muon suppression, (3) ensure collimator survival and machine protection against errand beam pulses, (4) not be excessively long, and (5) not amplify incoming trajectory fluctuations via the collimator wake fields. Two optical systems have been designed — the first linear, the second non-linear —, which promise to meet all these requirements for the design beam energy of 1.5 TeV. We describe the various design criteria, a preliminary performance assessment, and outstanding questions.

## INTRODUCTION

CLIC is designed to deliver electron-positron collisions of $10^{35}$ cm$^{-2}$s$^{-1}$ luminosity at a centre-of-mass energy of 3 TeV. Its total length is about 35 km. The CLIC rf power source is based on two-beam acceleration comprising 22 drive beams for each of the two main linacs. The beam-delivery system (BDS) includes collimation and final focus, and it occupies a total length of about 5 km (counting both sides). The CLIC BDS can be scaled to 500 GeV by varying the strengths of sextupoles and bending magnets, while maintaining the same total length. Table 1 compiles key parameters for the 500-GeV and 3-TeV systems. The BDS accommodates a single-stage momentum collimation, followed by betatron collimation and a 500-m long compact final focus with nonzero dispersion in the final low-beta quadrupoles. A comprehensive review of the CLIC BDS can be found in Ref. [1]. The five conflicting design requirements for the collimation system were listed in the abstract.

**TABLE 1.** Parameters of final focus (FF), collimation system (CS), and the beam for 3 TeV and 500 GeV cm energy. Emittance numbers refer to the entrance of the BDS. The spot sizes at the interaction point (IP) are rms values obtained by particle tracking and are up to 3 times larger than the 'effective' beam sizes which determine the luminosity

| parameter | symbol | 3 TeV | 500 GeV |
|---|---|---|---|
| FF, CS length [km] | | 0.5, 2.0 | 0.5, 2.0 |
| tot. BDS length [km] | | 2.5 | 2.5 |
| emittances [$\mu$m] | $\gamma\varepsilon_{x,y}$ | 0.68, 0.01 | 2.0, 0.01 |
| beta functions [mm] | $\beta^*_{x,y}$ | 6.0, 0.07 | 3.0, 0.05 |
| rms spot sizes [nm] | $\sigma^*_{x,y}$ | 67, 2.1 | 180, 4.2 |
| bunch length [$\mu$m] | $\sigma^*_z$ | 35 | 35 |
| crossing angle [mrad] | $\theta_c$ | 20 | 20 |
| repetition rate [Hz] | $f_{\text{rep}}$ | 100 | 200 |
| no. bunches/train | $n_b$ | 154 | 154 |
| particles per bunch | $N_b$ | $4 \times 10^9$ | $4 \times 10^9$ |
| lum. w/o pinch [$10^{34}$ cm$^{-2}$s$^{-1}$] | $L_0$ | 4.0 | 1.9 |

## COLLIMATION DEPTH AND FAILURES

The transverse collimation depth for CLIC is set by synchrotron radiation and beam loss in the final quadrupoles on the incoming side of the IP. The rms IP divergence for a single beam is less than 10 $\mu$rad. After the collision the outgoing spent beam and the generated electron-positron pairs extend up to 10 mrad [2], i.e., to 1000 times larger angles. Therefore, the much narrower cone of photons emitted by the incoming beam is of no concern for the exit line. Computing synchrotron-radiation fans at 3 TeV shows that the available beam stay-clear, limited by the aperture of the final permanent-magnet quadrupoles, is about $14\sigma_x$ and $83\sigma_y$ in the two transverse planes [1]. Since the dispersive component of the horizontal beam divergence is about equal to the betatron component, the actual limiting envelope for horizontal betatron oscillations is about $\sqrt{2}$ smaller, or $10\sigma_x$.

If the beam impacts on a collimator, the latter can be destroyed either by melting or by surface fracture. Already at the SLC, the beam damaged the gold-plated Ti-alloy collimators located at the end of the SLAC linac [3]. The particle flux of the CLIC beam is about $10^5$ times more intense. All proposed linear colliders envision a sequence of thin spoilers (0.5–1.0 radiation length), which are followed in some distance by thick absorbers (10 r.l.). The purpose of the spoilers is to increase the beam angular spread due to multiple scattering, in case of beam impact due to a failure. This increases the beam size at the absorbers and reduces the risk of melting. The spoiler-absorber scheme has been employed in the SLAC linac since many years [4]. The acceptable spot size at the spoilers is limited by surface fracture due to beam impact. This limit can roughly be estimated from the local temperature rise $\Delta T = (1/C_p)(\Delta E/\Delta m)$, where $C_p$ denotes the heat capacity and $\Delta E/\Delta m$ the ionization-energy deposition per unit mass, which is $(\Delta E/\Delta m) = (dE/dx)/(2\pi\sigma_x\sigma_y\rho)$. Comparing the ensuing tension $\sigma = E\alpha\Delta T$ (where $E$ denotes the Young's modulus and $\alpha$ the thermal expansion coefficient) with the ultimate tensile strength $\sigma_{UTS}$ yields the minimum spot size required to avoid surface fracture: $\sigma_x\sigma_y > (1/2\pi\rho)(dE/dx)N_b n_b E\alpha/(C_p\sigma_{UTS})$. The stability limit obtained from a more precise calculation [5] is displayed in Fig. 1. For carbon, the computation was done twice, with and without the additional heating due to beam image currents (i.c.). The plotting symbols superposed on the stability curves indicate the nominal design beam sizes at the spoilers for the momentum collimation, for the betatron collimation, and for a 2nd collimation stage in the final focus proper, respectively. The momentum spoiler will survive, if it is made from carbon, and possibly in the case of beryllium. The dedicated betatron collimators will always be destroyed, in case they are directly hit by the nominal beam. Most of the final-focus collimators survive, if made from carbon. Copper and titanium cannot withstand the beam impact anywhere.

**TABLE 2.** Collimator parameters

| cm energy | 3 TeV | 500 GeV |
|---|---|---|
| $\delta$ spoiler gap | ±4 mm | ±4.8 mm |
| $\beta_x$ spoiler gap | ±80 μm | ±300 μm |
|  | (10 $\sigma_x$) | (9 $\sigma_x$) |
| $\beta_y$ spoiler gap | ±104 μm | ±215 μm |
|  | (80 $\sigma_y$) | (69 $\sigma_y$) |
| spoiler material | Be | Be |
| spoiler length | 177 mm (0.5 r.l.) | 177 mm (0.5 r.l.) |
| absorber mat. | Ti (Cu coated) | Ti (Cu coated) |
| absorber length | 712 mm (10 r.l.) | 712 mm (10 r.l.) |
| no. $\delta, \beta_{x,y}$ spoil. | 1, 4, 4 | 1, 4, 4 |

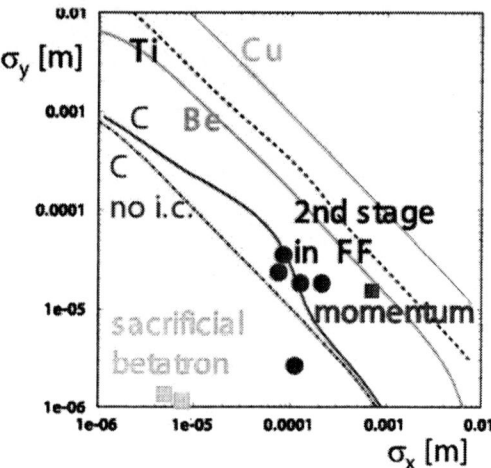

**FIGURE 1.** Nominal beam sizes at the CLIC spoilers superposed on a damage threshold diagram according to [5]

Large betatron oscillations are not easily generated from pulse to pulse. In addition, such oscillations filament rapidly in the linac due to the beam energy spread (partly unavoidable and partly introduced for BNS damping), increasing the beam emittance by at least two orders of magnitude [6]. Momentum errors are expected to occur much more frequently. Our design philosophy has been to demand passive survival for momentum errors, but to allow for sacrificial betatron collimators (which saves length). The likelihood of beam impact on the betatron collimators must then be minimized. Momentum errors in the linac can be caused, *e.g.*, if one of the 22 drive beams is missing, or if the beam is injected at the wrong phase, or with the wrong charge. Simulations using PLACET and MAD show that every energy error likely generates significant tails, at the 1% level, in the beam distribution incident on the momentum collimator [7]. Most of the momentum errors studied result in a beam-size increase by not more than a factor 2 or 3 [7]. Therefore, only carbon and possibly beryllium can withstand the beam impact on the momentum collimators for these failures, confirming our earlier finding for the nominal beam. The momentum errors may be accompanied by significant horizontal betatron oscillations. Indeed, failures which cause momentum errors smaller than 4% can induce betatron amplitudes of 10–45 $\sigma_x$ [7]. Therefore, in order to protect the horizontal betatron spoilers, the momentum collimation was tightened to ±1.5%, so that all dangerous beam pulses are intercepted by the momentum spoiler. Table 2 summarizes the collimator parameters. The slightly marginal Be was chosen as spoiler material in view of its better wake-field properties (see below). At 3 TeV the rms beam size at the momentum absorber for a beam scattered by the spoiler is 1.1 mm.

# HALO AND MUON BACKGROUND

Muons are generated at a rate of roughly $10^{-4}$ per lost electron. GEANT-3 and GEANT-4 simulations for CLIC [8] have shown that 1–10% of the muons produced by beam loss on the collimators reach a detector at the IP of 7.5 radius. If the halo amounts to $10^{-3}$ of the beam, about 25000 muons may pass through the detector per bunch train. Simulations performed by the CLIC physics study group suggest that even with this magnitude of background it is still possible, and in fact straightforward, to discern supersymmetric events emerging from the primary IP [9]. Details of the muon simulations are important. For example, increasing the size of the magnets from 20 cm to 50 cm reduces the number of muons by an order of magnitude. Also the full shower evolution needs to be considered [8].

The wave length required for beam-size measurements by a laser wire depends not only on the size to be measured, but also on the beam size in the orthogonal plane, as [10] $\lambda < \pi \sigma_y^2/(2\sigma_x)$. Inserting, for example, $\sigma_y \approx 2\ \mu$m $\sigma_x \approx 8\ \mu$m, we obtain $\lambda < 800$ nm, which is easily achieved. However, a severe concern is the magnitude of the laser-wire signal compared with the background from lost halo particles. Detailed simulations with the code BDSIM for an older version of the CLIC beam delivery show that at most locations downstream of the collimators, the background from beam loss is up to 7 orders of magnitude higher than the laser-wire signal [11]. This may imply the need for a dedicated diagnostics section upstream of the first collimator.

Simulations of the beam particle transport at large amplitudes are important for estimating the expected background levels. A comparison of tracking results for the nominal Gaussian beam by the four codes DIMAD, MAD, MERLIN and PLACET has revealed significant differences in the rms IP spot sizes, especially if a nonzero energy spread or synchrotron radiation are also included [12].

# WAKE FIELDS

The collimators consist of a central flat part and they are tapered at a shallow angle on either side. The flat part at the centre gives rise to a resistive wall wake field, the two tapers to a combination of resistive-wall and geometric wakes. The beam centroid deflection by a tapered circular collimator is [13, 14]:
$\Delta y' = (2N_b r_e/(\gamma \sigma_z)) \left[ \frac{(4\lambda \sigma_z)^{1/4}}{g^{3/2}} + L_F \frac{(\lambda \sigma_z)^{1/2}}{2\sqrt{\pi}g^3} \right] y$, with $\lambda[\text{m}] = \rho[\Omega\text{m}]/(120\pi)$, and assuming the optimum taper angle at which the wake is minimum [14]: $\theta_{\text{opt}} \approx 1.1(\lambda \sigma_z/g^2)^{1/4}$. The above formula is applicable for $\sqrt{\sigma_z \lambda} \ll g \ll \sqrt{\sigma_z \lambda}\sigma_z/\lambda$ [14]. For example considering a resistivity of $\rho \approx 6 \times 10^{-8}$ $\Omega$m (beryllium), it is valid for half gaps $g$ between 70 nm and 13 mm. Taking again Be and a gap $g = 100\ \mu$m, the optimum taper angle is 30 mrad. However, for the CLIC rms bunch length of 35 $\mu$m the anomalous skin effect may become important [15]. This has not yet been accounted for.

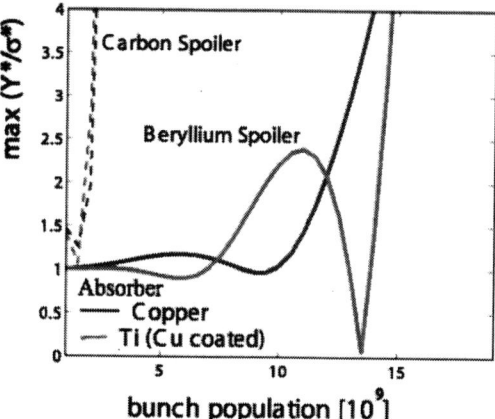

**FIGURE 2.** Maximum vertical jitter enhancement due to collimator wake fields for 4 vertical spoilers and 4 absorbers made from a variety of materials

Assuming the wake field for circular collimators, and considering the combined effect of 4 spoilers and 4 absorbers, we can compute the vertical centroid jitter enhancement due to the wake field. For an incoming betatron oscillation of a certain amplitude, the maximum position offset at the IP (*i.e.*, maximum as a function of the oscillation phase) equals the initial amplitude multiplied by a jitter enhancement factor. This enhancement is computed after normalizing both the initial amplitude and the final IP position to the ideal (linear) rms beam sizes at either position. The jitter enhancement due to wakes in the CLIC collimation system is displayed in Fig. 2 as a function of bunch intensity, for a few absorber and spoiler materials. We observe that a carbon spoiler is ruled out by its enormous wake field. A beryllium spoiler promises an acceptable performance, and so do absorbers consisting either of Cu-coated Ti or pure Cu. So far we have treated circular collimators only. This may be a good approximation also for square collimators. A theory for rectangular collimators of half width $h$ and half height $g$ was developed by G. Stupakov in 1996 [16] and 2001 [17]. According to [17], for $\sqrt{\theta h^2/(g\sigma_z)} < 3.1$, the deflection from the geometric wake field of two rectangular shallow tapers is $\Delta y'_G = (N_b r_p/\gamma)\sqrt{\pi}\theta h/(2\sigma_z g^2)y$. This contains an additional factor $h/b$, which can be significant. For comparison, the deflection by two circular tapers of radius $g$ is $\Delta y'_G = (N_b r_p/\gamma)2\theta/(\sqrt{\pi}\sigma_z g)y$, which is $4/\pi \approx 1.27$ times larger than that for a square aperture with half gaps $h = g$. At SLAC, wake fields of four different prototype collimators were measured [18], but, unfortunately, none of these has probed the regime which is relevant for CLIC.

## ALTERNATIVE NONLINEAR SYSTEM

As an alternative to the conventional linear collimation system described above, we have designed a nonlinear system, which employs three skew sextupoles. The main skew sextupole is placed at a position with large dispersion and blows up the vertical beam size at a single spoiler downstream, profiting from the large beam energy spread. This ensures a large beam size at the spoiler and, thus, its survival in case of a failure. Also the collimator gaps are much increased, and the strength of the wake fields is reduced. The vertical beam size at the spoiler is controlled by the dispersion and by the strength of the skew sextupole. The value of the dispersion is limited by the desired maximum length of the system and by emittance growth from synchrotron radiation. A second skew sextupole separated by an optical $-I$ transform from the first and located behind the spoiler cancels the nonlinear aberrations induced by its upstream counterpart. This pair of skew sextupoles is positioned so as to remove large-amplitude particles in the final-doublet (FD) phase. To take care of the orthogonal IP phase, where we may collimate at much larger amplitudes, we place a weak (pre-)skew sextupole upstream of the entire system, at an appropriate betatron phase. This pre-sextupole will nonlinearly deflect IP-phase particles, entering at large amplitudes, into the final-doublet phase, so that they will be deflected by the following much stronger FD-phase skew sextupole and impact on the same spoiler as the FD-phase particles. The horizontal and vertical rms beam sizes at the spoiler are 69 $\mu$m and 209 $\mu$m, respectively, which should be compared with the corresponding much smaller values of 8 $\mu$m and 1 $\mu$m for the conventional linear system. Likewise, the vertical spoiler half gap is 16.7 mm for the nonlinear system, instead of about 100 $\mu$m for the linear one. However, the collimation in the nonlinear system is not perfect. 'Holes' exist in the 6-dimensional phase space, where particles may escape collimation. More details on this system can be found in reference [19].

## CONCLUSIONS AND THANKS

Two collimation systems for CLIC at 3 TeV were presented. Both have a reasonable length of about 2 km. Simulations so far indicate a promising performance. Only minor changes are required to lower the centre-of-energy from 3 TeV to 500 GeV. On the other hand, the collimation depths and collimator materials are heavily constrained by various requirements. The nonlinear system is a promising alternative to the conventional system, with several potential advantages, but also a few drawbacks. A modified version of the nonlinear system could be applied in storage rings, which might be a possible solution for collimation at a future LHC upgrade.

A number of open questions remain: (1) Do the codes accurately describe the particle transport at large amplitudes? (2) Do we believe the predicted wake field for flat collimators? (3) How important is the anomalous skin effect for CLIC bunch lengths? (4) What is the best material for spoilers or collimators? (5) Do the absorbers survive the full beam impact? (6) Are 'holes' in phase space a showstopper for the nonlinear system? (7) Should we devote time and resources to study more exotic schemes, like crystals, plasmas, lasers, liquid metals, or wake-free collimators? Many of these questions could be addressed in a dedicated test facility.

We would like to thank K. Bane, F. Caspers, A. Drozhdin, P. Emma, S. Hertzbach, J. Irwin, T. Markiewicz, O. Napoly, P. Raimondi, T. Raubenheimer, M. Ross, M. Seidel, A. Seryi, G. Stupakov, N. Walker, and P. Tenenbaum for helpful discussions, and A. Bay, G. Guignard, J.-P. Riunaud, F. Ruggiero, and I. Wilson for supporting this study.

## REFERENCES

1. Aleksa, M., et al., *Proc. Nanobeam 2002*, Lausanne, CERN-Proceedings-2003-001, 2002, pp. 35–50.
2. Schulte, D., High-Energy Beam-Beam Effects in CLIC, CERN-CLIC-NOTE-391 (1999).
3. Decker, F.-J., et al., "Design and Wake Field Performamce of the New SLC Collimators," *Linac 96*, Geneva (1996).
4. De Staebler, H., Walz, D., private communication (1996).
5. Fartoukh, S., et al., Heat Deposition by Transient Beam Passage in Spoilers, CERN-SL-2001-012-AP (2001).
6. Assmann, R., et al., *EPAC 2000*, Vienna (2000) p. 522.
7. Schulte, D., et al., *PAC 2001*, Chicago (2001) p. 4068.
8. Burkhardt, H., *Proc. Nanobeam 2002*, Lausanne, CERN-Proceedings-2003-001, 2002, pp. 57–60.
9. Battaglia, M., and Burkhardt, H., private communications.
10. Frisch, J., "Technical Challenges Outstanding," *Web Proc. Nanobeam 2002*, Lausanne, at http://www.cern.ch/nanobeam (2002).
11. Blair, G., *Proc. Nanobeam 2002*, Lausanne, CERN-Proceedings-2003-001, 2002, pp. 169–173.
12. Redaelli, S., et al., *Proc. Nanobeam 2002*, Lausanne, CERN-Proceedings-2003-001, 2002, pp. 51–56.
13. Yokoya, K., Impedance of Slowly Tapered Structures, CERN-SL/90-88 (1990).
14. Irwin, J., Chapter 9 of the Zeroth Order Design Report for the Next Linear Collider, SLAC-R-0474-VOL-2 (1996).
15. Tenenbaum, P., Seryi, A., comments at this workshop.
16. Stupakov, G.V., Part. Acc. **56**, 83 (1996).
17. Stupakov, G.V., *PAC 2001*, Chicago (2001) p. 1859.
18. Tenenbaum, P., et al., *PAC 2001*, Chicago (2001) p. 418.
19. Faus-Golfe, et al., *8th EPAC*, Paris (2002) p. 533.

# Design and Performance of the Tesla Test Facility (TTF) Collimation System

## H. Schlarb

*Deutsches Elektronen Synchrotron (DESY), D-22607 Hamburg, Germany*

**Abstract.** To perform a-proof-of principle experiment of a SASE based Free Electron Laser a permanent magnet undulator has been installed in the TTF linac phase 1. The magnets used (NdFeB) are known to be sensitive to radiation damages if exposed to high energy electrons. Already beam losses in the order of $10^{-6}$ of the nominal TTF beam current (64 $\mu$A) are critical for the undulator and could have cause an irreversible damage of its magnets after a few months of operation. The protection of the undulator against radiation has been attacked twofold: first, a two stage spoiler-absorber collimation system has been installed and second, a beam loss detection system of high sensitivity has been added to the machine protection system. The performance and the operational experience of the active and the passive protection system are discussed.

## INTRODUCTION

The TESLA Test Facility linac is a superconducting linear accelerator whose purpose it is to demonstrate that a linear collider based on superconducting cavities can be built and operated reliably. The linac is constructed from two 12 m long acceleration modules, each is comprised of a string of eight 9-cell cavities operated at 1.3 GHz frequency. A laser driven photoinjector provides the TESLA 500 large bunch spacing ($\approx \mu$s) and high bunch charge (few nC). The low emittance beam produced in the RF gun of the injector opened up the possibility to drive a Free-Electron Laser (FEL). To perform a proof-of-principle experiment for a FEL operated in VUV-wavelength range, three permanent magnet undulators with an overall length of 15 m have been installed in summer 1999.

NdFeB permanent magnets have been used to achieve the high magnetic field quality required to initiate the FEL process. Investigations on the radiation threshold of the magnets predict 1% reduction of the magnet remanent field at an absorbed dose of 70 kGy deposited by charged particles with energies above 20 MeV [1]. At 230 MeV beam energy the average beam power amounts to 15 kW. In worse case scenario, operation with full intensity can cause critical magnet damages within one hour at rather small losses of 0.1% of the beam current.

To protect the undulator, an active and a passive protection system have been developed. The active system is based on beam loss detectors with sufficient high sensitivity [2]. The passive system was comprised of five collimators installed in a 5.5 m long section upstream the undulator [3].

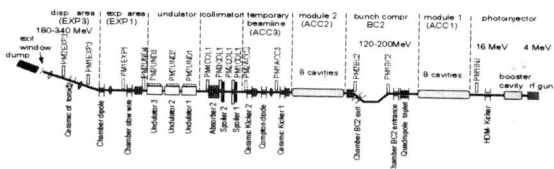

**FIGURE 1.** Scheme of the the Tesla Test Facility linac. In the scheme the location of loss monitors along the machine are indicated.

**TABLE 1.** Beam parameters

|  | TTF | FEL | Used |
|---|---|---|---|
| Bunch spacing | 1 $\mu$s | 0.11 $\mu$s | 0.44/1 $\mu$s |
| Beam current | 8 mA | 9 mA | 3-7 mA |
| Bunch charge | 8 nC | 1 nC | 2-4 nC |
| norm. emittance | 20 $\mu$m | 2 $\mu$m | 3-10 $\mu$m |
| Beam Energy | 150 – 500 MeV |  | 150 – 340 MeV |
| Beam duration | 800 $\mu$s |  | 10 – 800 $\mu$s |
| Repetition rate | 10 Hz |  | 1 Hz |
| Beam power | 10 – 36 kW |  | < 1.4 kW |

The TTF linac has an overall length of 100 m, sketched in Fig. 1. A magnetic chicane between the first and second acceleration module allows to compress the electron beam longitudinally. The design parameters for TTF and FEL operation are summarized in Tab. 1. Caused by various reasons, the parameters listed in the right column have been realized during the three years operation, until summer 2002. Thus the operation conditions were somewhat relaxed mainly due to the reduced average power (only 1 Hz operation) and the increased bunch spacing (factor of four).

## Beam core, beam halo and dark current

The high brightness beam is produced by impinging a laser beam of 7 ps rms duration on a CsTe photo-cathode installed in an RF gun operated at gradients of 40 MV/m. The high bunch charge and the short bunch length cause strong space charge forces. Caused by the non-linearities of the forces large beam tails can be produced while the beam core remains of high quality ($\gamma\varepsilon = 2$-$4\,\mu$m). For instance, at an electron beam propagated through a waist the repulsive forces acting on the electrons of the beam core are strong enough that they can not cross the bunch center. But electrons with large radial or longitudinal displacements can cross the beam center and start to populate a volume in the transverse phase space very different from the core electrons. The halo generating process is very sensitive to the machine setting and small changes in the charge or the magnetic fields can causes large changes in the halo distribution.

To the FEL process only electrons of the beam core contribute. An optimized FEL operation requires the matching of the core to the undulator optics. In this case, simulations predict that the miss-matched beam tails can be lost in the undulator. Thus, the specific feature of the electron source is threefold: first, the high charge density of the beam core can destroy the vacuum chamber within a few microseconds in case of an accidental beam loss. Second, the large beam halo can easily cause losses in the linac, and third, the beam halo distribution is very unstable and changes frequently.

In addition, caused by the high gradient in the RF-gun and the large RF-duty cycle, high dark currents can be accelerated through the entire linac. The relevant dark currents are emitted from the photo-cathode or the gun back-plane. The time structure is give by the 1.3 GHz RF periods and according to Fowler-Nordheim theory it is spread in phase by about 15-20° rms. The central emission is delayed by 50° with respect to electron beam. A part of the dark current survives the magnetic chicane and reaches the collimation section. Dark currents of more than $100\,\mu$A have been observed, more than 1% of the nominal beam current.

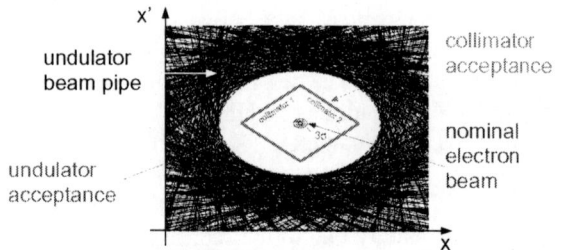

**FIGURE 2.** Phase space acceptances of the collimator and undulator at 230 MeV (x-direction).

# DESIGN AND PERFORMANCE OF THE COLLIMATION SYSTEM

## Basic layout

The undulator dipole field has superimposed strong focusing quadrupoles with a FODO-cell length of 0.96 m. The phase advance across the undulator is larger than $\pi/2$ which requires to collimate both betatron phases. Two spoilers separate by a drift space of 2.5 m length define the phase space acceptance of the collimator section. Figure 2 illustrate the properties of the beam distribution and the phase space acceptance of collimator and undulator. At 230 MeV, the clearance between the collimated beam (mono-energetic) and the undulator vacuum chamber is 1.1 mm (23% safety margin).

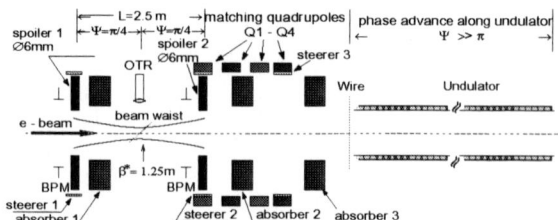

**FIGURE 3.** Basic layout of the collimation section.

The gradient of the undulator quadrupole is fixed. Four quadrupoles Q1-Q4, shown in Fig.3, match the collimated phase space to the energy dependent undulator FODO-cell optics. Steerer magnets are foreseen to place the beam in the center of the spoilers and to compensate for quadrupole displacements. High resolution BPMs are located in front of the spoilers. Both are mechanically aligned to the spoilers within $200\,\mu$m. They also provide the measurement of the dark currents with a precision of 500 nA. The beam current is measured before and after the first spoiler and behind the undulator by toroids. To detect losses secondary emission multipliers are mounted near to the collimators.

The spoilers are made of aluminum (AlMg$_3$). They can withstand a head on collision of the beam for $2.5\,\mu$s[1] (reaction time of beam interlock $2.2\,\mu$s). Three absorbers are installed to remove secondary particles produced at the spoilers. The purpose of the first absorber is to protect diagnostics components. All absorbers are made of copper (SE-Cu 2.0070).

The beam optics is shown in Fig. 4. The beam exiting the second acceleration module is matched using two quadrupole doubles into the collimator section. The transverse beam profile can be measured by inserting an OTR-screen. To simplify the matching conditions the

---

[1] Worst case scenario with $60\,\mu$m rms beam size at 9 MHz rep. rate

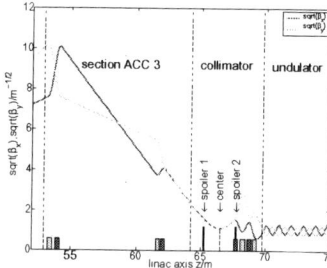

**FIGURE 4.** Optics at the collimator section (230 MeV).

screen is located between the spoilers at the beam waist.

## Energy bandwidth

Due to space limitations an energy collimation could not been added to the beam line. Caused by chromaticity particles with large energy deviation are improperly matched to the undulator. Figure 5 shows the energy bandwidth of the collimation system. For the simulation the incoming beam has been increased in emittance such that 50% of the particles pass the spoilers. The various colors shows the fraction of the collimated beam lost in the undulator (primary particle losses). The blue area ($\delta E/E_0 = +7.5\%$ to $-10\%$ at $E_0 = 230$ MeV) indicate the energy range with full collimation performance, hence no electrons are lost after the spoilers. Towards lower and higher energies $E_0$ the bandwidth decreases. For the entire energy range between 150 MeV and 350 MeV the bandwidth is sufficiently wide for off-crest beam acceleration mandatory for bunch compression in the magnetic chicane.

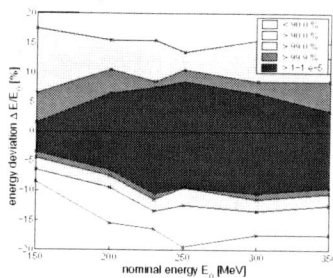

**FIGURE 5.** Energy bandwidth of the collimation system.

Operation of the linac in the red area is tolerable for short periods (weeks), assuming the beam halo does not exceed 1% of the nominal beam current. Linac operation in the green area is only possible with a significant reduced duty cycle, i.e. 1 Hz operation and small number of bunches in a pulse-train. Larger energy deviations (yellow or white area) cause significant exposure of magnets mainly in the first undulator module. The active protection system inhibits such operation.

*Dark current.* The spread in the emission phase of the dark currents causes a spread in energy at the entrance of the collimator which exceeds the bandwidth of the system. The phase spread is further increased due to the bunch compressor as shown in Fig.6(a) before the chicane and Fig.6(b) after ACC2. The particle energies are distributed from 90 MeV (deceleration in ACC2) to 240 MeV above the beam energy of 230 MeV. Obviously, the collimation system can not deal with such distribution and during time periods with high dark currents large dose values have been observed [4].

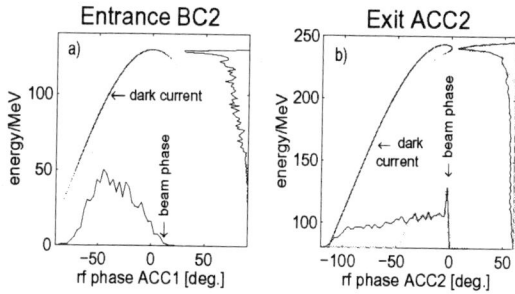

**FIGURE 6.** Distribution in the longitudinal phase space.

This dark current problem has been solved twofold: first, a new generation of polished photo-cathodes were used showing much lower dark currents emission and second, with a low energy scraper installed in BC2 particles exceeding the energy bandwidth of the collimator can be removed, as indicated by the green curves in Fig 6(b).

## Quadrupole displacements

The quadrupoles of the collimation section have been aligned by the survey within a precision of 200 $\mu$m. Particle tracking simulations predict that full collimator performance is achieved for quadrupole displacements below 50 $\mu$m. Consequently, the proper settings of the correctors superimposed to the first and forth matching quadrupole to eliminate the effect of quadrupole displacements is mandatory. Using beam based alignment techniques, the required orbit correction at the entrance of the undulator has been determined to be 1.36 mm and 1.51 mrad in the x-plane and 0.44 mm and 0.55 mrad in the y-plane.

## Secondary particle removal efficiency

Secondary particles produced at the spoilers escaping the absorber system contribute to the absorbed dose in the undulator and limits the amount loss at the collimator section. Using EGS4 [5], the removal efficiency for secondary particles has been determined to be 99.83%. About 0.17% of the energy incident on the spoilers is dumped mainly in the first undulator module. Figure 7 shows the expected energy deposition in the magnets along the undulator for different vertical position ($x \equiv 0$). The dose values are calculated for an electron incidence of 1 kJ on the spoilers. According to this simulation, the removal efficiency limits the acceptable beam losses at the spoilers to 1% of the beam current if 1000 hours nominal beam operation is assumed. In the simulations, control dosimeters are added to the geometry. The simulated dosimeter values are used to estimated the secondary efficieny experimentally and to distinguish undulator exposure caused by primary electrons (failure operation) or secondary particles from the collimators.

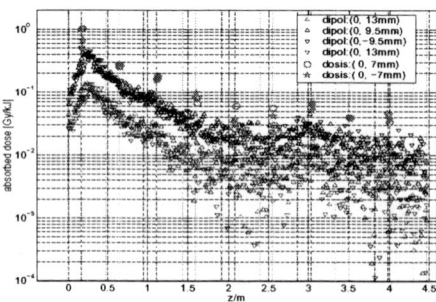

**FIGURE 7.** Absorbed dose along the first undulator module.

Experimentally, the removal efficiency has been found to be typically a factor of 3 to 5 worse than simulated. Most likely, the orbit correction to compensate for quadrupole displacements caused beam losses at the $2^{nd}$ absorber which explains the observed discrepancy.

## BEAM LOSS MONITOR SYSTEM

The machine protection system based on current measurements inhibit beam operation at losses in the percent range. To avoid damage of the undulator magnets losses of $I_{loss}/I_{nom} = 10^{-6}$ had to be detected and if exceeded the linac duty cycle automatically reduced. A second system has been developed, based on photomultipliers and secondary emission multipliers which detect electromagnetic showers initiated by beam losses. Each detectors controls either a parts of a beam line or a single beam line device only. The system provides the possibility to adjust individually the allowable loss threshold. To inhibit very exotic operation condition such that 100% of the beam is lost at a location without a detector the toroid based system was still active.

Since loss monitors detect secondary particles the calibration is non-trival: first, the detected signal height may depend on the incidence position and angle of primary electrons (geometry factor) and second, the difference

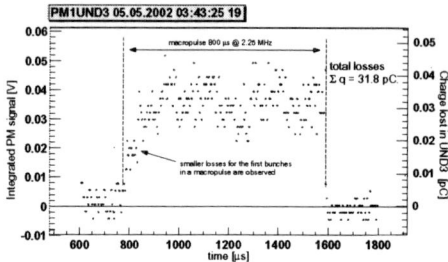

**FIGURE 8.** PM signal recorded by the control system.

signals of two toroids is too inaccurate to calibrated the photomultipliers.

The dependence of the geometry has been simulated with EGS4. To calibrate the photomultipliers of high sensitivity, small beam losses have been induced in a controlled way, i.e. using wire scanner. Figure 8 shows the beam loss in the third undulator recorded within a macropulse with 1800 bunches. The losses amount to 31.8 pC or $I_{loss}/I_{nom} = 4.4 \cdot 10^{-6}$ acceptable for 1 Hz linac operation. The lower threshold for the interlock system was limited by back ground radiation in the tunnel. The system fulfill the requirement to inhibit beam losses of $10^{-6}$, but has been operated typically at the level of $10^{-4} - 10^{-5}$ due to the reduce beam operation duty cycle.

## CONCLUSION

The new developed active and passive protection system provided sufficient safety for the linac operation during the entire operation of TTF phase I and opens up the possibility for high beam power experiments. After disassebling the undulators modules no magnetic field degradation has been observed. Difficulties were mainly related to the missing energy collimation and the tight tolerances on quadrupole displacements. The displacements have been corrected using dipole magnets, but the required orbit correction most likely reduced the removal efficiency for secondary particles by a factor of 3–5. Due to the limited energy bandwidth of the system higher dose values during time periods with high dark current emission has been observed.

## REFERENCES

1. Schlarb, H., Ph.D. thesis (2001), dESY-THESIS-2001-055.
2. Schlarb, H., *EPAC'02*, pp. 1966–1968 (2002).
3. Schlarb, H., *EPAC'02*, pp. 2706–2708 (2002).
4. Schlarb, H., *EPAC'02*, pp. 1190–1192 (2002).
5. Nelsen, W. R., Tech. Rep. SLAC-265, Stanford, CA 94305 (1985).

**Beam-Beam '03**

# Beam-Beam'03 Summary

W. Fischer[*] and T. Sen[†]

[*]*Brookhaven National Laboratory, Upton, New York 11973*
[†]*Fermi National Accelerator Laboratory, Batavia, Illinois 60510*

**Abstract.** This paper summarizes the presentations and discussions of the Beam-Beam'03 workshop, held in Montauk, Long Island, from May 19 to 23, 2003. Presentations and discussions focused on halo generation from beam-beam interactions; beam-beam limits, especially coherent limits and their effects on existing and future hadron colliders; beam-beam compensation techniques, particularly for long-range interactions; and beam-beam study tools in theory, simulation, and experiment.

## INTRODUCTION

The Beam-Beam'03 workshop was held in Montauk, Long Island, from May 19 to 23, 2003. It was attended by 15 participants from 9 institutions. Beam-Beam'03 was held in conjunction with with the 29th ICFA Advanced Beam Dynamics Workshop HALO'03. Part of the program, registration, abstract submission, and proceedings were shared with HALO'03. The workshop concentrated on beam-beam effects in circular colliders, with emphasis on hadron colliders, and followed earlier workshops on this subject [1, 2]. After a plenary talk on halo formation due to the beam-beam interaction, three main topics were discussed:

1. Beam-beam limits,
   especially coherent limits and their effects on existing and future hadron colliders
2. Beam-beam compensation techniques,
   particularly for long-range interactions
3. Beam-beam study tools
   in theory, simulation, and experiment

In the following we summarize the presentations and discussions for each of these topics.

## BEAM-BEAM HALO FORMATION

F. Zimmermann, CERN, summarized the measurements, simulations, and analytical models for the halo formation due to beam-beam interactions for both lepton and hadron colliders.

In lepton colliders, two beam-beam limits are observed: the first limit restricts the beam-beam parameter $\xi$, the second limit is due to the formation of transverse tails (Seeman, 1983). Tails in lepton colliders reach a 'steady state' due to radiation damping. They cause experimental background, reduce the beam lifetime, and often limit the luminosity. Dramatic increases of both core and tails were observed with increasing beam currents.

With no radiation damping in hadron colliders, the betatron amplitudes of particles in the tails are not reduced. Tails not only cause background, they can damage collimators and quench a superconducting machine.

A number of mechanisms for halo generation were considered in the past, among them beam-beam bremsstrahlung (Burkhardt et al., 1997), stochastic diffusion (Cornelis, 1993), Arnold diffusion (Chirikov, 1979), resonance trapping (Chao, Month, 1974), phase convection (Gerasimov, 1990), resonance streaming (Tennyson, 1980), and modulational diffusion (Chirikov, 1979). Transverse tails were most often measured with the help of collimators. In LEP, beam-beam bremsstrahlung was found to be the dominant tail generating process.

Halo generation in lepton colliders was studied with a number of computer codes. Self-generated boundary conditions were proposed by Irwin in 1989, and subsequently implemented in two codes. In addition, macro particle and PIC codes were developed. Typically $10^7$ to $10^9$ particle turns are tracked, and a good predictive power of these codes has been demonstrated.

Diffusive rates with beam-beam interactions in HERA and RHIC show similar values. However, the Tevatron in Run II and the LHC enter a new regime where long-range collisions dominate. These have caused fast beam losses in simulations and they may ensure that no tails develop.

Various tools are available to manipulate tails. Matched beam sizes, centered collisions, zero crossing angles and optimized tunes (with tolerances of approximately 0.001) were shown to be beneficial. Octupoles were used in VEPP-4, VEPP-2M, and DAΦNE.

**TABLE 1.** Comparison of maximum beam-beam parameters in hadron colliders [4, 5, 6, 7]. Note that machine configurations change over time and that parameters in routine operation may be different.

| quantity | ISR | SPS | Tevatron Run II (design) | HERAp | RHIC p 2003 | LHC (design) |
|---|---|---|---|---|---|---|
| bunches per beam | coasting | 3 | 36 | 174 | 55 | 2808 |
| experiments (head-on interactions) | 6 | 2 | 2 | 2 | 4 | 4 |
| long-range interactions | ... | 4 | 70 | — | 2 | 120 |
| beam-beam parameter per IP $\xi$ | 0.001 | 0.009 | 0.01 | 0.0007 | 0.004 | 0.003 |
| total beam-beam tune spread $\Delta Q_{bb}$, max | 0.008 | 0.028 | 0.024 | 0.0014 | 0.015 | 0.010 |

## BEAM-BEAM LIMITS

In circular colliders the beam-beam interaction is one of the most limiting effects. The maximum beam-beam parameters achieved in hadron colliders are shown in Tab. 1. A table comparing lepton colliders can be found in Ref. [3].

Y. Alexahin, FNAL, reviewed the theory and observations of coherent beam-beam effects. The eigenmodes of coherent dipole oscillations can be found by solving the Vlasov equation in first order perturbation theory. At intensity ratios greater than 0.6, the discrete $pi$-mode lies outside the continuous spectrum and therefore may not be Landau damped. Multi-bunch modes with $36 \times 36$ bunches in the Tevatron were considered. The tune spread induced by the head-on and long-range interactions is large enough to damp the multibunch modes provided the tunes of both beams are the same. If the anti-proton tunes are lower than the proton tunes, the coherent modes shift by less than the incoherent tunes and may not be damped. A number of mechanisms were proposed to suppress the $\pi$-mode: a split of bare lattice tunes (A. Hoffman), redistribution of phase advances between interaction points (A. Temnykh, J. Welch), different integer parts of tunes in separate rings (W. Herr), and long bunches (due to the overlapping of synchrotron sidebands). During discussions, Alexahin suggested separating "collective" from "coherent", the former applying to purely intensity dependent phenomena, the latter applying to phenomena where particle phases are correlated.

L. Jin, University of Kansas, showed a case of collective instability in HERA. When the $e^+$-beam approached a fourth order resonance, a 30% emittance growth was observed in the proton beam. This observation could be well reproduced in a simulation. Later he discussed the importance of tune spread to the collective beam-beam instability.

W. Fischer, BNL, gave a presentation of strong-strong and other beam-beam observations in RHIC. With the current bunch spacing, bunches in RHIC experience only two long-range interactions. It is intended to accommodate a total tune spread as large as has been achieved in the past. Furthermore, RHIC is the first bunched beam hadron collider in which strong-strong effects are observed. Beam-beam generated $\sigma$- and $\pi$-modes were seen with a frequency difference that matches expectations from calculations [8]. The coherent modes observed could be suppressed by separating the tunes of the two rings. This may not be sufficient if the beam-beam parameter is doubled and the triplets are better corrected, leaving the beam-beam interaction as the dominant source of transverse nonlinearities.

In two talks the performance of the B-factories were reviewed. W. Kozanecki, CEA-Saclay/SLAC, showed the recent performance of the SLAC B-factory. A strong interplay between electron cloud and beam-beam effects is observed. With changing parameters along the bunch train, luminosity and background optimization relies on a delicate balance between currents, tunes, beam-beam and e-cloud parameters. Long-range interactions have an observable negative impact on the luminosity. As of May 2003, the tunes were moved closer to the half integer and were found to improve machine performance. Beam-beam simulations show encouraging agreement with experiments although not all relevant phenomena were included. The beam-beam parameters $(\xi_x, \xi_y)$ achieved in the LER and HER respectively are $(0.065, 0.048)$ and $(0.075, 0.060)$, the luminosity reached $6 \cdot 10^{33} \text{cm}^{-2}\text{s}^{-1}$.

K. Ohmi, KEK, reported on the experience with finite crossing angles at KEKB. With a crossing angle of 2x11 mrad a luminosity of $1 \cdot 10^{34} \text{cm}^{-2}\text{s}^{-1}$ was achieved. No problems were encountered with the crossing angle up to a beam-beam parameter of $\xi \approx 0.05$. The beam-beam parameters $(\xi_x, \xi_y)$ achieved in the LER and HER respectively are $(0.097, 0.066)$ and $(0.067, 0.050)$. Electron cloud effects in the positron ring (LER) are mitigated (both in KEKB and PEP II) by wrapping solenoidal coils around most of the machine. As in PEP II, a day-by-day fine tuning of the machine parameters is required to maintain the highest luminosities. Simulations helped with the choice of the tuning parameters.

The use of crossing angles and long bunches is also under consideration for hadron colliders [9]. W. Fischer showed an example of a possible luminosity increase at the incoherent beam-beam limit with six superbunches in RHIC. Assuming that the incoherent tune shift is the limiting effect and neglecting a number of other effects, a luminosity increase of about two orders of magnitude was estimated.

T. Sen, FNAL, reviewed the theory and observations of beam-beam interactions in the Tevatron. One of the key observations is that a small tune footprint by itself does not guarantee good beam lifetime. At injection, the long-range beam-beam interactions (which create a small tune footprint) limit the anti-proton beam lifetime to 1-5 hours compared to 25 hours without the beam-beam interactions. No significant effect on the protons is seen. On the ramp, about 10% of the anti-protons are lost and the observed anti-proton emittance growth is suspected to be caused by beam-beam. During the beta-squeeze, anti-proton losses are low while proton losses are occasionally high enough to cause quenches. At collision, beam lifetimes are mainly determined by the $p - \bar{p}$ interactions at the detectors. Bunch dependent emittance growth of anti-protons due to beam-beam effects at collision is sometimes observed. This can usually be corrected by a change of tune. Changes to the helices, realignment of the Tevatron, cleaner IR optics, different bunch patterns and active beam-beam compensation are among the several methods under development to mitigate the effects of the beam-beam interactions.

B. Erdelyi, FNAL, compared simulations with experimental studies in the Tevatron. Until the recent commissioning of the vertical dampers, the vertical chromaticity was set to a high value to keep the protons stable at injection energy. This however lead to a low anti-proton lifetime and the emittance was found to decrease initially before reaching a constant value. From these observations, the dynamic aperture of anti-protons could be measured and was found to be in good agreement with the simulation results. At collision, lifetimes observed at different tunes were compared with dynamic aperture calculations at these tunes and found to be in qualitative agreement.

W. Fischer showed how the beam-beam interaction and unequal rf frequencies can generate tune modulation. This effect leads to a reduction of the beam lifetime in RHIC when the rf frequencies of the two rings are not locked, a situation typically encountered during the RHIC energy ramp.

## BEAM-BEAM COMPENSATION

Two approaches are currently pursued to compensate the long-range beam-beam interactions: electric wires and electron lenses. Attempts to compensate the direct space charge forces through four-beam schemes were not successful in the past, but are under investigation again. Also under investigation are the compensation of the beam-beam multipole effect with magnets. A short summary of earlier compensation schemes can be found in Ref. [10].

## Wire compensation at the SPS

The idea of compensating the long-range beam-beam interactions by the magnetic field of a current carrying wire was proposed for the LHC by Koutchouk. In the LHC the long-range interactions are clustered around each IP and occur at nearly the same betatron phase. Simulations showed that two wires placed around each IP reduced the tune footprints and increased the diffusive dynamic aperture by about $(1-2)\sigma$.

F. Zimmermann reported on recent experiments performed at the SPS to observe the effects of a single wire on a beam. A 1m long wire supported on a rigid structure and carrying 267 A of current was placed in the vacuum chamber. Water flow through the hollow wire was required for cooling. Orbit bumps were used to change the transverse separation of the beam and the wire. Beam lifetime dropped and background rates increased at separations smaller than $9\sigma$ - close to the predictions from simulations. Orbit distortions and tune shifts due to the wire were also close to predictions. Diffusion rates could not however be measured.

These initial observations are indeed encouraging and suggest that the idea is worth pursuing. The next critical step is to demonstrate that the wire can compensate the effect of another field on the beam. The plan in the next stage of the experiment is to install two wires in the SPS. The second wire will be powered to cancel the effect of the first wire on the beam. If the experiment succeeds, the wire compensation idea will likely be pursued seriously not only for the LHC but also for the Tevatron and future hadron colliders.

## Multiple wires and modeling for the Tevatron

The wire compensation principle is also being tested at the Tevatron. The long-range interactions occur at different phases all around the ring and both beams traverse the same beam pipe. This necessarily makes the application of the wire compensation more complicated. One advantage is that the wire needs to operate only in a DC mode since the average effect on all bunches needs to be compensated.

B. Erdelyi discussed a fast and accurate model of the field of a finite length wire that allows misalignments and is now implemented in the codes COSY Infinity and Six-Track. First simulation results at injection energy with four wires placed in the Tevatron are encouraging. The maximum current required in each wire is estimated to be 232 Amps, a value close to the current used in the SPS measurements. At suitably chosen distances and angles of the wire relative to the anti-proton beam, the resonance structure excited by the wires resembles that generated by the long-range interactions. However the resonance structure depends sensitively on the placement of the wires suggesting a more robust compensation is necessary. One possibility is to place several wires in a cylindrical cage at each location. Initial investigations of the multiple-wire scenario show that the nonlinear components of the field created can be chosen with greater flexibility. Nevertheless, several issues with the wire compensation principle in the Tevatron need to be resolved before it can proceed to an experimental test.

## TEL results

V. Shiltsev (FNAL) reported on the status of the beam-beam compensation at the Tevatron with an electron lens. The Tevatron electron lens (TEL) was designed to counteract mainly the effects of the tune spreads between anti-proton bunches and the large tune footprint due to the beam-beam interactions at top energy. Initial observations showed that the tune shift due to the electron beam was as expected but the action of the lens usually worsened the lifetime. Unexpectedly the TEL found use as a resonant kicker in clearing the DC beam that circulates in the machine.

Recently the situation improved when the electron gun that generated a uniform rectangular profile was replaced by a gun that generates a smooth Gaussian profile. At good working points the electron lens preserves the beam lifetime. During stores the TEL has occasionally been used in an attempt to reduce emittance growth of selected anti-proton bunches. A recent attempt was successful but two other attempts had no influence or slightly negative effects. Several upgrades are planned to improve the performance of the lens - perhaps the most important will be reducing the orbit jitter of the electron beam.

## Multipole compensation

J. Shi, University of Kansas, proposed a method for compensating the nonlinearities of the beam-beam interactions with multipoles. This is achieved by minimizing the coefficients in a Taylor map of the nonlinear fields order by order. It was applied to a model of the LHC using either correctors locally in the IR sections or distributed globally in the arcs. It was demonstrated that the tune footprint could be reduced and the dynamic aperture increased using only up to third and fourth order nonlinearities of the map. The sensitivity of this compensation to lattice and orbit errors was not addressed.

## Four beam compensation

K. Ohmi reported on a new simulation study of the four beam neutralization scheme as a possible luminosity upgrade for KEKB. This scheme where beam-beam forces are canceled by virtue of no net charge at the collision points was first tried at DCI (Orsay) in the 1980s but did not succeed because of coherent instabilities. The DCI performance was compared with simulations earlier [11]. Two schemes were investigated in the present study. One scheme uses the present KEKB rings for two beams and two external beams are provided by linacs. In the other scheme two additional rings are built to have four circulating beams. Active feedback systems to damp the coherent dipole motion were included. However both schemes are plagued by higher order coherent and incoherent motion and the available tune space is very limited.

## BEAM-BEAM STUDY TOOLS

J. Ellison, University of New Mexico, showed averaging techniques in the weak-strong case with only head-on interactions, pointing to areas of high and low stability of particle motion in the tune plane. Averaged Hamiltonians were derived to describe motion in the vicinity of two low order resonances: the 4th order resonance $2\nu_x + 2\nu_y = p$ and the linear coupling resonance $\nu_x - \nu_y = 0$. The conjecture is that motion is generically chaotic in this neighborhood. He also presented a new model for the two degrees of freedom collective beam-beam interaction.

J. Rogers, Cornell University, reviewed beam-beam simulation methods for lepton machines. A key motivation for the simulations is to understand whether coherent or incoherent motion or some combination of the two is responsible for the two beam-beam limits observed in $e^+ - e^-$ machines. Weak-strong simulation methods require tricks to follow particle distributions long enough to calculate lifetimes, typically of the order of an hour or $10^9$ turns. These include the leap frog method of Irwin (1989) and inclusion of scattering processes by Kim and Hirata (1998). Weak-strong simulations have proven useful for accelerator design, the choice of operating parameters, and the investigation of beam halos (second beam-beam limit). Self-consistent strong-strong simulations are necessary to understand coherent effects but at present are able to follow particle distributions only for several damping times. Each $e^+ - e^-$ collider has developed its own PIC style code. These include CESR: Krishnagopal and Siemann (1996), Anderson (1999); PEPII: Cai et al. (2001); KEKB: Ohmi (2000). The luminosities calculated from these codes for their respective machines are found to be within 10% of observed luminosities when the machine is well tuned. A comparison of these codes is desirable.

J. Shi reviewed the simulations for hadron machines. Strong-strong methods currently employed include the soft Gaussian approximation, direct multi-particle tracking, Particles-In-Cell (PIC), Hybrid Fast Multipole Method (HFMM), and canonical perturbation theory for solving the Vlasov equation. It is important to check that convergence is achieved with respect to simulation parameters such as the number of macro-particles and the grid-size. Currently only fast processes (within $10^6$ turns) can be analyzed. Slow particle loss, emittance growth and the formation of tails cannot be predicted with confidence. Using a PIC style code, he reported chaotic motions of the centroid in a model of the LHC at ten times the design value of the beam-beam parameter. This is an interesting prediction but observation of this phenomena may be unlikely in the near future.

J. Qiang, LBNL, discussed the computational challenges in modeling beam-beam and space charge simulations. These include efficient Poisson solvers on parallel computers, large particle numbers, long tracking times, and stable direct solvers. He also discussed a parallel computational tool for strong-strong and weak-strong beam-beam modeling. The code is based on shifted Green functions and models efficiently the long-range parasitic collisions. The code was used to investigate the emittance growth caused by modulated transverse offsets in RHIC and the LHC. For the Tevatron, the antiproton lifetime at injection has been simulated. The calculated lifetime is of the order of a few hours (close to observations) when the physical aperture chosen is small enough.

A. Sobol, University of New Mexico, presented numerical calculations of the phase space density for the strong-strong beam-beam interaction that addressed the problem of storing a large amount of data into a computers cache.

Weak-strong simulation tools are useful standard tools for both lepton and hadron colliders. But while strong-strong simulations have gained predictive power for lepton colliders, their use for hadron colliders so far is limited. This should only encourage further development of codes and new methods such as the direct integration of the non-linear Vlasov equation.

In a discussion with the HALO diagnostics groups it was pointed out that for operational observations of the beam-beam effect, it would be desirable to have most beam quantities available on a bunch-by-bunch basis. Due to abort or other gaps in the bunch fill patterns, parameters such as closed orbit, tune, linear coupling, chromaticity and emittance vary from one bunch to another. Currently, bunch-by-bunch coupling, chromaticity and emittance measurements are not easily available.

# CONCLUSIONS

While there are a number of beam-beam phenomena, in both lepton and hadron colliders, that are not completely understood, the three major questions currently relevant to collider operation may be the following.

1. Are coherent modes dangerous in hadron colliders? They could be if the modes are outside the incoherent tune spread. However the spectrum of these modes and their relationship to the incoherent spectra depends on several factors including the intensity ratio of the beams, long-range interactions, synchrotron tune, chromaticity, tune splits etc. Until now the presence of these beam-beam driven modes has not limited the operation of any collider - either lepton or hadron. Damping mechanisms, e.g. changing the tune split or an increase in chromaticity, seem to be available to render these modes innocuous. That may change in the future so theoretical and experimental studies of these modes need to be vigorously pursued.

2. Can beam-beam compensations techniques be made to work? This is being actively studied experimentally and theoretically at the Tevatron, and at the SPS for application in the LHC. Both the electron lens and wires may be used in the Tevatron. The lens would be used to reduce the tune shifts between bunches and the wires to reduce the average effect of the long-range interactions on all bunches. The accelerator physics challenges are many: ensuring the proton beam is not affected, and coherent instabilities are not excited, to name a few.

3. What can analysis and simulations predict in hadron machines? Solutions of the linearized Vlasov equation with beam-beam interactions have been successfully used to predict the frequencies of *pi*-modes. Analytically it would be desirable to develop a weakly nonlinear theory that exhibits coupling of the modes and perhaps other features. Numerical tools to analyze the nonlinear Vlasov equation also need to be developed. Lifetime simulations for the Tevatron at injection energy are now yielding results, of the order of an hour, close to observations. Longer lifetimes are at present out of reach. Further improvements in the modeling and the use of the latest advances in computing technology are

greatly needed to run both weak-strong and strong-strong simulations for the time scales of interest.

## REFERENCES

1. J. Poole and F. Zimmermann (editors), "Proceedings of the workshop on beam-beam effects in large hadron colliders – LHC99", Geneva, Switzerland, CERN-SL-99-039 AP (1999).
2. T. Sen and M. Xiao (editors), "Proceedings of a workshop on beam-beam effects in circular colliders", Fermilab, FERMILAB-Conf-01/390-T (2001).
3. J.T. Seeman, "Luminosity and the beam-beam interaction", Joint US-CERN-Japan-Russia Accelerator School on High Quality Beams, St. Petersburg and Moscow, Russia, AIP Conference Proceedings, Vol. 592, Melville, New York (2001).
4. W. Schnell, "Report on the ISR", proceedings of the 1975 Particle Accelerator Conference, Washington D.C. (1975).
5. W. Herr, "Beam-beam issues in the LHC and relevant experience from the SPS proton antiproton collider and LEP", proceedings of the Workshop on Beam-beam Effects in Circular Colliders, Fermilab, FERMILAB-Conf-01/390-T (2001).
6. M. Bieler et al., "Recent and past experiences with beam-beam effects at HERA", proceeding of the Workshop on Beam-beam Effects in Large Hadron Colliders – LHC99–, Geneva, CERN-SL-99-039 AP (1999).
7. The LHC Study Group, "The Large Hadron Collider conceptual design", CERN/AC/95-05 (LHC) (1995).
8. K. Yokoya and H. Koiso, "Tune shift of coherent beam-beam oscillations", Part. Accel. Vol. 27, pp. 181-186 (1990).
9. K. Takayama, J. Kishiro, M. Sakuda, Y. Shimosaki, and M. Wake, "Superbunch hadron colliders", Phys. Rev. Lett. Vol. 88 No. 14 (2002).
10. S. Peggs, "Beam-beam compensation schemes", in "Handbook of accelerator physics and engineering" edited b A.W. Chao and M. Tigner, World Scientific (1999).
11. B. Podobedov and R.H. Siemann, "Coherent beam-beam interaction with four colliding beams", Phys. Rev. E **52** No 3, pp 3066 (1995).

# Coherent Beam-Beam Effects, Theory and Observations

Y. Alexahin

*FNAL, Batavia, IL 60510*

**Abstract.** Current theoretical understanding of the coherent beam-beam effect as well as its experimental observations are discussed: conditions under which the coherent beam-beam modes may appear, possibility of their resonant interaction (coherent resonances), stability of beam-beam oscillations in the presence of external impedances. A special attention is given to the coherent beam-beam modes of finite length bunches: the synchro-betatron coupling is shown to provide reduction in the coherent tuneshift and - at the synchrotron tune values smaller than the beam-beam parameter - Landau damping by overlapping synchrotron satellites.

## 1 INTRODUCTION

Let us start with definition of the subject of the present report. By **coherent** effects we understand those arising from correlated in phase motion of particles (not necessarily with equal amplitudes as in a rigid-body motion).

There is a wider class of **collective** phenomena arising from mutual influence of two strong beams, e.g. the flip-flop effect, which are beyond the scope of this report. Here we just assume that a (quasi) equilibrium state does exist on a sufficiently long time scale.

The interest in the coherent beam-beam effect is twofold: it is useful in diagnostics of colliding beams but it is also a source of potential instability.

Though the coherent modes were routinely seen since the early days of $e^+e^-$ colliders, there had been a long-standing issue of how the coherent tuneshifts are related to the beam-beam parameter until the work by K.Yokoya et al. [1]. In that paper an exhaustive answer was given on the basis of the Vlasov perturbation theory which was afterwards successfully used in the studies of coherent beam-beam resonances [2, 3], Landau damping by the beam-beam tunespread [4], diverse effects of the synchro-betatron coupling [5].

In the present report we give an overview of the Vlasov perturbation theory of coherent beam-beam effect, compare some of its results with numerical simulations and experimental observations.

## 2 COHERENT BEAM-BEAM MODES

Let us make a conventional choice of the generalized azimuth $\theta = s/R$ as the independent variable and describe particle motion with the help of action-angle variables

$$\underline{I} = \{I_x, I_y, I_s\}, \quad \psi_{x,y} = \psi_{x,y}^{(original)} + \frac{v'_{x,y}}{\alpha_M R}z, \quad (1)$$

where the angle variables were renormalized to take into account chromaticity $v'_{x,y}$, $\alpha_M$ being the momentum compaction factor, $R$ the average machine radius and $z$ the longitudinal displacement w.r.t. the reference particle.

We choose the (quasi) equilibrium distribution function (see [6] for mathematical proof of existance) to be Gaussian:

$$F_0 = \frac{1}{(2\pi)^3 V} \exp(-\underline{\varepsilon}^{-1} \cdot \underline{I}), \quad (2)$$

$$\underline{\varepsilon} = \langle \underline{I} \rangle, \quad V = \varepsilon_x \varepsilon_y \varepsilon_s, \quad \underline{\varepsilon}^{-1} = (\varepsilon_x^{-1}, \varepsilon_y^{-1}, \varepsilon_s^{-1})$$

and study its small perturbations.

### 2.1 Vlasov Perturbation Theory

Evolution of the perturbation is governed by the Vlasov equation

$$\frac{\partial}{\partial \theta} F_1^{(k)} + \underline{v}^{(k)}(\underline{I}) \cdot \frac{\partial}{\partial \underline{\psi}} F_1^{(k)} = -F_0 \underline{\varepsilon}^{-1} \cdot \frac{\partial}{\partial \underline{\psi}} K_1^{(k)}(\underline{I}, \underline{\psi}; \theta) \quad (3)$$

where $k = 1, 2$ is the beam number, the r.h.s. describes the beam-wall and the beam-beam interaction; in the case of a finite bunch length the latter with the help of the **synchro-beam transformation** [7] can be presented in the form

$$K_1^{(k)}(\underline{I}, \underline{\psi}) = \sum_{IP} \frac{r_p N_{3-k}}{\gamma} \delta_p(\theta - \theta_{IP}) \int G^{(k)} F_1^{(3-k)}(\underline{I}', \underline{\psi}') d^3 I' d^3 \psi'.$$

By virtue of this transformation the interaction of a particle with the whole of the opposing bunch is lumped to the nominal interaction point, as a result the Green function explicitly depends on the momenta [5]:

$$G = -\ln\left\{[x - x' + (\alpha + \frac{p_x + p'_x}{2})(z - z')]^2 + [y - y' + \frac{p_y + p'_y}{2}(z - z')]^2\right\} \quad (4)$$

where $\alpha$ is half crossing angle. Coordinates in eq.(4) include the constant offset (if any) and the synchrotron part:

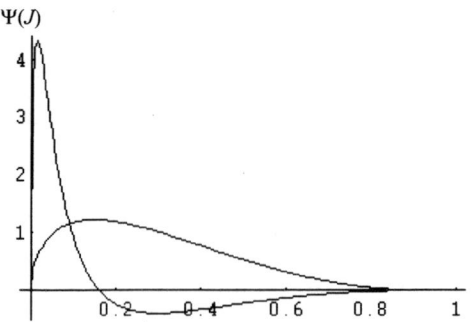

**FIGURE 1.** The first two discrete eigenmodes of horizontal oscillations in flat beams

$$x = (-1)^{k-1}d_x/2 + D_x\delta_p + \sqrt{2\beta_x I_x}\sin(\psi_x + \phi_x^{(k)}) \quad (5)$$

where $\phi_{x,y}^{(k)}$ is the betatron phase advance.

Eqs.(1,4,5) show how such factors as chromaticity, finite bunch length, crossing angle and dispersion enter the theory. Also, difference in intensity, bare lattice tunes and distribution in phase advances for the two beams can be taken into account.

It should be noted that by lumping the interaction in one point we exclude the possibility of the head particles in a finite-length bunch to talk to the tail particles via the opposing beam; thus we leave aside such important question as the beam-beam contribution to driving the head-tail instability.

## 2.2 Angle and Radial Modes

Performing Fourier expansion in the angle variables

$$\mathbf{f} = e^{\underline{\varepsilon}^{-1}\cdot\underline{I}/2}\begin{pmatrix}\sqrt{r_\xi}F_1^{(1)}\\F_1^{(2)}\end{pmatrix} = \sum_{\underline{m}}\exp(i\underline{m}\cdot\underline{\psi})\,\mathbf{f}_{\underline{m}}(\underline{I},\theta) \quad (6)$$

where $\underline{m} = \{m_x, m_y, m_s\}$ is the **angle mode** index, $r_\xi = N_1/N_2 \leq 1$, we cast the Vlasov equation into the form

$$i\frac{\partial}{\partial\theta}\mathbf{f}_{\underline{m}} = \sum_{\underline{m}'}\hat{A}_{\underline{m},\underline{m}'}\cdot\mathbf{f}_{\underline{m}'} \quad (7)$$

with the matrix integral operators

$$\hat{A}_{\underline{m},\underline{m}'} = \begin{pmatrix}\underline{m}\cdot\underline{v}^{(1)} & 0\\ 0 & \underline{m}\cdot\underline{v}^{(2)}\end{pmatrix}\delta_{\underline{m},\underline{m}'} + \\ + \underline{m}\cdot\underline{\varepsilon}^{-1}\sum_{IP}\frac{r_p}{\gamma}\sqrt{N_1 N_2}\delta_p(\theta-\theta_{IP})\begin{pmatrix}0 & \hat{G}^{(1)}_{\underline{mm}'}\\ \hat{G}^{(2)}_{\underline{mm}'} & 0\end{pmatrix} \quad (8)$$

For one-dimensional oscillations (e.g. horizontal for definiteness) $m_x = 1$ is usually referred to as the dipole mode, $m_x = 2$ as the quadrupole mode and so on. In fact each such mode presents a family of modes with different dependence on the action variables, called by B.Zotter the **radial modes** in contradistinction to the angle modes.

## 2.3 Discrete & Continuous Spectra

If the tunes are chosen so that for a given $\underline{m}$ the coherent resonance condition does not hold for any relatively low $\underline{m}'$,

$$\underline{m}\cdot\underline{v}^{(1)} + \underline{m}'\cdot\underline{v}^{(2)} \neq n, \quad (9)$$

then we may consider the mode $\underline{m}$ uncoupled and formulate the eigenvalue problem for the corresponding family of radial modes:

$$\hat{A}_{\underline{m},\underline{m}}\cdot\Psi_\lambda = \lambda\Psi_\lambda \quad (10)$$

For uncoupled modes the periodic $\delta$-function in eq.(8) can be replaced with $1/2\pi$.

Generally operator $\hat{A}_{\underline{m},\underline{m}}$ has mixed spectrum. Due to the first multiplicative part it necessarily has continuous spectrum with $\lambda$ spanning the range of variation of the proper combinations of the incoherent tunes in both beams, $\underline{m}\cdot\underline{v}^{(1)}\cup\underline{m}\cdot\underline{v}^{(2)}$. The integral part of $\hat{A}_{\underline{m},\underline{m}}$ produces by itself purely discrete spectrum, however the total operator may or may not have discrete eigenvalues.

Eigenfunctions can be normalized so that

$$(\Psi_\lambda, \Psi_\mu) = \delta_{\lambda\mu} \quad (11)$$

where the r.h.s. should be understood as the Kronecker symbol for $\lambda$ belonging to the discrete spectrum, and as the Dirac $\delta$-function otherwise.

In the case of equal intensities and tunes the eigenmodes split into two classes: $\pi$-modes: $f^{(1)} = -f^{(2)} = f^{(-)}$, and $\Sigma$-modes: $f^{(1)} = f^{(2)} = f^{(+)}$.

The spectrum of dipole $\pi$-modes was found to be [1,3] (in units of the beam-beam parameter):

- round beams: $\lambda = 1.214$
- flat beams (horizontal): $\lambda = 1.330, 1.026, 1.002$
- flat beams (vertical): $\lambda = 1.239$

Fig.1 shows the first two radial modes of horizontal oscillations in flat beams as functions of $J = I_x/\varepsilon_x$.

For all geometries there is just one discrete $\Sigma$-mode with unshifted tune ($\lambda = 0$) corresponding to the rigid-body oscillations:

$$\Psi_0 = \sqrt{J_x}\,e^{-(J_x+J_y+J_s)/2} \quad (12)$$

where $J_i = I_i/\varepsilon_i$.

In all these cases the spectra of both $\pi$- and $\Sigma$-modes include continuum (0, 1).

The ratio of the split in tunes of dipole $\pi$- and $\Sigma$-modes to the beam-beam parameter was dubbed the **Yokoya factor**, $Y$.

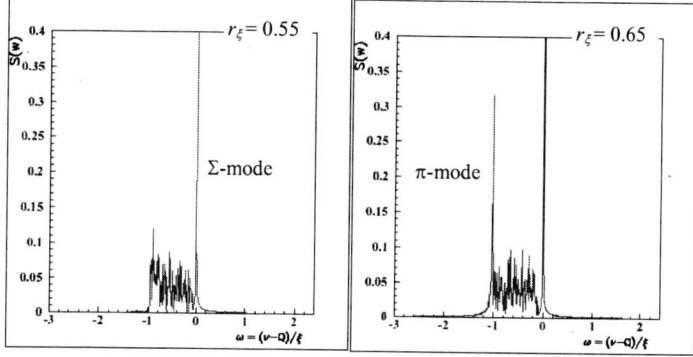

**FIGURE 2.** Spectra of oscillations in round *p-p* beams with the indicated values of intensity ratio.

There is a popular one-dimensional "slab" model in which the Yokoya factor was found to be as large as $Y = 1.5$ [6]. It should be stressed that there is a basic difference between this model and vertical oscillations in flat beams: in the latter case the problem is intrinsically two-dimensional [1], the vertical tune depends on the horizontal amplitude no matter how small the aspect ratio is. Higher dimensionality reduces coherence, hence smaller tuneshift than in the "slab" model.

Quadrupole π-mode also may have discrete eigenvalues [2,3], for horizontal oscillations in flat beams two such eigenvalues were found: $\lambda_1 = 2.044$, $\lambda_2 = 2.002$.

Important characteristics of the dipole eigenmodes are the ***spectral coefficients***

$$c_k(\lambda) = (\Psi_0, \Psi_\lambda^{(k)}), \quad k = 1, 2 \tag{13}$$

satisfying the relations

$$\int [c_1^2(\lambda) + c_2^2(\lambda)] d\lambda = 2, \quad \int c_1(\lambda) c_2(\lambda) d\lambda = 0 \tag{14}$$

where the integral is understood in the Stieltjes sense: sum over the discrete eigenvalues and integral over the continuum.

Squares of coefficients (13) give the relative spectral weight of the mode in oscillations excited by a dipole kick at the corresponding beam [5].

For horizontal π-modes in flat beams $c_1 = -c_2 = 0.724, 0.188, 0.064$. Small values of the spectral coefficients of the second and third radial modes (and their proximity to the continuum boundary) explain why they were not seen in experiment.

## 2.4 Transition from the Weak-Strong to the Strong-Strong Regime

An important question is under what conditions the discrete eigenvalues may exist. There are a number of factors which affect coherence of oscillations, the basic one being the intensity ratio $r_\xi$.

It was shown that in the round beams the discrete mode emerges from the continuum at $r_\xi \approx 0.6$ [4]. Fig.2 presents results of simulations by the Hybrid Fast Multipole Method [8] which confirm this conclusion.

Another important factor is difference in tunes of the two beams [9]. It was found that the tunesplit $\geq Y\xi$ is necessary to damp both π and Σ discrete modes [3].

Discussion of these and other factors (and their possible interference) can be found in Ref.[5].

## 2.5 Experimental Observations

Dedicated studies were performed at CESR for the vertical plane [10] and at Tristan accumulator ring for both transverse planes [1]. Measured values of the Yokoya factor coincide with theoretical values within a few percent.

The only observation of coherent beam-beam modes in hadron beams was made at RHIC [11]. There was also found a good agreement with theoretical predictions.

## 3 COHERENT RESONANCES

If the condition of a coherent resonance is met,

$$\underline{m}_1 \cdot \underline{\nu}^{(1)} + \underline{m}_2 \cdot \underline{\nu}^{(2)} = n \tag{15}$$

coupling between the modes $\underline{m} = \underline{m}_1$ and $\underline{m}' = -\underline{m}_2$ in eq.(8) should be taken into account. If parity of $m_{1x}$ and $m_{2x}$ or of $m_{1y}$ and $m_{2y}$ is different, then respectively horizontal or vertical offset is needed for the beam-beam interaction to produce the coupling.

Analysis shows that this coupling may lead to instability only in the case

$$(\underline{m}_1 \cdot \underline{\varepsilon}^{-1})(\underline{m}_2 \cdot \underline{\varepsilon}^{-1}) > 0 \tag{16}$$

Coherent beam-beam resonances were observed experimentally [12] and in simulation [13]. Fig.3 shows the measured dependence of the horizontal dipole π-mode tune on the Σ-mode tune in LEP at 46GeV. The red square data points mark the region of spontaneous excitation of the π-mode. This excitation was explained in [3] as a resonance of the dipole π-mode ($m_{1x} = 1$) and the quadrupole Σ-mode ($m_{2x} = 2$) in the presence of a moderate offset.

The possibility of such a resonance was confirmed in [13] by tracking with the use of the soft-Gaussian approximation for beam-beam kick. It was found however that the instability saturates at relatively small

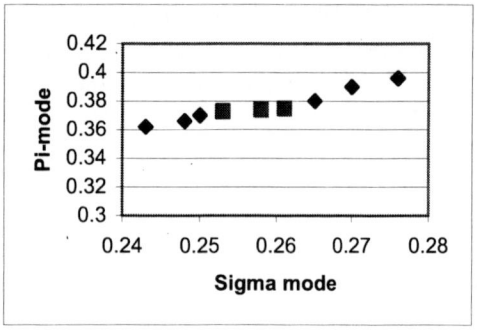

**FIGURE 3.** Spontaneous excitation of horizontal π-mode observed in LEP at 46GeV with beams colliding at 4 IPs (courtesy of K.Cornelis)

dipole amplitudes ($<0.04\sigma_x$) but at the expense of an unceasing emittance growth (Fig.4)

In absence of offsets higher order resonances (quadrupole-octupole in particular) can produce the emittance growth, but at a much lower rate [13].

## 4 FINITE BUNCH LENGTH EFFECTS

There are various sources of the Hamiltonian synchro-betatron coupling, we will consider here the betatron phase variation along the interaction region ("finite length effect") and chromaticity, the two in the case of short bunches combine in parameter (for horizontal oscillations)

$$\kappa = (\nu_x'/\alpha_M R - 1/\beta^*)\sigma_s \tag{17}$$

There are three distinct regimes depending on the ratio of the synchrotron tune to the beam-beam parameter [5]:

- Small beam-beam parameter (high $\nu_s$).

The effect of coupling can be quantified by factor $\lambda_\parallel$ (longitudinal eigenvalue) which can be extracted from the integral operator in the second term of operator (8); in the case $\kappa^2 \ll 1$ it is [5]

$$\lambda_\parallel = e^{-\kappa^2} I_{m_s}(\kappa^2) \tag{18}$$

where $I_m(x)$ is the modified Bessel function of order $m$.

Due to this factor the tunes of the π- and Σ-modes in finite-length bunches are shifted towards the center of the continuum, the Yokoya factor can be estimated as $Y \sim \lambda_\parallel Y_0$ with $Y_0$ being the value for infinitely short bunches.

Eq.(18) holds for the synchrotron modes $m_s \neq 0$ as well and shows that the coherent contribution to their spectra is strongly suppressed, the tunes of both π and Σ synchrotron modes being determined by the average incoherent tunes (~$\xi/2$ for head-on collisions)

$$\nu_{m_s} \approx \overline{\nu_{\text{inc}}} + m_s \nu_s \tag{19}$$

- Large beam-beam parameter (low $\nu_s$).

In this limit $\lambda_\parallel \approx 1$ so that the Yokoya factor is not affected, but the oscillations are not purely dipole, their phase varies along the bunch.

- Comparable values of the beam-beam parameter and the synchrotron tune.

The effect of synchro-betatron coupling is most dramatic in this case, the synchrotron sidebands of the continuum modes can overlap spectral lines of discrete π- and Σ-modes thus providing their Landau damping [5]. This prediction was confirmed by tracking in the soft-Gaussian approximation [14].

The described dependence of the coherent modes on the synchrotron tune can be compared with experimental results obtained at VEPP-2M [15]. Fig.5 shows the measured vertical tunes as functions of the

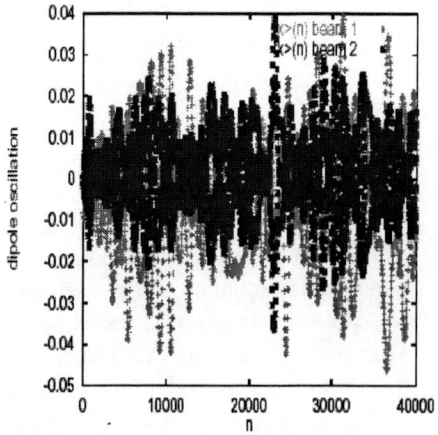
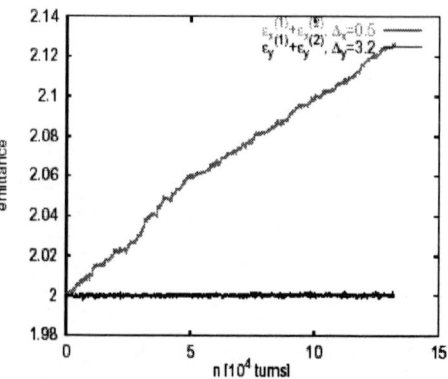

**FIGURE 4.** Tracking simulation of the dipole-quadrupole resonance at an offset of $0.3\sigma_x$ [13]. Left: center-of-mass oscillations in the two beams, right: combined horizontal emittance growth.

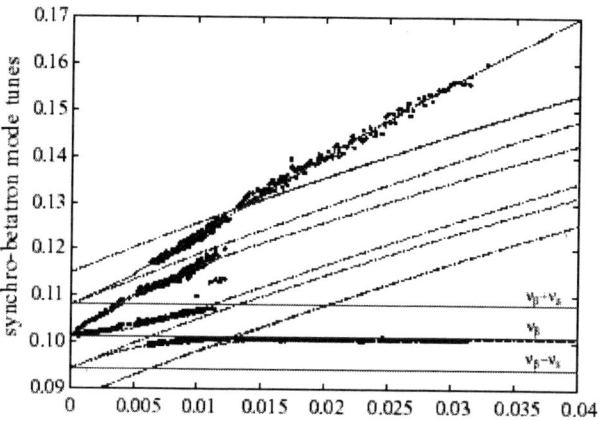

**FIGURE 5.** Measured tunes (dots) of vertical oscillations of one $e^+$ and one $e^-$ bunches colliding at two IPs in VEPP-2M as functions of the beam-beam parameter / IP; $\nu_y = 3.101$, $\nu_s = 0.007$, $\sigma_s = 3.5$cm, $\beta^* = 6$cm.

beam-beam parameter at fixed value of the synchrotron tune $\nu_s = 0.007$ and $\sigma_s/\beta^* \approx 0.6$. At small values of the beam-beam parameter ($\xi \leq 0.005$/IP) the Yokoya factor appears as small as $Y \approx 0.65$. In the opposite limit ($\xi = 0.03$/IP) taking into account the dynamic focusing effect it can be estimated as $Y \approx 1.14$.

Although numerically the Yokoya factor occurs smaller than expected in both limits, its qualitative behavior w.r.t. the ratio of the beam-beam parameter to the synchrotron tune is close to the prediction.

## 5 BEAM-BEAM EFFECT AND IMPEDANCE DRIVEN INSTABILITIES

Interplay of beam-beam and beam-wall interactions is the major reason for the continuing interest in coherent beam-beam effect.

### 5.1 Landau Damping of the Beam-Beam Modes

It was first suggested by J.Gareyte [16] that the large gaps between the coherent and incoherent tunes may switch off Landau damping in the strong-strong regime thus leaving the beams liable to instability.

As the further studies have shown, there is possibility to damp the discrete coherent modes by tunesplit and/or overlapping synchrotron sidebands.

The analytical theory of Landau damping by synchrotron sidebands was extended in [17] on the case of large bunch length, $\sigma_s \sim \beta^*$. Computed with its help (in the simplified case of flat beams at IP) beam-beam spectra in Tevatron at three values of chromaticity are presented in the upper row in Fig.6. When the chromaticity is close to the value $\nu_x' = 8$ which renders $\kappa = 0$ in eq.(17) the discrete modes are clearly seen, but are completely submerged into the continuum when the chromaticity is sufficiently far from this value.

The lower plot in Fig.6 demonstrates that in the case of unfavorable values of chromaticity Landau damping can be restored by splitting the bare lattice tunes in two beams by an amount $\geq \xi_x^{(pbar)}$ as discussed in Section 2.

### 5.2 Aggravation of TMCI by LR Interactions

It was observed at injection energy in LEP that the TMCI threshold in 8×8 operation was ~25% lower than in 1×1 case [18]. This reduction was caused mainly by the midarc long-range interactions where

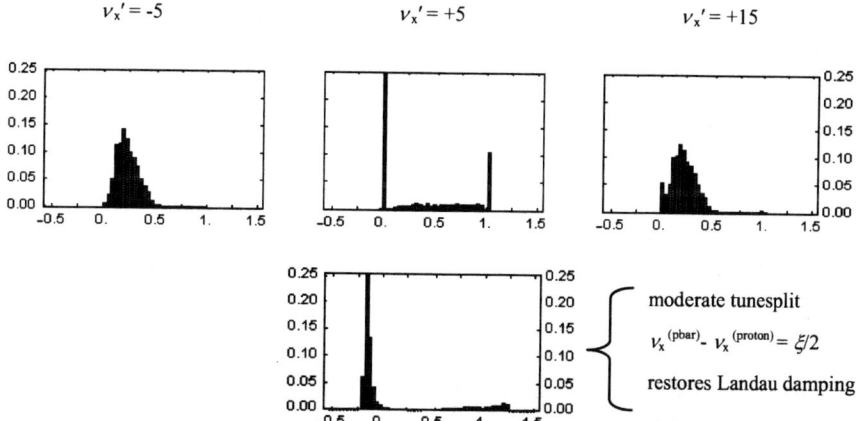

**FIGURE 6.** Effect of chromaticity and tunesplit on the beam-beam oscillations spectrum at the Tevatron upgrade parameters $\sigma_s/\beta^* = 50/35$, $\nu_s/\xi_x^{(pbar)} = 0.035$, $r_\xi = N_{pbar}/N_p = 1/2$, abscissa values in $\xi_x^{(pbar)} = 0.02$.

the beams were separated horizontally.

In the case of long-range interactions the coherent π-mode is shifted twice as much as the average incoherent tune and, according to eq.(19), the tunes of the synchrotron modes. In the result at some value of the beam current the tunes of the dipole and $m_S = -1$ π-modes in the plane of separation collide (Fig.7) creating potential for instability.

In the other plane (vertical in this case) the unstable situation is created by the $m_S = -1$ Σ-mode being shifted upwards to the dipole Σ-mode.

The instability itself was driven by the non-Hamiltonian coupling via the wake-fields, not by the beam-beam interaction.

However, there is still an open question whether the interaction of very long bunches can drive the head-tail instability. Some indication of such a possibility can be seen in the results of simulations with the ODYSSEUS code [19].

## ACKNOWLEDGEMENTS

The author is grateful to P.M.Ivanov for discussion of beam-beam experiments at VEPP-2M.

## REFERENCES

1. K.Yokoya et al., "Tune Shift of Coherent Beam-Beam Oscillations", KEK Preprint 89-14 (1989); Particle Accelerators, v.27, p.181 (1990).
2. P.Zenkevich, K.Yokoya, "Landau Damping of Coherent Beam-Beam Oscillations", KEK Preprint 92-116 (1992); Particle Accelerators, v.40, p.229 (1992).
3. Y.Alexahin, "Eigenmodes of Coherent Oscillations in Colliding Beams", in: Proc. LHC'99 Workshop, CERN-SL-99-039 AP, Geneva, 1999, pp. 41-44.
4. Y.Alexahin, "On the Landau Damping and Decoherence of Transverse Dipole Oscillations in Colliding Beams", CERN SL-96-064 AP (1996); Particle Accelerators, v.59, p.43 (1998).
5. Y.Alexahin, "A Study of Coherent Beam-Beam Effect in the Framework of the Vlasov Perturbation Theory", LHC Project Report 461 (2001); NIM A480 (2002) pp. 253-288.
6. J.Ellison, this Workshop.
7. K.Hirata, H.Moshammer and F.Ruggiero, "Synchro-Beam Interaction", CERN SL-AP/90-92 (1992), *Particle Accelerators*, v.40, p.205 (1993).
8. W. Herr, M.P. Zorzano and F. Jones, "A hybrid fast multipole method applied to beam-beam collisions in the strong-strong regime", in: Proc. Workshop on Beam-Beam Effects in Circular Colliders, Fermilab, 2001, pp.139-148.
9. A.Hofmann, "Beam-Beam Modes for Two Beams with Unequal Tunes", in: Proc. LHC'99 Workshop, CERN-SL-99-039 AP, Geneva, 1999, pp. 56-58.
10. D.H.Rice, in: Proc. 3rd Advanced ICFA Beam Dynamics Workshop on Beam-Beam Effects in Circular Colliders, Novosibirsk, 1989, pp. 17-25.
11. W.Fischer, "Strong-Strong and Other Beam-Beam Observations in RHIC", this Workshop.
12. K.Cornelis, private communication.
13. Y.Alexahin, M.P.Zorzano, "Excitation of Coherent Beam-Beam Resonances for Beams with Unequal Tunes in the LHC", LHC Project Note 226 (2000).
14. W. Herr and R. Paparella, "Landau damping of coherent beam-beam modes by overlap with synchrotron side-bands", LHC Project Note 304 (2002).
15. P.M.Ivanov et al., in: Proc. Workshop on Beam-Beam Effects in Circular Colliders, Fermilab, 2001, p.36.
16. J.Gareyte, in: Proc. 3rd Advanced ICFA Beam Dynamics Workshop on Beam-Beam Effects in Circular Colliders, Novosibirsk, 1989, pp.135-139.
17. Y.Alexahin, in: Proc. Workshop on Beam-Beam Effects in Circular Colliders, Fermilab, 2001, pp. 117-120
18. K.Cornelis, "TMC threshold as function of beam-beam interaction", SL Note 93-39 OP (1993)
19. J.Rogers, "Beam-Beam Simulations for Lepton Machines", this Workshop.

# Beam-Beam Instability in Case of Strong-Weak Beam-Beam Interactions

L. Jin and J. Shi

*University of Kansas, Lawrence, KS 66049*

**Abstract.** Beam-beam effects in HERA were studied with a self-consistent beam-beam simulation by using the particle-in-cell method. A remarkable agreement between the experimental measurement and the simulation result was observed on the emittance growth, luminosity reduction, and coherent beam-beam tune shifts. The simulation study also showed that the chaotic coherent beam-beam instability could occur in HERA and this collective beam-beam instability could be avoided with a slightly different tune.

## INTRODUCTION

To examine any possible luminosity reduction due to beam-beam effects, several beam experiments were performed on HERA [1, 2, 3]. In order to have a better understanding of those experimental data and to evaluate the beam-beam effect in the HERA upgrade, the dynamics of beam-size growth and the stability of the coherent beam-beam oscillation in HERA were studied with a self-consistent beam-beam simulation. A remarkable agreement between the experimental measurement and the simulation result was observed on the emittance growth, luminosity reduction, and coherent beam-beam tune shifts. The simulation study also showed that the chaotic coherent beam-beam instability [4] could occur in HERA. It was found that when the beam-beam parameter of the lepton beam exceeds a threshold that corresponds to an overlap of the lepton beam with a low-order single-particle beam-beam resonance, the onset of the collective beam-beam instability results in an enhanced growth of the proton-beam emittance. After the onset of this collective beam-beam instability, the phase-space area near the origin (closed orbit) becomes unstable for beam centroids and two initially centered beams could develop a spontaneous chaotic coherent oscillation due to beam-beam interactions. A study of the dynamics of beam particle distributions showed that after the onset of the beam-beam instability, the distributions significantly deviate from Gaussian distribution due to beam halo. The formation of the beam halo is a result of chaotic transport (chaotic diffusion) of particles from beam cores to beam tails. In the HERA Upgrade, the beam-beam parameter of the lepton beam is over 20 and 100 times larger than that of the proton beam in the horizontal and vertical direction, respectively, and the two rings have a very different working point. Traditionally, the beam-beam effect in such situation is considered as a typical strong-weak or very unsymmetrical case. For the strong-weak beam-beam interactions, it is commonly believed that the coherent beam-beam effects is not important. This study showed that the traditional boundary between the strong-strong and strong-weak beam-beam interactions is no longer valid in the non-integrable regime of beam-beam interactions. For high-intensity beams, the non-integrable beam-beam perturbation (phase-dependent perturbative Hamiltonian) could dominate the beam dynamics and the collective beam-beam instability could therefore occur in both cases of strong-strong (symmetrical or nearly symmetrical) and strong-weak (very unsymmetrical) beam-beam interactions. In the simulation, the beam-beam interaction at each IP was represented by a kick in transverse phase space and the beam-beam kick was calculated by using particle-in-cell method with $5 \times 10^5$ macro-particles in each beam. For lepton beams, the quantum excitation and synchrotron damping are treated as kicks during the tracking.

## NON-COLLECTIVE BEAM-BEAM EFFECTS

In HERA Accelerator Study 2000, the beam-beam effect at a very large beam-beam tune shift of the $e^+$ beam was studied [2]. In the experiment, the vertical beam-beam parameter $\xi_{e,y}$ of the $e^+$ beam was varied from 0.068 to 0.272 by changing the vertical beta-function $\beta^*_{e,y}$ of the $e^+$ beam at IPs from 1.0 m to 4.0 m. The emittance of the $e^+$ beam and the luminosity were measured as func-

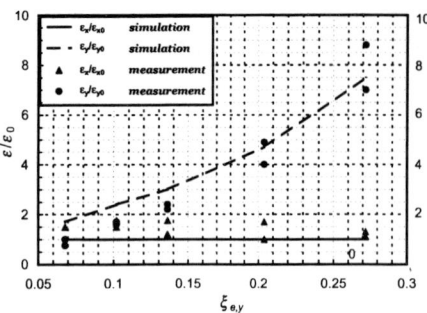
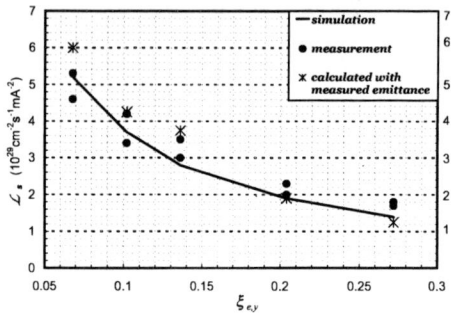

**FIGURE 1.** Emittance of the $e^+$ beam (upper figure) and specific luminosity (lower figure) v.s. $\xi_{e,y}$. $\varepsilon_0$ is the emittance without collision. Discrete points are from the experiment and continuous curves from the simulation. In the emittance plot, the dash (solid) curve is the vertical (horizontal) emittance.

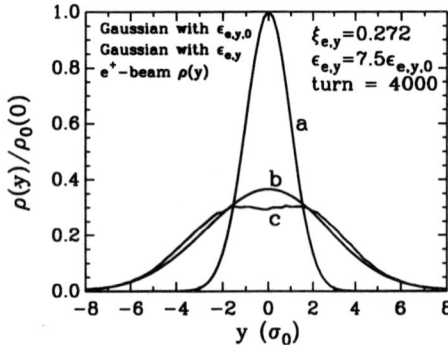

**FIGURE 2.** The vertical projection of the $e^+$-beam distribution at $\xi^*_{e,y} = 0.272$. (a) The initial beam distribution (Gaussian); (b) the Gaussian distribution with enlarged $e^+$-beam emittance; and (c) the $e^+$-beam distribution.

tions of $\xi_{e,y}$ at both HERA IPs. To have a better understanding of those measured data, we have reconstructed the HERA beam experiment with the beam-beam simulation. This study also served a detailed benchmark of our beam-beam simulation code with the experimental measurement. Figure 1 plots the emittance growth of the $e^+$ beam and the specific luminosity as a function of $\xi^*_{e,y}$ measured in the experiment and calculated by the beam-beam simulation. For each $\xi^*_{e,y}$ where the measurement was performed, two data points correspond to the measurement at H1 and ZEUS, respectively. Both the emittance and the luminosity plot show a remarkable agreement between the experiment and the simulation. In the experiment, the coherent beam-beam tune shift associated with the $e^+$-beam was also measured as $\Delta v_x = 0.009$ and $\Delta v_y = 0.013$ for the case of $\xi^*_{e,y} = 0.272$. From the simulation, $\Delta v_x = 0.007$ and $\Delta v_y = 0.014$, which is in a very nice agreement with the experimental result. Compared with the incoherent beam-beam tune shift of 0.082 and 0.544 (two IPs) in horizontal and vertical direction, respectively, the coherent beam-beam tune shifts in this case are extremely small. To understand such

the small coherent beam-beam tune shifts, the eigen-frequencies of the coherent oscillation were derived for the un-symmetrical case of beam-beam interaction based on the assumption of rigid Gaussian beams [1]. It was found from the derived eigen-frequency formula that the small coherent beam-beam tune shift in this case is due to a severe mismatch between two beams as a result of the emittance blowup of the $e^+$ beam. With the enlarged $e^+$-beam emittance in Fig. 1, however, the coherent tune shift calculated from the derived formula are $\Delta v_x = 0.013$ and $\Delta v_y = 0.019$ that do not agree well with the experimental or simulation results. The discrepancy here is due to a non-Gaussian distribution of the $e^+$ beam. A study of the dynamics of particle distributions during the beam-beam simulation showed that the distribution of the $e^+$ beam deviated from a Gaussian distribution with a significant drop at beam core and a growth of beam tails (Fig. 2). Compared with the distribution of the $e^+$ beam, a Gaussian beam has more particles in the core. The coherent beam-beam tune shift calculated based on Gaussian beams is therefore larger than that of the $e^+$ beam.

## COLLECTIVE BEAM-BEAM INSTABILITY

Simulation has been conducted for the HERA Upgrade with two $e$-$p$ collisions. When the working point of the $e$ beam is at $\vec{v}_e = (54.14, 51.21)$, the onset of the chaotic coherent beam-beam instability results in an emittance blowup on both the beams (Fig. 3) and a significant luminosity reduction. Moreover, the two initially centered beams developed a spontaneous chaotic coherent oscillation (Fig. 4). A study of the dynamics of beam-beam tune spread of the $e$ beam showed that this collective beam-beam instability is due to an overlap of the $e$ beam and the 4th-order resonance (Fig. 5). It can be seen in Fig. 5 that many particles in the $e$ beam are trapped inside the resonance. A simulation with a slightly different working point $\vec{v}_e = (54.072, 51.107)$ of the $e$ beam was therefore performed. At this new working point, the $e$-beam is away from the 4th-order resonance. The beam centroid motion is stable and no significant emittance growth was observed on both the beams. Consequently, the luminosity is recovered to the design value. The collective beam-beam instability in this case can therefore be avoided by eliminating the crossings of major beam-beam resonance. To determine the threshold of the onset of the coherent beam-beam instability, the emittance growth of the $p$ beam was studied as a function of bunch current of the $p$ beam in the case of one interaction point. It was found that the threshold is at 50% design $p$-beam current. This further confirms the effect of the 4th-order resonance on the collective beam-beam instability since at 50% design $p$-beam current the $e$ beam avoids the crossing of the 4th-order resonance.

The prediction of the coherent beam-beam instability in HERA was confirmed in a recent beam experiment in HERA [3]. In the experiment, $\xi_e = (0.0156, 0.0236)$, $\xi_p = (0.0014, 0.0004)$ and only one $e^+$-$p$ collision was used. The emittance of the $p$ beam and the luminosity were measured and compared at two different working points of the $e^+$ beam. When the working point of the $e^+$ beam is at $\vec{v}_e = (0.140, 0.210)$, the $e^+$ beam does not cross any major beam-beam resonance. In this case, no significant emittance growth of the $p$ beam and the luminosity reduction were observed. When the working point of the $e^+$ beam was moved to the measured coherent tunes of $\vec{v}_e = (0.215, 0.296)$, on the other hand, the $e^+$ beam overlaps with the 4th-order beam-beam resonance of $2v_x + 2v_y = 1$ and a $\sim$30% emittance growth was observed. In both cases, the working point of the $p$ beam is kept at $\vec{v}_p = (0.294, 0.298)$. The stability of the strong ($p$) beam is therefore severely affected by the dynamics of the weak ($e^+$) beam due to the onset of the collective beam-beam instability. We were able to reconstruct this experiment with the self-consistent beam-beam simulation. Fig. 6 plots the beam tune spread of the $e^+$ beam calculated during the simulation when the $e^+$ beam overlaps with the resonance of $2v_x + 2v_y = 1$. It clearly shows

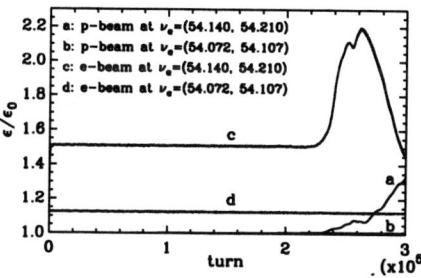

**FIGURE 3.** Evolution of the average of horizontal and vertical emittances for $p$ and $e$ beam.

a strong resonance effect on the beam. The emittance growth of both the beams calculated from the simulation was plotted in Fig. 7 and it remarkably agree with the phenomena observed in the experiment. It should be emphasized that the collective beam-beam instability observed here occurs in a situation of a typical strong-weak beam-beam interaction.

## COMMENTS ON BEAM-BEAM INSTABILITY OF HADRON BEAMS

To understand coherent beam-beam instabilities, many efforts have been made to formulate the beam-beam problem self-consistently by using various methods of approximations or perturbation expansions of the nonlinear Vlasov equation. In lepton storage-ring colliders, because of the radiation damping, the time scale for a lepton beam to reach the equilibrium distribution is much less than the storage time. Consequently, the study of beam dynamics can be focused on the behavior of the distribution near its steady states [5, 6, 7]. Moreover, a fast damping of high-order fluctuations permits the truncation of the moment expansion at fairly low orders. For lepton storage-ring colliders, therefore, methods of perturbation are usually effective in the study of beam-beam effects. For high-energy hadron beams, on the other hand, the damping time scale is usually larger than the storage time so that the motion of beam particles is determined by Hamiltonian dynamics. In the presence of nonlinear perturbations due to beam-beam interactions, the particle distribution may not reach any steady state within a fraction of the storage time. Without the dissipation, moreover, it is not even clear mathematically if the nonlinear Vlasov equation has any steady state close to Gaussian distribution. In the near-integrable regime of the beam-beam interaction, beam distributions change very little due to the beam-beam interaction in the time scale of interest. In this case, quasi-stationary states of the Vlasov equation could be considered and methods of perturbation could be employed to study the beam-beam effects. In the non-integrable regime of the beam-beam interaction, no stationary distribution for the nonlinear Vlasov equation has been found theoretically or observed experimentally. Note that the Gaussian distribution is not or not even close to an equilibrium distribution when the beam-beam parameter is large. Computer simulations have shown that the beam distribution could deviate significantly from its initial Gaussian distribution due to the formation of beam halo [4, 8]. Consequently, the truncation of the moment expansions or the linear stability analysis of equilibrium distributions of the nonlinear Vlasov equation is no longer valid in this case. Moreover, it has been well recognized in the field of

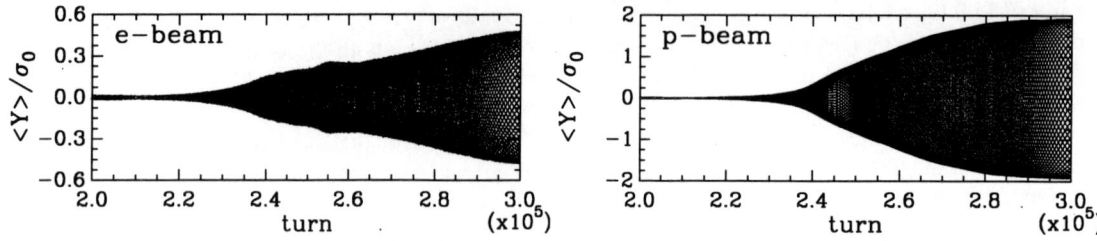

**FIGURE 4.** Chaotic coherent beam oscillation in HERA Upgrade with two $e$-$p$ collisions at $\vec{\nu}_e = (54.14, 51.21)$.

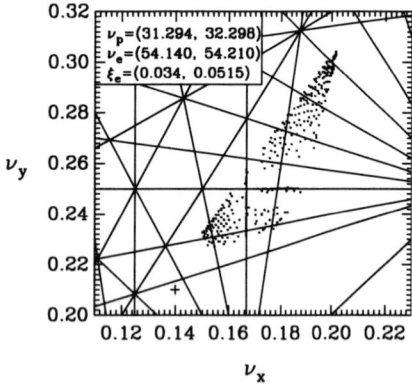

**FIGURE 5.** Tune spread of the $e$ beam in HERA Upgrade with two $e$-$p$ collisions when $\vec{\nu}_e = (54.14, 51.21)$.

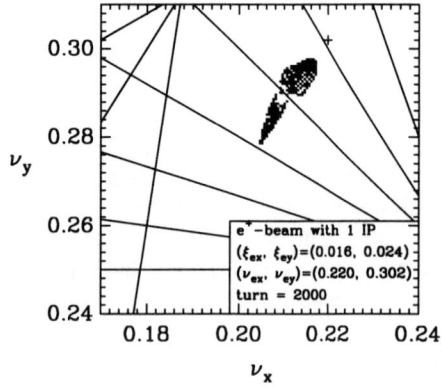

**FIGURE 6.** Tune spread of the $e^+$ beam in HERA with one $e^+$-$p$ collision when the lattice tune of the $e^+$ beam is $\vec{\nu}_e = (54.220, 51.302)$.

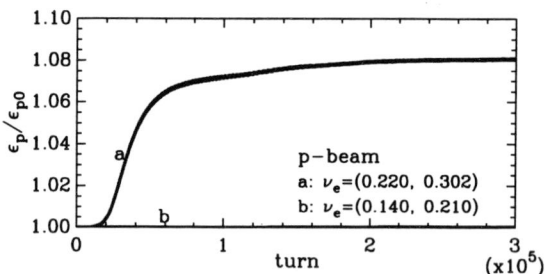

**FIGURE 7.** The evolution of the average of horizontal and vertical emittance of the $p$ beam. The lattice tune of the $e^+$ beam is at (a) $\vec{\nu}_e = (54.220, 51.302)$ and (b) $\vec{\nu}_e = (54.140, 51.210)$.

nonlinear dynamics that in the non-integrable regime of a Hamiltonian system, the use of perturbation expansions such as various canonical perturbation methods usually distorts the dynamics of the system and may result in incorrect conclusions. In order to understand the beam-beam effects of hadron beams, one has therefore to study transient states of the nonlinear Vlasov equation. For the transient state in the nonlinear regime of beam-beam interactions, only validated method currently available for a theoretical understanding of the beam-beam effects is the self-consistent numerical simulation.

## ACKNOWLEDGMENTS

This work is supported by the US Department of Energy under Grant No. DE-FG03-00ER41153. The authors would like to thank the Center for Advanced Scientific Computing at the University of Kansas for the use of the Supercomputer.

## REFERENCES

1. G.H. Hoffstaetter, *HERA Accelerator Studies 1999*, DESY HERA 00-02, (2000).
2. G.H. Hoffstaetter, *HERA Accelerator Studies 2000*, DESY HERA 00-07, (2000).
3. M. Minty, " Summary of Recent Beam-Beam Experiments After the Luminosity Upgrade", preprint, (2003).
4. J. Shi and D. Yao, *Phy. Rev. E* **62**, 1258 (2000).
5. A.W. Chao, M.A. Furman, and K.Y. Ng, in *Proc. of the European Part. Accel. Conf.*, Rome, 1988, edited by S. Tazzari (World Scientific, Singapore, 1989), p. 175.
6. N.S. Dikansky and D.V. Pestrikov, *Part. Accel.* **12**, 27, (1982).
7. A.W. Chao and R.D. Ruth, *Part. Accel.* **16**, 201 (1985).
8. L. Jin and J. Shi, "Importance of Beam-Beam Tune Spread to the Collective Beam-Beam Instability in Hadron Colliders" submitted to Phys. Rev. E, (2003).

# Summary of Beam-beam Observations during Stores in RHIC[1]

Wolfram Fischer

*Brookhaven National Laboratory, Upton, New York 11973*

**Abstract.** During stores, the beam-beam interaction has a significant impact on the beam and luminosity lifetimes in RHIC. This was observed in heavy ion, and even more pronounced in proton collisions. Observations include measurements of beam-beam induced tune shifts, lifetime and emittance growth measurements with and without beam-beam interaction, and background rates as a function of tunes. In addition, RHIC is currently the only hadron collider in which strong-strong beam-beam effects can be seen. Coherent beam-beam modes were observed, and suppressed by tune changes. In this article we summarize the most important beam-beam observations made during stores so far.

## INTRODUCTION

The beam-beam interaction is a major consideration in the operation of RHIC. It can lead to emittance growth and particle loss, and is a source for experimental background. Machine parameters, close to the maximum parameters achieved so far, are presented in Tab. 1, a full parameter list can be found in Ref. [1].

RHIC consists of two superconducting rings, Blue and Yellow, and has produced gold-gold, proton-proton and deuteron-gold collisions [2]. With RHIC's interaction region design (see Fig. 1) and with 4 experiments most bunches experience 4 head-on, and 2 long-range collisions per turn. However, due to the abort gaps, some bunches experience only 3, and some only 2 head-on collisions. The long-range interactions are with at least 7 rms beams sizes separation. With 120 or less bunches per ring (the current limit), sets of 3 bunches in one ring and 3 bunches in the other ring are coupled through the beam-beam interaction (see Ref. [3]).

Even small tune shifts due to the beam-beam interaction can be observed directly with a high precision tune measurement system [4]. In Fig. 2 a tune shift measurement is shown that was taken 3 hours into a gold-gold store.

Two beam splitting DX dipoles are the magnets closest to the interaction point (IP). They are each 10 m away from the IP (see Fig. 1). Beams collide nominally without a crossing angle. With rf manipulations, the crossing point can be moved longitudinally. If the bunch spacing is large enough (with 60 or less bunches per ring), it is possible to separate the beams longitudinally and switch off all 6 beam-beam interactions. If the crossing point is moved within the DX magnets, an observed tune shift is a sign of crossing angles (Figs. 1 and 2). The sum of all residual crossing angles is typically about 0.5 mrad.

Beam-beam phenomena observed in other hadron colliders [5] can also be seen in RHIC [6]. In addition, with bunches of equal intensity the beams are subject to strong-strong effects. To accommodate acceleration of different species, the two RHIC rings have independent rf systems. With different rf frequencies the beam-beam interaction is modulated and can have a visible impact on the beam lifetime [7].

**TABLE 1.** Machine parameters relevant to beam-beam interactions, for Au-Au and p-p collisions. The beam-beam parameters in deuteron-gold collisions are close to those in gold-gold collisions.

| parameter | unit | Au-Au | p-p |
|---|---|---|---|
| relativistic $\gamma$, injection | ... | 10.5 | 25.9 |
| relativistic $\gamma$, store | ... | 107.4 | 106.6 |
| no of bunches $n_b$ | ... | 55 | 55 |
| ions per bunch $N_b$ | $10^9$ | 1 | 100 |
| emittance $\varepsilon_{Nx,y95\%}$ | $\mu$m | 10 | 20 |
| chromaticities ($\xi_x, \xi_y$) | ... | (+2,+2) | |
| harmonic no. $h$, store | ... | $7 \times 360$ | 360 |
| synchrotron tune $Q_s$ | $10^{-3}$ | 3.0 | 0.5 |
| rms bunch length $\sigma_z$ | m | 0.3 | 0.7 |
| rms momentum spread $\sigma_p/p$ | $10^{-3}$ | 0.15 | 0.3 |
| envelope function at IP $\beta^*$ | m | 1–10 | |
| beam-beam $\xi$/IP | ... | 0.0023 | 0.0037 |
| crossing angle $\theta$ | mrad | 0.0 | |
| head-on collisions | ... | 2–4 | |
| parasitic collisions | ... | 4–2 | |

---

[1] Work supported by US DOE, contract No DE-AC02-98CH10886.

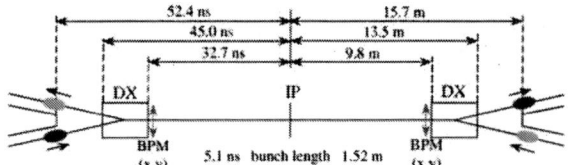

**FIGURE 1.** RHIC interaction region. Beams share a common beam pipe between the beam splitting DX dipoles. The bunch spacing shown corresponds to a fill pattern of 120 symmetrically distributed bunches.

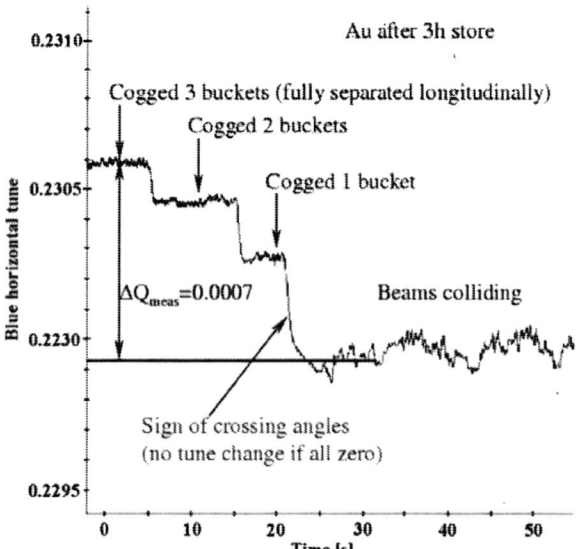

**FIGURE 2.** Tune as a function of longitudinal beam crossing position. Cogging by one acceleration bucket moves the crossing point by 5 m (tune measurement courtesy P. Cameron).

The beam-beam tune shift depends on the bunch intensity and emittance. Since the bunch intensities can be measured with good precision, the beam-beam tune shift measurement also provides an emittance estimate.

## LIFETIME AND EMITTANCE GROWTH

The beam-beam interaction is most pronounced for proton-proton collisions with larger $\beta^*$ values (Tab. 1). In lattices with small $\beta^*$ uncorrected nonlinear field errors in the triplets have a significant impact on the beam lifetime. Fig. 3 shows the distributions of the bunched beam lifetimes for polarized proton collisions during the 2001 run, with $\beta^* = 3$ m at all IPs. The distributions are relatively wide with a mean of about 15 hours. The average bunch intensity for all stores in the plot is $N_b = 0.4 \cdot 10^{11}$ with an 95% normalized emittance of $\varepsilon_N = 25$ $\mu$m, corresponding to a beam-beam parameter of $\xi = 0.0015$/IP. Gold beam lifetimes are generally lower due to intra-

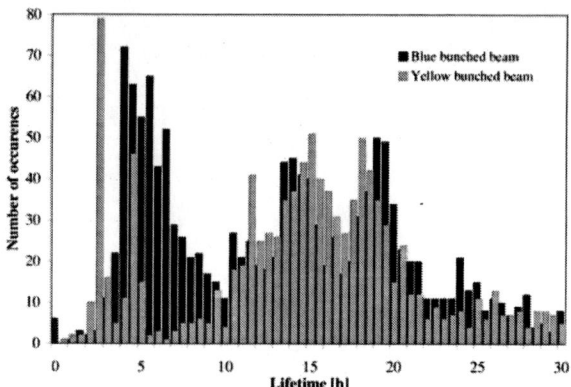

**FIGURE 3.** Blue and Yellow bunched beam lifetimes with proton-proton collision during one month of operation (lifetime fits courtesy J. van Zeijts).

beam scattering. A deuteron beam without collisions, in a lattice with $\beta^* = 2 - 10$ m, provides a good comparison to see the impact of the beam-beam interaction. Its lifetime is only marginally influenced by intra-beam scattering or nonlinear triplet errors. 55 bunches with $N_b = 0.5 \cdot 10^{11}$, stored for an hour, showed a lifetime of 830 h.

From the change in the beam intensity and the observed luminosity an estimate for the emittance growth can be obtained, assuming the same emittance in both beams. For proton-proton collisions we find $\Delta \varepsilon / \varepsilon = 4\%$ in the first store hour (with an rms value of 5%). For comparison, no emittance growth was observed with the ionization profile monitor in the deuteron beam measurement without beam-beam interaction.

While most bunches experience 4 head-on collisions, some bunches experience only 3 or 2 head-on collisions. In proton-proton collisions, these have visibly larger beam lifetimes (see Fig. 4). For this reason, beams were only collided in two experiments in the latter part of the last proton run.

## WORKING POINT AND BACKGROUND

Both transverse RHIC fractional tunes $(Q_x, Q_y)$ are kept between 0.2 and 0.25, and during stores close to the coupling resonance $Q_x = Q_y$. In this area the lowest order resonances are of order 9, 13, 14 and 17 (see Fig. 6). If the nonlinear dynamics are dominated by the beam-beam interaction and the crossing angles are all zero, no odd-order resonances are driven.

Experimental background rates were observed as a function of the tunes with deuteron-gold collisions (see Figs. 5 and Fig. 6), and proton-proton collisions. The tunes were moved parallel to the $Q_x = Q_y$ line, scanning

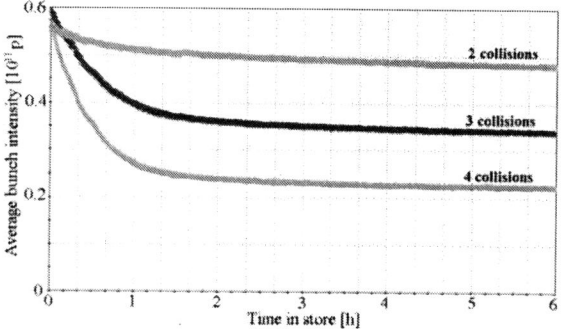

**FIGURE 4.** Average bunch intensity of 5 bunches each with 2, 3 and 4 head-on collision respectively during a proton store. The initial beam-beam parameter is $\xi = 0.002/\text{IP}$.

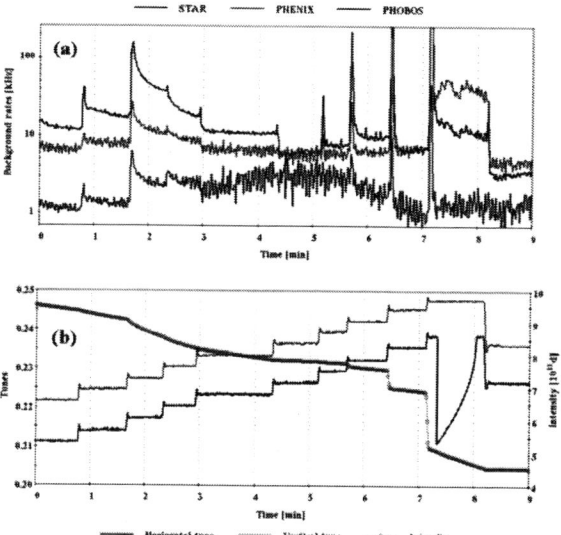

**FIGURE 5.** (a) Experimental background rates during a working point scan. (b) Deuteron beam intensity and transverse tunes during the scan.

the area considered for operation. With both beams, high background rates were found near $9^{th}$ order resonances, and low background rates near $13^{th}$ order resonances. The working points with low background rates are used in operation. High background rates near $9^{th}$ order resonances are another sign of residual crossing angles (cf. Fig. 2). Increased background rates were also found with a transverse offset [8].

## STRONG-STRONG OBSERVATIONS

RHIC sees strong-strong beam-beam effects. In addition to the tune ($\sigma$-mode) a new transverse oscillation mode ($\pi$-mode) occurs. For a single collision per turn the $\pi$-

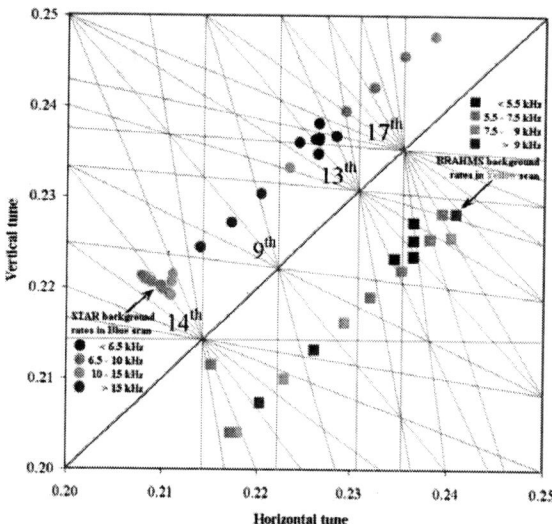

**FIGURE 6.** Experimental background rates as a function of fractional coherent tunes. For these measurements deuterons in the Blue ring collide with gold ions in the Yellow ring. The total beam-beam tune spread due to the beam-beam interaction is about $\Delta Q_{tot} = 0.005$ in both measurements.

mode is at a tune $Y\xi$ below the $\sigma$-mode, where $Y \approx 1.2$ for round beams [9]. If the beam-beam interaction is the dominant nonlinear effect, the $\pi$-mode can be outside the continuous spectrum and thus be undamped [10].

Coherent beam-beam modes were observed in an experiment with proton beams, with a beam-beam parameter $\xi = 0.003$ and a single collision per turn (see Fig. 7). The measured difference between the $\sigma$- and $\pi$-modes is consistent with a Yokoya factor of $Y \approx 1.2$. The locations of the $\pi$-modes were reproduced in a strong-strong simulation [12]. $\pi$-modes were also observed in routine operation with a beam-beam parameter $\xi = 0.0015$, four collisions per turn and linear coupling (see Fig. 8). The $\pi$-modes could be suppressed by small changes in one of the tunes (see Fig. 8).

## SUMMARY

The beam-beam interaction has a significant impact on lifetime and emittance of the RHIC beams, as well as the experimental background. In addition to beam-beam effects observed in other hadron colliders, coherent beam-beam modes were seen for the first time. So far, coherent modes could be suppressed by small tune changes.

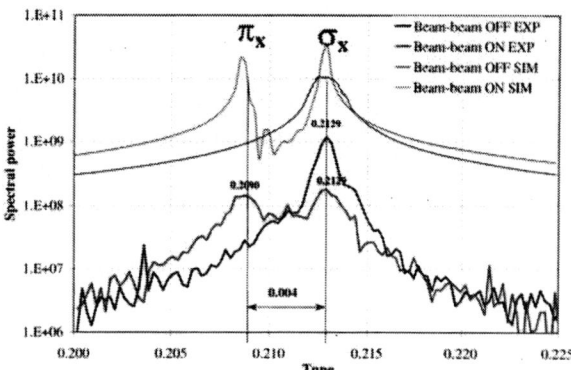

**FIGURE 7.** Coherent dipole modes in an experiment with a single proton bunch per beam, and in a corresponding simulation [12]. $\xi = 0.003$, spectra from 4096 turns.

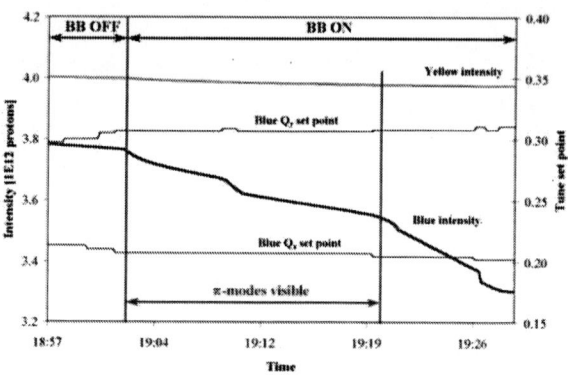

**FIGURE 9.** Blue and Yellow beam intensity and Blue tune set points during tuning for lifetime at the beginning of a proton store. $\pi$-modes were only visible for certain tune set points.

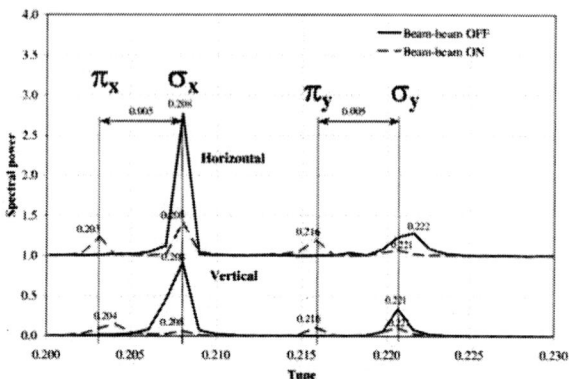

**FIGURE 8.** Coherent dipoles modes in operation with 4 collisions per turn. $\xi = 0.0015$/IP, spectra from 1024 turns.

## ACKNOWLEDGMENTS

The author is thankful for help and discussions to J.M. Brennan, M. Blaskiewicz, P. Cameron, H. Huang, W. MacKay, S. Peggs, F. Pilat, V. Ptitsyn, T. Roser, S. Tepikian, D. Trbojevic, and J. van Zeijts.

## REFERENCES

1. H. Hahn (editor), "RHIC Design Manual", revision of October 2000, http://www.rhichome.bnl.gov/NT-share/rhicdm/00_toc1i.htm.
2. T. Roser, "RHIC status and plans", proceedings of the 2003 Particle Accelerator Conference, Portland, Oregon (2003).
3. W. Fischer and S. Peggs, "RHIC as a test bench for beam-Beam Studies", proceedings of the Beam-Beam Workshop at Fermilab, BNL C-A/AP/61 (2001).
4. P. Cameron et al., "RHIC third generation PLL tune system", proceedings of the 2003 Particle Accelerator Conference, Portland, Oregon (2003).
5. S. Saritepe, G. Goderre, and S. Peggs, "Observations of the beam-beam interaction in hadron colliders", in "Frontiers of particle beams: intensity limitations", Springer-Verlag, Lecture Notes in Physics (1991).
6. W. Fischer, M. Blaskiewicz, J.M. Brennan, P. Cameron, R. Connolly, C. Montag, S. Peggs, F. Pilat, V. Ptitsyn, S. Tepikian, D. Trbojevic, and J. van Zeijts, "Observation of strong-strong and other beam-beam effects in RHIC", proceedings of the 2003 Particle Accelerator Conference, Portland, Oregon (2003).
7. W. Fischer, P. Cameron, S. Peggs, and T. Satogata, "Tune modulation from beam-beam interaction and unequal radio frequencies in RHIC", these proceedings.
8. A. Drees, Z. Xu, H. Huang, "Results from vernier scans at RHIC during runs 2000-2003", proceedings of the 2003 Particle Accelerator Conference, Portland, Oregon (2003).
9. K. Yokoya and H. Koiso, "Tune shift of coherent beam-beam oscillations", Part. Accel. Vol. 27, pp. 181-186 (1990).
10. Y. Alexahin, "On the landau damping and decoherence of transverse dipole oscillations in colliding beams", *Part. Accel.*, **V59**, p. 43; CERN-SL-96-064 (AP) (1996).
11. W. Fischer, L. Ahrens, M. Bai, M. Blaskiewicz, P. Cameron, R. Michnoff, F. Pilat, V. Ptitsyn, T. Sen, S. Tepikian, D. Trbojevic, M. Vogt, and J. van Zeijts, "Observation of coherent beam-beam modes in RHIC", BNL C-A/AP/75 (2002).
12. M. Vogt, J.A. Ellison, W. Fischer, and T. Sen, "Simulations of coherent beam-beam modes at RHIC", proceedings of the 2002 European Particle Accelerator Conference, Paris (2002).
13. W. Fischer, P. Cameron, S. Peggs, and T. Satogata, "Tune modulation from beam-beam interaction and unequal radio frequencies in RHIC", BNL C-A/AP/72 (2002).
14. M. Brennan et al., "Operation of the RHIC rf system", proceedings of the 2003 Particle Accelerator Conference, Portland, Oregon (2003).

# Beam-Beam Performance Of The SLAC B-Factory[*]

W. Kozanecki[§]
Y. Cai[¶], F.-J. Decker[¶], R. Holtzapple[¶†], J. Seeman[¶], M. Sullivan[¶], U. Wienands[¶]

[§]*DSM-DAPNIA-SPP, CEA-Saclay, 91191 Gif-sur-Yvette (France)*
[¶]*Stanford Linear Accelerator Center, Stanford, CA 94309 (USA)*
[†]*now at Alfred University, Alfred, NY 14802 (USA)*

**Abstract.** The beam-beam performance of PEP-II has been monitored by parasitic measurements recorded during routine physics running, and by a few dedicated accelerator physics experiments. These measurements indicate that in some cases, the beam-beam interaction and the electron-cloud-induced blowup of the low-energy positron beam are somehow coupled and enhance each other. Tailoring the bunch pattern to carefully balance these effects has proven very effective in maximizing the integrated luminosity. The comparison of recent simulation results with experimental data is encouraging.

## INTRODUCTION

The main parameters of the SLAC B-Factory [1] are listed in Table 1. In contrast to what is naturally enforced in single-ring colliders, the emittances and IP $\beta$-functions can be quite different in the two rings. The best performance has so far been achieved with rather different $e^+$ and $e^-$ beam-beam parameters. High luminosity also reproducibly favours an $e^+/e^-$ current ratio of 1.5 to 2.0, where one would naively expect 2.9 from the simplified energy-transparency condition [2]:

$$I^+ E^+ = I^- E^-$$

Steady luminosity and background improvements have relied on maintaining a delicate empirical balance between the currents, tunes, beam-beam parameters, and ECI-induced blowup as these quantities vary along each bunch train. Spot-size, beam-current and luminosity diagnostics (both bunch-by-bunch and averaging over the entire train) have proven essential to unravel these coupled phenomena.

**TABLE 1.** PEP-II Collision Parameters. The symbols "LER" and "HER" refer to the low-energy ($e^+$) and high-energy ($e^-$) ring, respectively.

| IP Parameter | Design | Recent peak performance | Units |
|---|---|---|---|
| Beam energy $E$ (LER/HER) | 3.1 / 9.0 | 3.1 / 9.0 | GeV |
| Crossing angle | 0 | < 1 | mrad |
| Luminosity | 3.00 | 6.57 | $10^{33}$ cm$^{-2}$s$^{-1}$ |
| Number of bunches | 1658 | 1034 | |
| Beam current $I$ (LER/HER) | 2146 / 750 | 1550 / 1175 | mA |
| $\beta^*_y / \beta^*_x$ | 1.5 / 50 | 1.2 / 40 ($e^+$), 1.2 / 28 ($e^-$) | cm |
| Emittances $\varepsilon_y / \varepsilon_x$ (low $I$) | 1.5 / 49 | 1.8 / 30 ($e^+$), 1.8 / 49 ($e^-$) | nm-rad |
| LER tunes $v_x / v_y$ | 38.64 / 36.57 | 38.52 / 36.57 | |
| HER tunes $v_x / v_y$ | 24.62 / 23.64 | 24.52 / 23.62 | |
| $\xi_y$ ($e^+/e^-$) | 0.03 / 0.03 | 0.082 / 0.040 | |
| $\xi_x$ ($e^+/e^-$) | 0.03 / 0.03 | 0.109 / 0.040 | |

---

[*] Work supported in part by the Department of Energy under Contract No. DE-AC03-76SF00515.

# INTERPLAY BETWEEN ELECTRON-CLOUD & BEAM-BEAM ISSUES

The build-up of the electron cloud along the positron bunch, and its impact on the luminosity, exhibit a strong dependence on the bunch pattern. An example from early PEP-II running is shown in Fig. 1. The first bunches of each mini-train have a high luminosity, which drops to 40 % of its initial value at the end of the longest train. The long gaps clear the electron cloud, which slowly builds up again along the next mini-train.

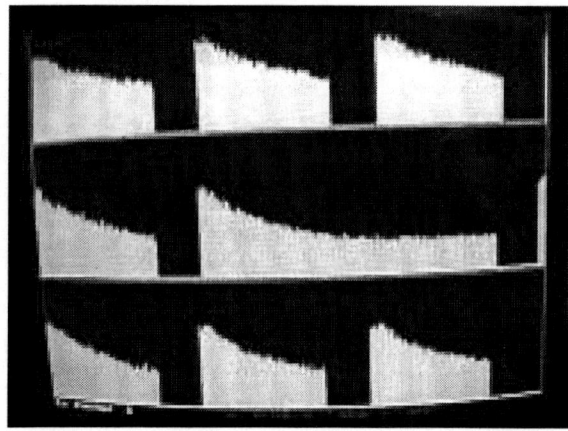

**FIGURE 1.** Luminosity vs. position along the bunch train. Pattern: 8.4 ns bunch spacing with 7 large gaps (July 2000). This data was acquired with the solenoids turned on in part of the straight sections only.

A related (and so far unexplained) observation is that of self-blowup of the $e^+$ beam during high-current collisions. This is illustrated in Fig. 2. With the $e^-$ current kept constant at 625 mA while increasing the $e^+$ current from 100 to 1300 mA, the transverse $e^+$ beam size grew rapidly, both horizontally and vertically, once the LER current exceeded a 900 mA threshold. In contrast, the $e^-$ horizontal and vertical beam sizes (not shown) remained constant. In the absence of an electron beam, the $e^+$ horizontal size was, at low current, 15% smaller than in collision, and remained constant with increasing LER current (Fig. 2). When a similar sequence of measurements was performed by varying the $e^-$ current instead, the $e^+$ beam size was independent of HER current (above 150 mA), but its value depended on the LER current itself. These surprising observations were interpreted as the electron-cloud instability (ECI) and the beam-beam interaction mutually "enhancing" each other at high positron current.

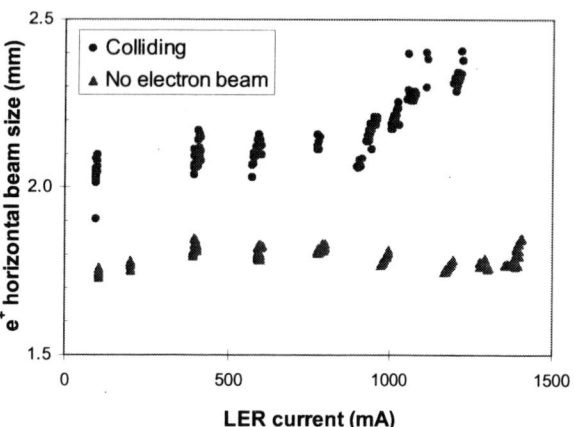

**FIGURE 2.** LER-current dependence of the $e^+$ horizontal beam-size averaged over the bunch train, in collision (circles) and with no $e^-$ beam circulating (triangles). The beam size is measured by a synchrotron-light monitor (SLM). Pattern: 597 bunches, 10.5 ns spacing (July 2000).

More recently, direct measurements of the $e^+$ beam size along the bunch train, using a gated synchrotron-light camera, revealed both a short-range blowup (within each mini-train), and a long-range buildup of the cloud over the entire length of the train. Such $e^+$ beam-size variations directly impact the luminosity *via* the positron charge density at the IP. But they also induce variations of the $e^-$ beam-beam tune shift along the train, resulting in an effective increase of the $e^-$ tune footprint and/or in $e^-$ spot-size variations. In addition, the PEP-II luminosity and beam lifetimes are quite sensitive to minor adjustments in the tunes of either beam. Because the tunes can only be optimized in an average sense (*i.e.* for the train as a whole), specific bunches (at the head or tail of the train, or at the front of individual mini-trains) occasionally exhibit significantly lower luminosity and/or poor lifetime.

**FIGURE 3.** Standard luminosity pattern in 2003 (6.3 ns spacing), alternating mini-trains of 10 and 11 bunches. Each

mini-train has constant luminosity except for bunches 1 and 3. During this measurement, solenoids covered most of the LER vacuum chamber in all arcs and straight sections.

These observations lead to an empirical optimization [3] of the bunch pattern to maximize the luminosity, taking into account:
- the total-current budget (limited by RF power and by beam-induced heating);
- the spacing between mini-trains (the longer the gaps, the better the $e^-$-cloud suppression);
- the mini-train length (the shorter the mini-train, the less the cloud builds up);
- the number of mini-trains (fewer mini-trains mean fewer "fragile" bunches).

The severity of the $e^-$-cloud effects, for a fixed bunch pattern, has been steadily decreasing over the years. The winding of low-field solenoids (25-35 Gauss) around all accessible sections of the LER vacuum chamber, combined with synchrotron-radiation scrubbing of vacuum surfaces, has progressively allowed operating PEP-II with increased beam currents and a denser bunch pattern (Fig. 3). Several electron-cloud-related phenomena are no longer apparent, such as the single-beam $e^+$ blowup at high current or the mutual enhancement of the ECI- and beam-beam-induced blowup. During recent running, the effects of the cloud typically reach equilibrium after the first two or three bunches in each mini-train.

Exploratory measurements, aimed at improving the luminosity, suggest that in the design bunch pattern (4.8 ns spacing), the ECI may again turn into a significant limitation as currents are further increased. Recent simulations suggest that doubling the solenoid field strength may prove an effective countermeasure: this upgrade is in progress.

## BEAM-BEAM FLIP-FLOP

The strong beam-beam forces between colliding bunches can result in a "flip-flop" of the transverse size of some bunches. The flip-flop occurs when the transverse size of the positron bunch shrinks and the corresponding $e^-$ bunch size grows. This phenomenon accounts for several concomitant observations: increase in IP beam size (evidenced by a reduced luminosity), shorter $e^+$-beam lifetime, and higher background in the BABAR detector.

Flip-flop has been studied [4] in the LER using a 2-ns gated camera. Measurements have shown that when an $e^+$ bunch flips: (i) the horizontal and vertical $e^+$ bunch sizes decrease by ~30% (Fig. 4); (ii) the luminosity drops by ~50% (Fig. 5), presumably because the corresponding $e^-$ bunch size(s) increase substantially; (iii) the LER lifetime drops when the bunch goes from its "flipped" (shrunk) state to the "flopped" state (Fig. 4); (iv) the transition is rather fast (~ 0.5 sec.) and usually occurs at the top of a store, where the LER x-tune is optimized for high luminosity; and (v) flip-flop primarily affects bunches in the front of a mini-train, which might be due to such bunches experiencing a different tune shift because the ECI has been suppressed there by the mini-gap.

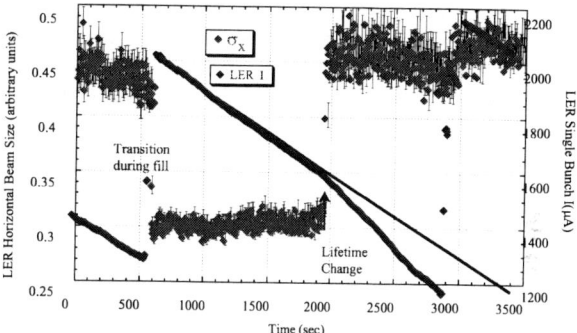

**FIGURE 4.** Single-bunch $e^+$ current and horizontal beam size vs. time, illustrating the flip-flop transition.

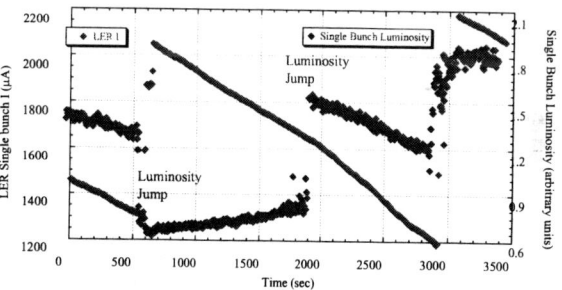

**FIGURE 5.** Single-bunch luminosity and $e^+$ current vs. time, illustrating the flip-flop transition.

A more detailed study is planned using two gated cameras, one per ring, to gain further insight into flip-flop dynamics.

## BEAM-BEAM LIMITS NEAR $\nu_x=1/2$

All of the above observations were carried out with the horizontal LER tune, and the vertical HER tune, near the 2/3 resonance (Table 2). Both rings were recently moved closer to the 1/2-integer to take advantage of the eventual luminosity enhancement associated with the dynamic-$\beta$ effect.

**TABLE 2. History of PEP-II working points.**

| Fractional tune | 1999-2003 | Since 5/2003 |
|---|---|---|
| LER ($v_x / v_y$) | 0.64 / 0.56 | 0.52 / 0.57 |
| HER ($v_x / v_y$) | 0.57 / 0.64 | 0.52 / 0.62 |

The new working point yields significantly improved luminosity performance, but also a qualitatively different beam-beam blowup pattern. The positron beam size now depends only on the $e^-$ current (and vice-versa): "self-blowup" of the $e^+$ beam, in particular, is no longer observed. Over the current range of a typical store, the vertical $e^-$ beam size shrinks by 30-40% as the LER current decays (Fig. 6). Similarly, the horizontal positron beam size is a linear function of the HER current, and is typically 50-60% larger at full $e^-$ current than in single-beam mode. These mutual blowup effects are reflected in the beam-current dependence of the luminosity, which exhibits clear saturation at high intensity (Fig. 7).

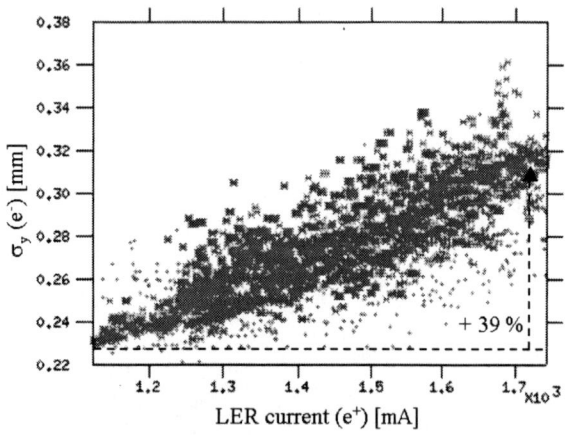

**FIGURE 6.** Dependence of the vertical $e^-$ SLM beam size on the LER current, during a typical store ($v_x^{+,-} \sim 0.52$).

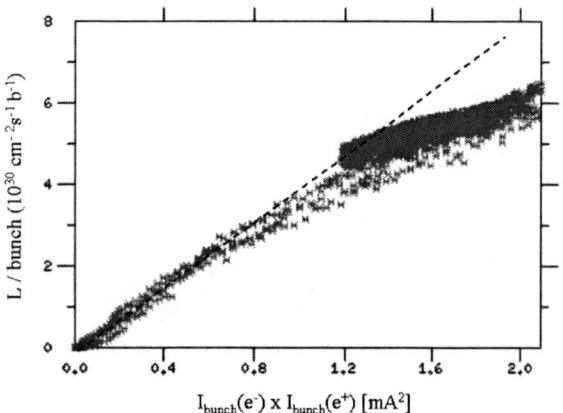

**FIGURE 7.** Beam-current dependence of the luminosity during a typical store.

Combining the measured beam-current dependence of the $e^+$ and $e^-$ beam sizes and of the luminosity with (measured or assumed) emittances and IP $\beta$-functions, provides an estimate of the beam-beam parameters $\xi_{x,y}^{+,-}$ (Table 1). Although approximate, the results constitute a measure of the actual beam-beam performance of the machine, as well as a guide towards further luminosity improvements.

## SIMULATIONS

As a result of dramatic increases in available computer speed, it has become feasible to use the particle-in-cell method for beam-beam simulations. Such an approach is self-consistent, because the electromagnetic field is computed by solving the Poisson equation with the charge distributions being updated as the beams collide. Recently, it was found that the extent of the x-y grid can be much reduced if an inhomogeneous potential is assigned on the boundary [5]. A smaller spatial extent allows for a denser mesh, thereby increasing the resolution of the Poisson solver.

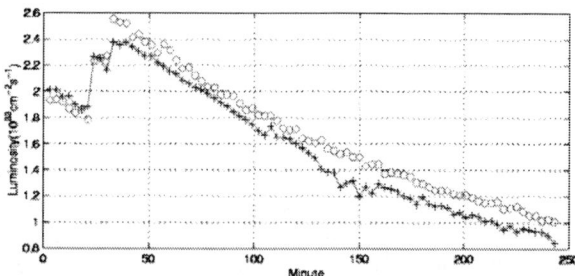

**FIGURE 8.** Time evolution of the measured luminosity on 10/1/2000 (crosses), and comparison with that predicted by a two-dimensional simulation (circles).

Fig. 8 compares the PEP-II luminosity measured during physics running, with that computed using a two-dimensional simulation [6] at the same beam currents. The prediction is absolute in that none of the simulation input parameters were fitted: all of them had been chosen close to their nominal value except for the vertical emittance, which is not so important because of the vertical beam-beam blowup at high intensity. The agreement between simulation and measurement is surprising, and is remarkable given the simplicity of such a 2-D model. This success is largely attributed to the fact that the operating tunes were well optimized and that many resonances (including synchro-betatron lines) had been carefully avoided.

In general however, three-dimensional effects such as hourglass effect and phase averaging must be included in beam-beam simulations. Achieving the necessary numerical convergence in three dimensions requires the use of parallel supercomputers.

One of the most important aspects of parallel computing is how to minimize the communication among processors. Each application may have a different optimal solution. For beam-beam simulations, we have developed an efficient strategy utilizing dual processors (Fig. 9). Macro-particles are evenly distributed across many processors. The processors are divided into two groups, one for the positron beam and the other for the electrons. Before the collision, the beam distribution on the grid is summed within each group, and the resulting distribution is distributed back to all processors in the group. Then the total distribution is exchanged between the groups. That allows us to solve the Poisson equation and compute the force on the macro-particles in every local processor.

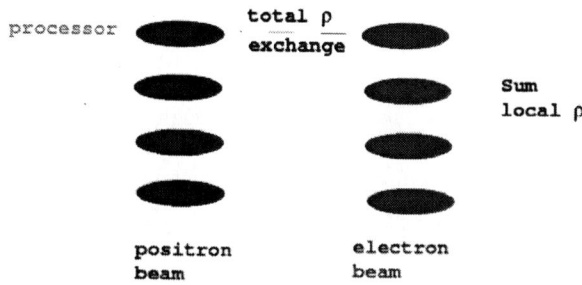

**FIGURE 9.** Illustration of how the processes in the parallel computer communicate with each other.

In this scheme, the macro-particles always remain confined to the same computing process. Only the beam distribution on the grid is exchanged among the processors. The division into two groups essentially allows to double the speed without much penalty. With a parallel supercomputer comprising 32 processors in all, we increased the speed of the simulation by a factor of 24. This enabled us to achieve numerical convergence in a three-dimensional simulation using 15 longitudinal slices, for a typical set of symmetric PEP-II parameters at high beam intensities.

## SUMMARY

The impact of the electron-cloud instability on the PEP-II luminosity has been minimized by a combination of scrubbing, installation of solenoid windings around the LER, and bunch-pattern optimization. But electron-cloud effects still play an ubiquitous role in the beam-beam performance of the B-Factory. Even though they appear less important at the new working point, they may still constitute a significant limitation at high luminosity.

The PEP-II horizontal tunes were recently moved close to the 1/2 integer. At this working point, the machine achieved a peak luminosity of $6.57 \cdot 10^{33} cm^{-2} s^{-1}$, with estimated beam-beam parameters $\xi_x/\xi_y$ of about 0.109/0.082 (0.040/0.040) in the LER and HER respectively.

Beam-beam simulations show encouraging agreement with experiment, but more extensive validation of 3-dimensional codes is urgently needed.

## ACKNOWLEDGMENTS

This work was supported in part by the U.S. Department of Energy, under Contract Number: DE-AC03-76SF00515. We gratefully acknowledge Robert Ryne's generosity in providing the computing resource at NERSC. One of us (W.K.) thanks SLAC for its hospitality.

## REFERENCES

1. Seeman, J., *et. al.*, SLAC-PUB-9332 (2002).

2. *PEP-II, an Asymmetric B Factory: A Conceptual Design Report*, SLAC report SLAC-R-418, 1993, Sec. 4.4.

3. Decker, F.-J., *et. al.*, SLAC-PUB-9272 (2002); SLAC-PUB-9360 (2002).

4. Holtzapple, R. L., *et. al.*, SLAC-PUB-9238 (2002).

5. Cai, Y., Chao, A. W., Tzenov, S. I., and Tajima, T., *Phys. Rev. ST Accel. Beams* **4**, 011001 (2001).

6. Cai, Y., SLAC-PUB-8811 (2001).

# Collision with finite crossing angle at KEKB

Kazuhito Ohmi*, Masafumi Tawada* and Yoshihiro Funakoshi*

*KEK, 1-1 Oho, Tsukuba, 305-0801, Japan

**Abstract.** KEK B factory (KEKB) achieved the luminosity of $10^{34}$ cm$^{-2}$s$^{-1}$ with a finite crossing angle scheme. The crossing angle made easy the design of the interaction region for a narrow bunch spacing, with the result that it gave a gain of high collision repetition. The crossing angle did not degrade the beam-beam performance up to the beam-beam parameter of $\xi \approx 0.05$. We review the beam-beam effect of KEKB as a successive case of collision with finite crossing angle.

## INTRODUCTION

KEK B factory (KEKB) is an asymmetric $e^+e^-$ collider with finite crossing angle scheme. The beam energies are 3.5 GeV and 8 GeV for $e^+$ and $e^-$, respectively, and they collide with each other with a half crossing angle of 11 mrad in the horizontal plane. KEKB had been designed to deliver the luminosity $10^{34}$ cm$^{-2}$s$^{-1}$ in the energy region of $\bar{B}B$ mesons production. The commissioning had started at October in 1999, and the design luminosity was achieved at May 2003. The crossing angle did not have serious effects for the luminosity. The beam-beam parameter was achieved the design, 0.05, though many operation parameters were different from the design.

The parameters are shown in Table 1. The differences, which make the status of KEKB very clear, are summarized as follows [1].

- Number of bunch ($N_b$) could not increase to the design value. This is due to electron cloud effect. Before 2002, more than 1150 bunches (8 ns spacing) did never contribute the luminosity. 1280 bunches and more (7.5-7 ns in averaged spacing) contribute the luminosity this year (2003). Winding solenoid and scrubbing improved the luminosity for more bunches.

- Bunch population ($N_e$) was much higher than the design. The beam-beam parameter calculated by the design beam size was more than 0.1. Actually beam size enlargement occurred due to the beam-beam effect, with the result that the beam-beam parameter was reduced 0.05.

- Beta functions were squeezed as possible as we could.

- Horizontal tune was approached to half integer as possible as we could. The luminosity increased with approach to half integer. Vertical tune was also optimized.

- Synchrotron tune was higher than the design. It seemed to contribute to suppress the head-tail instability caused by electron cloud.

We come up against two fronts from the viewpoint of the beam dynamics. One is the electron cloud effect. Solenoid magnets have been winded to remove electron cloud every shutdown periods [2]. The luminosity increased for higher current after winding the solenoid, though it had hit the ceiling before. We still have a luminosity reduction for increasing the number of bunches.

The other is the beam-beam effect. We put much more current in each bunch than the design, since the number of bunches is limited by the electron cloud effect. Since the beam size is enlarged by the beam-beam effects, the beam-beam parameter does not increase any longer in this current region. Nevertheless the luminosity increases for putting more current, though it does not depend on square of current. Increasing the current is one of method to get more luminosity.

The beam-beam performance is strongly affected by orbit distortion and optics parameters at the collision point. They are used as tuning knobs, which are controlled and/or scanned to get higher luminosity day by day at KEKB. Efficient tuning knobs depended on the machine condition. We had discussed the luminosity dependence on the parameters [3] at the early stage of commissioning in 2000. The luminosity was an order lower at that time. We discuss again the parameter dependence for the present parameters.

**TABLE 1.** Basic parameters of KEKB

|  | Design | | Operation | |
|---|---|---|---|---|
|  | HER | LER | HER | LER |
| $C$ | 3016 m | | 3016m | |
| $E$ | 8 GeV | 3.5 GeV | 8 GeV | 3.5 GeV |
| $N_b$ | 5000 | | 1280 | |
| $N_e$ | $1.4 \times 10^{10}$ | $3.3 \times 10^{10}$ | $7.5 \times 10^{10}$ | $5.5 \times 10^{10}$ |
| $\beta_x/\beta_y$ | 80 cm / 8 mm | | 60 cm / 7 mm | |
| $\varepsilon_x/\varepsilon_y$ | 18 nm / 0.18 nm | | 24 nm / 0.24 nm | 18 nm / 0.18 nm * |
| $\sigma_z$ | 7 mm | | 4 mm | |
| $\nu_x/\nu_y/\nu_s$ | 0.52/0.58/0.015 † | | 0.513 / 0.586 / 0.02 | 0.506 / 0.545 / 0.025 |
| $\tau_{xy}/T_0$ | 4,000 | 4,000 | 4,000 | 4,000 |
| $\theta_c$ | $2 \times 11$ mrad | | $2 \times 11$ mrad | |

\* The emittance coupling was assumed to be 1%.
† In first design $\nu_y$ was chosen 0.08. The beam-beam effect is the same for $\nu_y = 0.58$ and 0.08 [4].

## BEAM-BEAM TUNING

The beam-beam performance should depend only on the one turn map at the collision point.

$$x(s) = \mathcal{M}_{bb} \circ \mathcal{M}_{rev} x(s-C) \quad (1)$$

where $\mathcal{M}_{bb}$ and $\mathcal{M}_{rev}$ are maps of the beam-beam interaction and lattice transfer. Though the beam-beam limit is determined by $\mathcal{M}_{bb}$ finally, the beam-beam performance is controlled by $\mathcal{M}_{rev}$. The zero-th and linear components of the map $\mathcal{M}_{rev}$, which are closed orbit and optics of two beams at the collision point, are considered to be leading terms for the luminosity performance.

### Offset of two beams

We first discuss error of the closed orbit $x_0^{\pm} = (x, p_x, y, p_y, z, p_z)_0^{\pm}$ at the collision point. $x_0^{\pm}$ and $y_0^{\pm}$ are transverse orbit offset. $z_0^{\pm}$, which is deviation of collision timing, should be fixed by measurement of the timing of bunches. $p_{x,0}^{\pm}$ and $p_{y,0}^{\pm}$ correspond to a small crossing angle. In our case, since beams collide with a finite horizontal crossing angle $\theta_c = 2 \times 11$ mrad, small error of $p_x \ll \theta_c$ is negligible. It is not necessary to take notice of $p_{z,0}^{\pm}$, which does not affect the beam-beam force. Therefore it is enough to control the transverse offset and vertical crossing angle for the zero-th order map.

When two beams collide with an offset, the geometrical luminosity is reduced as follows, without effects of crossing angle and hour glass.

$$L \propto \exp\left(-\frac{\Delta x^2}{4\sigma_x^2} - \frac{\Delta y^2}{4\sigma_y^2}\right). \quad (2)$$

Fig.1 shows the luminosity reduction due to orbit distortions. The luminosity without orbit distortion was al-

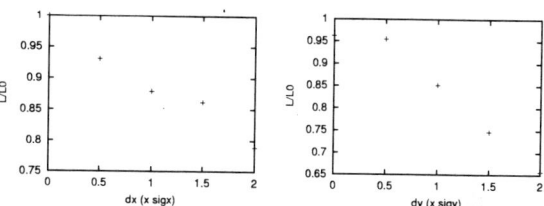

**FIGURE 1.** Luminosity reduction due to orbit offset.

ready reduced about half of geometrical one. These results were obtained by weak-strong simulation. Actually orbit distortion has strong-strong effect. A strong-strong simulation will be done [5]. The beam size was enlarged by the offset collision, with the result that the reduction for no offset was somewhat loose in this simulation.

The aspect ratio of vertical size for the bunch length is $\sigma_y/\sigma_z = 5 \times 10^{-4}$. Vertical crossing angle of the order of 0.5mrad can affect the luminosity geometrically. The effect was weak for the low beam intensity [3]. The simulation using recent parameters will be done [5].

### Optics at the collision point

We discuss how optics parameters affect the beam-beam effects. The one turn map including the beam-beam interaction is multiplication of beam-beam map and the transfer map between $s^* + 0$ to $s^* - 0 + C$. The transfer map is the same as the one turn map at $s^*$ in the case without the beam-beam interaction. The linear part of the transfer map is characterized by the optics parameters at the collision point. Though the optics is distorted by the beam-beam interactions, the distorted optics parameters are connected with the bare optics parameter. The revolution matrix is parameterized as [6]

$$M(s) = V^{-1}(s)UV(s) \quad (3)$$

241

where
$$V(s) = B(s)R(s)H(s). \tag{4}$$

The matrix $B$ normalizes the betatron motion,
$$B = \begin{pmatrix} B_x & 0 & 0 \\ 0 & B_y & 0 \\ 0 & 0 & B_z \end{pmatrix} \tag{5}$$

where
$$B_i = \begin{pmatrix} 1/\sqrt{\beta_i} & 0 \\ \alpha_i/\sqrt{\beta_i} & \sqrt{\beta_i} \end{pmatrix} \tag{6}$$

The matrices, $R$ and $H$ resolve $x-y$ and longitudinal couplings, respectively.
$$R = \begin{pmatrix} b & 0 & -r_4 & r_2 & 0 & 0 \\ 0 & b & r_3 & -r_1 & 0 & 0 \\ r_1 & r_2 & b & 0 & 0 & 0 \\ r_3 & r_4 & 0 & b & 0 & 0 \\ 0 & 0 & 0 & 0 & 1 & 0 \\ 0 & 0 & 0 & 0 & 0 & 1 \end{pmatrix}, \tag{7}$$

where $b^2 + |R| = 1$.
$$H = \begin{pmatrix} 1 - \frac{|H_x|}{1+a} & \frac{H_x S_2 H_y S_2}{1+a} & -H_x \\ \frac{H_y S_2 H_x S_2}{1+a} & 1 - \frac{|H_y|}{1+a} & -H_y \\ -S_2 H_x^t S_2 & -S_2 H_y^t S_2 & a \end{pmatrix}, \tag{8}$$

where
$$H_{x(y)} = \begin{pmatrix} \zeta_{x(y)} & \eta_{x(y)} \\ \zeta'_{x(y)} & \eta'_{x(y)} \end{pmatrix} \tag{9}$$

and $a^2 + |H_x| + |H_y| = 1$.

The matrix $U$ are characterized by tunes of three modes,
$$U = \begin{pmatrix} U_x & 0 & 0 \\ 0 & U_y & 0 \\ 0 & 0 & U_z \end{pmatrix}, \tag{10}$$

where
$$U_i = \begin{pmatrix} \cos\mu_i & \sin\mu_i \\ -\sin\mu_i & \cos\mu_i \end{pmatrix}. \tag{11}$$

The beam envelope matrix is expressed by
$$\Sigma(s) = \langle x(s) x^t(s) \rangle = V(s) \Sigma_n V^t(s). \tag{12}$$

$\Sigma_n$ is a diagonal matrix represented by emittances,
$$\Sigma_n = diag(\varepsilon_x, \varepsilon_x, \varepsilon_y, \varepsilon_y, \varepsilon_z, \varepsilon_z), \tag{13}$$

where $\beta_z \varepsilon_z = \sigma_z^2$ and $\varepsilon_z/\beta_z = \delta$.

Optics parameters are designed to be finite values for $\beta_{x,y,z}$ and zero for others at the collision point. In actual operation, these parameters are deviated from the design: the beta may be larger than the design and other parameters become nonzero. These distorted optics parameters affect the beam-beam performance [3].

The vertical beam size is critical for the flat beam. The vertical size is enlarged by coupling from other freedom. The geometrical beam size is roughly expressed by
$$\sigma_y^2 \approx \beta_y \varepsilon_y + \sigma_x^2 r_1^2 + r_2^2 \sigma_{p_x}^2 + \eta_y^2 \sigma_\delta^2 + \sigma_z^2 \zeta_y^2. \tag{14}$$

The optics errors affect the luminosity further strongly considering dynamical effect of the beam-beam interaction. Figure 2 shows the luminosity degraded by the optics errors of some parameters.

In KEKB operation, these parameters are optimized everyday. The dispersions and coupling parameters are controlled by using a closed orbit bump at sextupole magnets. The beta functions, exactly speaking, waist position of beta function, are controlled by several quadrupole magnets near the collision point. The parameters are scanned to optimize the luminosity.

## Tune

We had performed tune survey since the design stage of 1994 [7] by using a weak-strong simulation. Tune survey has been done by strong-strong simulation since starting the commissioning. The design operating point was $(v_x, v_y) = (0.52, 0.08)$. The operating point is changed to $(0.52, 0.58)$ [4] at 2000. The change made possible fine tuning of beam-beam performance: that is, the closed orbit becomes controllable, with the result that the optics parameters and others tuned by orbit bumps also become controllable.

Figure 3 shows result of recent tune survey obtained by the strong-strong simulation. For the tune survey of LER and HER, tune of HER and LER was fixed as (.513,.582) and (.511,.553), respectively.

Recent operating point tends to be lower in vertical for LER. The optimal tune of LER is near 0.54-0.55 in vertical. This behavior is consistent with the simulation result.

Tune of the colliding beam is ambiguous, because it was spreaded by the strong nonlinear beam-beam force. Therefore tune of a non-colliding bunch is monitored with a gated time window. The non-colliding bunch is called the pilot bunch [8]. The tune, which is very sensitive for the luminosity and the beam life time controlled along a pattern depending on the current [9].

A possibility toward more luminosity is remained for the HER tune from the simulation. The best operating point of HER is slightly lower in horizontal and near 0.54 in vertical, which is the same as those of LER. However when we try to go down the horizontal tune, the pilot bunch is lost due to shortening of the life time.

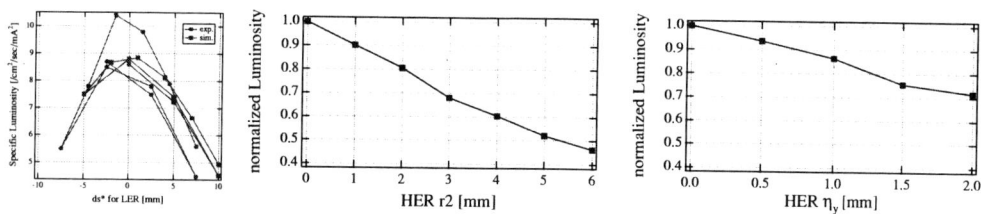

**FIGURE 2.** Typical luminosity behavior for changing the beta waist, x-y coupling and dispersion.

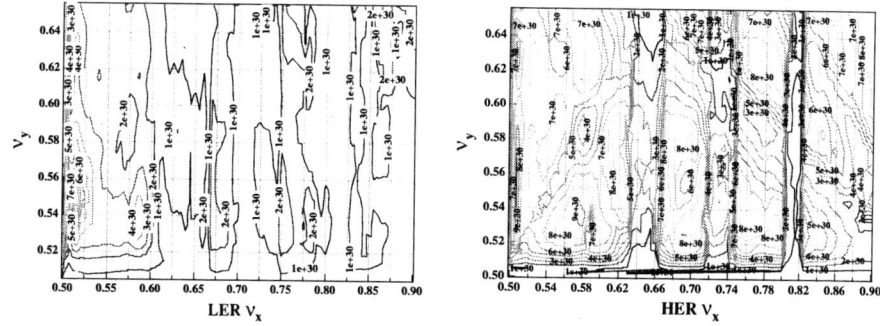

**FIGURE 3.** Luminosity for the tunes of LER and HER. Red line corresponds to the highest luminosity.

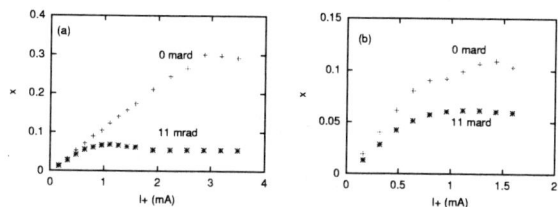

**FIGURE 4.** Beam-beam parameters obtained by weak-strong and strong-strong simulations for the crossing angles of zero and 11 mrad.

## CROSSING ANGLE

The crossing angle is equivalent to an optics error with $\zeta_x$ at the collision point, except a nonlinear transformation with a small effect. Figure 4 shows the beam-beam parameter for zero and finite (11 mrad) crossing angle. The results obtained by strong-strong and weak-strong simulations are coincide for crossing angle of 11 mrad, while they are no coincide for zero crossing angle. The beam-beam parameter is no limit for the weak-strong simulation, though is around 0.1 for the strong-strong simulation.

## SUMMARY

Optics functions and tunes, which are defined by the $6 \times 6$ revolution matrix, control the beam-beam performance. We have studied how the parameters affect the luminosity using beam-beam simulations based on the weak-strong and strong-strong model. The luminosity is scanned for the parameters at KEKB and optimized everyday. The simulations are compared and examined with the luminosity optimization in KEKB.

The crossing angle can be regarded as a kind of optics error. Luminosity and beam-beam parameter at zero crossing angle is much better than those at finite crossing angle. Crab cavities generates $\zeta_x$ at the collision point, which can cancel the effect of crossing angle. The crab cavities improve the luminosity for KEKB.

The authors thank members of KEKB commissioning group for their help.

## REFERENCES

1. Akai, K., et al., *Nucl. Instrum. Meth. A*, **499**, 191–227 (2003).
2. Fukuma, H., et al., *AIP Conf. Proc.*, **642**, 357–359 (2003).
3. Ohmi, K., *Proceeding of EPAC2000*, pp. 433–435 (2000).
4. Wu, Y. Z., Funakoshi, Y., Tawada, M., and Ohmi, K., A new working point for the KEKB, Tech. rep., KEK-report-2001-5 (2001).
5. Tawada, M., Ohmi, K., and Funakoshi, Y. (to be published).
6. Ohmi, K., Hirata, K., and Oide, K., *Phys. Rev. E*, **49**, 751–765 (1994).
7. KEKB design group, KEKB design report, Tech. rep., KEK-report-95-7 (1995).
8. Ieiri, T., et al., *Phys. Rev. ST-Accelerators and Beams*, **5**, 094402 (2002).
9. Ohnishi, Y., *private communications* (2003).

# Luminosity Increase at the Incoherent Beam-Beam Limit with Six Superbunches in RHIC[1]

W. Fischer and M. Blaskiewicz

*Brookhaven National Laboratory, Upton, New York 11973*

**Abstract.** By colliding bunches of greater length under a larger angle, the tune spread caused by the beam-beam interaction can be reduced. Assuming a constant limit for the beam-beam tune shift, the bunch intensity can then be raised. In this way, a luminosity increase is possible. We review this strategy for proton beams in RHIC, with two collisions and consider six long bunches. Barrier cavities are used to fill every accelerating bucket of the machine, except for an abort gap, and to create the superbunches bunches at store. Resonances driven by the beam-beam interaction and coherent effects are neglected in this article.

## INTRODUCTION

Luminosity limits set by the incoherent beam-beam tune shift were discussed for unbunched beams by Keil [1]. He showed that an increase in the crossing angle reduces the beam-beam tune shift and allows a higher line density, which in turn leads to an increased luminosity. Recently, Ruggiero and Zimmermann extended this analysis to bunched beams [2]. With one horizontal and one vertical collision under the same angle, the beam-beam tune spread in both planes is the same for round beams.

Extremely long bunches, called superbunches, are the basis of a recently proposed hadron collider concept [3]. In this proposal, beam is stacked in very long bunches using barrier cavities, and accelerated with an induction device [4].

In this article we estimate the luminosity for six very long bunches in RHIC given a certain limit for the incoherent beam-beam tune spread. With six symmetrically distributed superbunches any two of the RHIC experiments can be served with luminosity. For the scheme under investigation here, barrier cavities are needed for injection and for the gap maintenance at store. Acceleration is done with the existing 28 MHz system with harmonic number $h = 360$ [5]. In an earlier article [6] we considered bunches in the RHIC accelerating and storage buckets, as well a superbunches that fill the whole circumference except for an abort gap.

Basic parameters are summarized in Tab. 1. We assume that a total tune spread of $\Delta Q_{max} = -0.03$ can be accomodated, caused by one horizontal and one vertical crossing. This is consistent with the maximum values achieved in the SPS and Tevatron, but challenging for routine operation.

The crossing angle $\theta$ is measured as the full angle from one beam to the other. With the current vertical corrector strength, a crossing angle of 0.84 mrad can be implemented at store [7]. However, some of this strength may be needed to correct for unwanted orbit effects. We therefore assume that vertical crossing angles of 0.5 mrad can be implemented with the existing hardware. Larger horizontal crossing angles were used in the past.

We take for the length, in which the beam-beam force is active, the distance between the DX beam splitting magnets. Once the beams reach these magnets they are quickly separated. The effective detector length, the region in which collisions are recorded, is the largest length currently used by any one of the RHIC detectors [8].

We neglect here resonances driven by the beam-beam interaction, coherent effects and end effects of the superbunches. However, we note that large crossing angles can be beneficial in damping coherent beam-beam modes [9]. Furthermore, it is assumed that the long-range beam-beam interactions during the energy ramp do not lead to significant emittance increases or beam losses.

## BEAM PREPARATION

At injection a long bunch that almost fills the circumference, except for an abort gap, is maintained by a barrier cavity. New bunches are injected into buckets that are then merged with the existing single superbunch. In this way, an amount of beam can be injected much larger than currently possible.

When the injection is finished the accelerating system is turned on, and the beam is captured in all the 28 MHz buckets, except for the abort gap. We assume that $4 \cdot 10^{11}$

---

[1] Work supported by US DOE, contract No DE-AC02-98CH10886.

**TABLE 1.** Parameters for acceleration and superbunches.

| quantity | unit | accel. bunch | super bunch |
|---|---|---|---|
| circumference $C$ | m | 3833 | |
| beam-beam limit $\Delta Q_{max}$ | ... | −0.03 | |
| crossing angle $\theta$ | mrad | 0.5 | |
| lattice $\beta^*$ at store | m | 1.0 | |
| relativistic $\gamma$ at store | ... | 260 | |
| emittance $\varepsilon_N$, 95% | $\mu$m | 20 | 20 |
| interaction region length $l$ | m | 20 | 20 |
| eff. detector length $l_{det}$ | m | 0.7 | |
| particles per bunch $N_b$ | $10^{11}$ | 4.0 | 215 |
| number of bunches $n_b$ | ... | 320 | 6 |
| bunch area $S$, 95% | eV·s | 1.0 | ... |
| rf frequency $f_{rf}$ | MHz | 28 | ... |
| gap voltage $V_{gap}$ | MV | 0.3 | ... |
| rms bunch length $\sigma_z$ | m | 0.45 | ... |
| luminosity $L$ | $10^{33}\text{cm}^{-2}\text{s}^{-1}$ | 1.5 | 2.3 |

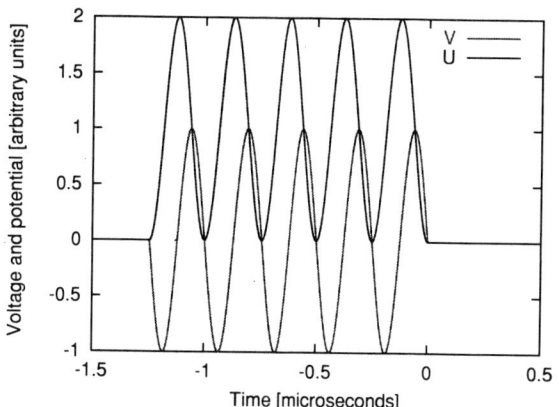

**FIGURE 1.** Voltage $V$, and potential $U$ waveforms of a barrier cavity with $f_{rf} = 4$ MHz.

protons can be accelerated in 320 of the 360 accelerating buckets. During acceleration the beams are vertically separated in the interaction regions. Bunches experience 5 parasitic collisions in every interaction region [10], a total of 30 per turn. It is possible to provide a separation of at least 7 transverse rms beam sizes. Operational experience so far has shown that the beam-beam effects cannot be completely suppressed in this way. Compared to current running conditions, the beam-beam effect may be mitigated by a larger transverse separation and better tune control along the ramp. We assume here that the beam can be accelerated without significant emittance growth or beam loss.

At store the beam is then transferred into six long bunches that are maintained by barrier cavities. The length of the superbunches is determined by the maximum line density that can be sustained at the beam-beam limit given a certain crossing angle.

## SUPERBUNCH GAP MAINTENANCE

In this section the maintenance of the six superbunches with barrier cavities is discussed. Experience with and plans for barrier cavities are reported in Refs. [4, 12, 13, 14].

Let $\varepsilon = E - E_s$ denote the energy deviation for a given particle and let $\tau$ denote its arrival time with respect to the synchronous particle. Using turn number $n$ as the time-like variable the equations for $\tau$ and $\varepsilon$ are

$$\frac{d\varepsilon}{dn} = -qV_s + qV_{rf}(\tau), \quad (1)$$

$$\frac{d\tau}{dn} = T_{rev}\eta \frac{\varepsilon}{\beta^2 E_s}, \quad (2)$$

where $q$ is the particle charge, $V$ the rf voltage, $T_{rev}$ the revolution time, $\eta$ the slip factor and $\beta$ the relativistic beam parameter. The subscript $s$ denotes the synchronous particle. Eqs. (1) and (2) correspond to the Hamiltonian

$$H(\tau,\varepsilon) = \frac{T_{rev}\eta}{2}\frac{\varepsilon^2}{\beta^2 E_s} + qV_s\tau - q\int_0^\tau V_{rf}(\tau_1)d\tau_1. \quad (3)$$

For adiabatic processes the phase space density is constant on curves of constant $H(\varepsilon,\tau)$. For these a dimensionless potential energy $U(\tau)$ can be defined by

$$U(\tau) = \frac{2\beta^2}{\eta T_{rev}(E_s/q)}\left[V_s\tau - \int_0^\tau V_{rf}(\tau_1)d\tau_1\right] \quad (4)$$

with which the maximum energy deviation on a given contour $\hat{\varepsilon} = (E - E_s)_{max}$ can be written as

$$\frac{\hat{\varepsilon}^2}{E_s^2} = \frac{\varepsilon^2}{E_s^2} + U(\tau). \quad (5)$$

We choose $V_{rf}$ so that $U(\tau) \geq 0$. With Eqs. (4) and (5) the potential and rf voltage for a given energy deviation $\hat{\varepsilon}$ can be determined for a given waveform of the barrier cavity voltage.

For gap maintenance we have $V_s = 0$. A gap between the bunches of 1 $\mu$s length can be created, for example, by one waveform of a $f_{rf} = 1$ MHz rf system [13]. In this way about half of the RHIC circumference can be filled with beam in six superbunches. A gap of 1 $\mu$s length would also be sufficient as an abort gap. For shorter gaps between bunches, a higher frequency is needed. The voltage and potential waveforms for such a system are illustrated in Fig. 1, where a sinusoidal waveform for the voltage is assumed, $V(t) = -\hat{V}\sin(2\pi f_{rf}t)$. The peak voltage $\hat{V}$ as a function of the energy spread $\hat{\varepsilon}$ can be obtained from Eq. (4) as

$$\hat{V} = \frac{\pi}{2}\frac{\eta T_{rev}f_{rf}}{\beta^2}\frac{E_s}{q}\frac{\hat{\varepsilon}^2}{E_s^2}. \quad (6)$$

**TABLE 2.** Rf parameters at injection and storage.

| quantity | unit | injection | storage |
|---|---|---|---|
| relativistic $\gamma$ | ... | 26 | 260 |
| kinetic energy $E_k$ | GeV | 23.4 | 243.0 |
| slip factor $\eta$ | ... | 0.00044 | 0.00191 |
| energy spread $\hat{\varepsilon}$ | ... | $10^{-3}$ | $10^{-3}$ |
| barrier frequency $f_{rf}$ | MHz | 1.0 | 1.0 |
| gap voltage $\hat{V}$ | kV | 0.2 | 9 |

With an energy spread of $\hat{\varepsilon}/E_s = 10^{-3}$ and a frequency of $f_{rf} = 1$ MHz the peak voltage needed at injection and storage is 0.2 kV and 9 kV respectively (see Tab Tab. 2). Previous barrier cavity work has created 10 kV single period sine waves using a single cavity [13]. Thus gap maintenance appears possible.

## LUMINOSITY

In Ref. [2] formulas are given for the incoherent tune shift due to the beam-beam interaction for particles in the beam center, and for the luminosity. For the conditions given in Tab. 1 the luminosity per interaction point is $L = 2.3 \cdot 10^{33} \text{cm}^{-2}\text{s}^{-1}$ with six superbunches. This about two orders of magnitude larger than the luminostiy under current running conditions $L = 2.7 \cdot 10^{31} \text{cm}^{-2}\text{s}^{-1}$ ($N_b = 10^{11}, n_b = 55$). In Tab. 1 also given is the luminosity for colliding the 320 acceleration bunches, $L = 1.5 \cdot 10^{31} \text{cm}^{-2}\text{s}^{-1}$. In this case the large number of parasitic collisions needs to be analyzed. With six superbunches the luminosity is about 50% higher than with the bunched beam.

We now show the change of the superbunch length and luminosity per interaction point under variation of the crossing angle $\theta$, intensity of the acceleration buckets $N_b$, and the sustainable total beam-beam tune spread $\Delta Q_{min}$.

In Fig. 2 the variation is shown for the crossing angle $\theta$. With small crossing angles the superbunches become very long. With crossing angles below 0.2 mrad the whole ring would be filled. With crossing angles larger than 0.5 mrad the luminosity increase slowes down.

At the beam-beam limit the achievable luminosity is proportional to the bunch intensity and the beam-beam tune shift $\Delta Q_{max}$. It is not dependent on the emittance since both the beam-beam tune shift and the luminosity are inversely proportional to the emittance.

For superbunches and crossing angles $\theta \ll 1$ one has [1, 2]

$$L = \frac{\gamma N_b n_b}{\beta^*} |\Delta Q_{max}| F(\theta, l, l_{det}) \quad (7)$$

where the form factor $F(\theta, l, l_{det})$ is fixed for a certain configuration of $(\theta, l, l_{det})$. The linear dependence of the luminosity on the bunch intensity $N_b$ can be seen in Fig. 3, and on the beam-beam tune shift $\Delta Q_{max}$ in Fig. 4.

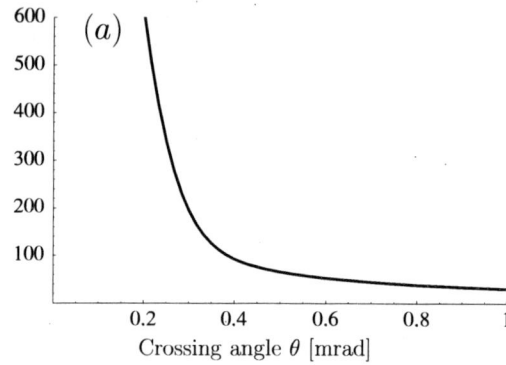

**FIGURE 2.** Superbunch length and luminosity per interaction point as a function of the crossing angle $\theta$ in parts (a) and (b) respectively. Other parameters are given in Tab. 1.

## SUMMARY

We estimated the achievable luminosity with six superbunches in RHIC for the incoherent beam-beam spread of $\Delta Q_{max} = -0.03$. The estimated luminosity of $L = 2.3 \cdot 10^{33} \text{cm}^{-2}\text{s}^{-1}$ is about two orders of magnitude larger than the luminosity under current running conditions, and about 50% higher than for bunches with the same total intensity. For the preparation of six superbunches at store, barrier cavities are needed with parameters close to those that were demonstrated in the past.

A number of effects were neglected in this study. Among those are resonant effects, coherent effect, end effects of the superbunches, and long-range beam-beam interactions on the energy ramp. These effects will reduce the estimated luminosity. Furthermore, a number of system changes will be needed [15].

## ACKNOWLEDGMENTS

The authors are thankful for discussions with J.M. Brennan, M. Harrison, W. MacKay, T. Roser, F. Ruggiero, S. Peggs, V. Ptitsyn, K. Takayama, and F. Zimmermann.

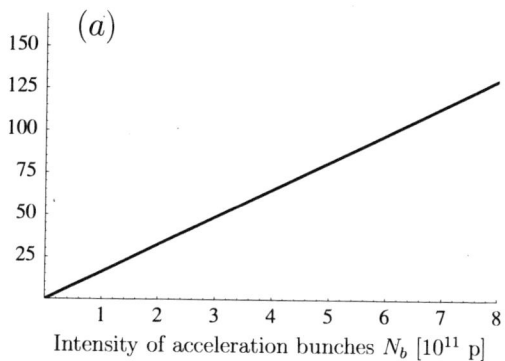
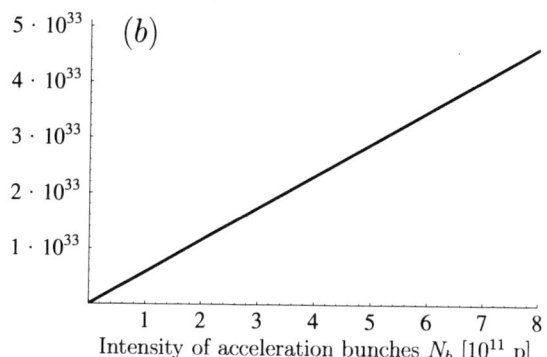

**FIGURE 3.** Superbunch length and luminosity per interaction point as a function of the bunch intensity $N_b$ of the accelerated bunches in parts (a) and (b) respectively. Other parameters are given in Tab. 1.

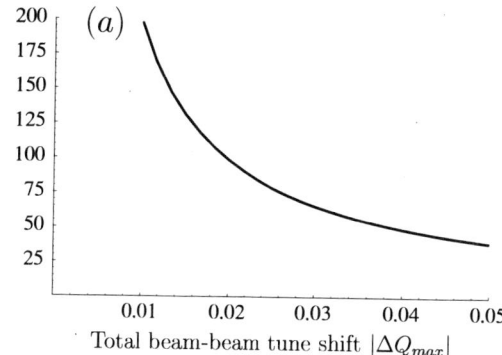
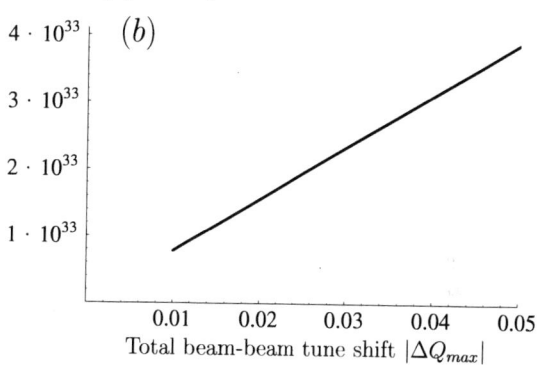

**FIGURE 4.** Superbunch length and luminosity per interaction point as a function of the total beam-beam tune shift $|\Delta Q_{max}|$ in parts (a) and (b) respectively. Other parameters are given in Tab. 1.

## REFERENCES

1. E. Keil, "Luminosity optimization for storage rings with low-$\beta$ sections and small crossing angles", Nucl. Instrum. Methods **113**, 333 (1973).
2. F. Ruggiero and F. Zimmermann, "Luminosity optimization near the beam-beam limit by increasing bunch length or crossing angle", Phys. Rev. ST Accel. Beams **5** 061001 (2002).
3. K. Takayama, J. Kishiro, M. Sakuda, Y. Shimosaki, and M. Wake, "Superbunch hadron colliders", Phys. Rev. Lett. Vol. 88 No. 14 (2002).
4. J. Kishiro and K. Takayama, "Induction synchrotron", Nucl.Inst. Meth. A451, 304-317 (2000).
5. H. Hahn (editor), "RHIC design manual", revision of October 2000, http://www.rhichome.bnl.gov/NT-share/rhicdm/00_toc1i.htm.
6. W. Fischer and M. Blaskiewicz, "Luminosity increase at the incoherent beam-beam limit in RHIC", BNL C-A/AP/94 (2003).
7. V. Ptitsyn, private communication (2002).
8. D. Barton, PHOBOS; W. Cristie, STAR; M. Purschke, PHENIX; D. Beavis, BRAHMS; private communication (2002)
9. W. Herr and R. Paparella, "Landau damping of coherent beam-beam modes by overlap with synchrotron sidebands", CERN LHC Project Note 304 (2002).
10. S. Peggs, "Beam-beam collisions and crossing angles in RHIC", proceedings of the LHC Beam-Beam Workshop at CERN 1999, BNL RHIC/AP/169 (1999).
11. M. Blaskiewicz and J.M. Brennan, "A barrier bucket experiment for accumulating de-bunched beam in the AGS", proceedings of the 1996 European Particle Accelerator Conference, Sitges, Spain (1996).
12. K.Y. Ng, "Multiple Injections with Barrier Buckets", proceedings of the 1998 European Particle Accelerator Conference, Stockholm, Sweden (1998).
13. M. Blaskiewicz, J.M. Brennan, T. Roser, K. Smith, R. Spitz, A. Zaltsman, M. Fujieda, Y. Iwashita, A. Noda, M. Yoshii, Y. Mori, C. Ohmori, Y. Sato, "Barrier Cavities in the Brookhaven AGS", proceedings of the 1999 Particle Accelerator Conference, New York (1999).
14. T. Bohl, T. Linnecar and E. Shaposhnikova, "Barrier buckets in the CERN SPS", proceedings of the 2000 European Particle Accelerator Conference, Vienna, Austria (2000).
15. M. Harrison, "High luminosity p-p operation at RHIC", BNL RHIC/AP/8 (1993).

# Theory and Observations of Beam-beam effects at the Tevatron

T. Sen

*Fermi National Accelerator Laboratory, Batavia, Illinois 60510*

**Abstract.** Long-range beam-beam interactions in Run II at the Tevatron are the dominant sources of beam loss and lifetime limitations of anti-protons, especially at injection energy. I discuss observations and theoretical understanding of these beam-beam effects.

## INTRODUCTION

The Tevatron is currently delivering luminosities close to $4 \times 10^{31}$ cm$^{-2}$ sec$^{-1}$ to the CDF and D0 experiments. In a record store on May 17th, 2003, (average initial luminosity = $4.49 \times 10^{31}$ cm$^{-2}$sec$^{-1}$), the average bunch intensities at the start of collisions were $N_p = 2.43 \times 10^{11}$, $N_{\bar{p}} = 0.25 \times 10^{11}$. These are to be compared with design values for Run II of $2.7 \times 10^{11}$ and $1.35 \times 10^{11}$ for protons and anti-protons respectively.

After 36 bunches of protons are injected and placed on the proton helix, anti-protons are injected four bunches at a time. After all bunches are injected, acceleration to top energy takes about 85 seconds. After reaching flat top, the optics around the interaction regions (IRs) is changed to lower $\beta^*$ from 1.6 m to 0.35 m at B0 and D0 and the beams are brought into collision. During a high energy physics store each bunch experiences two head-on collisions with bunches in the opposing beam and seventy long-range interactions. At all other stages, each bunch experiences only long-range interactions - seventy two in all.

Beam-beam forces change the linear and nonlinear dynamics of particles in fundamental ways. The long-range beam-beam force has in general both quadrupolar and skew-quadrupolar components. The magnitude of the tune and coupling shift depends on the beam separation but also on the amplitude of the particle experiencing the force. Since there is dispersion at the locations of the long-range interactions in the Tevatron, the separation between the test particle and the other beam and therefore the tune shift depends on the momentum deviation of the particle. An amplitude dependent tune shift implies that the chromaticity shift is also amplitude dependent. These amplitude dependences introduce the familiar tune footprint but also coupling and chromaticity footprints within a bunch.

Theoretical expressions have been developed for the amplitude dependent tune shifts, chromaticities and coupling for arbitrary aspect ratios [1]. Some conclusions are easily drawn however from the round beam case. At large distances, both the tune shift and the coupling fall as $1/d^2$ while the chromaticity falls off more rapidly as $1/d^3$. The tune shift for round beams vanishes if the plane of the helix is at 45°. Minimizing the chromaticity requires the plane of the helix to be either at 30° or vertical. The coupling vanishes if the separation is either in the horizontal or the vertical plane. Since each of these parameters has a different dependence on the helix angle, it is not possible to minimize them simultaneously with a choice of the angle.

Each anti-proton bunch sees a different sequence of long-range interactions with different optics functions. This leads to a significant spread of tune shifts, coupling and chromaticities between the bunches. Thus the working point for example cannot be optimized for all bunches at once. Due to the amplitude dependence of these quantities, the spread within a bunch can be comparable to the spread between the bunches.

## INJECTION ENERGY

There are 138 different locations of parasitic interactions around the ring. The smallest separation is about 4$\sigma$ but there is a large variation between 5-12 $\sigma$ at most of the other parasitics. At most parasitics the beta functions are larger in the horizontal plane. This asymmetry in the beta functions is reflected in parameters like the tune shifts.

Figure 1 shows the small amplitude tune shifts of the anti-protons, bunch by bunch. The tune footprint of anti-proton bunch 1 with (a) only the beam-beam interactions and (b) with beam-beam and machine nonlinearities is

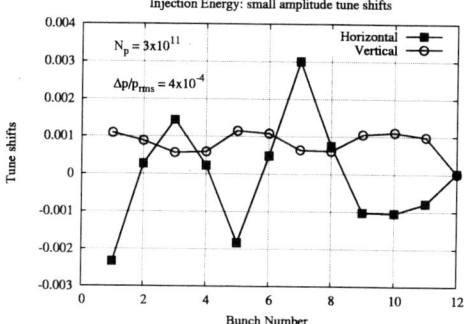

**FIGURE 1.** Analytically calculated beam-beam induced tune shifts of anti-protons at small amplitude, bunch by bunch.

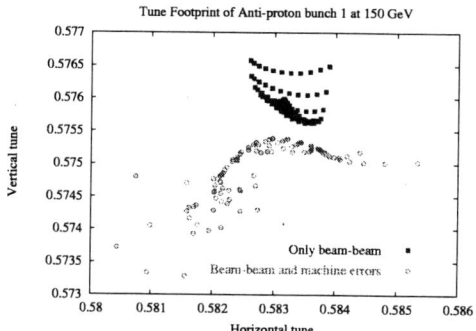

**FIGURE 2.** Tune footprint of pbar bunch 1 with and without machine nonlinearities.

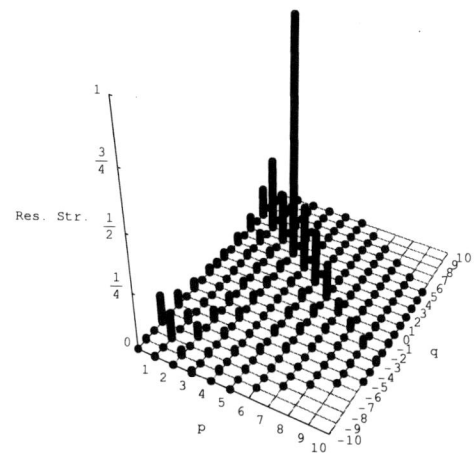

**FIGURE 4.** Resonance strengths at injection at an amplitude of $2\sigma$ with lattice and beam-beam nonlinearities. The resonance condition is $p\nu_x + q\nu_y =$ integer.

shown in Figure 2. The tune footprint at large amplitudes is determined by the machine nonlinearities which change the sign of the detuning. The tune spread within a bunch is about 0.005, about five times smaller than at collision energy.

The beam-beam induced global coupling at small amplitude is of the same magnitude as the lattice induced global coupling. Figure 3 shows the small amplitude

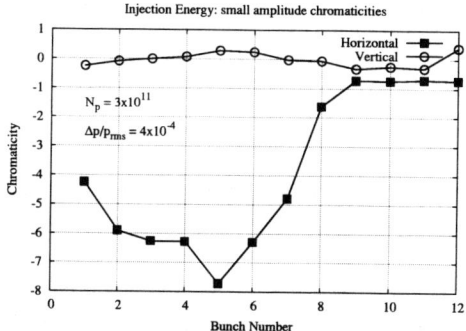

**FIGURE 3.** Beam-beam induced chromaticity at small amplitude, bunch by bunch (150 GeV).

chromaticity. The large spread in chromaticity could make some bunches more sensitive to synchro-betatron resonances and to coherent instabilities at high anti-proton intensities.

Analytically calculated resonance driving terms due only to beam-beam interactions show that seventh order resonances dominate. With lattice nonlinearities included, the resonance driving terms have been calculated from a nonlinear map. The lattice nonlinearities include the correction sextupoles and the error fields in the arc and IR magnets. The beam-beam driven resonances (mainly 7th order) are considerably stronger than the purely lattice driven resonances (mainly 4th order). Figure 4 shows the resonances driven with both nonlinearities. Bunches 10, 11 and 12 (at the end of a train) suffer the least changes to their tunes, coupling and chromaticities and the resonant widths from the beam-beam interactions are also small for these bunches. This is in qualitative agreement with the higher lifetimes observed in bunches towards the end of a train at injection.

The beam-beam interactions have a strong influence on the anti-proton lifetime at 150 GeV. During a study in September 2002 with only anti-protons injected into the Tevatron, the lifetimes at 150 GeV varied between 10-25 hours compared to lifetimes of 1-10 hours in stores. The losses during acceleration were also small about 2%, compared to typical losses around 10% in stores.

The dynamic aperture of a few anti-proton bunches at 150 GeV has been calculated with extensive simulations. The nonlinearities in the model include the measured multipoles in the magnets, the chromaticity and feed-down sextupoles together with the beam-beam interactions. The results from two different codes, MAD and

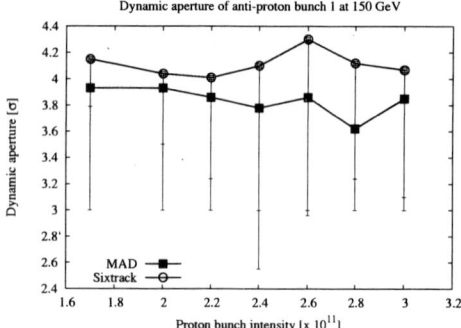

**FIGURE 5.** Dynamic aperture after $10^6$ turns of anti-proton bunch 1 vs proton intensity at injection. The average value (over all angles in coordinate space) along with one-sided error bars to represent the minimum value at each intensity are shown.

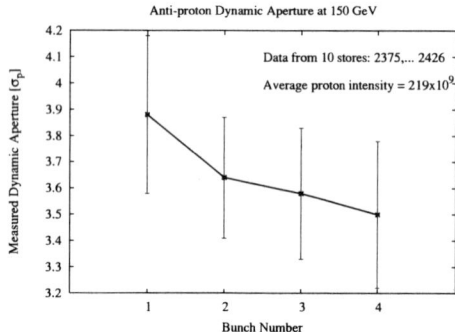

**FIGURE 6.** Dynamic aperture of the first four anti-proton bunches at injection calculated from the measured drop in emittance in 10 recent stores. The error bars represent the variation over the stores.

Sixtrack, typically agree to within 15%. Figure 5 shows a plot of the average dynamic aperture (after $10^6$ turns or 2 seconds in the Tevatron) as a function of the proton bunch intensity. This plot predicts that the dynamic aperture at 150 GeV is about $4\sigma_p$ and nearly independent of the proton intensities over this range.

The lifetimes of most bunches varies between 1-10 hours under typical conditions but is observed to be relatively independent of the proton intensities so far - in agreement with the result expected from Figure 5. In several recent stores, the emittance of the anti-protons was observed to decrease with time after injection. We consider the first four bunches injected since their emittances are measured 10 time with the flying wires before acceleration. The dynamic aperture was calculated from the asymptotic emittance for 10 recent stores where there was an observed emittance reduction. The dynamic apertures, shown in Figure 6, are in good agreement with the simulation results in Figure 5.

A. Kabel at SLAC has developed a six-dimensional code called PlibB for lifetime simulations. An interesting prediction from one of his simulations is a significant increase in the lifetime when the vertical chromaticity is reduced below 4 units. Another parallel code called Beam-Beam3D has been developed by J. Qiang at SLAC. As with Kabel's code, the only nonlinearities included so far are those due to the beam-beam interactions. Lifetimes from simulations using this code can be in the range of 1-2 hrs if the physical aperture is small enough. Further development will include the nonlinearities of the magnets.

## RAMP AND SQUEEZE

Anti-proton losses during the ramp were measured to be $\sim$2% during a dedicated anti-proton only store in September 2002. However during regular stores with protons present, anti-proton losses averaged around 11% in March 2003. These losses occur during the entire ramp. The anti-proton loss during the ramp is also well correlated with the anti-proton vertical emittance. Reducing the emittance blow-up while injecting anti-protons in the Tevatron onto the anti-proton helix would therefore also reduce the loss during the ramp.

## COLLISION

The smallest separations at the parasitic collisions occur at the ones immediately upstream and downstream of the head-on collisions at B0 and D0. The beta functions at these four parasitics are also the largest. Consequently the tune shifts and resonance driving terms (for example) contain the largest contributions from these parasitics amongst all the parasitics.

At collision energy, bunch by bunch orbit positions are observed at the synchrotron light monitor. The long-range interactions are largely responsible for the variations seen in the anti-proton bunches. Comparison of these observed orbits with analytical calculations shows good agreement. The bunch by bunch variation in tunes is quite different from the variation at injection [cf. Figure 1]. Only the head (bunch 1) and the tail (bunch 12) of the train have tune shifts significantly different from the others. Figure 7 shows the tune footprint (up to $6\sigma$) of these bunches and that of bunch 6 which is representative of all the other bunches. The extent of the footprint (about 0.024) is largely contributed (0.020 units) by the head-on collisions for all bunches. The beam-beam induced coupling at small amplitudes is smaller than at injection. The beam-beam induced chromaticity (both the variation over the bunches and the magnitude) is larger than at injection.

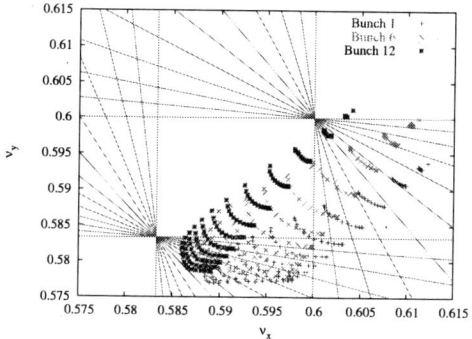

**FIGURE 7.** Tune footprints of bunches 1, 6 and 12 superposed on the 5th and 12th order sum resonances.

The head-on collisions create the strongest nonlinear fields but drive only even order resonances, mainly the 12th order resonances. The effects of these resonances is considerably weakened (by two orders of magnitude at some amplitudes) as a consequence of phase averaging [1]. However with increasing chromaticity the number of synchrotron sidebands with significant width increases as well which increases the effective width of the resonances. The long-range interactions drive the odd order 5th and 7th order resonances. The relatively large chromaticity generated by these interactions create synchrotron sidebands of these resonances which overlap at relatively small amplitudes.

A nonlinear Taylor map has been used to calculate the resonances with both lattice and beam-beam nonlinearities. As at injection, we find that at an amplitude of $2\sigma_p$, the 7th and 5th order resonances driven by the beam-beam dominate. The largest of these are the (3, 4) and (3, 2) resonances. Compared to injection, there are now many more resonances of comparable strength but the strengths themselves are smaller.

Extensive dynamic aperture and other tracking calculations at collision optics and Run II design parameters lead to several conclusions. For example, the head-on interactions largely determine the tune footprint but they have very little influence on the dynamic aperture. Of the seventy long-range interactions, the four interactions nearest to the two IPs are the most important in determining the DA but the synchro-betatron resonances driven by the other long-range interactions are important. These calculations also predict that the DA will drop by about $1\sigma$ from present intensities ($N_p \sim 210 \times 10^9$) to design intensity ($N_p = 270 \times 10^9$). Analysis of data shows that at present the anti-proton lifetime during stores is not strongly influenced by the beam-beam effects.

In several dedicated studies we have explored the possibility of improving the anti-proton lifetime by changing the collision helix. Increasing the helix separation at two of the four nearest parasitics in dedicated experiments and at the start of some stores did not result in a significant increase in the lifetime.

Emittance growth at the start of a store has occasionally been a concern. In most of these cases the anti-proton emittance growth rate was large at the start, then dropped with falling beam intensities. This emittance growth is strongly bunch dependent; typically bunches 1 and 12 have lower growth rates than the others. Small changes to the tune usually suffice to lower the growth rate. More details about experimental beam-beam studies can be found in [2].

## SUMMARY

At injection energy the long-range beam-beam inteactions limit the anti-proton lifetime to under 10 hrs but have not influenced proton lifetimes much. During the ramp anti-proton losses (and perhaps emittance growth) are largely due to beam-beam effects. Occasional proton losses, large enough to quench the Tevatron, during the final cogging and squeeze are attributed to beam-beam effects. Proton losses during luminosity are occasionally much higher than observed during machine studies with only protons of similar intensities. Lifetimes of both anti-protons and protons at collision however are largely determined by luminosity with a comparable contribution from residual gas scattering to the proton lifetime.

New helical solutions that increase the separations are being developed using the present set of electrostatic separators. There are also plans to install additional separators around the IPs to increase the separations at the nearest parasitics. Other planned measures that will reduce beam-beam limitations include (a) Improving the alignment in the Tevatron, (b) Injecting beams with smaller emittances, (c) Operating with lower chromaticities, (d) Testing different bunch patterns, (e) Improving the IR optics, (f) Searching for better working points, (g) Active compensation of beam-beam effects with the Tevatron electron lens [3] and perhaps in tandem, compensation with current carrying wires [4].

## REFERENCES

1. T. Sen, B. Erdelyi, M. Xiao, *Theoretical studies of beam-beam effects in the Tevatron at collision energy*, PAC 2003
2. X.L. Zhang et al., *Experimental studies of beam-beam effects at the Tevatron*, PAC 2003
3. V. Shiltsev, this workshop
4. B. Erdelyi and T. Sen, this workshop

# Tune Modulation from Beam-Beam Interaction and Unequal Radio Frequencies in RHIC[1]

W. Fischer, P. Cameron, S. Peggs and T. Satogata

*Brookhaven National Laboratory, Upton, New York 11973*

**Abstract.** The two RHIC rings have independent rf systems to accommodate different species. Thus, the radio frequencies can differ when the phase and radial loops are closed, and the rf frequencies of the two rings are not synchronized. A radio frequency difference leads to longitudinally moving beam crossing points. When the crossing points are between the beam splitting dipoles, the beams experience the beam-beam interaction. Outside the interaction region the beam-beam interaction is switched off. In this way the tune is modulated. A computation of the tune modulation depth, pulse shape and frequency is presented. Tune modulation measurements are shown.

## INTRODUCTION

During the early RHIC operation it has been observed that beams with 56 bunches suffer lifetime degradation as soon as the rf phase and radial loops are closed at injection. In Fig. 1 an example for such an observation is shown.

Beams are injected with equal radio frequencies in both rings and longitudinal separation [1]. Before the ramp starts the phase and radial loops are closed, which leads to a small difference in the radio frequencies since both rings have independent rf systems. A typical difference is between zero and 100 Hz, while the radio frequency at injection is 28.022 MHz for gold beams. The difference in the radio frequencies results in a longitudinally moving beam crossing point.

When the beam crossing point moves through the interaction region (IR), one beam will experience the field of the other beam when the crossing point is located between the DX beam splitting magnets (see Fig. 2). Outside this region the beams are well separated and do not see the fields of the other beam. Thus, the difference in the radio frequencies can lead to a tune modulation.

In the 2001/2002 run a total beam-beam tune spread of $\Delta Q_{bb} = 0.005$ was achieved in gold operation with 4 collisions. In proton operation the tune spread reached $\Delta Q_{bb} = 0.01$. This tune spread corresponds to the tune modulation depth of small amplitude particles. In the following we compute the tune modulation frequency and wave form taking into account crossing angles. Tune modulation measurements with a phase locked loop (PLL) are presented.

**FIGURE 1.** Blue and Yellow lifetime deterioration due to beam-beam interaction with unequal radio frequencies.

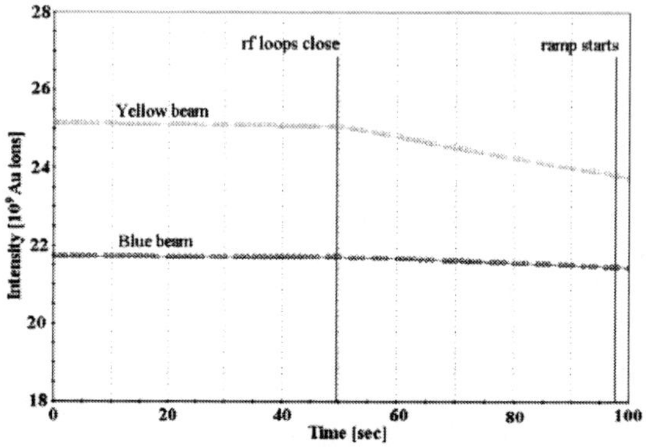

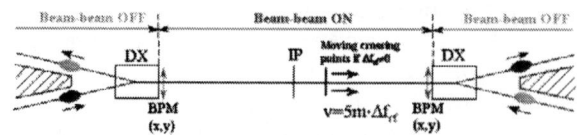

**FIGURE 2.** Schematic of RHIC interaction regions layout indicating regions where the beam-beam interaction is on and off.

## TUNE MODULATION PARAMETERS

Denoting the two radio frequencies with $f_{rf,1}$ and $f_{rf,2}$ and the harmonic number with $h$, the difference in the revolution time of the two rings is

$$\Delta T = T_1 - T_2 = \frac{h}{f_{rf,1}} - \frac{h}{f_{rf,2}} = \frac{h\Delta f_{rf}}{f_{rf}^2}. \quad (1)$$

[1] Work supported by US DOE, contract No DE-AC02-98CH10886.

CP693, *Beam Halo Dynamics, Diagnostics, and Collimation*, edited by J. Wei, W. Fischer, and P. Manning

The crossing point moves by $c\Delta T/2$ per turn, where $c$ is the particle velocity. The velocity of the crossing point can then be computed as

$$v_{CP} = \frac{c}{2}\frac{\Delta f_{rf}}{f_{rf}}. \qquad (2)$$

**Example RHIC:** With $f_{rf} = 28.022$ MHz and $\Delta f_{rf} = 5$ Hz, it follows $v_{CP} = 27$ ms$^{-1}$. The speed of light is $c = 2.998 \cdot 10^8$ ms$^{-1}$,

We now compute the wave form of the tune modulation as the crossing point moves from one crotch to the crotch of the other side of the interaction region. Assuming round beams and head-on collisions, particles with small betatron amplitudes will experience horizontal and vertical tune shifts $\Delta Q_{bb}$ between the DX magnets. The tune shifts will become negligible when the crotches are approached. If $L$ is the distance between the crotches, the crossing point moves in time

$$T_{IR} = \frac{L}{v_{CP}} = \frac{2L}{c}\frac{f_{rf}}{\Delta f_{rf}} \qquad (3)$$

from one crotch to the other. Particles with larger betatron amplitudes show a smaller tune modulation depth.

**Example RHIC:** With the values in the above example and $L = 31.4$m, it follows $T_{IR} = 1.16$ s.

The tune shift, rms beam size and beam separation as a function of the longitudinal position are shown in Fig. 3 assuming no crossing angle, and in Fig. 4 assuming a half crossing angle of 0.22 mrad. Without a crossing angle the beam-beam interaction is effectively switched off when the crossing point reaches the DX magnets.

With a crossing angle the beam-beam interaction can be switched off much earlier. With 0.22 mrad half crossing angle (see Fig. 4) the beam-beam interaction is switched off after only a few meters. In 2001 the Brahms experiment was running with a half crossing angle of up to 0.9 mrad due to concerns about the quench performance of an DX magnet in this interaction region.

How often the tune modulation waveform computed above is executed depends on the fill pattern. We assume here rings that are symmetrically filled with $N$ bunches. The distance between IPs is then $C/(2N)$, where $C$ is the circumference. The time between two waveform executions is then

$$T_C = \frac{C}{2Nv_{CP}} = \frac{C}{Nc}\frac{f_{rf}}{\Delta f_{rf}}. \qquad (4)$$

**Example RHIC:** With the values in the above examples and $N = 6$, it follows $T_C = 11.9$ s. With $N = 60$ it follows $T_C = 1.19$ s.

In Fig. 5 the tune modulation wave form is shown schematically. Figs. 6 and 7 show the situation for RHIC

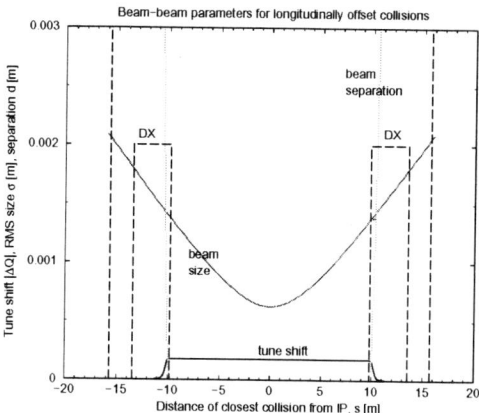

**FIGURE 3.** The tune shift (black), rms beam size (red) and beam separation (green) as a function of the longitudinal position without a crossing angle. Assumed are one crossing point, $\xi = \Delta Q_{bb} = 0.0002$, $\beta^* = 5$ m, $\varepsilon_{N,95\%} = 50$ $\mu$m. The objects starting at 10 m from the IP are the DX magnets, the crotches are at 16 m from the IP.

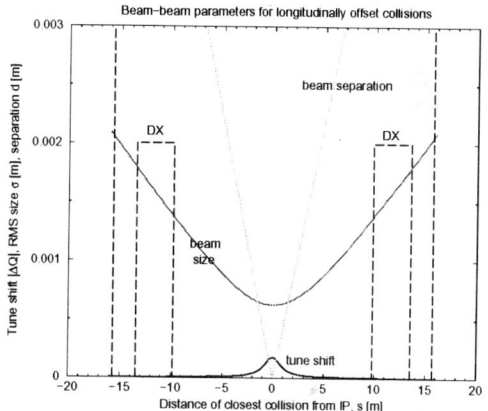

**FIGURE 4.** The tune shift (black), rms beam size (red) and beam separation (green) with the same parameters as in Fig. 3, except for the a half crossing angle of 0.22 mrad.

with 6 and 60 bunches respectively. With 60 bunches and the numbers in the examples, one has a tune modulation with the frequency $f_{mod} = 0.9$ Hz and a modulation depth of up to a few $10^{-3}$. The modulation depth can thus reach values that are orders of magnitudes larger than modulation depth from power supply ripple. Since the wave form is not sinusoidal, higher tune modulation harmonics are also created.

The effect of tune modulation on the dynamic aperture has been studied with experiments and simulations for the Tevatron [3, 4], SPS [5], HERA [6, 7, 8], RHIC [9] and the LHC [10, 11]. A detrimental effect of tune modulation on the dynamic aperture and lifetime was found even with tune modulation depth well below $10^{-3}$.

To avoid the reduction in beam lifetime and dynamic

aperture, the RHIC beams were separated vertically by 10 mm during injection and on the ramp [12]. Furthermore, a scheme was implemented to synchronize the two beams during the ramp with proton beams [13].

## TUNE MODULATION MEASUREMENTS

In Figs. 8 and 9 measurements are shown that illustrate tune modulation mechanism. All measurements were done with a phase locked loop (PLL), which allows to measure the tune to better than $10^{-4}$.

In Fig. 8 the tune shift is initially measured when the bunches are separated longitudinally, i.e. the crossing point is beyond the DX magnet. The crossing point is then moved in 3 steps until it is at the nominal interaction point. A cogging step by one bucket moves the crossing point by 5 m longitudinally. The tune shift corresponding to the last crossing point is a sign of crossing angles (compare with Fig. 4). With beams colliding the tune is depressed through the beam-beam interaction. From the measured beam-beam tune shift the emittance can be inferred since the bunch currents are known.

For Gaussian round beams the beam-beam parameter is

$$\xi = \frac{3 N_b r}{2\pi \varepsilon_N}. \quad (5)$$

where $N_b$ is the bunch intensity, $r$ the classical particle radius, and $\varepsilon_N$ the normalized 95% emittance. The observed tune shift is approximately

$$\Delta Q = \frac{1}{2} N_x \xi \quad (6)$$

where $N_x$ is the number of beam-beam interactions per turn. The factor 1/2 stems from the the observation of the coherent tune shift [14].

In Fig. 9 the Blue vertical tune and the Yellow horizontal tune are shown at the end of the ramp and during rf manipulations at flat top. In each ring there are 55 bunches. In the Blue ring there are, on average, $6.36 \cdot 10^{10}$ protons per bunch, in the Yellow ring there are $4.27 \cdot 10^{10}$ protons per bunch. During the energy ramp both tunes are moving. In addition, the tunes are modulated since the rf frequencies are unlocked. There is a modulation period of $T_c = 3.3$ s during the ramp and, according to Eq. (4), the difference between the radio frequencies is $\Delta f_{rf} = 1.8$ Hz. The rf frequency is $f_{rf} = 28.149$ MHz.

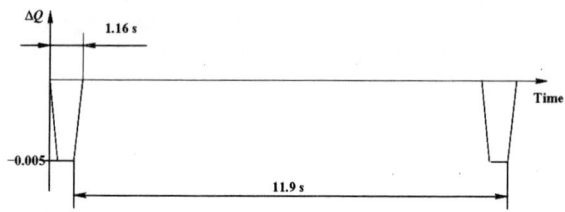

**FIGURE 6.** Tune modulation, with parameters as in the RHIC examples and 6 symmetrically distributed bunches in each ring.

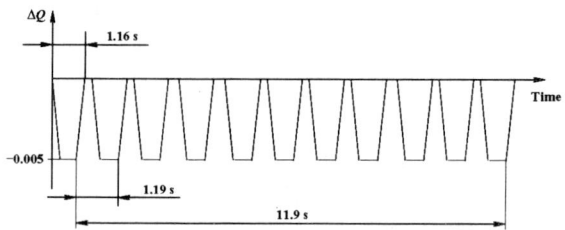

**FIGURE 7.** Tune modulation with parameters as in the RHIC examples and 60 symmetrically distributed bunches in each ring.

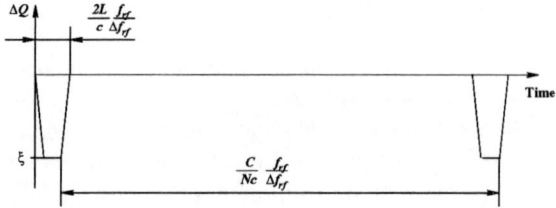

**FIGURE 5.** Tune modulation due to beam-beam interaction with unequal radio frequencies. Shown are the length of the beam-beam interaction and the time between beam-beam interactions.

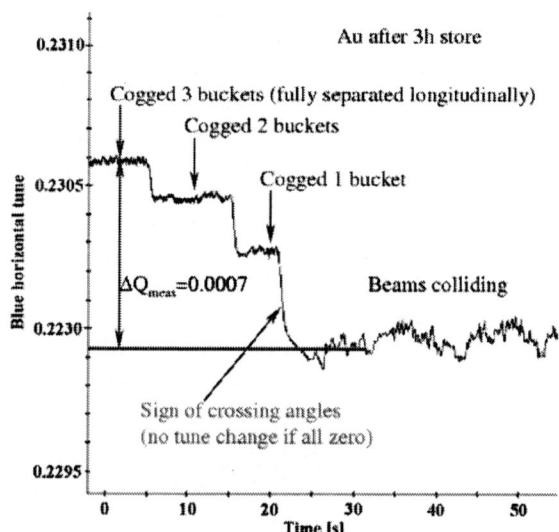

**FIGURE 8.** Tune shift due to beam-beam interaction with different longitudinal locations of the beam crossing points. The measurement was taken after gold beam was stored for three hours.

When reaching the flat top, the tune modulation stops once the rf frequencies are synchronized. Then the abort gaps are aligned longitudinally ("cogging"). During this process the rf frequency of one ring is slightly changed temporarily to allow one bunch pattern to move against the other one. The maximum frequency difference in the process is $\Delta f_{rf} = 10$ Hz. This leads to a modulation period of $T_c = 0.6$ s that is also measured with the PLL. When the cogging is finished the tunes are depressed by the beam-beam interaction. The Blue beam probes the Yellow beam and thus its tune depression depends on the Yellow emittance and bunch intensity. The Yellow beam probes the Blue beam. If both beams have the same emittance, it follows from Eq. (5)

$$\frac{\Delta Q_{Blue}}{\Delta Q_{Yellow}} = \frac{N_{b,Yellow}}{N_{b,Blue}} \quad (7)$$

in accordance with the observation shown in Fig. 9.

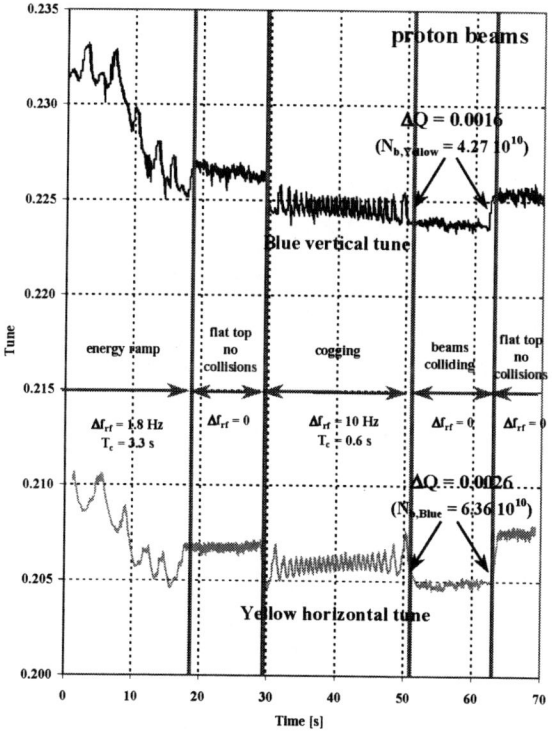

**FIGURE 9.** Blue vertical and Yellow horizontal tune, measured with the PLL, at the end of the ramp and during rf manipulations at flat top.

## SUMMARY

When the two RHIC beams have slightly different radio frequencies the tune is modulated with frequencies of typically 5 to 100 Hz with modulation depths of up to a few $10^{-3}$, the total beam-beam tune shift. Measurements and simulations for existing hadron machines have shown that such conditions are likely to affect the dynamic aperture and beam lifetime severely. This has also been observed in RHIC.

## ACKNOWLEDGMENTS

The authors would like to thank for help and discussions: M. Bai, M. Blaskiewicz, J.M. Brennan, J. DeLong, T. Hayes, V. Ptitsyn, S. Tepikian, T. Roser, and D. Trbojevic.

## REFERENCES

1. J. DeLong, J.M. Brennan, private communication (2001).
2. W. Fischer and S. Peggs, "RHIC as a Test Bench for Beam-Beam Studies", proceedings of the Beam-Beam Workshop, Fermilab (2001).
3. A. Chao et al., "Experimental investigations on nonlinear dynamics in the Fermilab Tevatron", Phys. Rev. Lett. 61, pp. 2752-2755 (1988).
4. T. Satogata, "Nonlinear resonance islands and modulational effects in a proton synchrotron", PhD Thesis, Northwestern University (1993).
5. W. Fischer, M. Giovannozzi and F. Schmidt, "Dynamic Aperture Experiment at a Synchrotron", Phys. Rev. E, Vol. 55, Number 3, p 3507 (1997).
6. F. Zimmermann, "Emittance Growth and Proton Beam Lifetime in HERA", PhD thesis, Hamburg University, DESY 93-059 (1993).
7. O. Brüning, "An Analysis of the Long-term Stability of the Particle Dynamics in Hadron Storage Rings", PhD thesis, Hamburg University, DESY 94-085 (1994).
8. W. Fischer, "An Experimental Study on the Long-term Stability of Particle Motion in Hadron Storage Rings", PhD thesis, Hamburg University, DESY 95-235 (1995).
9. W. Fischer and T. Satogata, "A Simulation Study On Tune Modulation Effects in RHIC", BNL RHIC/AP/109 (1996).
10. W. Fischer and F. Schmidt, "Long-term Tracking for the LHC including ripple", CERN SL/Note 94-75 (1994).
11. M. Böge, A. Faus-Golfe, J. Gareyte, H. Grote, J.-P. Koutchouk, J. Miles, Q.Qing, T. Risselada, F. Schmidt, S. Weisz, "Overview of the LHC dynamic aperture studies", proceedings of the 1997 Particle Accelerator Conference, Vancuver (1997).
12. V. Ptitsyn, private communication (2001).
13. J.M. Brennan, private communication (2001).
14. E. Keil, "Beam-beam Dynamics", CERN Accelerator School, Rhodes, Greece, CERN 95-06 (1995), CERN SL/94-78 (1994).

# Progress Report on Beam-Beam Compensation with Electron Lenses in Tevatron

V. Shiltsev[1], Y. Alexahin[1], K. Bishofberger[2], G. Kuznetsov[1], N. Solyak[1], M. Tiunov[3], X. Zhang[1]

[1]*FNAL, Batavia, IL 60510, USA*
[2]*UCLA, Los Angeles, CA 90095, USA*
[3]*BINP, Novosibirsk, Russia*

**Abstract.** We discuss the original idea of beam-beam compensation (BBC) in Section I, sequence of events in 2001-2002 and use of the Tevatron Electron Beam (TEL) for DC beam removal in Section II, (anti)proton lifetime improvement in Section III, experimental data on the BBC attempts in Section IV and, conclusively, Section V is devoted to discussion on important phenomena, needed improvements and future plans.

## ORIGINAL GOALS OF BBC

The idea of beam-beam compensation (BBC) in the Tevatron proton-antiproton collider [1] originally assumed installation of a single low-energy high-current DC electron beam device which would create nonlinear space charge force acting on antiprotons and compensating in average electromagnetic forces due head-on collisions with protons in the two collision points. Later, it was realized that because of non-uniform bunch loading scheme in the Tevatron Run II (each beam contains 3 trains of 12 bunches spaced by 396 ns, the trains are separated by 2.6 μs gaps), antiproton-bunch dynamics depends on bunch position in the bunch train, and two electron lens with pulsed and variable currents can be used to compensated bunch-by-bunch differences, e.g. tune variation [2,3].

Figure 1 shows antiproton tune diagram for design Run II parameters without BBC (Figure 1a), with a single TEL installed at the location with $\beta_x \gg \beta_y$ and compensating variations of tuneshift in horizontal plane (Figure 1b). If the second linear lens set at a location with $\beta_Y \gg \beta_X$ then antiproton footprint can be reduced as depicted in Figure 1c, while optimization of electron beam current density profile may result in even further reduction of the antiproton tunespread, Figure 1d. Analytical calculations and numerical tracking [4,5] showed that the BBC should lead to significant improvement of lifetime of some bunches (outliers in Figure 1, bunches #1 and #12 in each train)

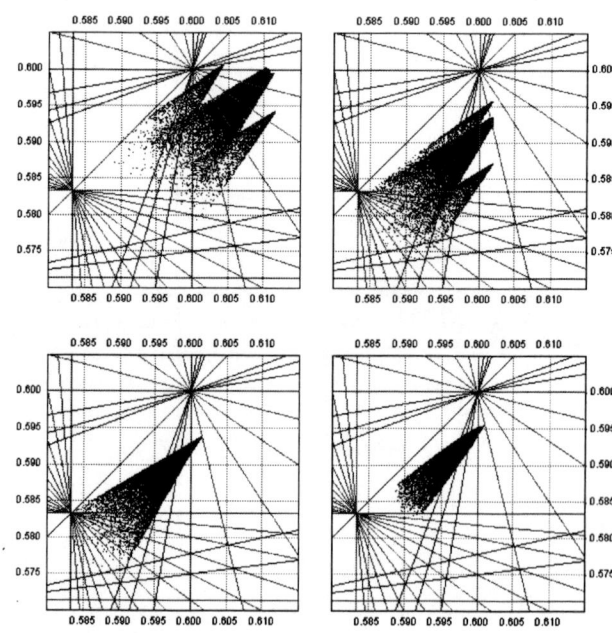

Figure 1: Tevatron tune diagram (a) and various BBC (b, c, and d).

and, thus, should improve integrated luminosity by 5-10%.

That was the ultimate justification for design and construction of the first TEL, which started in 1998. In Spring 2001 the first TEL has been installed in the Tevatron tunnel and commissioned (see Figure 2).

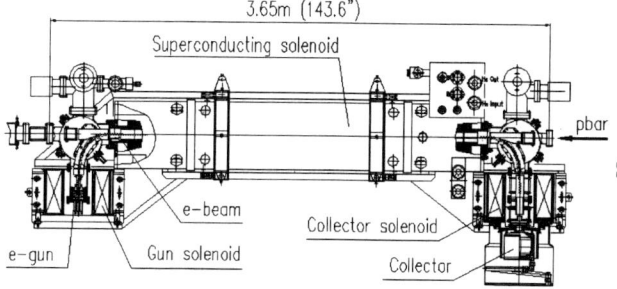

Figure 2: The first Tevatron Electron Lens.

Detailed description of magnetic, vacuum and electron beam system of the TEL, its diagnostics and operation can be found in [6], see also references therein.

## PROJECT PROGRESS IN 2001-2002

In the first series of beam studies in 2001-2002, we achieved tuneshifts of 980 GeV protons of about $dQ$=+0.008 with –3 A of the electron beam current [7]. The original 10kV electron gun generated constant current density distribution in 3.4 mm diameter beam over 2 m long interaction region. Schottky detectors in the Tevatron are used to measure the tunes of the proton bunches. During one test of the lens, three proton bunches (without antiprotons) were injected into the Tevatron and ramped to 980 GeV, and the observed (fractional) horizontal tune of all three bunches was 0.5795. Then the lens was pulsed in order to interact with only one of the three bunches. The spectra associated with the other two bunches remained unaltered, but the third shifted by 0.0082 to 0.5877. Figure 3 shows the resulting spectra; the two untouched bunches produced the set of peaks on the left, and only after turning the TEL on did the third bunch produce the set on the right.

The tuneshift dependence on electron current and energy, on electron beam position and timing was found in good agreement with theoretical formula [1,6].

$$dQ_{x,y} = \mp \frac{\beta_{x,y}}{2\pi} \cdot \frac{1 \pm \beta_e}{\beta_e} \cdot \frac{J_e L_e r_p}{e \cdot c \cdot a_e^2 \cdot \gamma_p} \quad (1)$$

The proton lifetime was in the range of 10 hours (some 24 hours at the best). At first, it was not clear what was limiting it: electron–beam current/position fluctuations or nonlinear beam–beam effects complicated by inaccurate electron beam alignment with respect to (anti)protons.

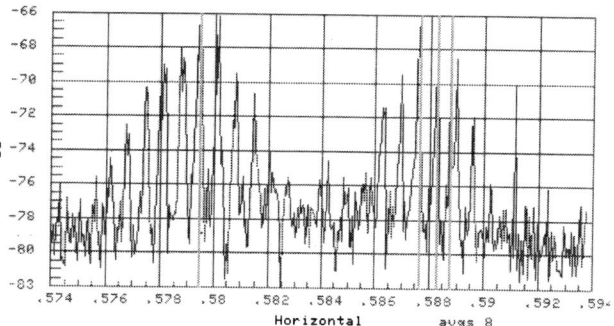

Figure 3: Horizontal tune of 980 GeV protons shifted by the TEL.

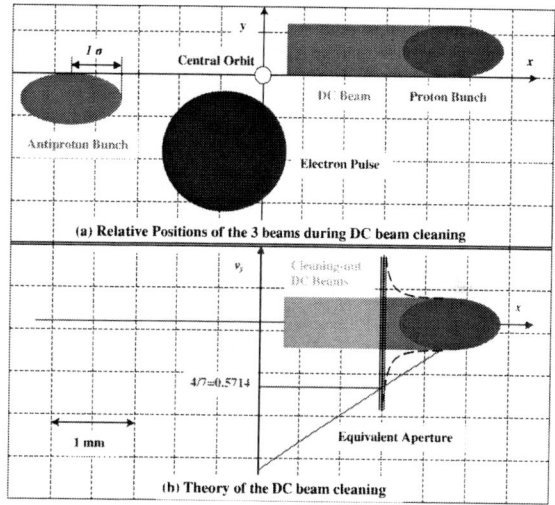

Figure 4: DC beam cleaning: (a) beam positions (b) physics of the abort gap cleaning.

In 2002 the TEL was found to be an invaluable instrument for cleaning DC beam in the Tevatron—an application which was not foreseen at the start of the project. The DC beam consists of particles slowly leaking from RF buckets at 980 GeV and circulating around the ring unsynchronized with RF, including within the abort gaps between bunch trains. A few $10^9$ particles are enough to cause a quench on beam abort. Betatron tunes of Tevatron beams—0.583 in horizontal and 0.575 in vertical plane—are close to $4/7^{th}$ resonance line at 0.5714. The TEL current is fired in the gaps between bunch trains every $7^{th}$ turn and thus excites the DC-beam particles to very large amplitudes until they are lost; Figure 4 depicts the transverse positions of the three beams and the technique of cleaning. Since early 2002, the TEL has been used operationally for DC-beam cleaning in every Tevatron HEP store [8].

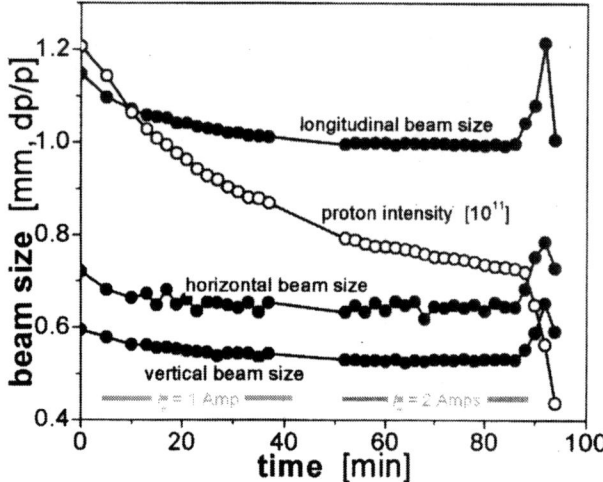

Figure 5: TEL as a "soft collimator."

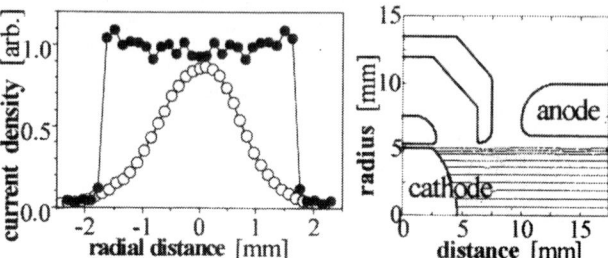

Figure 6: Beam profiles of the flattop gun and Gaussian gun and a cross-section of the latter.

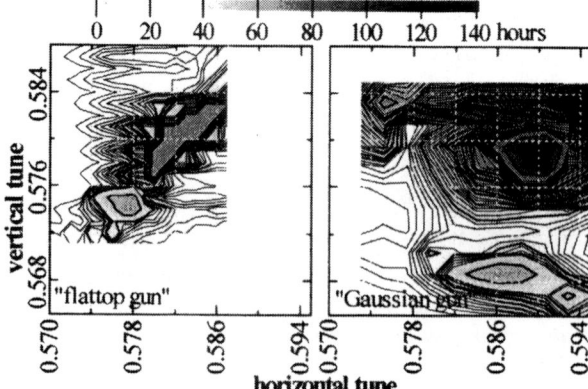

Figure 7: Tune scans with flattop and Gaussian e-beams.

## LIFETIME IMPROVEMENT

Our studies in 2001-2002 showed that poor steering of the electron beam is by far the most important factor affecting the (anti)proton lifetime $\tau=(dN/dt/N)^{-1}$. It can affect $\tau$ even at relatively small electron currents. Lifetime dependence on the electron current with fixed steering correctors was roughly $\tau \propto 1/J^2$.

Eventually, we realized that the edges of the electron beam act as a "soft collimator." For example, Figure 5 shows the size of a particular bunch while it was collimated in this manner. One amp of electron-beam current was applied initially. Many particles were quickly lost, decreasing the beam size; however, the loss rate began to level off because the remaining core bunch was stable. To confirm our understanding, the beam current was doubled to two amps, but the beam size was still secure. Also shown is the bunch intensity (open circles) in units of $10^{11}$ particles, and the linear attrition rate indicates that there was a uniform, slow diffusion of particles in phase space, which caused a small amount of continuous losses. At the very end of the study, the electron beam was misaligned purposefully. The bunch, now passing through the highly nonlinear beam edge, quickly gained emittance and lost particles.

This unfortunate effect spurred the design of a new gun with a very smooth, almost Gaussian-shaped profile. The perveance of this "Gaussian" gun is only 1.8 µP versus 5.6 for the flat-profile gun, but the central current density is about the same than that of the flattop gun. Figure 6 shows transverse distribution of the electron current density generated in the Gaussian and flattop beams [5].

Figure 7 supplies cogent evidence that a smoother beam profile can preserve the bunch lifetime. Two working-point scans (measuring lifetime at various horizontal and vertical tunes) were conducted—the first with the flattop gun, the second with the Gaussian gun. While the two scans did not cover exactly the same regions of tune space, most of each scan overlaps.

The plots have identical boundaries and color scales, and contours are drawn every 20 hours. The flattop gun could not surpass 70 hours, and its highest lifetimes were confined to a small diagonal region. On the other hand, the Gaussian gun offered lifetimes exceeding 120 hours over a much broader area. Again, these values are indistinguishable from typical Tevatron lifetimes. The TEL-induced tune shift in both scans was set to about 0.004.

## BBC: SUPPRESSION OF "SCALLOPS"

The very first evidence of successful BBC was suppression of vertical emittance growth of antiproton bunches tuneshifted by the TEL.

The transverse antiproton emittance growth, after 980-GeV beams are brought to collision, is caused by

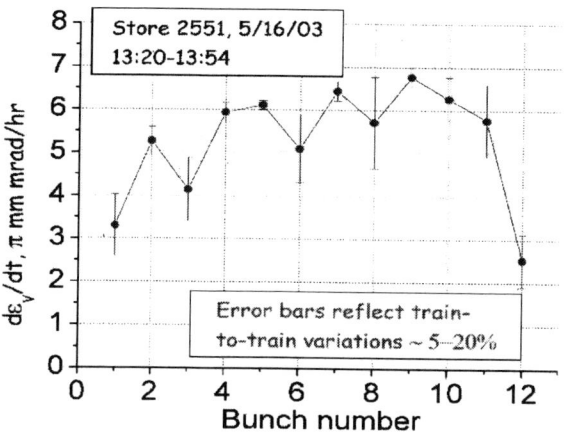
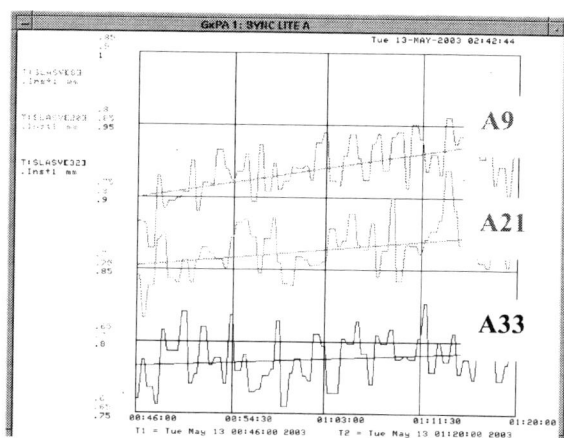

Figure 8: Emittance growth of pbar bunches

Figure 9: Vertical size of three pbar bunches in store 2540.

beam-beam interaction and occurs in the Tevatron when proton bunch intensity exceeds 180e9 [9]. Figure 8 shows the emittance growth rate of twelve antiproton bunches during the first 34 minutes of Tevatron HEP store #2551. Because of 3-fold symmetry of proton loading, the emittance growth rates are the same within 5-20% for corresponding bunches in different trains (e.g. for #1,13,25 or for #2,14,26, etc), which is indicated by error bars in Figure 8. One can see that blowup rates are smaller for bunches closer to the end or start of the train. For comparison, the emittances of all of the bunches before collisions are very similar, in the range of 18-22 $\pi$ mm mrad (95% normalized).

After about an hour, the blowup flattens out, and the distribution of the emittances over different pbar bunches looks like three "scallops." The scallops, though, do not appear in every store because the effect is dependent on antiproton tunes—particularly how close one of them is to some important resonance. For a typical working point of $Q_x$=0.582, $Q_y$=0.590, 5th order (0.600), 7th order (0.5714) and 12th order (0.583) resonances play major role in the pbar beam dynamics [9]. In April-May 2003 it was observed that vertical tune changes as small as -0.002 often resulted in a reduction of the amplitude of the scallops. Smaller, but still visible, scallops were observed also in the proton bunches.

The TEL was used at the beginning of several HEP stores in attempt to reduce the "scallops." First, it was demonstrated that the TEL can be transferred from DC beam removal regime to the BBC regime. This switch includes adjusting the gun's cathode voltage from 6kV to 4.5kV and increasing the cathode filament power from 39 W to 46 W (these increase the electron space charge), changing triggering from 3 pulses every 7th turn (DC cleaning) to 1 pulse every turn (for BBC), changing electron pulse timing (from abort gap to one

of bunches), shortening the e-pulse width, and, finally, using strong TEL dipole correctors to move the electron beam several millimeters in order to interact with the pbars. All these steps with zero electron current produced no significant effect on colliding beams or detector backgrounds, thereafter the TEL with about 0.6A of current was timed on a single pbar bunch at the beginning of the Tevatron stores and *we observed that the TEL can slow vertical emittance growth of the antiproton bunch it was timed on.*

Figure 9 presents evolution of vertical rms sizes of three antiproton bunches #9, 21 and 33 over the first 34 minutes after "initiating collisions" in store #2540 (May 13, 2003). The TEL was acting only on bunch #33. The size has been measured with use of SyncLite Monitor [10]. Corresponding emittance growth was 4.1 $\pi$ mm mrad/hr for bunch #9, 2.2$\pi$ mm mrad/hr and only 1.0 $\pi$ mm mrad/hr for #33. We consider that as evidence of the improvement due to the TEL At the beam parameters: current 0.6 A, energy 4.5kV, rms e-beam size 0.8 mm, interaction region length 2.05 m – expected maximum horizontal pbar tune shift was about –(0.003-0.004), vertical -0.001 (estimated). After 34 minutes the TEL was turned off, and emittances of all three bunches leveled.

During four weeks in April and May, there were eight attempts to attempt BBC at the beginning of the HEP stores. There were no "scallops" in three stores: #2445, #2490, and #2495, and though the TEL was acting on antiprotons, we observed no effect on emittance growth, as well as pbar losses and lifetime. Only Schottky power detector channel SHPWR responded to the TEL current by 0.5dB rise. A faulty TEL pulse generator led to emittance excitation by noise and quick (one minute) loss of corresponding antiproton bunches in stores #2487 and #2502, but it did not lead to loss of the store. After that was fixed, we had scallops and the TEL on bunch #33 in three stores and we suppressed the vertical emittance growth in #2540, effect was neutral in #2546, and somewhat negative (faster emittance growth) in #2549. Table 1 summarizes the emittance growth rates for three "equivalent" pbar bunches (namely, bunch #9 for each of the three trains) in three stores with the TEL off and three stores with the TEL on bunch #33 only.

TABLE 1. Vertical RMS sizes of various bunches.

| store | duration | A9 | A21 | A33 |
|---|---|---|---|---|
| 2536 | 40 min | 9.9 | 9.2 | 9.3 |
| 2538 | 35 min | 1.9 | 1.7 | 2.8 |
| **2540** | 34 min | 4.1 | 2.2 | **1.0** |
| **2546** | 30 min | 3.9 | 1.9 | **4.0** |
| **2549** | 26 min | 4.5 | 3.6 | **7.1** |
| 2551 | 34 min | 6.7 | 6.6 | 7.0 |

One can see that without the TEL, emittance growth rates over the first 30-40 minutes of the stores for the three "equivalent" bunches were the same.

Again, the effect of the TEL is obvious, though not well controlled since it can be negative as well as positive. We believe that the uncertainty is due to imprecise centering of the electron beam on the antiprotons. The pbar orbit at F48 can migrate by up to 0.5 mm over a time scale of twelve hours and up to 1 mm over a scale of several days to a week [11]. Unfortunately, the electrical centers of the TEL BPMs are dependent on the signal bandwidth, and the difference between the short pbar pulse position and the long electron pulse position can not be determined with accuracy better than 0.5-1.5 mm (though, resolution of the BPMs for any of the beams alone is about 20-40 microns). Such errors in positioning of $\sigma=0.8$ mm electron beam with respect to $\sigma=0.5$ mm pbar bunch may result, for example, in significant variation of the TEL-induced tuneshift and even in changing sign of the tuneshift. We plan to improve the TEL BPMs [8].

## CONCLUSIONS, FUTURE STEPS

For the past 18 months, the TEL has been needed to clean the abort gap of residual particles. Recently we have indications that the TEL can compensate beam-beam effects in the Tevatron—it reduces the scallops.

We will continue experimental studies of BBC using the TEL at F48, which can be used not only for suppression of the scallops but also at other stages of the Tevatron cycle (injection, ramp, squeeze, collisions). We may choose to act on protons as well to do BBC or suppress coherent instabilities.

Also, we will study effects of coherent longitudinal [12] and transverse waves in the electron-(anti)proton interaction and explore the need of a better high frequency stabilization of the electron current and position.

We plan to improve the TEL BPMs and commission bunch-by-bunch tune diagnostics with a 1.7 GHz Schottky detector [13]. Fabrication of the second electron lens in collaboration of IHEP (Protvino) is underway and will be finished in the summer of 2004.

Future hardware changes are focused on having wider electron beam with higher current in the TEL. They include: a) adding solenoids in the bends of the TEL in order to allow propagation of electron lens with smaller field in the main superconducting solenoid; b) constructing a new 15 kV HV modulator; c) installing a new electron gun combining flattop and smooth, Gaussian-like tails.

The number of people involved with the TEL has evolved and increased over the past few years. Appreciation goes to H. Pfeffer, G. Saewert, A. Semenov, D. Wildman, D. Wolff, and M. Olson, all of who have contributed considerable effort.

## REFERENCES

1. V. Shiltsev, D. Finley, FNAL-TM-2008 (1997).
2. V. Shiltsev, FNAL-TM-2031 (1997).
3. V. Shiltsev et.al., Phys. Rev. ST-AB, 2,071001(1999).
4. D. Shatilov et.al., Proc. PAC'01, p. 2002.
5. Y. Alexahin et.al., Proc. PAC'01, p. 2005.
6. V. Shiltsev et.al., Proc. PAC'01, p. 158.
7. K. Bishofberger et.al., Proc. Advanced Accelerator Concepts Workshop '02, p. 821.
8. X. Zhang et.al., Proc. PAC'03.
9. T. Sen et.al., Proc. PAC'03.
10. H. Cheung et.al., Proc. PAC'03.
11. V. Shiltsev et.al., FNAL-Conf-02/250 (2002).
12. V. Parkhomchuk et.al., Zh.Tech.Phys, **73**, 8 (2003).
13. R. Pasquinelli, Proc. PAC'03.

# Wire Map and Applications to Long-Range Beam-Beam Compensation

B. Erdelyi and T. Sen

*Fermi National Accelerator Laboratory, P.O. Box 500, MS-220, Batavia, IL, 60510*

**Abstract.** Long range beam-beam effects play an important role in the Tevatron. Active compensation is envisaged by current wires, as proposed recently for the LHC. Here, we present the first steps in this program, namely the principle of the compensation, the derivation of the transfer map of the wire, potentially increasing robustness by utilizing wire cages, and some very preliminary results of application to the Tevatron at injection energy.

## INTRODUCTION

The idea of compensating the long-range beam-beam effects with current carrying wires was considered recently for the LHC [1]. The principle is simple; the strong beam can be regarded as a current, and its effect on the weak beam can be alleviated by another current placed in such a way that the electric field of the strong beam and the magnetic field of the weak beam cancel each other. The advantage of such an approach consists of the simplicity of the method and the possibility to deal with all multipole orders at once.

We initiated a similar study for the Tevatron. However, the Tevatron is different from the LHC in the sense that both (proton and antiproton) beams share the same beam tube, and the long-range interactions take place all over the ring's circumference and phase space. This requires careful studies since the wires affect both beams, and new optimization strategies need to be formulated. For an overview of up to date beam-beam studies at the Tevatron see [2]. Also, in parallel, there is another, complimentary, approach to long-range beam-beam compensation pursued at the Tevatron by utilizing electron lenses [3].

To simulate the effect of the wires on the nonlinear dynamics, one needs to implement it in beam dynamics codes. The implementation should be fast, accurate, symplectic, and allow misalignments. Moreover, it should be facilitating optimization with ease. We chose to implement the wire in differential algebra used by codes such as Cosy Infinity, and for tracking in Sixtrack. To this end one needs to compute the transfer map of the wire. The first step is to look at the equations of the motion. This leads to the necessity of computing the magnetic field generated by a finite piece of straight wire at an arbitrary point in space, and its integral over the length of the region occupied by the wire.

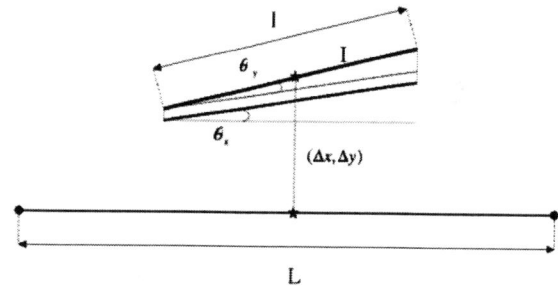

**FIGURE 1.** The wire is embedded in a drift, and may have arbitrary length and orientation.

## DERIVATION OF THE WIRE TRANSFER MAP

This section contains the steps needed for the derivation and implementation of the transfer map of the wire. A schemtaic of the model utilized for the wire is shown in Figure 1.

### The Equations of the Motion

In canonical coordinates $\left(x, a = \frac{p_x}{p_0}; y, b = \frac{p_y}{p_0}\right)$, where $p_0$ is the momentum of the reference particle, the equations of the motion take the following form (assuming no longitudinal fields):

$$x' = \frac{a}{\sqrt{1-a^2-b^2}}, \; y' = \frac{b}{\sqrt{1-a^2-b^2}}, \quad (1)$$

$$a' = -\frac{B_y}{\chi_0}, \; b' = \frac{B_x}{\chi_0}, \quad (2)$$

where $\chi_0$ is the magnetic rigidity of the reference particle. In the Hamiltonian formulation the first two equations are recognized as the equations of the drift, while the last two as the equations of a kick. This leads to the possibility of a second order symplectic integration of the system by an operator splitting method (which is equivalent to the so-called thin lens approximation), i.e. if the transfer map of a wire, embedded in a drift of length $L$, is denoted by $\mathcal{M}_w$, the map of the drift by $\mathcal{M}_d$, and the map of the kick by $\mathcal{M}_k$, the relation between them is

$$\mathcal{M}_w^{(L)} =_2 \mathcal{M}_d^{(L/2)} \circ \mathcal{M}_k^{(L)} \circ \mathcal{M}_d^{(L/2)}. \quad (3)$$

The map of a drift is readily available, and the solution for the kick is given by

$$a_f = a_i - \int_{-L/2}^{L/2} \frac{B_y}{\chi_0} dz, \quad (4)$$

$$b_f = b_i + \int_{-L/2}^{L/2} \frac{B_x}{\chi_0} dz. \quad (5)$$

Therefore, to compute the kick one needs the integrated field generated by the wire. We note that this approach takes into account the wire's fringe field region too.

## The Magnetic Field of a Finite Straight Wire

The magnetic field can be computed using the Biot-Savart law. Assume that in an arbitrary coordinate system $(x,y,z)$ we have a wire of length $l$, such that the start of the wire is at a distance $\vec{r}_P$ from a point $P$ where we want to compute the field. If the length of the wire is parametrized by $\lambda$ such that $\vec{l}(\lambda) = \lambda \vec{l}$ with $\lambda \in [0,1]$, the field, integrated over $L$, is given by

$$\left\langle \vec{B}_P \right\rangle = 10^{-7} I \int_0^1 d\lambda \int_{-L/2}^{L/2} \frac{\vec{l} \times \vec{r}_P}{\left|\vec{r}_P - \lambda \vec{l}\right|^3}, \quad (6)$$

where $I$ is the current.

The integrals can be done analytically. However, they simplify if the coordinate system is assumed parallel to the wire, that is, if $\vec{l} = (l_x, l_y, l_z)$, the coordinate system is oriented such that $l_x = l_y = 0$. In this case the result is given by

$$\left\langle \vec{B}_P \right\rangle = 10^{-7} I \frac{u-v}{x^2+y^2} \begin{pmatrix} x \\ y \\ 0 \end{pmatrix}, \quad (7)$$

where $u = \sqrt{\left(\frac{L}{2}+l\right)^2 + x^2 + y^2}$ and $v = \sqrt{\left(\frac{L}{2}-l\right)^2 + x^2 + y^2}$. Here $x$ and $y$ are regarded as the sums of the wire distance from the longitudinal axis and the particle's betatron amplitude. In case the wire is tilted with respect to the coordinate system, first the tilt of the coordinate system is performed, then the map of the wire applied.

## The Full Map of a Wire

Combining the maps of the pieces we can get the full transfer map of the wire, which allows for arbitrary length, current and placement, including pitch and yaw. It is given by

$$\mathcal{M}_w = \mathcal{S}_{\Delta x, \Delta y} \circ \mathcal{T}_{\theta_x, \theta_y}^{-1} \circ \mathcal{M}_d^{(L/2)}$$
$$\circ \mathcal{M}_k^{(L)} \circ \mathcal{M}_d^{(L/2)} \circ \mathcal{T}_{\theta_x, \theta_y}, \quad (8)$$

where $\mathcal{T}_{\theta_x,\theta_y}$ represents the tilt of the coordinate system by horizontal and vertical angles $\theta_x, \theta_y$ to orient the coordinate system parallel to the wire, and $\mathcal{S}_{\Delta x, \Delta y}$ represents a shift of the coordinate axes to make the coordinate systems after and before the wire agree.

Specifically, the map that tilts the coordinate axes with angle $\theta_x$ is

$$\begin{cases} x_f = x_i \left(\cos\theta_x - \sin\theta_x \tan(\alpha - \theta_x)\right), \\ a_f = \sqrt{(1+\delta)^2 - b_i^2} \sin(\alpha - \theta_x), \\ y_f = y_i - x_i \sin\theta_x \frac{b_i}{\sqrt{(1+\delta)^2 - b_i^2}\cos(\alpha-\theta_x)}, \\ b_f = b_i. \end{cases} \quad (9)$$

The (non-commuting) tilt in the other transverse plane is obtained by interchanging $x \leftrightarrow y$ and $a \leftrightarrow b$. The inverse tilts are just tilts with opposite signs of the tilt angles. The shift needs to be performed with

$$\Delta x = -L \tan(\theta_x), \quad (10)$$

$$\Delta y = -\frac{L \tan(\theta_y)}{\cos(\theta_x)}. \quad (11)$$

## APPLICATIONS TO THE TEVATRON

At injection energy there are long-range interactions at 138 different locations, spread all over the circumference of the ring and phase space. There is practically no possible way to correct each interaction with an individual wire. We are also limited by the small number of drift spaces where wires could be installed.

As a first step in this study we looked at a set of four wires, placed in drifts that are at least 1 m long, where the horizontal and vertical beta functions are not too different, where the proton and antiproton closed orbits are well separated (in order for the wire to not affect the proton beam), and at a reasonable distance from the beam pipe. The wire needs to be at some comfortable distance from the antiproton beam to allow for orbit drifts and manipulations. We chose to focus on resonance strengths as a criterion for correction. The long-range beam-beam interactions drive mostly seventh order resonances at injection, with the (3,4) resonance dominating, making it a natural candidate for correction by wires.

The wires have been installed at the following sections: A17, F0, E0, and C0. The length of the wires was fixed to 1 m. To obtain a rough estimate for the current in the wires, we assumed round proton beams with design parameters and that the beam-beam kicks add up linearly. This leads to

$$I_w = (\frac{r_p m_p c^2}{0.2998}) N_b \times 10^7 = 12.89 [\text{Amp}] \qquad (12)$$

per interaction, where $N_b = 2.7 \times 10^{11}$ is the number of protons per bunch. This multiplied by 72 (interactions) / 4 (wires) gives $I_w = 232$ A in each wire. This is a large current and the wire may require cooling, as also envisaged for the LHC. Since the wire is placed on the side of the antiproton beam that is opposite to the proton beam, the sign of the current should be in the direction of the proton beam. These constraints fix the longitudinal positions, length, and currents. There is no obvious constraint that fixes the transverse positions. If there were one wire per interaction it is obvious that for the correction to not depend on the particle amplitudes, the wires should be placed at the same distance from the antiproton beam as the proton beam is from the antiproton beam. In our case with 72 interactions per bunch (and varying separations) and only four wires, there is no such clear criterion. From the map of the wire it is clear that by changing the antiproton beam - wire distance, or even the orientation of it, the map will be different, and might drive different resonances.

Indeed, the resonance structure changed with the transverse placement in simulations. Here we present some preliminary results. In the first set of runs we fixed some high currents and distance of the wires from the antiproton beam to 10 mm, and varied the orientation in the first quadrant, namely at angles 0°, 22.5°, 45°, 67.5° and 90° with respect to the horizontal. We computed the resonances up to order seven. Figure 2 shows the resonance

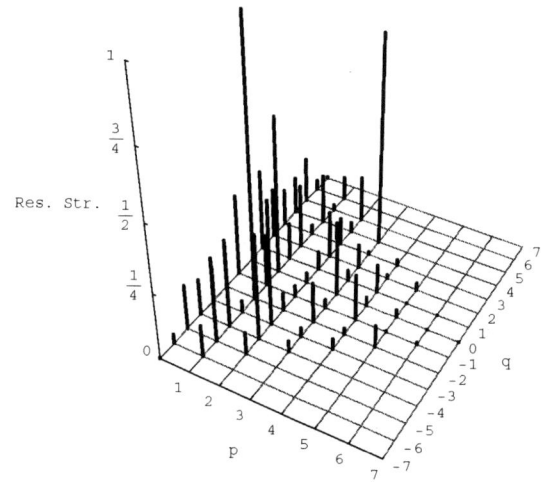

**FIGURE 2.** Resonance strengths at injection generated by the wires placed along the diagonal in the first quadrant in the $x - y$ plane.

structure when the angle was fixed at 45°. The (3,4) resonance is strongly driven at this angle.

In another simulation, we kept the angle of the wire the same as the plane of the helix but varied the distance of the wire from the anti-proton beam. If the wires are too far, the resonance structure in this geometry did not resemble the one generated by the beam-beam interactions. However as the distance decreased, the resonance structures become more alike with mostly the seventh order resonances driven. It is clear nonetheless that the resonance structure depends sensitively on the exact placement of the wire and a more robust compensation may be necessary.

## MULTIPLE WIRES AT A SINGLE LOCATION

One way to increase the robustness and flexibility of the compensation method is to place several wires around the circumference of a cylindrical cage at each location [4]. We consider here the fields produced by such a structure.

For simplicity we assume that the wires are infinitely long and each carries the same current $I_W$. The geometry of the arrangement is shown in Figure 3. We assume that there are $N_W$ such wires along the circumference. $\alpha_j$ is the angle of the $j$th wire (w.r.t the $x$ axis) from the center of the beam pipe. From the field of an infinitely long wire it follows that the field at an anti-proton with coordinates

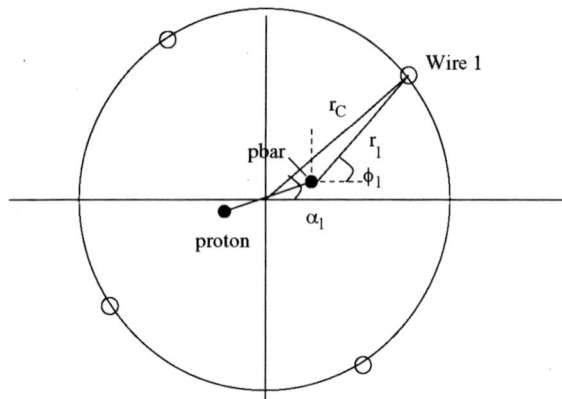

**FIGURE 3.** Cross section of the wire cage showing the wires placed on a circle of radius $r_C$ with the 1st wire at an angle of $\alpha_1$ w.r.t the $x$ axis.

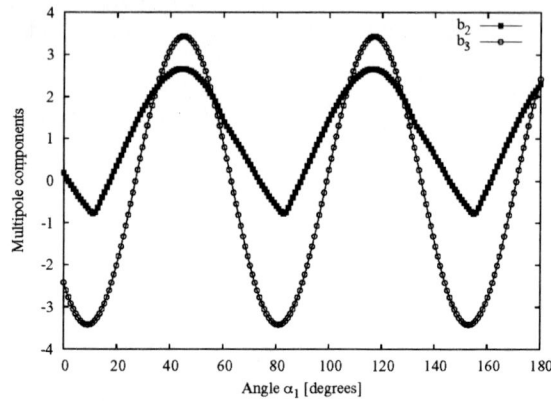

**FIGURE 4.** Normal sextupole and octupole coefficients $(b_2, b_3)$ due to 5 wires placed uniformly around the beam vs the orientation angle of the wires.

$(x, y)$ due to these wires is

$$B_y + iB_x = \frac{\mu_0}{2\pi} I_W \sum_{j=1}^{N_W} \sum_{n=0}^{\infty} [-\cos(n+1)\phi_j + i\sin(n+1)\phi_j] \left[\frac{(x+iy)^n}{r_j^{(n+1)}}\right], \quad (13)$$

where

$$r_j = [(r_C \cos\alpha_j - r_{\bar{p}}\cos\theta_{\bar{p}})^2 + (r_C \sin\alpha_j - r_{\bar{p}}\sin\theta_{\bar{p}})^2]^{1/2}, \quad (14)$$

$$\phi_j = \arctan\left[\frac{(r_C \sin\alpha_j - r_{\bar{p}}\sin\theta_{\bar{p}})}{(r_C \cos\alpha_j - r_{\bar{p}}\cos\theta_{\bar{p}})}\right]. \quad (15)$$

Here $r_C$ is the radius of the cage measured from the center of the beam-pipe, $(r_{\bar{p}}, \theta_{\bar{p}})$ are the distance and angle respectively of the anti-proton beam also from the center of the beam pipe. Comparing this with the usual multipole expansion

$$B_y + iB_x = B_0 \sum_{n=0}^{\infty} (b_n + ia_n) \left[\frac{x+iy}{R_{ref}}\right]^n \quad (16)$$

we can write the multipole coefficients as

$$b_n = -\sum_{j=1}^{N_W} \cos(n+1)\phi_j \left[\frac{\langle r_W \rangle}{r_j}\right]^{n+1}, \quad (17)$$

$$a_n = \sum_{j=1}^{N_W} \sin(n+1)\phi_j \left[\frac{\langle r_W \rangle}{r_j}\right]^{n+1}, \quad (18)$$

where the main field $B_0$ and reference radius $\langle r_W \rangle$ are

$$B_0 = \frac{\mu_0}{2\pi} \frac{I_W}{\langle r_W \rangle}, \quad \langle r_W \rangle = \frac{1}{N_W} \sum_j r_j. \quad (19)$$

We use these expressions to calculate the multipole coefficients as a function of the orientation of the wires. We let $\alpha_1$ be the smallest angle of the wires w.r.t the anti-proton beam and assume (again for simplicity) that all other wires are distributed uniformly in azimuth, i.e. spaced at an angle $= 2\pi/N_W$ apart.

Figure 4 shows the variation in the normal sextupole and octupole components $(b_2, b_3)$ as a function of the angle $\phi_1$. We used $N_W = 5$ and chose the plane of the helix to be at $45°$. We remark on some features: (i) the magnitude of each component varies by a factor of two to three as the angle is varied, (ii) the ratio of these components also changes with the angle. Another feature not shown here is that the ratio of these components also depends on the number of wires, e.g. with 3 wires, the sextupole components are larger.

These aspects of the fields with multiple wires will allow greater flexibility in compensating selected resonances. We envisage a cage with several wires but powering only those wires at each location (depending on the orientation of the helix for example) which improve the lifetime. This investigation is still in a preliminary stage. Many more issues, such as the influence on the proton beam etc. need to be considered in a more detailed study.

## REFERENCES

1. J.P. Koutchouk, *Principle of a correction of the long-range beam-beam effect in LHC using electromagnetic lenses*, LHC Project Note 223 (2000)
2. T. Sen, *Theory and observations of beam-beam interactions in the Tevatron*, these proceedings
3. V. Shiltsev, *Status of beam-beam compensation with electron lenses in Tevatron*, these proceedings
4. J.P. Koutchouk, personal communication

# Multipole Compensation of Long-Range Beam-Beam Interactions

J. Shi, L. Jin, and O. Kheawpum

*University of Kansas, Lawrence, KS 66049*

**Abstract.** In cases of multi-bunches operation in storage-ring colliders, serious long-range beam-beam effects could be due to many parasitic collisions that are localized inside interaction regions or/and distributed around the ring. To reduce the long-range beam-beam effects, the compensation of long-range beam-beam interaction with magnetic multipole correctors based on minimizations of nonlinearities in one-turn or sectional maps of a collider has been proposed. With LHC as a test model, the effectiveness of the multipole compensation of long-range beam-beam interactions was studied in terms of improvement of dynamic aperture and reduction of emittance growth. The study showed that both the local and global compensation of long-range beam-beam interactions with magnetic multipole correctors is very effective in increasing the dynamic aperture and improving the linearity of the phase-space region relevant to the beams.

## INTRODUCTION

The principle of the multipole compensation of long-range beam-beam interactions is based on the idea of global compensation of nonlinearity in a storage-ring collider by using maps [1, 2]. Without beam-beam interactions, the nonlinear beam dynamics in a storage ring can be described by a one-turn map that contains all global information of nonlinearities in the system. By minimizing nonlinear terms of one-turn maps order-by-order with a few groups of magnetic multipole correctors, one can reduce the nonlinearity of the system globally. To include long-range beam-beam interactions into the map for the global compensation, one should recognize that a large beam separation is typical at parasitic collision points. In both LHC and Tevatron, for example, the beam separation is in a scale of 6–14 $\sigma$ where $\sigma$ is the nominal beam size. In the phase-space region relevant to the beam, the long-range beam-beam interactions can thus be expanded into a Taylor series around the beam separation and be included into the one-turn map for the global compensation of the nonlinearities of the system [3]. In the case of localized parasitic collisions, one can use a group of local multipole correctors to minimize nonlinear terms of a local sectional map that contains the Taylor expansion of long-range beam-beam interactions. With a few groups of multipoles correctors, therefore, nonlinear terms in one-turn and/or sectional maps including the long-range beam-beam interactions can be minimized order-by-order and, consequently, the nonlinearity of the system in the phase-space region of interest can be significantly reduced.

In order to minimize nonlinear terms of a one-turn or sectional map with a few parameters of multipole correctors, we postulate that the $n$th-order undesirable nonlinearity in a one-turn or sectional map can be characterized by the magnitude of its $n$th-order undesirable coefficients which are defined by

$$\lambda_2 = \left[ \sum_{i+j+k+l=2} \left( u_{ijkl} - u^0_{ijkl} \right)^2 \right]^{1/2} \quad (1)$$

where $u_{ijkl}$ is the coefficient associated with the term of $x^i p_x^j y^k p_y^l$ in the map. $u^0_{ijkl}$ of $i+j+k+l=2$ denote the sextupole components from the chromaticity correctors and should be excluded from the minimization and $u^0_{ijkl} = 0$ of $i+j+k+l > 2$. For convenience, we define the $n$th-order global/local compensation when all $\lambda_i$ with $i = 2,...,n$ are minimized order-by-order using the multipole correctors up to the $n$th order.

To implement the global or local compensation of the nonlinearities with the one-turn and/or sectional maps, the maps can be obtained by using the method of differential algebra [4] or Lie algebra [5] based on the measured field errors of magnets and the long-range beam-beam interactions at designed parasitic beam crossings. During the operation, the beam-beam interactions and thus the maps change with the beam sizes. The strength of the multipole correctors for the compensation could be adjusted dynamically with the beam-size growth if necessary. If a one-turn map can be extracted with desired accuracy directly from beam-dynamics measurements [6, 7], moreover, the global compensation of the nonlinearities of a storage-ring colliders could be further optimized during the operation [2]. Such a beam-based global compensation is especially important when there is a significant uncertainty on the linear lattice and/or in the field measurement of magnets.

## TESTING RESULTS

With LHC as a model, the effectiveness of the multipole compensation of long-range beam-beam interactions was evaluated in terms of the improvement of dynamic aperture and the reduction of emittance growth. Dynamic aperture was calculated with $10^5$-turns tracking and the emittance growth was examined with a self-consistent beam-beam simulation by using the particle-in-cell method with one million particles [8]. The fractional parts of horizontal and vertical tunes of LHC are $(\nu_x, \nu_y) = (0.31, 0.32)$. Head-on and long-range beam-beam interactions at IP1 and IP5 as well as multipole field errors in the lattice were included in the tracking. For long-range beam-beam interactions, there are 15 parasitic collisions on the each side of an IP. The crossing angle of two counter-rotating beams is 300 $\mu$rad with vertical crossing at IP1 and horizontal at IP5. The separation of the two beams at parasitic collision points ranges from 7 to 13 $\sigma$, where $\sigma$ is the transverse beam size. To test the local compensation of both long-range beam-beam interactions and nonlinear field errors in the triplets, four groups of correctors in the triplets of IP1 and IP5 (one in each triplet) were used to minimize $\lambda_n$. To test the global compensation of the nonlinearities, we included four corrector packages symmetrically located in arcs. Each package of correctors contains normal and skew components of desired multipole correctors.

### a. Linear lattice

In order to isolate the effect of the multipole compensation on the long-range beam-beam interactions, we studied the case that contains only beam-beam interactions and otherwise a linear lattice. In Fig. 1, the ratio of $\lambda_n$ after to before the compensation was plotted as a function of the order of the compensation for $\xi = 0.0034$, 0.0068, 0.01, and 0.02. In all the cases, after the compensation the quadratic nonlinearity ($\lambda_2$) in the one-turn map was reduced to roughly 10% of the original and higher-order terms ($\lambda_n$, $n > 2$) were also reduced significantly. The compensation is therefore very effective in reducing the nonlinearities due to long-range beam-beam interactions. The reduction rate of $\lambda_n$ after the compensation, however, decreases with $n$ as shown in Fig. 1. As the number of monomials of a given order in the one-turn map increases quickly with the order, it is difficult computationally to locate an optimized minimum (semi-global minimum) of $\lambda_n$ when $n$ is large. Fig. 1 also shows that the reduction rate of $\lambda_n$ after the compensation is about the same for all $\xi$.

Figure 2 plots the DA as a function of $\xi$ without or with the multipole compensation of the long-range beam-beam interactions. When the DA without the compensation is smaller than 9 $\sigma$, where $\sigma$ is the nominal transverse beam size of LHC, the multipole compensation improves the DA significantly. Without the compensation, the DA also decreases with $\xi$ quickly. After the compensation, such reduction of the DA becomes much slower as the more nonlinear the system, the larger the increase of the DA after the compensation. In the case that the DA without the compensation is larger than 9 $\sigma$ when $\xi < 0.005$, the original system is already quite linear and the compensation fails to further improve the DA. Note that 9 $\sigma$ is the average beam separation at parasitic collision points. In the phase-space region that is close to or larger than the beam separation, the expansion of the long-range beam-beam interaction at the beam separation is invalid and the multipole correctors for the compensation could make the system even more nonlinear there. The phase-space region that is relevant to the beams is, however, near the closed orbit. In that region, the expansion of the long-range beam-beam interactions is always good and the compensation should therefore be effective in improving the dynamics of the beams. When the original DA is close to or larger than the beam separation, the DA is thus no longer a good quantity to characterize the benefit of the multipole compensation of the long-range beam-beam interactions. In Fig. 3, the increase of DA after the compensation was plotted as a function of the order of the compensation. It shows that as the order increases the further improvement of the DA becomes less pronounced. This indicates that the lower-order (cubic and 4th-order) nonlinearities

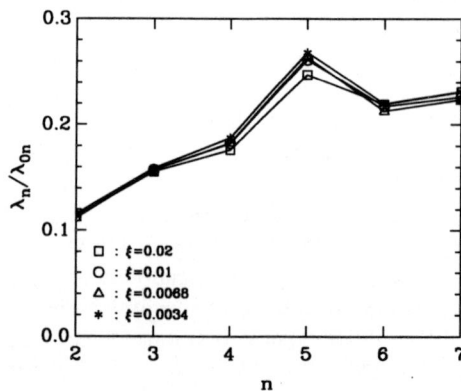

**FIGURE 1.** $\lambda_n$ after the multipole compensation of long-range beam-beam interactions vs. the order of the compensation. $\lambda_{0n}$ is the magnitude of nonlinear coefficients of the map without the compensation. The correctors for the compensation are in the arcs. Curves of different $\xi$ almost overlap each other.

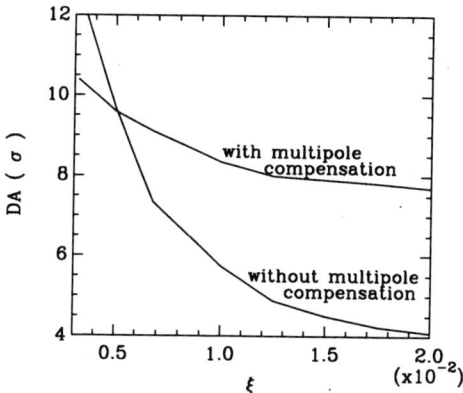

**FIGURE 2.** Dynamic aperture vs. beam-beam parameter without or with the multipole compensation of long-range beam-beam interactions in the linear lattice with head-on and long-range beam-beam interactions. $\sigma$ is the nominal transverse beam size of LHC.

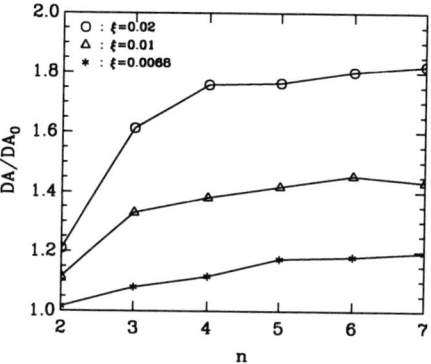

**FIGURE 3.** The increase of dynamic aperture after the multipole compensation of long-range beam-beam interactions vs. the order of the compensation. $DA_0$ is the dynamic aperture without the compensation. $\xi = 0.0068$, 0.01, or 0.02.

dominate the long-range beam-beam interactions.

To exam the effect of the multipole compensation of the long-range beam-beam interactions on the linearity of the phase-space region that is relevant to the beams, we studied the evolution of beam particle distributions by means of the strong-strong beam-beam simulation with $5 \times 10^5$ particles in each beam. During the tracking of the distributions, the transverse r.m.s. emittance was calculated in terms of $\varepsilon = \langle \xi_x^2 + \eta_x^2 + \xi_y^2 + \eta_y^2 \rangle / 2$ where $(\xi_x, \xi_y)$ and $(\eta_x, \eta_y)$ are the normalized transverse coordinates and their conjugate momenta, and $\langle \rangle$ is the average over particles in each beam. Fig. 4 plots the emittance growth without or with the multipole compensation as functions of the beam-beam parameter. It confirms that the multipole compensation of the long-range beam-beam interactions improves the linearity of the phase-space region nearby the closed orbit and, consequently, improves the dynamics of the beams even though the

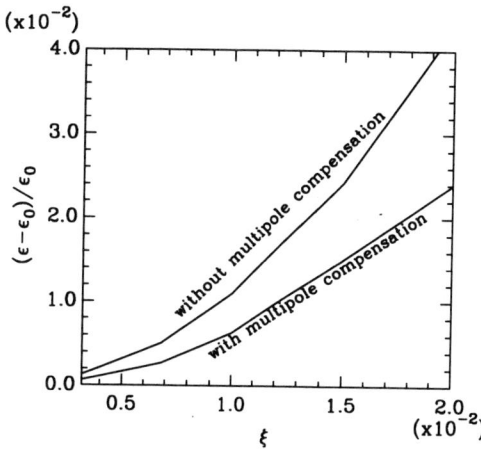

**FIGURE 4.** Emittance growth in $5 \times 10^4$ turns vs. beam-beam parameter without or with the multipole compensation of long-range beam-beam interactions in the linear lattice with head-on and long-range beam-beam interactions. The upper (lower) curve is without (with) the compensation. $\varepsilon_0$ is the initial emittance.

phase-space region close to or beyond the average beam separation at the parasitic crossings could become more nonlinear due to the multipole correctors for the compensation. With the multipole compensation, the evolution of the emittance with the long-range beam-beam interactions was found to be almost overlap with that without the long-range beam-beam interactions. The multipole compensation therefore eliminates the effect of long-range beam-beam interactions on the emittance growth.

### b. Nonlinear lattice

In the case of the nonlinear lattice with beam-beam interactions, the multipole compensation can reduce overall nonlinearities due to both the long-range beam-beam interactions and nonlinear field errors in the lattice. In order to improve the statistical significance of the simulation that involves random field errors, we studied 50 different cases of random multiple components of the field errors generated with different seed numbers in a random number generator routine. In this study, the beam-beam parameter was taken to be $\xi = 0.0068$. Note that the nominal beam-beam parameter of LHC is $\xi = 0.0034$. Because in the simulation we only included beam-beam interactions at two IPs instead of four in LHC, we intentionally used $\xi$ that is twice the LHC design value in order to be closer to the LHC situation. Without including any beam-beam interactions, the smallest and average DA of these 50 samples is 7.7 and 9.8 $\sigma$, respectively. With both the head-on and long-range beam-beam interactions but without the multipole compensation, the smallest and average DA of the fifty random cases is 5.7 and 6.5 $\sigma$. The dynamic aperture of LHC is thus dominated by the long-range beam-beam interactions. After the 5th-order multipole compensation with the correctors in the IRs, the smallest and average DA increase to 8.2 and 8.8 $\sigma$, respectively, which is a more than 40% gain in the DA. With the 5th-order multipole compensation by using four groups of global correctors in the arcs, the smallest and average DA are about the same as those when the correctors are in the IRs. Even though the correctors in the arcs are away from the sources of the nonlinearities, they are as effective as the correctors in the IRs for the global compensation of the long-range beam-beam interactions and nonlinear field errors.

To evaluate the effect of the compensation on beam-size growth, the evolution of the particle distributions of the beams were also studied by means of the strong-strong beam-beam simulation. Without the multipole compensation, the long-range beam-beam interactions result in a 57% increase of the emittance growth during $5 \times 10^4$ turns. With the multipole compensation, the increase of the emittance growth due to the long-range beam-beam interactions was again completely eliminated in the both cases when the correctors are close to or far away from the parasitic collisions. Moreover, due to a simultaneous compensation of nonlinear field errors in the lattice and the long-range beam-beam interactions, the emittance growth after the multipole compensation is smaller than that without long-range beam-beam interactions and the multipole compensation.

## SUMMARY

The global or local multipole compensation of long-range beam-beam interactions based on the minimization of nonlinearities in one-turn or sectional maps is an effective means to suppress long-range beam-beam effects due to localized or/and distributed parasitic collisions. With a few groups of multipoles correctors, nonlinear terms in one-turn or sectional maps including long-range beam-beam interactions can be minimized order-by-order and, consequently, the nonlinearity of the system in the phase-space region that is relevant to the beams is significantly reduced. In the case of localized parasitic collisions such as in LHC, both the wire compensation [9] and the multipole compensation are very effective in eliminating adverse long-range beam-beam effects. With the multipole compensation, however, the overall nonlinearities in the system including both the long-range beam-beam interactions and magnetic field errors in the lattice can be treated systematically with a same group of multipole correctors since the field errors and long-range beam-beam interactions can be simultaneously considered in the maps. In LHC, for example, there is already a group of multipole correctors inside each inner triplet for a local magnetic-field correction. The multipole compensation of long-range beam-beam interactions could thus be accomplished by using the same group of correctors. In the case of distributed parasitic collisions such as in Tevatron, no other viable solution is currently available for a compensation of nonlinear long-range beam-beam interactions. The multipole compensation scheme proposed here opens up a possibility for the reduction of nonlinear long-range beam-beam effects in this case. To apply the multipole compensation of long-range beam-beam interactions in Tevatron, one needs to consider that the two counter-rotating ($p$ and $\bar{p}$) beams of very different intensities have to pass through common multipole correctors because they share a single vacuum pipe. The minimization of nonlinearities of one-turn maps has therefore to be done simultaneously for both the beams. Since the beam-beam interaction on the $p$ beam is much weaker than that on the $\bar{p}$ beam, in the nominal condition of Tevatron RUN II, no compensation of the beam-beam interactions is needed for the $p$ beam. In order to satisfy this un-symmetrical requirement of the multipole strengths of the correctors for the compensation of the long-range beam-beam interactions, the correctors can be placed unsymmetrically with respect to the closed orbits of the two beams, i.e. the orbit of the $p$ beam is at the center of the correctors while the orbit of the $\bar{p}$ beam is away from the center so that the correctors exert much stronger nonlinear perturbations on the $\bar{p}$ than that on the $p$ beam. During the minimization of the nonlinearities of the maps, the off-center distance of the $\bar{p}$ beam will be optimized, within any hardware constraints, together with the strengths of the multipole correctors.

### ACKNOWLEDGMENTS

This work is supported by the U.S. Department of Energy under Grant No. DE-FG03-00ER41153. The authors would like to thank KCASC at the University of Kansas for the use of the Supercomputer. We would also like to thank Dr. T. Sen for many stimulating discussions.

## REFERENCES

1. J. Shi, Part. Accel. **63**, 235 (2000)
2. J. Shi, Nucl. Instr. & Meth. A**444**, 534 (2000).
3. J. Shi, O. kheawpum, and L. Jin, in *Proc. of the 8th European Part. Accel. Conf.*, Paris, June 2002.
4. M. Berz, Part. Accel. **24**, 109 (1989).
5. A. Dragt, in *Physics of High-Energy Particle Accelerators*, edited by R. A. Carrigan *et al.*, (AIP, new York, 1982).
6. C. Wang and J. Irwin, SLAC report, SLAC-PUB-7547, (1997).
7. S. Peggs and C.Tang, BNL report RHIC/AP/159, (1998).
8. J. Shi and D. Yao, Phy. Rev. E**62**, 1258 (2000).
9. J.-P. Koutchouk, in *Proc. of Workshop on Beam-beam Effects in Circular Colliders*, Edited by T. Sen and M. Xiao, FERMILAB-Conf-01/390-T, (2001).

# Beam-beam Simulations in Four-beam Scheme for High Luminosity $e^+e^-$ Colliders

Y. Ohnishi* and K. Ohmi*

*High Energy Accelerator Research Organization (KEK), Tsukuba, Japan

**Abstract.** Beam-beam interaction with four-beam scheme and beam-beam compensation is studied with computer simulations. The modes, phase space structures, and feasibility of bunch-by-bunch feedback system for coherent dipole motions with beam-beam interactions are discussed with four-beam scheme and beam-beam compensation.

## INTRODUCTION

For high luminosity at $e^+e^-$ collider, KEKB has achieved to be $10^{34}$ cm$^{-2}$s$^{-1}$ in 2003. When we try to achieve higher luminosity by a factor 10 or 100 than the present, non-conventional schemes such as space charge compensations and round beam scheme should be considered in the same way that a finite crossing angle at KEKB[1]. The finite crossing angle was guaranteed by computer simulations and has been operated successfully.

The luminosity of the storage ring is limited by the beam-beam interactions. Space charge compensation has been proposed to cure effects of the beam-beam interaction. One is a direct space charge compensation with four beams and the other is an indirect space charge compensation using secondary beams or plasmas(beam-beam compensation). Four-beam scheme was tested in DCI (Dispositif de Collisions dans l'Igloo) at the Laboratoire de l'Accélérateur Linéaire (Orsay, France) in 1975-1980[3] and a simulation study is reported[2]. Beam-beam compensation has been performed in Tevatron[4].

This paper reports results of strong-strong beam-beam simulations of space charge compensations. This simulation study includes radiation damping ($\tau_{x,y}$=40 msec) and bunch-by-bunch feedback system that has a similar performance at the present. The lattice is based on the Super B Factory as extremely high luminosity $e^+e^-$ collider for the future. The beam is flat-beam, beta functions are squeezed to be 3 mm at I.P as well as bunch length is 3 mm, and beam current more than 10 A should be stored.

## EIGENVALUE ANALYSIS

We consider beam-beam interactions in the one dimension for the simple case. The four-beam scheme is represented by beam 1:$e^-$, beam 2:$e^-$, beam 3:$e^+$, and beam 4:$e^+$. The beam 1 and the beam 3 collide the beam 2 and the beam 4. Revolution matrix includes the beam-beam interactions can be written by:

$$\tilde{R} = K_{34} \cdot K_{32} \cdot K_{14} \cdot K_{12} \cdot R, \quad (1)$$

where

$$R = \begin{pmatrix} c_1 & s_1 & 0 & 0 & 0 & 0 & 0 & 0 \\ -s_1 & c_1 & 0 & 0 & 0 & 0 & 0 & 0 \\ 0 & 0 & c_2 & s_2 & 0 & 0 & 0 & 0 \\ 0 & 0 & -s_2 & c_2 & 0 & 0 & 0 & 0 \\ 0 & 0 & 0 & 0 & c_3 & s_3 & 0 & 0 \\ 0 & 0 & 0 & 0 & -s_3 & c_3 & 0 & 0 \\ 0 & 0 & 0 & 0 & 0 & 0 & c_4 & s_4 \\ 0 & 0 & 0 & 0 & 0 & 0 & -s_4 & c_4 \end{pmatrix},$$

$c_i = \cos 2\pi v_i$, $s_i = \sin 2\pi v_i$, and $v_i (i=1,2,3,4)$ is lattice tunes. The beam-beam interaction matrix between beam 1 and beam 2 is represented by

$$K_{12} = \begin{pmatrix} 1 & 0 & 0 & 0 & \cdots & 0 \\ k_{12} & 1 & -k_{12} & 0 & \cdots & 0 \\ 0 & 0 & 1 & 0 & \cdots & 0 \\ -k_{12} & 0 & k_{12} & 1 & \cdots & 0 \\ \vdots & \vdots & \vdots & \vdots & \ddots & \vdots \\ 0 & 0 & 0 & 0 & \cdots & 1 \end{pmatrix},$$

where $k_{ij} = 2\pi \xi_{ij}$ and $\xi_{ij}$ is beam-beam parameter. The beam-beam interaction matrix, $K_{14}$, $K_{32}$, and $K_{34}$, are calculated by the similar way. The sign of $k_{ij}$ is different between the defocusing and the focusing case of the beam-beam interaction.

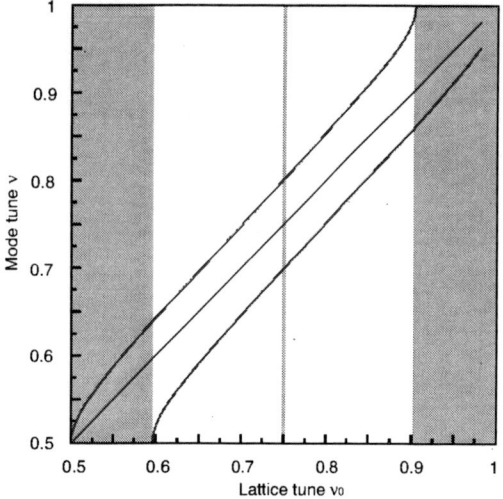

**FIGURE 1.** Mode tunes as a function of lattice tunes. Shaded region is unstable.

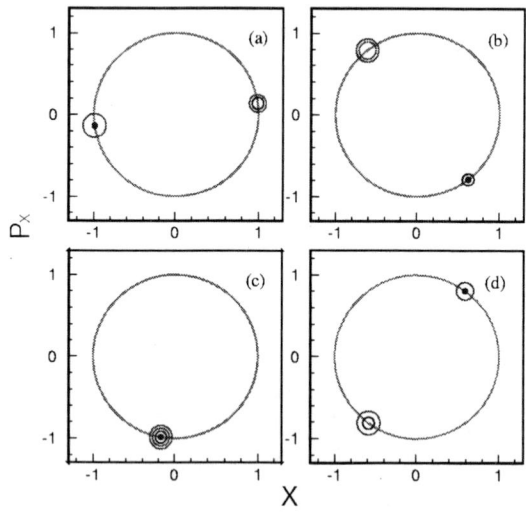

**FIGURE 2.** Eigenvectors corresponds to four modes. Circles from small to large size show beam 1 to beam 4, respectively. (a) negative tune shift, (b) positive tune shift, (c) and (d) no tune shift.

Four eigen modes are derived from the revolution matrix, $\tilde{R}$. Figure 1 shows mode tunes as a function of the lattice tune. There are a positive tune shift, a negative tune shift, and two no tune shifts. The beam-beam parameter, $\xi$, is 0.1. The bands with a width of the beam-beam tune shift at half integer and integer resonances are unstable region due to negative and positive tune shift. There is also unstable line at the 4th-order resonance. The eigen vectors corresponding to the eigenvalues are shown in Fig. 2.

# SIMULATION

Lattice used in this study is based on Super B Factory at KEK(SuperKEKB)[5]. The beam energy is asymmetric energy of 8 GeV(HER) and 3.5 GeV(LER). The HER is consists of beam 1:$e^-$ and beam 3:$e^+$ and the LER of beam 2:$e^-$ and beam 4:$e^+$. The machine parameters are as followings: $\beta_x^*/\beta_y^*$=30 cm/3 mm, emittance is 33 nm with 2% xy coupling, the number of particles for beam 1 and beam 3 is $5.14\times10^{10}$ per bunch, $11.75\times10^{10}$ per bunch for beam 2 and beam 4. The total beam current is 4.1 A per beam for HER and 9.4 A per beam for LER when the number of bunches is assumed to be 5000 per beam. All beams collide with head-on collision. The beam-beam parameter is $\xi_x/\xi_y$=0.1/0.07 in this case. In this study, each beam has the same betatron tunes.

Strong-strong beam-beam simulation code for four-beam scheme is used with modifications of strong-strong simulation for two-beam scheme which has been used to study beam-beam interactions at KEKB. We consider 4-dimensional phase space in collisions and 128×256 mesh in transverse plane. Each mesh size is 10 $\mu$m in horizontal and 0.3 $\mu$m in vertical. Number of macro-particles is $10^5$ in each beams to make the simulations reliable.

## Four-beam scheme

Several working points are tested to check stabilities of the colliding beams as shown in Fig. 3. Working points within beam-beam tune shifts from the half integer resonances and the 4th-order resonances such as $\nu_{x,y}=0.75$ are unstable region. When horizontal tune stays in the stable region, the beams are stable even though vertical tune is close to $\nu_y=0.75$. Coherent dipole motions and incoherent effects are observed in horizontal and vertical direction in the unstable region. The beam-centroid motion and beam size at I.P for $(\nu_x,\nu_y)=(0.77,0.72)$ are shown in Fig. 4. The beam-centroid motion and beam size are normalized by the initial beam size. The initial beam size is 100 $\mu$m in horizontal and 1.4 $\mu$m in vertical. The coherent dipole motion is occurred and is damped in the horizontal direction. Horizontal beam size becomes twice of the initial beam size rapidly, then the beam size is gradually increased. The growth of horizontal beam size is the result that seven islands appeared around the circle of the small amplitude move to larger amplitude. Coherent dipole motion also occurs and beam size is increased in the vertical direction.

Recently, bunch-by-bunch feedback system has been developed to make storage of high beam current with multi-bunches stable. Therefore, feasibility of bunch-by-bunch feedback system is investigated when coherent

**TABLE 1.** Luminosity at several working points. Feedback system is applied if it is necessary.

| $\nu_x$ | $\nu_y$ | FB | Horizontal | Vertical | $L \times 10^{30*}$ cm$^{-2}$s$^{-1}$ |
|---|---|---|---|---|---|
| 0.77 | 0.72 | off | unstable | unstable | < 2.0 |
| 0.77 | 0.72 | on | quads | stable | 4.0 |
| 0.72 | 0.65 | on | quads | stable | 10.5 |
| 0.64 | 0.72 | off | stable | stable | 7.0 |
| 0.515 | 0.55 | on | quads | stable | 7.9 |
| 0.64 | 0.62 | off | stable | stable | 7.8 |

* luminosity per a pair of colliding bunches

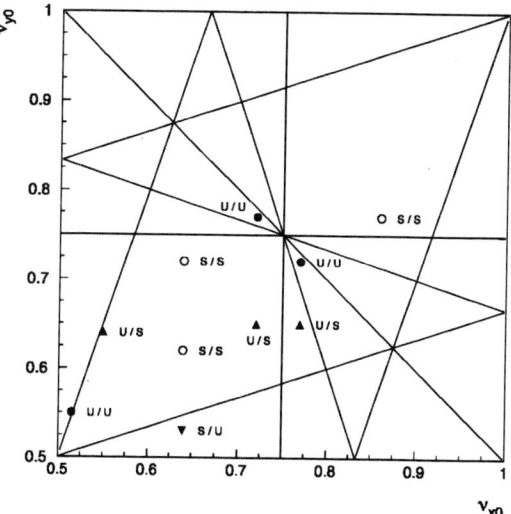

**FIGURE 3.** Tune diagram. Solid lines show 4th-order resonances. Closed circle shows unstable for both planes, triangle shows unstable in horizontal and stable in vertical, opposite triangle shows stable in horizontal and unstable in vertical, open circle shows stable in both planes. Beam-beam parameter is $\xi_x/\xi_y = 0.1/0.07$.

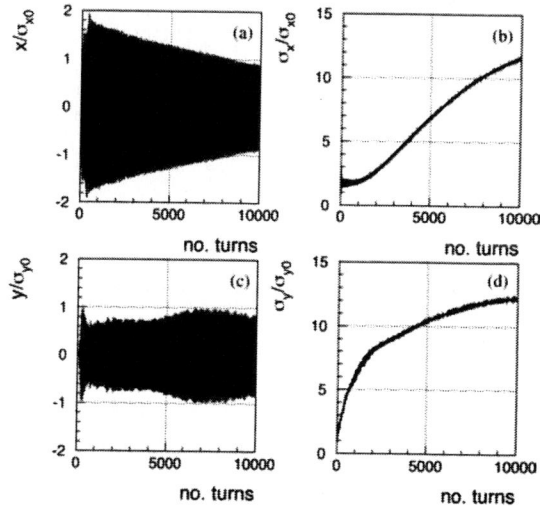

**FIGURE 4.** Normalized beam-centroid motion and beam size of beam 1 for $(\nu_x, \nu_y) = (0.77, 0.72)$. Similar behaviors are found in beam 2, beam 3, and beam 4.

dipole motion is occurred. We assume that 1 msec for the damping time of the transverse feedback system which is performed at the present machine. The beam-centroid motion and beam size at I.P for $(\nu_x, \nu_y) = (0.77, 0.72)$ with the feedback system are shown in Fig. 5. The coherent dipole motion can be damped with the feedback system after the beam size becomes large by a factor 2-5 of the initial beam size. Quadrupole motion is still left in the horizontal direction. Simultaneously, increase of the beam size can be suppressed with the damping of the coherent dipole motion. In the vertical direction, the beam size becomes flip-flop state between colliding beams after 4000 turns.

Figure 6 shows the beam-centroid motion and beam size at the working point, $\nu_x/\nu_y = 0.515/0.55$ with the feedback system. Quadrupole motion in horizontal occurs and is damped up to 3000 turns. Coherent dipole motion is observed in the both direction. The vertical dipole motion can be damped and the horizontal dipole motion can not be damped by the feedback system completely. Luminosity at several working points are listed in Table 1.

## Beam-beam compensation

We define beam 1:$e^-$ and beam 4:$e^+$ as circular beams and beam 2:$e^-$ and beam 3:$e^+$ as secondary beams provided from a linac collider. The beam 2 and the beam 3 are discarded after collision. Therefore, there is no coherent beam-beam effect at least multi-turn resonant type. The parameters of beams are same as those of the four-beam scheme. This is an ideal case since it is hard to realize this beam condition. The working point of beam 1 and beam 4 is the same tune in the horizontal and vertical direction, respectively. Stabilities of the beam-beam compensation are investigated at the working points near the 4th-order resonance and the half integer resonance. No coherent dipole motions in the horizontal and vertical directions occurs at the working point, $(\nu_x, \nu_y) = (0.77, 0.72)$. However, an octupole motion occurs in the horizontal direction. Amplitude of the octupole motion does not increase. Figure 7 shows the phase space plots of the beam 1 and beam 4 in the horizontal direction. The position $X$ and the momentum $P_X$ in the normal coordinate are defined as $X = x/\sqrt{\beta_x}$ and $P_X = (\alpha x + \beta p_x)/\sqrt{\beta}$, where $\beta$ and $\alpha$ are Twiss parameters. The vertical beam size is increased by a factor 7 of the initial beam size and saturated. The luminosity

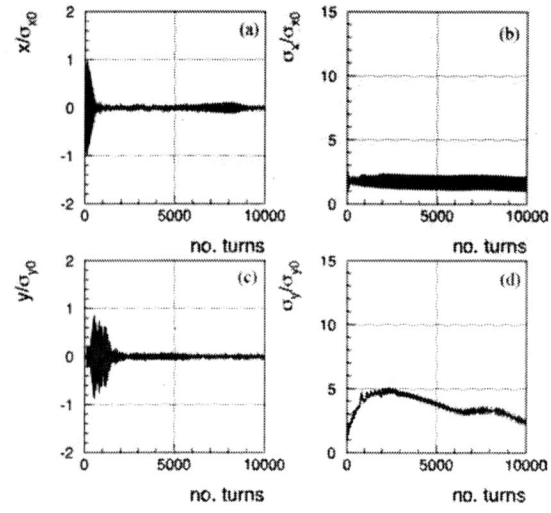

**FIGURE 5.** Normalized beam-centroid motion and beam size of beam 1 with feedback system for $(v_x, v_y) = (0.77, 0.72)$.

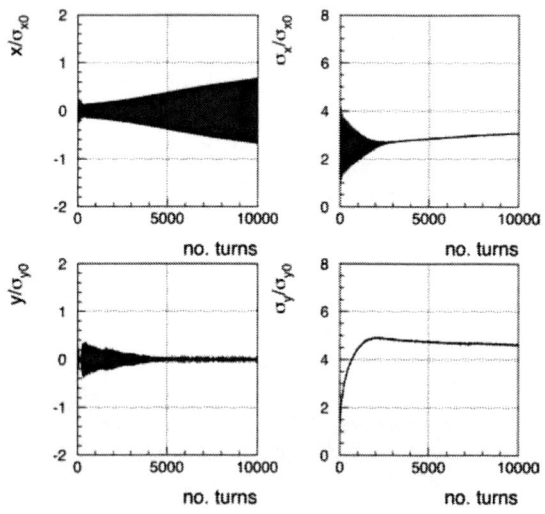

**FIGURE 6.** Normalized beam-centroid motion and beam size of beam 1 with feedback system for $(v_x, v_y) = (0.515, 0.55)$.

per a pair of colliding bunches is expected to be $3.4 \times 10^{30}$ cm$^{-2}$s$^{-1}$ which is smaller than that of four-beam scheme. On the other hand, the luminosity of $16.7 \times 10^{30}$ cm$^{-2}$s$^{-1}$ per a pair of colliding bunches is achieved at $(v_x, v_y) = (0.72, 0.65)$. This value is larger than the four-beam scheme with feedback system at the same tune. For the near half integer resonance, $(v_x, v_y) = (0.515, 0.55)$, quadrupole motion is observed in the horizontal direction and beam size blow up in the vertical direction. This behavior is similar to the four-beam scheme.

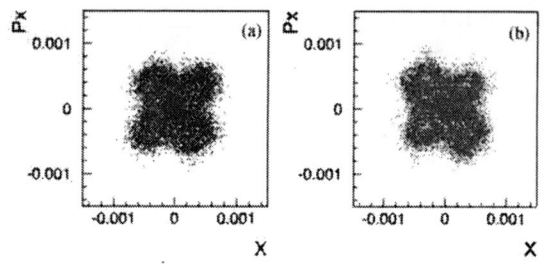

**FIGURE 7.** Phase space plots in beam-beam compensation, (a) beam 1 and (b) beam 4 for $(v_x, v_y) = (0.77, 0.72)$.

## SUMMARY

We perform strong-strong simulations in the four-beam scheme and the beam-beam compensation. Stabilities are investigated for several working points, especially near half integer and 4th-order resonances. Coherent dipole motions and incoherent effects are observed near the 4th-order resonance and half integer resonance in four-beam scheme. In the case of 4th-order resonance, the feedback system can work to damp the coherent dipole motion and suppress the incoherent effects as the result. Coherent dipole motions do not occur in the beam-beam compensation. However, quadrupole motions and octupole motions are observed. Space charge cancellation by means of the four-beam scheme and the beam-beam compensation, the stable region is small in principle. In addition, machine errors are not included in the simulations. When we try higher beam-beam parameters, the stable region becomes much smaller. The important point is how to cure higher-order motions even though coherent dipole motions can be controlled by the feedback system and to achieve higher beam-beam parameter in the stable.

## ACKNOWLEDGMENTS

We would like to thank the KEK Computing Research Center(CRC) for the supercomputer system.

## REFERENCES

1. KEKB Design Report, KEK Report 95-7, 1995.
2. B. Podobedov, R.H. Siemann (SLAC), Phys. Rev. E52, 1995.
3. J.Le Duff et al., Proc. of XIth International Conference on High Energy Accelerators, CERN, Geneve, July 7-11, 1980.
4. V.Shiltsev et al., Batavia 2001, Beam-beam effects in circular colliders 52-57.
5. EoI, January 2002, see http://belle.kek.jp/~yamauchi/EoI.ps.

# Beam-Beam Simulations for Lepton Machines

Joseph T. Rogers

*Laboratory for Elementary-Particle Physics*
*Cornell University, Ithaca, NY 14853 USA*

**Abstract.** We survey the range of techniques for numerical simulation of the beam-beam interaction in circular colliders in which synchrotron radiation is present. These include techniques used in weak-strong, quasi-strong-strong, and strong-strong simulations. Self-consistent strong-strong simulations usually use macroparticle-in-cell methods with a variety of refinements to increase the computational speed. Other approaches include macroparticle sampling methods and numerical solution of the Vlasov equation. We also describe and compare the existing beam-beam simulation codes for circular lepton colliders.

## INTRODUCTION

The luminosity of circular $e^+e^-$ colliders is usually limited by the beam-beam effect. There are two observed types of beam-beam limits when the bunch current is increased [1, 2]. In the first beam-beam limit the beam-beam parameter $\xi_V$ or $\xi_H$ reaches a limit as the core size of the beam starts to grow. In the second beam-beam limit a halo forms and beam loss increases. Circular $e^+e^-$ colliders may also be subject to the long-range beam-beam interaction where they pass each other in the same chamber.

There are two possible classes of causes for these beam-beam limits: incoherent (single-particle) and coherent particle motion. Particle loss due to incoherent motion of a particle subject to nonlinear fields is familiar in accelerators, and could be a cause of the beam-beam limits. There is also evidence that coherent modes of motion can cause a beam-beam limit. The charge neutralization experiment at DCI, in which an $e^+$ and an $e^-$ beam collided with another set of $e^+$ and $e^-$ beams, showed no increase in the limiting value of the beam-beam parameter $\xi$ [3]. This surprising result has been explained theoretically and through simulation [4]. At the beam-beam limit the coherent modes of the beams (which result in a separation of the positive and negative charge) become unstable. The implication is that the beam-beam limit might be due to coherent motion even when charge neutralization is not used. Other simulation studies have indicated the presence of coherent beam-beam instabilities [5, 6].

At this time we do not know whether the first beam-beam limit is caused by coherent or incoherent motion; neither do we understand the cause of the second beam-beam limit. We hope that simulation studies can help determine these causes and replace the question marks in Figure 1.

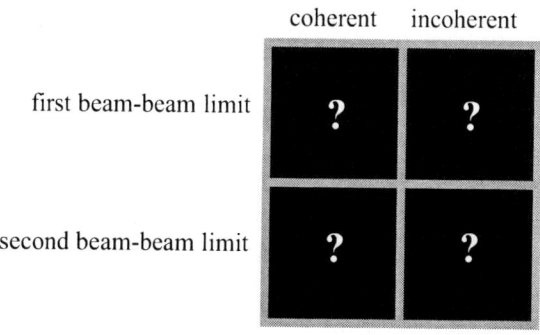

**FIGURE 1.** At present the causes of the first and second beam-beam limits are not known.

## SIMULATION METHODS

There are two main classes of simulation methods, with many possible variations. In "weak-strong" simulations, one beam ("strong") is modeled as a fixed charge distribution. It serves as the source of the electromagnetic field that perturbs the particle distribution in the other ("weak") beam. Weak-strong simulations can simulate only incoherent effects.

In "strong-strong" simulations both beams serve as the source of the field that perturbs the particle distribution in the other beam. This type of simulation is self-consistent. Strong-strong simulations can simulate both coherent and incoherent effects.

## Weak-Strong Methods

Weak-strong simulations have the advantage that the electromagnetic field calculation need only be done once, because the source distribution does not change from turn to turn. As a result, these codes are fast. Because weak-strong codes are fast, many macroparticles can be tracked for many turns. Thus these codes are particularly useful for halo and lifetime calculations, *i.e.*, the second beam-beam limit.

Weak-strong simulations have the disadvantage that they are sensitive only to incoherent effects, and are not self-consistent.

Computation speed is a concern. Even though weak-strong codes are relatively fast, it is difficult to calculate lifetimes of the order of hours. In practice, one cannot track the actual number of particles in the beam, so one represents them by a set of macroparticles that sample the phase space. In a typical $e^+e^-$ collider a beam-beam lifetime of one hour corresponds to a particle lifetime of the order of $10^9$ turns. Tracking for several times $10^9$ particle-turns can be too slow (although it may be practical in the near future with faster processors). Some codes resort to clever tricks involving variable particle number. Several examples are described in the next section.

### *Examples of Weak-Strong Codes and Methods*

The code BBC [7] is a widely used 6-D symplectic weak-strong code. It implements crossing-angle collisions by making a transverse Lorentz boost of the beams to a frame in which the beams collide head-on. After the beam-beam calculation the beams are given an inverse boost to return to the lab frame.

Halo distributions were investigated with a limited number of macroparticles by the use of a "leap-frog" method [8]. The method involves establishing a boundary in horizontal-vertical amplitude space in which, for example, 90% of the initial distribution of $n$ particles lies. The $n$ particles are then tracked for one damping time. During the tracking, (1) at every few turns, a particle outside the boundary is randomly selected and its coordinates saved, and (2) all coordinates of particles passing outward across the boundary are saved. The coordinates from (1) are used to generate a set of $n$ new particles in the region outside the boundary. These particles are then tracked. When a particle passes inward through the boundary, it is replaced with a particle with coordinates from (2). This method is extended to multiple boundaries, allowing the calculation of halo distributions far from the beam core.

The beam-beam interaction in the presence of rare scattering processes such as beam-gas scattering has been modeled [9]. In this method the distribution has macroparticles $i = 1, 2, ..., n$. Each macroparticle $i$ represents $N_i$ particles. The particles undergo scattering randomly. When a particle in macroparticle $i$ scatters, a new $(n+1)$-th macroparticle is created with $N_{n+1} = 1$ particle, leaving $N_i - 1$ particles in macroparticle $i$.

## Strong-Strong Methods

Strong-strong simulations have the advantage that they are sensitive to both coherent and incoherent effects. They are self-consistent, and are therefore particularly useful for calculating the first beam-beam limit, in which the core sizes of both beams grow.

Strong-strong simulations have the disadvantage that the electromagnetic calculation must be done repeatedly. These codes are much slower than weak-strong codes. Because they are relatively slow, fewer macroparticle-turns can be tracked, and strong-strong codes are of limited usefulness for halo and lifetime calculations.

Computation speed is a serious hurdle. Strong-strong simulations require clever tricks to track the beam for several radiation damping times. Some of these methods will be described below.

### *Types of Strong-Strong Simulations*

There are several types of strong-strong simulations.

In the **particle-particle method** the pairwise interaction between particles is calculated. The computation time scales as $n^2$, where $n$ is the number of macroparticles. This method is too slow to be practical, but can serve as a check on other calculation methods.

In **dynamic Gaussian models**, the macroparticle distribution is fit by a Gaussian charge distribution, which serves as the source of the field that perturbs the other beam. The 1$^{st}$ moments (rigid-Gaussian model) or 1$^{st}$ and 2$^{nd}$ moments (soft-Gaussian model) are free to evolve.

**Particle-in-cell (PIC) simulations** also track macroparticles moving under the influence of the beam-beam force. The electromagnetic field is calculated on a discrete grid after assigning the macroparticle charge to grid points. The field at the location of the counter-rotating macroparticles is determined by interpolation from the grid.

In **quasi-strong-strong simulations** the roles of the weak and strong beams are exchanged periodically. This type of simulation is self-consistent on long time scales, and has the speed advantages of a weak-strong simulation.

In the **fast multipole method**, the potential due to distant source particles is calculated as a multipole expansion.

**Numerical Vlasov equation solvers** evolve the phase space density directly without tracking particles.

We'll examine examples of these methods in the sections below.

*Examples of PIC Codes and Methods*

Particle-in-cell methods are perhaps the most widely used for strong-strong codes at present. In PIC codes the charge of the macroparticle must be assigned to points on the grid that is used to compute the fields. One such charge assignment, the "4-point cloud-in-cell" is shown in Figure 2. Other possibilities for charge assignment include assigning all the charge to the nearest grid point (NGP), and a 5-point or 9-point charge assignment.

After the field is calculated on the grid by some method, the field at the position of the perturbed particle must be determined by interpolation between grid points. The interpolation must use the same weighting as the charge assignment to conserve the transverse momentum of the beams. Figure 3 shows a 4-point cloud-in-cell interpolation of the fields.

The field on the grid can be determined using a fast Poisson solver [5]. This author solved the Poisson equation on a Cartesian grid by the Fast Forier Transform and Cyclic Reduction (FACR) method. Coherent phenomena were seen in the simulation results, including the flip-flop instability (a stationary equilibrium with unequal beam sizes) and period-$n$ anticorrelated beam size oscillations.

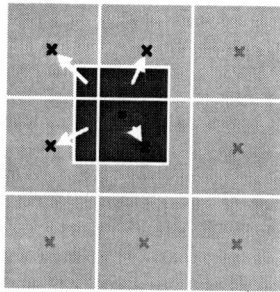

✶ = grid point

**FIGURE 2.** 4-point cloud-in-cell assignment of macroparticle charge to the grid. The fraction of the macroparticle charge that is assigned to a grid point is equal to the fraction of the square (centered on the macroparticle) that lies within a grid cell.

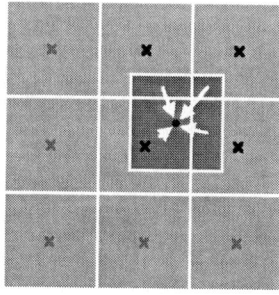

**FIGURE 3.** 4-point cloud-in-cell interpolation of the field from the grid points to the position of the perturbed particle. Again, the weighting is proportional to the fraction of the square (centered on the macroparticle) that lies within a grid cell.

Another PIC code, ODYSSEUS [10], uses adaptive methods to increase its speed so that longitudinal dynamics can be incorporated. Particles are sorted into longitudinal slices. The grid periodically adapts its size and aspect ratio to cover the beams (except the transverse tails). The fields experienced by the transverse tails are generated from a soft-Gaussian fit to the source beam. Longitudinal tails are treated with a weak-strong method. The convolution of the charge density with the Green's function is done in Fourier coefficient space. A sharpening function is included in the convolution to counter the "low-pass" effect of the charge assignment. ODYSSEUS is able to predict the luminosity of a well-tuned and well-understood machine to within 10%. Self-excited coherent motion is seen in the simulation results, including head-tail modes. Recently the crossing angle has been implemented by Hirata's method of a transverse boost to a frame where the collisions are head-on, and nonlinear transport through the arcs has been included.

In using a Green's function approach to calculating the fields, one must do a convolution of the Green's function with the charge density on the grid. A brute-force convolution has a computation time that scales as $N_g^2$, where $N_g$ is the number of grid points. However, cyclic convolution can be done as a multiplication of Fourier series coefficients. An economical approach is to calculate the fast Fourier transform (FFT) of the Green's function and charge density, multiply them, and then calculate the inverse FFT of their product. The FFT and inverse FFT dominate the computation time, but the time required to calculate the FFT scales as $N_g \log N_g$, which is faster than the brute-force convolution. A cyclic convolution requires a grid that is four times larger than the area occupied by the charge, as shown in Figure 4.

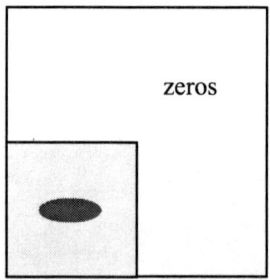

**FIGURE 4.** Cyclic convolution requires a grid that is four times larger than the area occupied by charge.

A Poisson solver with a grid much smaller than the beam pipe was developed [11] to reduce the number of grid points. Before solving the Poisson equation, the potential on the grid boundary is determined by a Green's function method. The Poisson solver uses FFT and cyclic reduction (FACR). This code runs on 32 parallel processors at NERSC. It has been recently extended to include longitudinal dynamics, with linear interpolation of the fields between adjacent slices. The simulated beam-beam limit I PEP-II is in excellent agreement (at the 10 to 15% level) with observations.

A code using a Green's function method with longitudinal dynamics is described in [12]. The convolution of the charge density with the Green's function is done in Fourier coefficient space. The potential is interpolated between longitudinal slices, so few slices are needed. A crossing angle is implemented by a boost to a frame with head-on collisions. The simulated luminosity in KEKB is in excellent agreement (at the 15% level) with observations.

The standard PIC method, in which the field domain is identical to the particle domain, is inefficient for separated beams (i.e., for calculation of the long range beam-beam interaction). Most of the grid is empty of particles. A shifted Green's function method has been developed [13] to overcome this inefficiency. In this method the field domain is made different from the particle domain by replacing the Green's function $G(\mathbf{r},\bar{\mathbf{r}}) = -\ln|\mathbf{r}-\bar{\mathbf{r}}|$ by a shifted Green's function $G_s(\mathbf{r},\bar{\mathbf{r}}) = -\ln|\mathbf{r}_c+\mathbf{r}-\bar{\mathbf{r}}|$, where the geometry is defined in Figure 5.

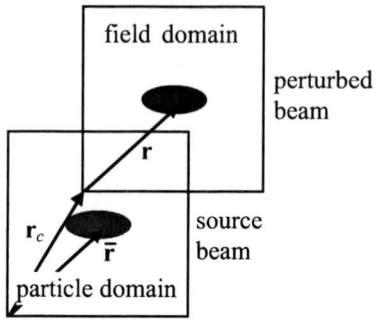

**FIGURE 5.** Geometry of the shifted Green's function method.

*Examples of Other Strong-Strong Codes and Methods*

A quasi-strong-strong code was developed [14] to incorporate long-term self-consistency, while maintaining the speed advantages of the weak-strong approach. The method is identical to the standard weak-strong simulation, but after a set number of collisions, the 1st and 2nd moments of the weak beam are calculated. The weak beam is replaced by a strong Gaussian beam with the same moments, and the strong beam is replaced by a weak beam. This exchange of weak and strong beams is performed periodically.

The Fast Multipole Method has not yet been applied to $e^+e^-$ colliders. In this method the force due to nearby source particles is calculated directly, and the potential due to distant source particles is calculated as a multipole expansion. The advantage of the Fast Multipole Method is that it is efficient for separated beams. Its disadvantage is that there are occasional exaggerated deflections due to close encounters of macroparticles. A Hybrid Fast

Multipole method has been applied to hadron colliders [15]. It uses both a grid and a multipole expansion of the fields.

Numerical Vlasov equation solvers evolve the phase space density without reference to particles. This technique has rarely been applied to $e^+e^-$ colliders. The advantage of this method is that it avoids "noise" due to poor sampling of phase space by macroparticles. A disadvantage is that the grid exists in phase space, and the number of grid points for a 4-D phase space may put too much demand on memory. Also, the phase space density is not automatically positive. A Vlasov-Fokker-Planck code for 2-D phase space with synchrotron radiation excitation and damping [16] demonstrated the existence of an equilibrium state. It conserves the integral of phase space density to 1 part in $10^5$ over several damping times.

## CONCLUSIONS

Weak-strong simulations are relatively fast and have provided useful guidance for accelerator design and the choice of tunes. Strong-strong simulations have gained real predictive power and are now fast enough to incorporate longitudinal dynamics. They regularly predict luminosities to ~10% from first principles. Extending a Fast Multipole Method code to include synchrotron radiation effects would be straightforward. FMM may be competitive with the shifted Green's function method. A number of 6-D strong-strong codes now exist, and should be systematically compared with each other.

## ACKNOWLEDGMENTS

The author wishes to thank E. Anderson, Y. Cai, J. Ellison, J. Irwin, S. Krishnagopal, K. Ohmi, J. Seeman, R. Siemann, and J. Urban for past discussions on beam-beam codes and assistance in producing this paper. The author gratefully acknowledges the support of the National Science Foundation.

## REFERENCES

1. R.H. Siemann, "The Beam-Beam Interaction in $e^+e^-$ Storage Rings", SLAC-PUB-6073 (1993).

2. J.T. Seeman, "Beam-beam interaction: luminosity, tails, and noise", SLAC-PUB-3182 (1983).

3. J. LeDuff, et al., Proc. 11$^{th}$ International Conference on High Energy Accelerators (1980) 707.

4. B. Podobedov and R. Siemann, "Coherent beam-beam interaction with four colliding beams", Phys. Rev. E **52** (1995) 3066.

5. S. Krishnagopal, "Luminosity-limiting coherent phenomena in electron-positron colliders", Phys. Rev. Lett. **76** (1996) 235.

6. J.T. Rogers, M.A. Palmer, A.P. Romano, C.R. Jones, "Beam-Beam Simulation Studies of CESR-c and Observations in CESR", Proc. Workshop on Beam-Beam Effects in Circular Colliders, FERMILAB-Conf-01/390-T (2001) 27.

7. K. Hirata, CERN SL Note 97-52 AP.

8. J. Irwin, Proc. 3$^{rd}$ Advanced ICFA Beam Dynamics Workshop (1989) 123 (SLAC-PUB-5743, 1992).

9. E.-S. Kim and K. Hirata, "Simulation of tail distributions due to random processes and beam-beam interaction in KEK-B", Proc. 1997 Part. Accel. Conf. (1997) 384.

10. E.B. Anderson, T.I. Banks, and J.T. Rogers, Proc. 1999 Part. Accel. Conf. (1999) 1686.

11. Y. Cai, A.W. Chao, S.I. Tzenov, and T. Tajima, "Simulation of the beam-beam effects in $e^+e^-$ storage rings with a method of reduced region of mesh", Phys. Rev. ST-AB **4**, 1 (2001) 011001-1.

12. K. Ohmi, "Simulation of beam-beam effects in a circular e+e- collider", Phys. Rev. E **62** (2000) 7287.

13. J. Qiang, M.A. Furman, and R.D. Ryne "Strong-strong beam-beam simulation using a Green function approach", Phys. Rev. ST-AB **5** (2002) 104402-1.

14. K. Ohmi, K. Hirata, and N. Toge, "Quasi-strong-strong simulations for beam-beam interactions in KEKB", Proc. 5$^{th}$ European Part. Accel. Conf., (1997) 1164.

15. W. Herr, M.P. Zorzano, and F. Jomes, "Hybrid fast multipole method applied to beam-beam collisions in the strong-strong regime", Phys. Rev. ST-AB 4 (2001) 054402-1.

16. R.L. Warnock and J.A. Ellison, Proc. 2$^{nd}$ ICFA Workshop on High Brightness Beams" (2000).

# Parallel Strong-Strong/Strong-Weak Simulations of Beam-Beam Interaction in Hadron Accelerators

Ji Qiang*, Miguel Furman*, Robert D. Ryne*, Wolfram Fischer[†], Tanaji Sen** and Meiqin Xiao**

*Lawrence Berkeley National Laboratory, Berkeley, CA 94720*
[†]*Brookhaven National Laboratory, Upton, NY 11973*
***Fermi National Laboratory, Batavia, IL 60510*

**Abstract.** In this paper, we present the results of using a parallel computational tool, BeamBeam3D, developed at Lawrence Berkeley National Laboratory, for strong-strong/strong-weak modeling of the beam-beam effects in three hadron accelerators: RHIC, Tevatron and LHC. This tool calculates self-consistently the electromagnetic beam-beam forces for arbitrary distributions and separation during each collision when a strong-strong beam-beam interaction model is used. When a strong-weak model is used, the code has the option of using a Gaussian approximation for the strong beam. Using the strong-strong model, we have studied the effect of time modulated offset beam-beam interaction on the emittance growth in the RHIC and LHC. We observed an extra 0.04% emittance growth after 300,000 turns in the RHIC where the time-averaged beam-beam offset is one transverse rms beam size and the modulation frequency is 10 Hz. There is no significant additional emittance growth in the LHC after one million turns where the time-averaged offset is zero. Using the strong-weak model, we have also studied the antiproton lifetime subject to 72 long range beam-beam interactions at 150 GeV injection energy in the Tevatron. The simulation shows a qualitative agreement with the experimental observation of the smaller antiproton emittance having a longer lifetime.

## INTRODUCTION

The beam-beam interaction puts a strong limit on the luminosity of the high energy hadron colliders. For example, in the Tevatron Run II experiment, long-range beam-beam forces significantly reduce the antiproton lifetime and is one of the major factors preventing the achievement of designed luminosity. To study the beam-beam interaction, an important approach is to use self-consistent macroparticle simulation. However, in the hadron accelerators, radiation damping is very weak for hadron particles. It requires to track the particles for many thousand turns to study the long-term emittance growth and particle losses. Using a parallel beam-beam computational tool will significantly reduce the study time. Meanwhile, it also provides the capability to include more complex physical process and to do the simulations with higher numerical accuracy.

## COMPUTATIONAL MODEL

In the computational beam-beam model, we have used six dimensional phase space coordinates $(x, p_x, y, p_y, \Delta z, \Delta p_z/p_0)$ to describe the particle motion in the accelerator. Here, $p_{x,y}$ is the transverse momentum normalized by the total momentum of a reference particle ($p_0 = E_0/c$), $\Delta z = s - ct(s)$ with $c$ the speed of light, and $\Delta p_z = |p| - p_0$. To calculate the electromagnetic force from the beam-beam interaction, we have used a multiple slice model. In this model, each beam bunch is divided into a number of slices along the longitudinal direction in the moving frame. Each slice contains nearly the same number of particles at different longitudinal locations $\Delta z$. The collision point between two opposite slices $i$ and $j$ is determined by

$$s_c = \frac{1}{2}(\Delta z_i^+ - \Delta z_j^-) \qquad (1)$$

The transverse coordinates of the particles at the collision point are given by

$$x^c = x + s_c p_x \qquad (2)$$
$$y^c = y + s_c p_y \qquad (3)$$

The slopes of the particles are updated using the beam-beam electromagnetic forces at the collision point following

$$p_{x_{new}} = p_x + \Delta p_x \qquad (4)$$
$$p_{y_{new}} = p_y + \Delta p_y, \qquad (5)$$

where

$$\Delta p_{x_2} = \frac{2q_1q_2N_1}{\gamma_2 4\pi\varepsilon_0 m_2 c^2} E_{x_1} \quad (6)$$

$$\Delta p_{y_2} = \frac{2q_1q_2N_1}{\gamma_2 4\pi\varepsilon_0 m_2 c^2} E_{y_1}. \quad (7)$$

In the above equations, the subscripts 1 and 2 pertain to each of the two beams, the corresponding equations for the other beam are obtained from the above by the exchange 1<->2, $\gamma = 1/\sqrt{1-\beta^2}$, $\beta_i = v_i/c$, $i = x,y,z$, $c$ is the speed of light, $\varepsilon_0$ is the vacuum permittivity, $q$ is the charge of the particle, $m$ is the rest mass of particle, $N$ is the number of particles in a bunch, and $E_x$ and $E_y$ are the transverse electric fields generated by the opposite moving beam. After the collision, the particles of each slice drift back to their original locations according to

$$x = x^c - s_c \, p_{x_{new}} \quad (8)$$
$$y = y^c - s_c \, p_{y_{new}} \quad (9)$$

The electric fields generated by the opposite moving beam can be obtained from the solution of Poisson's equation. The solution of Poisson's equation can be written as

$$\phi(x,y) = \int G(x,\bar{x},y,\bar{y}) \rho(\bar{x},\bar{y}) \, d\bar{x}d\bar{y} \quad (10)$$

where $G$ is the Green's function and $\rho$ is the charge density. For the case of transverse open boundary conditions, the Green's function is given by:

$$G(x,\bar{x},y,\bar{y}) = -\frac{1}{2}\ln((x-\bar{x})^2 + (y-\bar{y})^2) \quad (11)$$

The convolution for $\phi$ in Eq. 10 can be computed efficiently using an FFT in the doubled computational domain as described by Hockney and Eastwood. [1].

In the FFT-based algorithm, the particle domain and the electric field domain are contained in the same computational domain. Here, the particle domain is the configuration space containing the charged particles, and the field domain is the space where the electric field is generated by the charged particles. In the beam-beam interaction, the two opposite moving beams might not overlap with each other. For example, in the long-range interaction, the two colliding beams could be separated by more than several $\sigma$, where $\sigma$ is the rms size of the beam. Thus the field domain where the electric field is generated by one beam can be different from the particle domain containing the beam. In the beam-beam simulation, the origin of the field domain can be at an arbitrary location and varies from turn to turn. To apply Hockney's algorithm directly will require the computational domain to contain both the particle domain and the field domain, i.e. both beams. Since there is a large empty space between two beams, containing both beams in one computational domain will result in a poor spatial resolution of the beams. This is also computationally inefficient because the electric fields in the empty space between two beams are not used.

To avoid this problem, we have defined a *shifted Green function* as

$$G_s(x,\bar{x},y,\bar{y}) = -\frac{1}{2}\ln((x_c+x-\bar{x})^2 + (y_c+y-\bar{y})^2) \quad (12)$$

where $x_c$ and $y_c$ are the center coordinates of the field domain. The electric potential in the field domain is written as

$$\phi(x+x_c,y+y_c) = \int G_s(x,\bar{x},y,\bar{y}) \rho(\bar{x},\bar{y}) \, d\bar{x}d\bar{y}. \quad (13)$$

Using the shifted Green function, the center of the field domain is shifted to the center of the particle domain. The range of $x$ and $y$ cover both the particle domain and the field domain in one computational domain. The FFT can be used to calculate the convolution in Eq. 10 using the new Green function. A more detailed discussion of the shifted Green function method can be found in [2].

To summarize, using the shifted Green function:

- avoids the requirement that the particle domain and the field domain be contained in one big computational domain,
- leads to better numerical resolution for the charge densities and the resulting electric fields than the conventional method, because the empty space between the beams is not included in the calculation,
- is far more efficient, in terms of computational effort and storage, than the traditional approach of gridding the entire problem domain.

When the strong-weak beam-beam model is used, the program has the option to use the electric field from the numerical solution of Poisson's equation or from the Gaussian approximation of the strong beam. In the latter case, the electric field is calculated following a Gaussian code [3].

The effects of external fields can be represented, in the small-amplitude approximation, by a linear transfer map between collision points. A one-turn map is included to take into account the linear machine chromaticity effect. The effects of radiation damping and quantum excitation are represented using a localized stochastic map.

## STRONG-STRONG SIMULATIONS OF BEAM-BEAM INTERACTION IN RHIC AND LHC

In the hadron colliders, the closed orbit of each bunch at interaction point can be perturbed due to the vibration of focusing magnets or due to parasitic beam-beam

**TABLE 1.** RHIC Physical Parameters for the Beam-Beam Simulations

| | |
|---|---|
| beam energy (GeV) | 23.4 |
| protons per bunch | $8.4 \times 10^{10}$ |
| $\beta^*$ (m) | 3.0 |
| RMS spot size at the IP (mm) | 0.629 |
| betatron tunes ($v_x, v_y$) | (0.22, 0.23) |
| chromaticity ($q'_x, q'_y$) | (2, 2) |
| synchrotron tune $v_z$ | 3.7e-4 |
| RMS bunch length (m) | 3.6 |
| momentum spread | 1.6e-3 |
| offset (sigma) | 1 |
| oscillation frequency (Hz) | 10 |

**TABLE 2.** LHC nominal beam-beam parameters

| | |
|---|---|
| beam energy (TeV) | 7 |
| protons per bunch | $1.05 \times 10^{11}$ |
| $\beta^*$ (m) | 0.5 |
| RMS spot size at the IP ($\mu m$) | 15.9 |
| betatron tunes ($v_x, v_y$) | (0.31, 0.32) |
| synchrotron tune $v_z$ | 0.0021 |
| RMS bunch length (m) | 0.077 |
| momentum spread | 1.11e-4 |

interaction. This results in the bunches colliding at the interaction point with a small transverse offset. For example, in RHIC, there exists triplet vibration with frequency close to 10 Hz, which leads to offset of 5 − 10% of one $\sigma$ at the interaction point [4]. For practical purposes, it is important to know if offsets (static and modulated) cause additional emittance growth. To study this effect, we have carried out a strong-strong beam-beam simulation of two proton beams colliding in RHIC. The major physical parameters used in the simulation are given in Table 1. Here, we have chosen an offset of one $\sigma$ from the closed orbit in order to maximize its effect since the beam-beam force is strongest around one $\sigma$ for a Gaussian density distribution. The initial distribution was Gaussian but was allowed to evolve freely. We first studied the proton emittance growth in RHIC with static offset beam-beam collision. Fig. 1 shows a comparison of the emittance evolution (averaged over horizontal and vertical plane) with and without offset beam-beam collision. We see that emittance growth after 300,000 turns

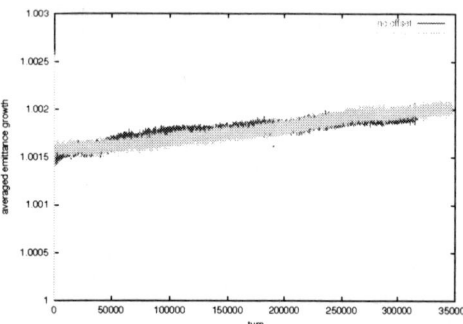

**FIGURE 1.** Emittance evolution (averaged over horizontal and vertical plane) with and without a one $\sigma$ offset beam-beam collision in RHIC.

is about the same with and without the offset. This suggests that the static beam-beam offset collision will not cause an extra emittance growth. The averaged emittance has grown by about 0.05% after 300,000 turns. This emittance growth is driven by the nonlinear electromag-

netic forces during the beam-beam interaction. Next, we studied the emittance growth with time modulated offset beam-beam collision. Here, we have assumed 10% oscillation amplitude around one $\sigma$ time-averaged offset in the horizontal direction. The oscillation frequency is set as 10 Hz to emulate the triplet vibration in RHIC. Fig. 2 shows a comparison of the averaged emittance evolution with and without offset beam-beam collision. With the time modulated offset beam-beam collision, the

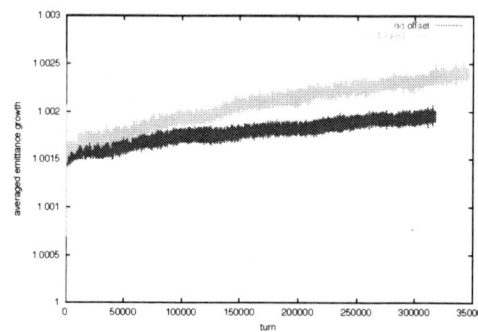

**FIGURE 2.** Emittance evolution (averaged over horizontal and vertical plane) with and without a time-modulated offset beam-beam collision in RHIC.

averaged emittance growth is about 0.09% after 300,000 turns. There is about 0.04% extra emittance growth due to the effect of time modulated beam-beam collision. In this simulation for RHIC, we have assumed that time-averaged beam-beam offset is one $\sigma$ to maximize the effect from time modulated offset beam-beam collision. In LHC, it is estimated that the time-averaged beam-beam offset is about zero but with about $0.1\sigma$ bunch-to-bunch closed orbit variation due to the parasitic beam-beam interaction [5]. We have done a strong-strong simulation using a set of nominal LHC physical parameters with $0.1\sigma$ oscillation amplitude and 1000 turn oscillation period. The nominal beam-beam parameters of the LHC are given in Table 2. Fig. 3 shows the emittance growth with and without the time modulated beam-beam collision. We see that there is about 0.05% emittance growth after one million turns with the nominal head-on collision. Even though the final emittance from the time modulated beam-beam collision is slightly higher than that without

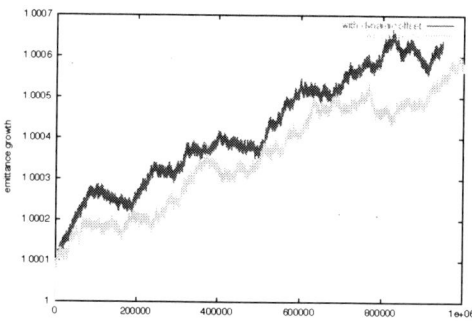

**FIGURE 3.** Emittance evolution (averaged over horizontal and vertical plane) with and without offset beam-beam collision in LHC.

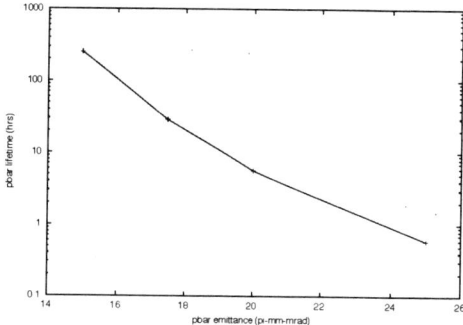

**FIGURE 4.** Antiproton lifetime as a function of antiproton emittance at 150 GeV Tevatron.

offset, the emittance growth rate is about the same. All the above simulations have been done using one million particles for each beam, 128 × 128 mesh points, and single slice model.

## STRONG-WEAK SIMULATION OF LONG RANGE BEAM-BEAM INTERACTION IN TEVATRON

We have assumed a strong-weak beam-beam interaction model in the calculation of the antiproton lifetime since the antiproton intensity is much smaller than the proton intensity (typically a factor of 10). There are 72 long range beam-beam interaction points at the 150 GeV injection stage of the Tevatron. The linear transfer maps between collision points for antiproton particles are obtained from the MAD simulation of the Tevatron lattice. At the collision point, every antiproton receives a kick from the beam-beam force generated by the proton beam. The beam-beam force is calculated assuming a Gaussian distribution for the proton beam. After each turn, a linear chromaticity map and a random diffusion map are applied to all antiprotons. The random kick was added to simulate the effect of gas scattering which has a sizable influence on the emittance growth at injection. The simulations are run for 100,000 turns using one million particles. In the simulation, we have assumed an "aperture" size of $3.25\sigma$, where the $\sigma$ is the horizontal or vertical rms size at each collision point. The aperture is such that a particle is lost with a unit probability when it hits it. The antiproton lifetime $\tau$ is estimated from fitting the antiprotron intensity with function $I_0 \exp(-t/\tau)$ using a least square method. Here, $I_0$ is the initial antiproton intensity.

Fig. 4 shows the antiproton lifetime as a function of the initial antiproton emittance. With a factor 2 increase of the antiproton emittance, the antiproton lifetime decreased drastically by more than a factor of 100. The strong antiproton emittance dependency of the lifetime may be due to the following two effects: On the one hand, the larger antiproton emittance gives a larger antiproton beam size and results in a faster loss to the aperture. On the other hand, the larger antiproton beam size reduces the distance between the proton beam and the antiproton particles, which results in stronger nonlinear beam-beam interactions.

## ACKNOWLEDGMENTS

This work was performed using NERSC and CCS supercomputers under the auspices of a SciDAC project, "Advanced Computing for 21st Century Accelerator Science and Technology, which is supported by the US DOE/SC Office of HENP and the Office of ASCR under contract DE-AC03-76SF00098.

## REFERENCES

1. R. W. Hockney and J. E. Eastwood, "Computer Simulation Using Particles," McGraw-Hill Book Company, New York, 1985.
2. J. Qiang, M. A. Furman, R. D. Ryne, Phys. Rev. ST Accel. Beams **5**, 104402 (2002).
3. M. A. Furman, "Beam-Beam Simulations with the Gaussian Code TRS," in Proc. ICAP98, Monterey, California, Sept. 14-18, 1998.
4. C. Montag, J. M. Brennan, J. Butler, and P. Koello, "Measurements of mechanical triplet vibrations in RHIC," Proc. of the 2002 EPAC, Paris, France (2002).
5. W. Herr, "Beam-beam issues in the LHC and relevant experience from the SPS proton antiproton collider and LEP," in Proceedings of a Workshop on Beam-beam Effects in Circular Colliders, Fermilab, June 25-27, 2001.

# Weak-Strong Beam-Beam: Averaging and Tune Diagrams

James A. Ellison*, H. Scott Dumas†, Marc Salas*, Tanaji Sen**, Andrey Sobol* and Mathias Vogt‡

*University of New Mexico, Albuquerque, NM, USA
†University of Cincinnati, Cincinnati, OH, USA
**Fermilab, Batavia, IL, USA
‡DESY, Hamburg, FRG

**Abstract.** We sketch the outlines of a perturbation scheme for the kick-lattice map in two degrees of freedom (4-d phase space) and apply it to the weak-strong beam-beam interaction. Our approach yields rigorous error bounds on the precision of our approximations (proofs will appear elsewhere), divides tune space into distinct regions where different approximations are valid, clarifies the dynamics near certain generically nonintegrable resonances, and generalizes to higher degrees of freedom and more complex models such as the strong-strong beam-beam.

## INTRODUCTION

In this paper we discuss a perturbation analysis of the kick-lattice map in two degrees of freedom (**DOF**):

$$w_{n+1} = R(v)\left(w_n + \varepsilon \, \text{Kick}(w_n)\right), \quad (1)$$

(here $\varepsilon$ represents the size of the kick), and apply it to the weak-strong beam-beam interaction (**WSBB**). We convert (1) to a standard form for the *method of averaging* by a transformation which leads to the map equation

$$x_{n+1} = x_n + \varepsilon \, \mathscr{F}(x_n, nv_0, a) + O(\varepsilon^2), \quad v = v_0 + \varepsilon a, \quad (2)$$

for which the associated averaged flow equation is

$$\dot{y} = \varepsilon \, \overline{\mathscr{F}}(y, v_0, a). \quad (3)$$

We show how to construct the map function $\mathscr{F}$ and its average $\overline{\mathscr{F}}$ (the latter being a vector field for an autonomous Hamiltonian). We summarize the procedure for obtaining rigorous error bounds, $x_n = y(n) + O(\varepsilon), 0 \leq n \leq T/\varepsilon$, which shows that we have the asymptotics right. We quantify the notions of far-from-low-order-resonance (**FFLOR**) and near-to-low-order-resonance (**NTLOR**) in the $v_h$–$v_v$ tune diagram. The dynamics of (3) is integrable in the FFLOR case and generically nonintegrable near the crossing of low-order resonance lines, but integrable near low-order resonance lines and far from low-order crossings. Finally, we present examples of the two types of NTLORs for the WSBB and illustrate the nonintegrability by finding numerical chaos.

Our main points are: (1) $\overline{\mathscr{F}}$ is easy to construct and gives the "correct" approximation to the dynamics in the sense of asymptotics, (2) the tune space naturally divides into four regions: FFLOR, NTLOR-I, NTLOR-II, and a (small) complementary set, (3) we clarify the dynamics near the crossing of two low-order resonance lines, and (4) the generalizations to more frequencies (e.g. tune modulation) and to the strong-strong beam-beam interaction [1] are fairly straightforward and are in progress. We hope that this work will complement the large body of previous research on the WSBB, which we do not attempt to discuss here.

## THE KICK-LATTICE MODEL OF WSBB

The basic equation of motion for the kick-lattice model in 2 DOF is given by

$$w_{n+1} = R(v)\left[w_n + \varepsilon \begin{pmatrix} 0 \\ -H'(A_n)w_{1,n} \\ 0 \\ -H'(A_n)w_{3,n} \end{pmatrix}\right] \quad (4)$$

where $R$ represents the lattice and depends on the tune vector $v$, and $H$ generates the kick with $A_n := (w_{1,n}^2 + w_{3,n}^2)/2$. Here, $w_n := (w_{1,n}, w_{2,n}, w_{3,n}, w_{4,n})^T$ and the first two components are scaled $x$ position and momentum and the last two components are scaled $y$ position and momentum. We treat the linear uncoupled lattice, where

$$R(v) := \text{diag}\left(e^{\mathscr{J}_2 2\pi v_h}, e^{\mathscr{J}_2 2\pi v_v}\right) \quad (5)$$

and where

$$\mathcal{J}_2 := \begin{pmatrix} 0 & +1 \\ -1 & 0 \end{pmatrix}, \quad e^{\mathcal{J}_2 t} = \begin{pmatrix} \cos t & \sin t \\ -\sin t & \cos t \end{pmatrix}. \tag{6}$$

Our analysis is a perturbation analysis in the parameter $\varepsilon$; thus $\varepsilon$ represents the size of the kick and is assumed to be small. For the WSBB, when the strong beam is a round Gaussian beam with $\sigma_h = \sigma_v =: \sigma$, we have

$$H'(A) := \frac{1}{A}\left(1 - e^{-A/\sigma^2}\right). \tag{7}$$

We now want to transform (4) to a *standard form for the method of averaging*. A usual approach is to transform $w \to x$ via $w = R(v)^n x$, which just transforms from the solution of the unperturbed problem. However, the analysis is simplified considerably by writing

$$v = v_0 + \varepsilon a = \begin{pmatrix} v_{0h} \\ v_{0v} \end{pmatrix} + \varepsilon \begin{pmatrix} a_h \\ a_v \end{pmatrix} \tag{8}$$

and transforming via

$$w_n = R(v_0)^n x_n. \tag{9}$$

In the NTLOR case, $v_0$ is the resonant tune vector and the vector $a$ measures the displacement from the resonance. In the FFLOR case, $a = 0$ and $v_0 = v$. This trick allows us to treat both cases simultaneously and simplifies the treatment of the NTLOR case. After some calculation, we have

$$x_{n+1} = x_n + \varepsilon \mathcal{J}_4 \left( \nabla_x \mathcal{H}_1(x_n, a) + \nabla_x \mathcal{H}_2(x_n, nv_0) \right) + O(\varepsilon^2) \tag{10}$$

$$\mathcal{H}_1(x, a) := 2\pi \left( a_h \frac{x_1^2 + x_2^2}{2} + a_v \frac{x_3^2 + x_4^2}{2} \right) \tag{11}$$

$$\mathcal{H}_2(x, nv) := H\left( \frac{w_{1,n}^2(x) + w_{3,n}^2(x)}{2} \right) \tag{12}$$

$$w_{1,n}(x) = x_1 \cos(2\pi n v_h) + x_2 \sin(2\pi n v_h)$$
$$w_{3,n}(x) = x_3 \cos(2\pi n v_v) + x_4 \sin(2\pi n v_v),$$

where $\mathcal{J}_4 := \mathrm{diag}(\mathcal{J}_2, \mathcal{J}_2)$. The averaged map problem becomes

$$y_{n+1} = y_n + \varepsilon \mathcal{J}_4 \left( \nabla_y \mathcal{H}_1(y_n, a) + \nabla_y \overline{\mathcal{H}}_2(y_n, v_0) \right), \tag{13}$$

and the associated averaged flow is

$$\dot{y} = \varepsilon \mathcal{J}_4 \left( \nabla_y \mathcal{H}_1(y, a) + \nabla_y \overline{\mathcal{H}}_2(y, v_0) \right). \tag{14}$$

We anticipate that for $0 \le n \le T/\varepsilon$ there exists a constant $C(T) > 0$ such that $|x_n - y_n| \le C(T)\varepsilon$ and $|x_n - y(n)| \le C(T)\varepsilon$. Our mantra is "the error is order $\varepsilon$ on order $1/\varepsilon$ time intervals." We stress that these are rigorous error bounds and not error estimates. Numerical evidence indicates that the error is $O(\varepsilon)$ on much longer intervals [2]. However, eventually phase error sets in, so from an accelerator physics point of view it is more interesting to find invariants of the averaged problem that are approximate invariants of the exact problem over very long times. We are investigating this.

We have carried out a detailed error analysis in the case of one frequency and applied it to the WSBB in 1 DOF [3]. We are working on the proof in the 2 DOF case [4]; it shows how to quantify the regions in tune space which are NTLOR and FFLOR interms of the perturbation parameter $\varepsilon$ and a "cut off Diophantine set" (**CODS**). In the next section we outline the proof in the FFLOR case; it can be skipped without much loss of continuity.

## ERROR ANALYSIS / FFLOR

Here we summarize the error analysis for the single frequency, FFLOR case (for details see [3]). Let $x_n$ be the solution of $x_{n+1} = x_n + \varepsilon f(x_n, nv)$, where $f(x, \theta)$ is $2\pi$-periodic in $\theta$, and let $y_n$ be the solution of $y_{n+1} = y_n + \varepsilon \overline{f}(y_n, v)$, where $\overline{f}$ is the $\theta$-average of $f$. Then

$$\|x_N - y_N\| =$$
$$\varepsilon \left\| \sum_{n=0}^{N-1} \underbrace{f(x_n, nv) - f(y_n, nv)}_{\clubsuit} + \underbrace{f(y_n, nv) - \overline{f}(y_n, v)}_{\spadesuit} \right\|$$

$$\le \varepsilon L \sum_{n=0}^{N-1} \overbrace{\|x_n - y_n\|}^{\clubsuit} + \varepsilon \underbrace{\left\| \sum_{n=0}^{N-1} \widetilde{f}(y_n, nv) \right\|}_{= M_N} \tag{15}$$

where $L$ is the Lipschitz constant of $f$ in its first argument, and $\widetilde{f} = f - \overline{f}$. If $M_N \le M$, then the Gronwall inequality gives $\|x_N - y_N\| \le \varepsilon M e^{\varepsilon LN} \le \varepsilon M e^{LT} = O(\varepsilon)$ for $0 \le N \le T/\varepsilon$. However, $M_N$ grows like $N$ on resonance (i.e., for $v$ a low-order rational). Thus the basic question is, For what $v$ is $M_N$ bounded on $[0, T/\varepsilon]$? The Fourier series for $f(x, \theta)$ gives

$$\widetilde{f}(y, \theta) = \sum_{k \ne 0} f_k(y) e^{ik\theta} = \sum_{0 < |k| \le R_\varepsilon} f_k(y) e^{ik\theta} + O(\varepsilon), \tag{16}$$

where in the second equality we truncate the Fourier series at $R_\varepsilon$ so that the tail is $O(\varepsilon)$. This leads to the following bound on $M_N$:

$$M_N = \left\| \sum_{n=0}^{N-1} \widetilde{f}(y_n, nv) \right\|$$

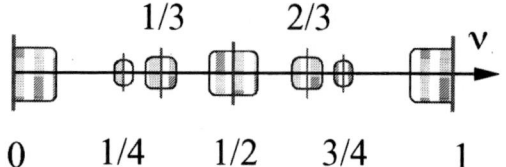

**FIGURE 1.** A possible cut off Diophantine set (CODS): The blue blobs designate the open sets removed from tune space; here the cutoff $R_\varepsilon = 4$.

$$\leq \left\| \sum_{n=0}^{N-1} \sum_{0<|k|\leq R_\varepsilon} f_k(y_n) e^{ikn\nu} \right\| + O(N\varepsilon)$$
$$= I_N + O(1) \text{ for } 0 \leq N \leq T/\varepsilon. \quad (17)$$

Summation by parts and the fact that $y_{n+1} - y_n = O(\varepsilon)$ (see (13)) give

$$I_N \leq \sum_{0<|k|\leq R_\varepsilon} \frac{c_k}{|1 - e^{i2\pi k\nu}|} \quad (18)$$

and we see the appearance of (the notorious) small divisors in the terms of this last sum. To overcome the small divisors, we use a CODS (see [3]) of $\nu$ values which avoid low-order rationals, so that $I_N = O(1)$. A schematic of the CODS is shown in Figure 1.

## THE AVERAGE OF $\mathcal{H}_2$

Since $\mathcal{H}_1$ does not depend on $n$, $\overline{\mathcal{H}}_1 = \mathcal{H}_1$. Thus to determine the averaged map (13) and averaged flow (14) we need to find

$$\overline{\mathcal{H}}_2(x, \nu_0) := \lim_{N\to\infty} \frac{1}{N} \sum_{n=0}^{N-1} \mathcal{H}_2(x, n\nu_0). \quad (19)$$

This is facilitated by introducing action-"angle" variables $J, \Phi$ defined by

$$\begin{pmatrix} x_1 \\ x_2 \end{pmatrix} = \sqrt{2J_h} \begin{pmatrix} \sin\Phi_h \\ \cos\Phi_h \end{pmatrix}, \begin{pmatrix} x_3 \\ x_4 \end{pmatrix} = \sqrt{2J_v} \begin{pmatrix} \sin\Phi_v \\ \cos\Phi_v \end{pmatrix} \quad (20)$$

which leads to

$$\mathcal{H}_2(x, n\nu_0) = H\left(J_h \sin^2(2\pi n\nu_h + \Phi_h) + h \to v\right). \quad (21)$$

Define

$$V(J, \theta) := H\left(J_h \sin^2\frac{\theta_h}{2} + J_v \sin^2\frac{\theta_v}{2}\right), \quad (22)$$

where the $1/2$ is inserted so that $V$ has minimal period $2\pi$ in $\theta$. Its associated Fourier series may be written:

$$V(J, \theta) = \sum_{m\in\mathbb{Z}^2} V_m(J) e^{im\cdot\theta}, \quad (23)$$
$$V_m(J) := \frac{1}{(2\pi)^2} \int_{\mathbb{T}^2} V(J, \theta) e^{-im\cdot\theta} d^2\theta.$$

It follows that

$$\mathcal{H}_2(x, n\nu_0) = \sum_{m\in\mathbb{Z}^2} V_m(J(x)) e^{i2m\cdot\Phi(x)} \left(e^{i4\pi(m\cdot\nu_0)}\right)^n. \quad (24)$$

At this stage it is not convenient to exploit the symmetry of $V_m$ because averages are most easily computed using the complex exponential. Thus

$$\overline{\mathcal{H}}_2(x, \nu_0) = \sum_{\substack{m\in\mathbb{Z}^2 \\ 2m\cdot\nu_0\in\mathbb{Z}}} V_m(J(x)) e^{i2m\cdot\Phi(x)} \quad (25)$$

since

$$\frac{1}{N} \sum_{n=0}^{N-1} \left(e^{i4\pi(m\cdot\nu_0)}\right) \xrightarrow{N\to\infty} 0 \text{ unless } 2m\cdot\nu_0 \in \mathbb{Z}. \quad (26)$$

## THE AVERAGED FLOW

The averaged flow is given by (14). In the FFLOR case ($\nu_0 = \nu \in$ CODS and $a = 0$), $\mathcal{H}_1 = 0$ and $\overline{\mathcal{H}}_2(y, \nu_0) = V_0(J(y))$. The dynamics is integrable and is just a rotation with tune spread. In the NTLOR case $\mathcal{H}_1(y, a) = 2\pi\left(a_h J_h(y) + a_v J_v(y)\right)$, $\overline{\mathcal{H}}_2(y, \nu_0)$ is given by (25), and we have two cases:

- Type I resonance: *close* to low-order resonance *line* but *far* from *crossings* of low-order resonance lines.
- Type II resonance: *close* to *crossing* of low-order resonance lines.

It is not difficult to show that the normal forms for resonances of type I are integrable, while the normal forms for resonances of type II are generically nonintegrable. To illustrate these basic features, we now discuss two examples of type I resonance and one of type II resonance.

### Example 1 (NTLOR-Type I)

Here we take $\nu_h = \frac{1}{2} + \varepsilon a_h$ but $\nu_v = \xi \in$ CODS (that is, $\xi$ is far from low-order rationals). A schematic of this set is illustrated by the blue region in Fig. 2 centered on the red line at $\nu_h = 0.5$. The averaged Hamiltonian may be written $\overline{\mathcal{H}}(x) = 2\pi a_h J_h + F(J_v, \frac{1}{2}x_1^2)$; thus $J_v$ is a constant of the motion and the $(x_1, x_2)$ dynamics is given by

$$\dot{x}_1 = 2\pi a_h x_2 \quad (27)$$
$$\dot{x}_2 = -2\pi a_h x_1 - \mathbf{D}_2 F(J_{v,0}, \frac{1}{2}x_1^2) x_1 \quad (28)$$

which is clearly integrable. The basic dynamics can be understood in terms of the phase plane portrait in the $(x_1, x_2)$-plane defined by $\pi a_h(x_1^2 + x_2^2) + F(J_{v,0}, \frac{1}{2}x_1^2) =$ const. $= E$.

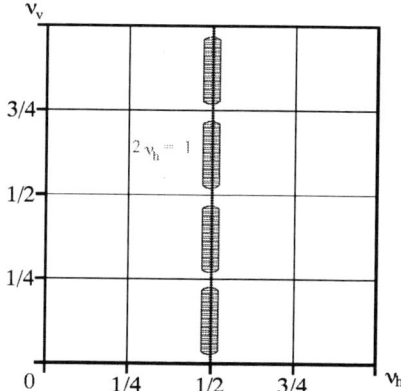

**FIGURE 2.** Tune Diagram : $\nu_h = \frac{1}{2} + \varepsilon a_h$, $\nu_v = \xi \in$ CODS.

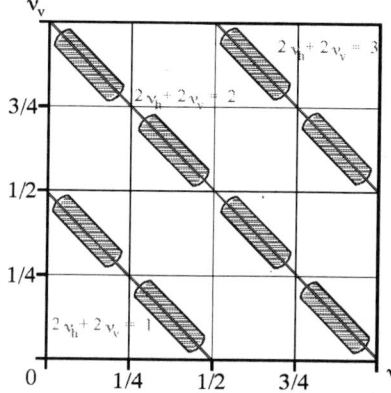

**FIGURE 3.** Tune Diagram : $2\nu_h + 2\nu_v = p + \varepsilon 4\alpha$, $2\nu_h - 2\nu_v = \xi \in$ CODS.

## Example 2 (NTLOR-Type I)

Here we consider the important $2\nu_h + 2\nu_v$ resonance. We set $2\nu_h + 2\nu_v = p + \varepsilon 4\alpha$, where $p = 1, 2, 3$ and $\alpha$ measures the distance from the resonance, and take $2\nu_h - 2\nu_v = \xi \in$ CODS. A schematic of this set is illustrated by the blue regions in Fig. 3. These regions have width $O(\varepsilon)$ and are centered on the three red diagonal lines in the unit square.

The averaged Hamiltonian is given by $K(\Phi, J) = 2\pi\alpha(J_h + J_v) + F(\Phi_h + \Phi_v, J)$ in action-angle variables, and we note that it does not depend on $p$. Thus the dynamics is the same in all eight regions shown in the figure. Since the Hamiltonian depends on the angles through their sum, $J_v - J_h$ is an integral of the motion and the system is integrable.

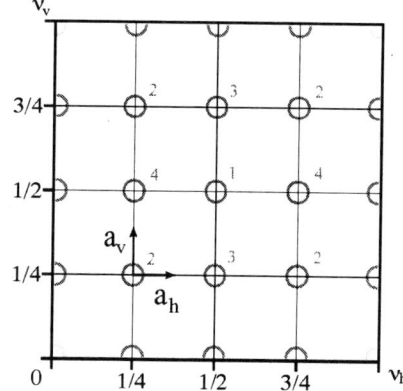

**FIGURE 4.** Tune Diagram : $\nu_0 = (\frac{q}{4}, \frac{p}{4})^T$.

## Example 3 (NTLOR-Type II)

Here we take $\nu_0 = (\frac{q}{4}, \frac{p}{4})^T$, $q = 1, 2, 3$, $p = 1, 2, 3$. This gives four distinct cases as indicated by the red $O(\varepsilon)$ circles in the tune diagram (Fig. 4) and labeled 1, 2, 3 and 4. Setting $A_{ij} = \frac{1}{2}(x_i^2 + x_j^2)$, the Hamiltonians in the four cases are:

1: $\overline{\mathcal{H}}(x) = 2\pi(a \cdot J) + H(A_{13})$,
2: $\overline{\mathcal{H}}(x) = 2\pi(a \cdot J) + (H(A_{13}) + H(A_{24}))/2$,
3: $\overline{\mathcal{H}}(x) = 2\pi(a \cdot J) + (H(A_{13}) + H(A_{14}))/2$, and
4: $\overline{\mathcal{H}}(x) = 2\pi(a \cdot J) + (H(A_{13}) + H(A_{23}))/2$.

We conjecture that for most values of the vector $a$ these systems are nonintegrable with chaotic dynamics. We have analyzed case 1, but $\mathcal{H}_2(x, n\nu_0)$ is actually independent of $n$, so it's not interesting from an averaging viewpoint. Here we discuss case 2 where $p = 1, 3$ and $q = 1, 3$. In this case the equations of motion are

$$\begin{aligned}
\dot{x}_1 &= \left(2\pi a_h + \frac{1}{2}H'(A_{2,4})\right)x_2 \\
\dot{x}_2 &= -\left(2\pi a_h + \frac{1}{2}H'(A_{1,3})\right)x_1 \\
\dot{x}_3 &= \left(2\pi a_v + \frac{1}{2}H'(A_{2,4})\right)x_4 \\
\dot{x}_4 &= -\left(2\pi a_v + \frac{1}{2}H'(A_{1,3})\right)x_3.
\end{aligned} \qquad (29)$$

We note that $x = 0$ is an equilibrium solution. We have carefully analyzed this case and run detailed numerical calculations in the so-called "surface-of-section" (**SOS**), also known as the Poincaré section. A discussion of the SOS may be found in for example [6]. Here we present just a few of the SOS plots; the interested reader will find more results in [5].

Figure 5 shows the SOS for the integrable case, $a_h = a_v = 0.0$, which has $x_1x_3 - x_2x_4$ as a 2nd integral of motion. Each point on the vertical axis inside the outer oval is a periodic solution. In configuration $(x_1, x_3)$-space,

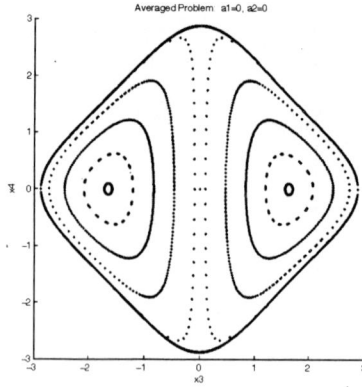

**FIGURE 5.** $v_0 = (\frac{1}{4}, \frac{1}{4})$, $\varepsilon = 0.01$, $E = 1$, $a_h = 0.0$, $a_v = 0.0$, $0 \leq N \leq 5000$.

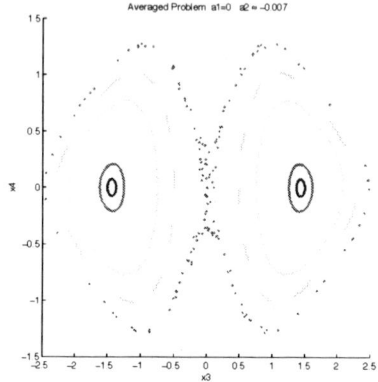

**FIGURE 6.** $v_0 = (\frac{1}{4}, \frac{1}{4})$, $\varepsilon = 0.01$, $E = 1$, $a_h = 0.0$, $a_v \approx -0.007$, $0 \leq N \leq 5000$.

these periodic motions occur on straight lines through the origin.

Figure 6 shows the SOS for $a_h = 0.0$ and $a_v \approx -0.007$. Here we have a clear indication of chaos with a stochastic layer which includes the origin. All the red points start from the initial condition $x_4 = 0$, $x_3 = 0.1$. Note the similarity to the integrable case in Fig. 5 and that the stochastic layer may be related to a break-up of the line of fixed points in the integrable case.

As $a_h$ is increased from zero, transition from the non-integrable behavior of Fig. 6 to "near-integrability" is to be expected, since we are moving from a type II resonance toward a type I resonance. This is illustrated by Fig. 7, which shows the SOS for $a_h = 0.5$ and $a_v \approx -0.007$ and which has the clear appearance of integrability. In fact the SOS appears integrable at $a_h = 0.05$.

In all cases there is excellent agreement between the averaged problem (14) and the exact simulation from (4) and (9), as can be seen in [5].

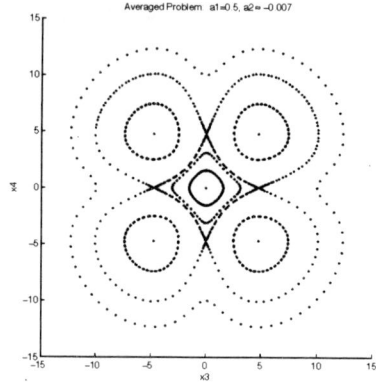

**FIGURE 7.** $v_0 = (\frac{1}{4}, \frac{1}{4})$, $\varepsilon = 0.01$, $E = 1$, $a_h = 0.5$, $a_v \approx -0.007$, $0 \leq N \leq 500$.

## ACKNOWLEDGMENTS

The work of J.A.E., M.S., and A.S. was supported by DOE grant DE-FG03-99ER41104. H.S.D. was supported by the C.P. Taft Foundation at U. Cincinnati.

## REFERENCES

1. Ellison, J., and Vogt, M., "A New Model for the Collective Beam-Beam Interaction," in *Slides from Beam-Beam Workshop 2003, Montauk*, AIP, 2003, URL http://www.rhichome.bnl.gov/AP/BeamBeam/Workshop03/BB03_program.html.
2. Salas, M., and Sobol, A., Private communication (2003).
3. Dumas, H., Ellison, J., and Vogt, M., *submitted* (2002).
4. Dumas, H., Ellison, J., and Vogt, M., in progress (2002).
5. Ellison, J., Dumas, H., Salas, M., Sen, T., Sobol, A., and Vogt, M., "Weak–Strong Beam–Beam : Averaging and Tune Diagrams," in *Slides from Beam-Beam Workshop 2003, Montauk*, AIP, 2003, same URL as above.
6. Tabor, M., *Chaos and Integrability in Nonlinear Dynamics*, John Wiley and Sons, New York, 1989.

# Numerical Calculation of the Phase Space Density for the Strong-Strong Beam-Beam Interaction

A. Sobol and J. A. Ellison

Mathematics and Statistics Department
University of New Mexico

**Abstract.** We developed a parallel code to calculate the evolution of the 4D phase space density of two colliding beams, which are coupled via the collective strong-strong beam-beam interaction, in the absence of diffusion and damping, using the Perron-Frobenius (PF) operator technique.

We restrict our study to the head-on interaction between two short, counter-rotating bunches at a given IP and a linear uncoupled lattice elsewhere. If $X$ represents some quantity of one beam, then $X^*$ represents the same quantity of the other beam. Index $n$ will refer to quantities right before the interaction on turn $n$ and index $n^+$ denotes quantities right after the interaction on turn $n$. Finally, $\Psi_n(x, p_x, y, p_y)$ denotes the phase space density at the phase space point $z = (x, p_x, y, p_y)$, and $\rho_n(x,y)$ denotes the corresponding spatial density.

The evolution at the IP is given by

$$\Psi_{n^+}(Kz) = \Psi_n(z),$$

where the kick is given by

$$K(z) = (x, p_x + k_x(x,y), y, p_y + k_y(x,y)),$$

$$k_x = \zeta \frac{\partial \Phi^*}{\partial x}, \; k_y = \zeta \frac{\partial \Phi^*}{\partial y},$$

$$\Delta \Phi^* = -2\pi \rho^*,$$

$$\rho^*(x,y) = \int \int \Psi^*(x, p_x, y, p_y) dp_x dp_y,$$

and the beam-beam parameter is $\zeta = \frac{2e^2 N^*}{\gamma 4\pi\varepsilon_0 mc^2}$. Note that it is trivial to calculate the inverse kick map $K^{-1}$:

$$K^{-1}(z) = (x, p_x - k_x(x,y), y, p_x - k_y(x,y)).$$

The evolution around the ring is given by

$$\Psi_{n+1}(Mz) = \Psi_{n^+}(z),$$

$$M = \begin{pmatrix} M_x & 0 \\ 0 & M_y \end{pmatrix}.$$

We do not impose any restriction on $M_x$ and $M_y$ other than invertibility.

This method, as originally proposed by Warnock [1], needs a 4D interpolation for the lattice map and a 2D interpolation for the kick map. Vogt discovered that this two-degree of freedom case was significantly more expensive than was anticipated from a scaling argument based on work in the one-degree of freedom case [2, 3] and that this discrepancy was probably a cache issue involving the 4D interpolation. Here we describe a solution of this problem by representing the 4D one-turn map as a composition of maps each requiring only 2D interpolations.

Let us introduce a rectangular mesh in 4D phase space. We will keep the values of densities at the meshpoints in four-dimensional arrays. The density $\Psi_{n^+}(z)$ after an interaction will be calculated at grid point $z$ by tracking the coordinate of the gridpoint backward $z' := M^{-1}(z)$ and assign $\Psi_{n^+}(z) := \Psi_n(z')$. Values of $\Psi_n(z')$, in between grid points, can be found via quadratic interpolation using known values of $\Psi_n$ at gridpoints.

The evolution of densities on the orbit is calculated in the similar way by tracking the coordinate of the gridpoint backward $z' := K^{-1}(z)$ and assign $\Psi_{n+1}(z) := \Psi_n(M^{-1}z)$.

To calculate the kick at the IP, we solve Poisson's equation using the conjugate gradient method [4]. We chose this method because we think it is one of the most suitable methods for a parallel computation.

Note that the amount of data involved in computation is proportional to the 4th power of grid size. For example, if we had only 100-point-per-dimension grid, we would have to store 100 million values. So much data cannot fit cache of any processor, which dramatically slows down computations. Using parallel computers doesn't solve this problem automatically because we need all values in the cache of the same processor for the interpolation. A special method of distributing data must be developed to perform calculations efficiently.

We believe that this problem can be solved by representing the kick-lattice map as a composition of three maps each of which changes only two coordinates out of four. Instead of the evolution equation

$$\Psi_{n+1}(z) = \Psi_n((MK)^{-1}z)$$

we consider the equivalent system of equations

$$\begin{cases} \Psi_{n+1/3}(z) = \Psi_n(K^{-1}z), \\ \Psi_{n+2/3}(z) = \Psi_{n+1/3}(\bar{M}_x^{-1}z), \\ \Psi_{n+1}(z) = \Psi_{n+2/3}(\bar{M}_y^{-1}z), \end{cases}$$

where

$$\bar{M}_x = \begin{pmatrix} M_x & 0 \\ 0 & I \end{pmatrix}, \quad \bar{M}_y = \begin{pmatrix} I & 0 \\ 0 & M_y \end{pmatrix}.$$

Since only two coordinates of vector $z$ change on each "1/3" step, we can fix other two coordinates and calculate a two-dimensional layer of the four-dimensional array. So, we keep only the values of one two-dimensional layer in the cache and calculate a 2D interpolation instead of keeping all the values of the array in cache and calculating a 4D interpolation. Since layers can be calculated independently, this process can be naturally performed in parallel.

We developed a C++ parallel code, which implements this idea. The code can be compiled such that the spatial density is displayed during program execution. This helps to debug the program, and we can see if the mesh resolution is reasonable. The graphics is implemented using Mesa3D (analogous to OpenGL) libraries.

However, while running on a large amount of data for many hours, displaying graphics would slow down the computation. So we made it possible to compile this program without graphics being displayed.

We made a test run of this program on a parallel computer (Los Lobos) with parameters given in the following table and we know that the time that calculations take scales according to the following numbers. It takes about 5 seconds to calculate one turn on 25-point-per-dimension grid on one processor.

| | |
|---|---|
| Beam energy (TeV) | 7 |
| Protons per bunch | $1.05 * 10^{11}$ |
| $\beta^*$(m) | 0.5 |
| rms spot size at IP ($\mu m$) | 16 |
| Betatron tunes ($v_x, v_y$) | (0.31, 0.32) |

Though we are quite sure that algorithm is implemented correctly, the test run of the program revealed some problems. We can observe some nonphysical effects, like negative probability. Also, we can see that the densities cease to be smooth very rapidly, so our approximation is not valid for large times. The last effect is probably due to the fact that there is no smoothing factor in the equation, and we are investigating this.

## ACKNOWLEDGMENTS

Our work was supported by DOE grant DE-FG03-99ER41104. Some simulations were performed at the Albuquerque High Performance Computing Center.

## REFERENCES

1. Warnock, R. L., and Ellison, J. A., "A General Method for Propagation of the Phase Space Destribution, with Application to the Saw-Tooth Instability," in *The Physics of High Brightness Beams*, 2000.
2. Ellison, J. A., and Warnock, R. L., "Existence and Properties of an Equilibrium State with beam-beam collision," in *Quantum Aspects of Beam Physics, 18th Advanced ICFA Beam Dynamics Workshop*, 2000.
3. Vogt, M., Ellison, J. A., Sen, T., and Warnock, R. L., "Two Methods for Simulating the Strong-Strong Beam-Beam Interaction in Hardron Colliders," in *Proceedings of Workshop on Beam-Beam Effects in Circular Colliders*, Fermilab, 2001.
4. Kelley, C., *Iterative Methods for Linear and Nonlinear Equations*, SIAM, Philadelphia, 1995.

# Comments and Discussions

# Halo Generation and Beam Cleaning by Resonance Trapping

## Alex Chao

*Stanford Linear Accelerator Center, Menlo Park, CA, 94025*

**Abstract.** One of the mechanisms of halo generation occurs when particles are trapped by a nonlinear resonance $\nu \approx m/p$ as the tune $\nu$ is being modulated. This note is intended to renew some attention to this mechanism, and hopefully to trigger new study to understand it more quantitatively. This same mechanism used in reverse could serve to clean beam halos and to relax requirements on the collimators.

## HALO MECHANISM

It was proposed that beam halos can be generated in a high-intensity storage ring by way of a particle trapping mechanism due to a modulating nonlinear resonance [1, 2]. A survey of this mechanism in relation to other nonlinear resonance effects has been made in [3, 4, 5], although a systematic quantitative evaluation of the resonance trapping mechanism to the application of halos seems not yet available. The hope of this note is to renew some attention to this halo generation mechanism.

In this mechanism, the tune of the accelerator is close to a nonlinear resonance $\nu \approx m/p$. The resonance will generate islands in the 2-D phase space. (We here consider 1-D dynamics only.) If the tune of a particle $\nu$ is slowly modulating in time, the islands will move in the phase space. They will move towards the phase space origin as $\nu$ approaches the resonant value $m/p$ and away from the phase space origin as $\nu$ moves away from $m/p$. As the tune modulates, the islands therefore move in and out in phase space. The speed at which they move is determined by how fast the tune modulates. In addition to the location of the islands, the island area also changes as the tune changes.

One reason the tune might be modulating is through synchrotron oscillation. If the storage ring has a finite chromaticity, a particle's tune will oscillate as its energy executes synchrotron oscillation. Another possible reason is power supply ripple causing the tune to oscillate. One might also imagine other possible reasons such as quantum radiation diffusion or intrabeam scattering although in that case the tune does not oscillate but wonders around when the particle energy wonders in random walks and the storage ring has a finite chromaticity.

As the islands move in and out in phase space, they may trap particles and carry the trapped particles with them as they move. The trapping efficiency, $\eta_{\text{trap}}$, a key quantity to be considered, depends on the island area $A$ and the island moving speed $S$. In fact, we may loosely write[1]

$$\eta_{\text{trap}} \sim A e^{-S}$$

Larger islands trap more particles, while islands moving too fast become ineffective traps. In fact, trapped particles may leak out of the islands if the islands move too fast, i.e. the islands become "leaky".

One possible halo generation mechanism therefore occurs as follows. As the tune modulates and the islands move in and out in phase space, the islands will be able to carry particles from small amplitudes to large amplitudes, and vice versa. However, if the islands have a greater trapping efficiency when they are closer to phase space origin ($\nu$ is closer to $m/p$) than when they are away from phase space origin ($\nu$ is away from $m/p$), then the modulating islands will act as a pumping mechanism to carry particles systematically from the beam core to the beam tail. The pumping rate is proportional to

$$\nu_{\text{mod}} [\eta_{\text{trap}}(core) - \eta_{\text{trap}}(tail)]$$

where $\nu_{\text{mod}}$ is the tune modulation frequency. A more quantitative analysis of this pumping rate is yet to be invented and carried out, although useful references exist, e.g. [3, 4, 5, 6].

One might also envision an emittance dilution mechanism if $\eta_{\text{trap}}(core) < \eta_{\text{trap}}(tail)$. In this case, the empty island buckets are brought into the beam core as empty filaments, not unlike the mechanism of phase displacement acceleration [7]. This effect however probably dilutes the effective beam core rather than causing a beam halo.

One might envision two qualitatively different types of nonlinear resonances as far as resonance trapping is

---

[1] The actual formula is more involved. For one special case, see [2].

concerned. One type, caused by head-on beam-beam or space charge effects, has the property that the driving force tends to decrease as the particle moves far into the beam tail. For the other type, caused by magnet nonlinearities or long-range beam-beam effects, the force tends to increase as the particle moves far into the beam tail. One would expect different halo generation behavior in these two types of nonlinear resonances.

## BEAM CLEANING

It turns out that the resonance trapping mechanism can in principle be used to clean the beam tails [8]. To do so, we will consider purposely modulating the tune in a pre-programmed manner, with the aim of bringing particles with modest amplitudes (particles to be cleaned) all the way out of the beam and into a collimator. In this application, the nonlinear resonance is used similarly to beam extraction [9, 10]. This arrangement of dynamic beam cleaning may be useful to avoid having collimators very close to an intense beam core. For application to storage ring colliders, whether a strong beam-beam interaction would dominate the phase space topology, and therefore ender the resonance trapping ineffective, is one issue of concern, unless the beams are separated during cleaning [11].

## ACKNOWLEDGEMENTS

I enjoyed the insightful discussions with Shinji Machida, Jean-Bernard Jeanneret, Frank Zimmermann, and Alexei Fedotov on this topic during and after the workshop. They educated me on the latest developments and also gave me some useful references.

## REFERENCES

1. A. Schoch, CERN Report 57-23 (1958).
2. Alexander W. Chao and Melvin Month, Nucl. Instru. Meth. 121, 129 (1974).
3. A. Fedotov, these proceedings.
4. F. Zimmermann, these proceedings.
5. S. Peggs, Handbook of Accelerator Physics and Engineering, Ed. Chao/Tigner, 2nd print, p.95 (2002).
6. O. Bruning, Ph.D. Thesis, U. Hamburg, DESY Report 94-085 (1994).
7. E.W. Messerschmid, CERN Report CERN/ISR-TH/73-31 (1973).
8. A.W. Chao and M. Month, Nucl. Instru. Meth. 133, 405 (1976).
9. H.G. Hereward, Proc. Vth Int. Conf. on High Energy Accelerators, Dubna (1963).
10. R. Cappi and M. Giovannozzi, Phys. Rev. Lett. 88, 104801 (2002).
11. J-B. Jeanneret, private communication, this workshop.

# List of Participants

| familyname | firstname | mi | institution | email |
|---|---|---|---|---|
| Aleksandrov | Alexander | V. | ORNL | sasha@ornl.gov |
| Alexahin | Yuri | I | FNAL | alexahin@fnal.gov |
| Assadi | Saeed | | Oak Ridge National Lab | saeed@sns.gov |
| Bai | Mei | | Brookhaven National Laboratory | mbai@bnl.gov |
| Becker | Jesse | | Brookhaven National Laboratory | jbecker@bnl.gov |
| Blaskiewicz | Michael | | Brookhaven National Laboratory | mmb@bnl.gov |
| Cameron | Peter | | Brookhaven National Laboratory | cameron@bnl.gov |
| Chao | Alex | | SLAC | achao@slac.stanford.edu |
| Connolly | Roger | | Brookhaven National Laboratory | connolly@bnl.gov |
| Danilov | Viatcheslav | V | ORNL SNS Project | danilovs@ornl.gov |
| Drees | Angelika | | Brookhaven National Laboratory | drees@bnl.gov |
| Ellison | James | A | University of New Mexico | ellison@math.unm.edu |
| Erdelyi | Bela | | Fermilab | erdelyi@fnal.gov |
| Fedotov | Alexei | | Brookhaven National Laboratory | fedotov@bnl.gov |
| Fischer | Wolfram | | Brookhaven National Laboratory | Wolfram.Fischer@bnl.gov |
| Fliller III | Raymond | P | Brookhaven National Laboratory | rfliller@bnl.gov |
| Franchetti | Giuliano | | GSI | G.Franchetti@gsi.de |
| Gassner | David | M | Brookhaven National Laboratory | gassner@bnl.gov |
| Gerigk | Frank | | CLRC/RAL | frank.gerigk@rl.ac.uk |
| Gilpatrick | John | D | Los Alamos National Laboratory | gilpatrick@lanl.gov |
| Giovannozzi | Massimo | | CERN | massimo.giovannozzi@cern.ch |
| Gluckstern | Robert | L | University of Maryland | rlg@physics.umd.edu |
| Hanke | Klaus | | CERN | klaus.hanke@cern.ch |
| HE | Ping | | Brookhaven National Laboratory | phe@bnl.gov |
| Henderson | Stuart | D | Oak Ridge National Laboratory | shenderson@ornl.gov |
| Hofmann | Ingo | | GSI | i.hofmann@gsi.de |
| Igarashi | Susumu | | KEK | susumu.igarashi@kek.jp |
| Ikegami | Masanori | | KEK | masanori.ikegami@kek.jp |
| Jeanneret | Jean-Bernard | | CERN | Bernard.Jeanneret@cern.ch |
| Jeon | Dong-o | | ORNL | jeond@ornl.gov |
| Jin | Lihui | | University of Kansas | jinlh@ku.edu |
| Kaltchev | Dobrin | | TRIUMF | kaltchev@triumf.ca |
| kewisch | jorg | b | Brookhaven National Laboratory | jorg@bnl.gov |

| | | | | |
|---|---|---|---|---|
| Kishek | Rami | A | University of Maryland | ramiak@ebte.umd.edu |
| Kostin | Mikhail | A | Fermilab | kostin@fnal.gov |
| Kozanecki | Witold | | CEA-Saclay | witold@slac.stanford.edu |
| Lagniel | Jean-Michel | | CEA-FRANCE | jean-michel.lagniel@cea.fr |
| Lee | Yong | Y | Brookhaven National Laboratory | yylee@bnl.gov |
| Luccio | Alfredo | U | Brookhaven National Laboratory | luccio@bnl.gov |
| Ludewig | Hans | | Brookhaven National Laboratory | ludewig@bnl.gov |
| Macek | Robert | J | LANL | macek@lanl.gov |
| Machida | Shinji | | KEK | shinji.machida@kek.jp |
| Markiewicz | Thomas | W | SLAC | twmark@SLAC.stanford.edu |
| Mokhov | Nikolai | V | Fermilab | mokhov@fnal.gov |
| Montag | Christoph | H | Brookhaven National Laboratory | montag@bnl.gov |
| MORI | YOSHIHARU | | KEK | yoshiharu.mori@kek.jp |
| Murdoch | Graeme | R | ORNL (SNS Project) | murdochgr@sns.gov |
| Noriaki | Nakao | | KEK | noriaki.nakao@kek.jp |
| Ohmi | Kazuhito | | KEK-ACC | ohmi@post.kek.jp |
| Ostroumov | Petr | N | Argonne National Laboratory, Physics Division | ostroumov@phy.anl.gov |
| Pichoff | Nicolas | | CEA-FRANCE | nicolas.pichoff@cea.fr |
| Pilat | Fulvia | C | Brookhaven National Laboratory | pilat@bnl.gov |
| Prebys | Eric | J | Fermilab | prebys@fnal.gov |
| Prior | Christopher | R | CLRC Rutherford Appleton Laboratory | c.prior@rl.ac.uk |
| Qiang | Ji | | LBNL | jqiang@lbl.gov |
| Rakhno | Igor | | Fermilab | rakhno@fnal.gov |
| Raparia | Deepak | | Brookhaven National Laboratory | raparia@bnl.gov |
| Raubenheimer | Tor | | SLAC | tor@slac.stanford.edu |
| Rogers | Joseph | T | Cornell University | jtr1@cornell.edu |
| Ruggiero | Alessandro | G | Brookhaven National Laboratory | agr@bnl.gov |
| Schlarb | Holger | | DESY | holgers@slac.stanford.edu |
| Schmidt | Rüdiger | | CERN-AB-CO | rudiger.schmidt@cern.ch |
| Sen | Tanaji | | Fermilab | tsen@fnal.gov |
| Seryi | Andrei | | SLAC | seryi@slac.stanford.edu |
| Shea | Thomas | | SNS/ORNL | shea@sns.gov |
| Shi | Jack | J | University of Kansas | jshi@ku.edu |
| Shiltsev | Vladimir | D | Fermilab | shiltsev@fnal.gov |
| Shobuda | Yoshihiro | | JAERI | shobuda@post.kek.jp |
| Simos | Nikolaos | | Brookhaven National Laboratory | simos@bnl.gov |
| Sobol | Andrey | V | University of New Mexico | andrey@math.unm.edu |
| Still | Dean | A | Fermilab | still@fnal.gov |
| Tenenbaum | Peter | | SLAC | quarkpt@SLAC.Stanford.EDU |
| Turchetti | Giorgio | | Dept. of Physics University of Bologna | turchetti@bo.infn.it |

| | | | | |
|---|---|---|---|---|
| Vretenar | Maurizio | | CERN | maurizio.vretenar@cern.ch |
| wang | lanfa | | Brookhaven National Laboratory | wangl@bnl.gov |
| Wangler | Thomas | P | Los Alamos National Laboratory | twangler@lanl.gov |
| Warsop | Christopher | M | Rutherford Appleton Lab., | c.m.warsop@rl.ac.uk |
| Wei | Jie | | Brookhaven National Laboratory | jwei@bnl.gov |
| Weng | Wu-Tsung | | Brookhaven National Laboratory | weng@bnl.gov |
| Witkover | Richard | L | Brookhaven National Laboratory | witkover@bnl.gov |
| Wittenburg | Kay | | DESY | Kay.Wittenburg@desy.de |
| Yakimenko | Vitaly | | Brookhaven National Laboratory | yakimenko@bnl.gov |
| Yip | Kin | | Brookhaven National Laboratory | kinyip@bnl.gov |
| Zimmermann | Frank | | CERN | frank.zimmermann@cern.ch |
| Zou | Yun | | University of Maryland | yunzou@glue.umd.edu |

# HALO'03 Program (DRAFT 2003-05-05)

| Session | Beam Dynamics (BD) | Diagnostics (DI) | Collimation (CO) | Beam-Beam (BB) |
|---|---|---|---|---|
| A (Tue AM) | | D. Lowenstein - welcome<br>A. Fedotov - Mechanisms of halo formation<br>T. Shea - halo diagnostics | | |
| B (Tue AM) | | N. Mokhov - Beam Collimation at Hadron Colliders<br>F. Zimmermann - Halo Formation from Beam-Beam Interaction<br>Y. Mori - Status of the 50 GeV proton synchrotron J-PARC project | | |
| C (Tue PM) | **(Mechanisms & Dynamics)**<br>Pichoff - halo study at CEA<br>Ostroumov - halo in HI SC linac<br>Jeon - halo in SNS linac<br>Gerigk - halo for ESS & Linac4 FE<br>(discussion) | (organization meeting) | **(High power proton machines I)**<br>Warsop - ISIS loss control<br>Prebys - FNAL booster collimation<br>Noriaki - MARS14 study for J-PARC RCS | **(BB limits)**<br>Alexahin - Coherent beam-beam effects<br>Lin - Beam-beam Instability in HERA<br>Fischer - Beam-Beam Obser. RHIC |
| D (Tue PM) | **(Mechanisms & Dynamics)**<br>Lagniel - equipartition & emittance exchange<br>Hofmann - resonance with space charge<br>Ruggiero - halo from mismatch in SCL<br>Franchetti - 4th order resonance & SC<br>(discussion) | | **(High power proton machines II)**<br>Simos - SNS collimation design<br>Ludewig - Dose & residual estimates in ring<br>Murdoch - SNS collimator handling | **(BB Limits)**<br>Kozanecki - Beam-Beam SLAC<br>Ohmi - Collision with finite crossing angle<br>Fischer - Six superbunches in RHIC |
| E (Wed AM) | | Working group progress report<br>M. Giovannozzi - Dynamic Aperture for Single-Particle Motion: Overview of Theoretical<br>T. Raubenheimer - Lepton linear collider halo and collimation issues | | |
| F (Wed AM) | **(joint session with DI)** | **(Experiments & Measurements)**<br>Wangler - LANL halo experiment<br>Igarashi - emittance growth at KEK<br>Cameron - tune-based measureme<br>Yakimenko - optical stochastic cool<br>(discussion) | **(Hadron Colliders I)**<br>Still - Tevatron Run-II halo removal<br>Jeanneret - collimation at LHC<br>Schmidt - loss & protection at LHC<br>Rakhno - protection at beam accidents | **(BB Limits)**<br>Sen - Beam-beam interactions at the Tevatron<br>Erdelyi - Sim. & Exp. Studies of beam-beam<br>Fischer - Tune modulation in RHIC |
| G (Wed PM) | **(Dynamics)**<br>Danilov - transverse instability<br>Turchetti - collective effects & Coulomb<br>Zeitlin - localization & stochastics<br>Fedorova - nonlinear effects<br>Weng - halo on AGS from 1.2 GeV linac | Gilpatrick - LEDA wires<br>Gassner - AGS wires<br>Macek - PSR wires<br>Assadi - SNS Laser wire | **(Hadron Colliders II)**<br>Fliller - crystal collimation at RHIC<br>Kostin - Simulation with MARS14 | **(BB Compensation)**<br>Shiltsev - Status of beam-beam with TEL<br>Shi - Global and local beam-beam in HERA |
| H (Wed PM) | **(Discussions)**<br>Discussion of dynamics topics<br>Discussion of experiments<br>Possible short contributions | **(Collimation and diganostics)**<br>Seryi - Non-linear Optics<br>Wittenburg - DESY wires<br>Drees - RHIC Collimation | **(joint session with DI)** | **(BB Compensation/BB Theory)**<br>Ohmi - Neutralized collisions<br>Ellison - Weak-strong averaging |
| I (Thu AM) | | Working group progress report<br>C. Prior - Space Charge Simulations<br>J. Wei - Beam Collimation in High Power Proton Accelerators | | |
| J (Thu AM) | **(Simulations)**<br>Kishek - single-turn simulation<br>Qiang - halo in 3D mismatched beam<br>Machida - halo simulation at KEK<br>Ikegami - simulation for J-PARC | **(Diagnostics and Beam-beam)** | **(Linear Colliders I)**<br>Kozanecki - detector background & requirements<br>Seryi - collimation for NLC, TESLA, CLIC<br>Zimmermann - Collimation for CLIC<br>Markiewicz - GEANT3 simulation for NLC<br>Markievicz - dealing with muons | **(joint session with DI)** |
| K (Thu PM) | **(Multi-particle simulations) (BD/BB)**<br>Rogers - Beam-beam for leptons<br>Shi - Beam-beam simulations for hadro<br>Luccio - Simulation of high-intensity<br>(discussion) | Gassner - SNS halo<br>Connolly - IPM<br>Macek - PSR electron cloud<br>Zou - Retarded Field Analyzer | **(Linear Colliders II)**<br>Schlarb - TTF collimation system<br>Markiewicz - renewable collimator technology | **(joint session with BD)**<br><br>Joint discussions with Dynamics WG |
| L (Thu PM) | **(joint session with CO)**<br><br>Joint discussions with Collimation WG<br>BD discussions: simulations, codes, benchmarking | Blaskiewicz - instabilities<br>Montag - tail shaping<br>Pilat - RHIC beam exp | **(Halo prevention) (BD/CO)**<br>Tennenbaum - collimation wakefields<br>Kaltchev - LHC cllimation with DIMAD | **(BB Discussion)**<br>Ellison - A new model for 2DOF<br>Qiang - Parallel computational tool<br>Sobol - Numerical calculations<br>Zeitlin - Efficient analysis of beam-beam |
| M (Fri AM) | discussion / summary draft | discussion / summary draft | discussion / summary draft | discussion / summary draft |
| N (Fri AM) | | Halo dynamics working group summary<br>Halo diagnostics working group summary<br>Halo collimation working group summary<br>Beam-beam workshop summary | | |

# 29th ICFA Advanced Beam Dynamics Workshop on
# BEAM-HALO DYNAMICS, DIAGNOSTICS, & COLLIMATION

## May 19 - 23, 2003   Montauk, Long Island, USA

### * in conjunction with 3rd Workshop on Beam-Beam Interactions

The workshop consists of working groups on halo dynamics (space charge, magnetic nonlinearities, resonance excitation, beam-beam, intra-beam scattering, instabilities and electron cloud, noise and diffusion processes, analytical and simulation techniques), halo diagnostics (diagnostics requirements, instrumentation design and performance, machine projection, experimental machine studies), and halo collimation (lattice design, betatron and momentum collimation design, set-up and collimator design, material tests, machine tests).

| Advisory Committee | Program Committee | Beam-Beam Committee* |
|---|---|---|
| R. Baartman (TRIUMF) | J.M. Brennan (BNL) | Y. Cai (SLAC) |
| W. Barletta (LBNL) | W. Chou (FNAL) | W. Fischer (BNL) |
| A. Chao (SLAC) | Y. Fedotov (IHEP) | M. Furman (LBNL) |
| I. Gardner (RAL) | R. Garoby (CERN) | W. Herr (CERN) |
| H. Haseroth (CERN) | S. Henderson (ORNL) | M. Minty (DESY) |
| S. Holmes (FNAL) | I. Hofmann (GSI) | F. Pilat (BNL) |
| N. Holtkamp (ORNL) | J.M. Lagniel (CEA) | T. Sen (FNAL) |
| R. Macek (LANL) | N. Mokhov (FNAL) | R. Talman (Cornell) |
| R. Maier (FZJ) | Y. Mori (KEK) | K. Yokoya (KEK) |
| H. Okamoto (U. Hiroshima) | A. Mosnier (CEA) | |
| C. Pagani (INFN) | C. Prior (RAL) | Local Committee |
| G. Rees (RAL) | T. Raubenheimer (SLAC) | P. Cameron |
| T. Shea (ORNL) | T. Roser (BNL) | M. Campbell |
| R.H. Siemann (SLAC) | F. Ruggiero (CERN) | A. Drees |
| A.N. Skrinsky (BINP) | H. Schmickler (CERN) | A. Fedotov |
| W.T. Weng (BNL) | H. Takayama (KEK) | N. Franco |
| H. Wiedemann (SLAC) | H. Thiessen (LANL) | J. Hauser |
| F. Willeke (DESY) | R. Wanzenberg (DESY) | L. Hoff |
| J. Wei (Chair, BNL) | R. Webber (FNAL) | P. Manning |
| J.Y. Xia (IMP) | J. Wei (Chair, BNL) | S. LaMontagne |
| ICFA Beam Dynamics Panel members | | W. McGahern |
| | | D. Raparia |
| | | N. Simos |
| | | J. Wei (Chair) |
| | | D. Zadow |

Sponsored by Brookhaven National Laboratory,
Spallation Neutron Source Project,
Brookhaven Science Association,
US Department of Energy,
ICFA Panel on Beam Dynamics

For further information, contact:
P. Manning, Brookhaven National Laboratory
Upton, NY 11973-5000, USA
Phone: +1 631-344-4072   Fax: +1 631-344-5729
E-mail: halo03@bnl.gov

www.sns.bnl.gov/halo03

Photo by Tom Donohue

A joint session between HALO Dynamics and HALO Diagnostics working groups while Pete Cameron and G. Franchetti were presenting.

A parallel session with the HALO Collimation working group while D. Still was presenting.

Workshop participants, seemingly completely absorbed by the speakers' presentations, are not to be distracted by the tempting sounds from the nature.

A big lunch after a hard morning's work …

… yet still more work at the lunch table (Ingo Hofmann, Massimo Giovannozzi, and …)

… and more (Dick Witkover and Tom Shea)

Joseph Rogers and his family after a meal at the Gurney's

Yun Zou and Jorg Kewisch

Mimi and Alfredo Luccio with Ingo Hofmann

Jean-Bernard Jeanneret, Angelika Drees, and Dean Still

… well, this is no longer allowed indoors… (Nick Simos, Jean-Bernard Jeanneret, and Rüdiger Schmidt)

Bob Macek and Shinji Machida during a break

Susumu Igarashi and Masanori Ikegami

Alex Chao is telling Ji Qiang and Lanfa Wang a story

Pete Cameron, xxxx, and Vitaly Yakimenko

Mei Bai, Ji Qiang, and Dong-o Jeon

Sandro Ruggiero and Rüdiger Schmidt in a coffee break with Tor Raubenheimer and Alfredo Luccio in the background

Mike Blaskiewicz, Nick Simos, Bill Weng, Fulvia Pilat, xxxx, and Vladimir Shiltsev

Massimo Giovannozzi and James Ellison

Y.Y. Lee chatting with Dick Witkover after a fresh morning exercise, with Rami Kishek in the background

Enough is enough, now it is time for the spa (Pete Cameron and Bill Weng)

Nick Franco the computer czar and his wife

Pam Manning the master coordinator and her husband

# AUTHOR INDEX

## A

Aberle, O., 180
Ajguirei, I. L., 180
Alexahin, Y., 221, 256
Annala, J., 176
Arutunian, S., 129
Assmann, R., 180, 184, 205

## B

Baishev, I., 180
Bazzani, A., 81
Beauvais, P.-Y., 49
Beebe-Wang, J., 85
Benedetti, C., 81
Bernal, S., 89
Biryukov, V., 192
Bishofberger, K., 256
Blair, G. A., 200, 205
Blaskiewicz, M., 244
Bojon, J.-P., 180
Brodowski, J., 162, 167, 172
Brown, K. A., 126
Bruno, L., 180
Burkhardt, H., 184, 205

## C

Cai, Y., 235
Cameron, P., 9, 103, 118, 137, 252
Carlier, E., 180, 184
Catalan-Lasheras, N., 162, 167
Chao, A., 291
Chapochnikova, E., 180
Chesnokov, Y., 192
Chiang, I. H., 126
Chiaveri, E., 180
Church, M., 176
Connolly, R., 140
Cousineau, S., 167

## D

Danilov, V. V., 77
Davino, D., 167
Decarlo, A., 172
Decker, F.-J., 235
Dehning, B., 180, 184
Drees, K. A., 133, 151, 192
Drozhdin, A. I., 158, 188, 200

Dumas, H. S., 282
Duperrier, R., 49

## E

Ellison, J. A., 282, 287
Erdelyi, B., 261

## F

Fartoukh, S., 180, 205
Faus-Golfe, A., 205
Fedotov, A. V., 3, 47
Ferrari, A., 180
Fischer, W., 215, 231, 244, 252, 278
Fliller, R. P., 133, 192
Franchetti, G., 65, 73
Funakoshi, Y., 240
Furman, M., 278

## G

Gassner, D., 126, 137, 192
Gerigk, F., 61
Gilpatrick, J. D., 122
Giovannozzi, M., 26, 73
Goddard, B., 180, 184
Grau, M., 140

## H

Haber, I., 89
Hammons, L., 192
Haouat, G., 49
Hendricks, B., 176
Hirst, J., 172
Hofmann, I., 47, 65, 73
Holmes, J. A., 77
Holtzapple, R., 235

## I

Igarashi, S., 114
Igarashi, Z., 96
Ikegami, M., 96
Irie, Y., 158

## J

Jeanneret, J. B., 180, 184, 205
Jeon, D.-o., 57
Jimenez, J. M., 180
Jin, L., 227, 265

## K

Kain, V., 133, 180, 184
Kaltchev, D., 180
Kato, T., 96
Keller, L., 200
Kheawpum, O., 265
Kishek, R. A., 89
Kondo, Y., 96
Kostin, M. A., 196
Kourotchkine, I., 180
Kozanecki, W., 200, 235
Kramper, B., 176
Kuznetsov, G., 256

## L

Lagniel, J.-M., 49
Legan, A., 176
Li, H., 89
Longo, C., 167
Ludewig, H., 162, 167, 172

## M

Machida, S., 93
Markiewicz, T., 200
Martini, M., 73
Maruyama, T., 200
McIntyre, G., 192
Metral, E., 73
Michnoff, R., 140
Miura, T., 114
Mokhov, N. V., 14, 151, 158, 188, 196, 200
Montag, C., 144
Mullany, B., 167
Murdoch, G., 162, 172

## N

Nakamura, E., 114
Nakao, N., 158
Napoly, O., 200

## O

Ohmi, K., 240, 269
Ohnishi, Y., 269
O'Shea, P. G., 89
Ostroumov, P. N., 53

## P

Pancin, J., 205
Peggs, S., 192, 252
Pichoff, N., 49
Potter, K., 172
Preis, H., 180
Prior, C. R., 32
Puccio, B., 184

## Q

Qiang, J., 65, 96, 278
Quinn, B., 89

## R

Rakhno, I. L., 188
Rambaldi, S., 81
Raparia, D., 85, 162, 167
Raubenheimer, T., 200
Redaelli, S., 205
Reiser, M., 89
Risselada, T., 205
Rogers, J. T., 273
Roseberry, T., 172
Ruggiero, A. G., 69, 85
Ruggiero, F., 180
Ryne, R. D., 65, 96, 278

## S

Salas, M., 282
Satogata, T., 252
Schlarb, H., 209
Schmidt, R., 180, 184
Schreiber, H.-J., 205
Schubert, J., 172
Schulte, D., 200, 205
Seeman, J., 235
Sen, T., 215, 248, 261, 278, 282
Seryi, A., 200
Shea, T., 9
Shi, J., 227, 265
Shiltsev, V., 256

Shimosaki, Y., 114
Shirakata, M., 114
Sievers, P., 180
Simos, N., 162, 167, 172
Sobol, A., 282, 287
Solyak, N., 256
Still, D., 176
Striganov, S. I., 196
Sullivan, M., 235

## T

Takayama, K., 114
Tawada, M., 240
Tenenbaum, P., 200
Tepikian, S., 140
Terekhov, V., 192
Tiunov, M., 256
Toyama, T., 114
Trbojevic, D., 133, 192
Tropin, I. S., 196
Tsoupas, N., 85
Tuozzolo, J., 167, 172
Turchetti, G., 81

## U

Ueno, A., 96
Uriot, D., 49
Uythoven, J., 180

## V

Vlachoudis, V., 180
Vogt, M., 282
Vos, L., 180
Vossenberg, E., 180

## W

Walker, N., 200
Walter, M., 89
Wangler, T. P., 108
Warsop, C. M., 154
Wei, J., 38
Weng, W. T., 85
Wenninger, J., 184
Werner, M., 129
Wienands, U., 235
Witkover, R., 137
Wittenburg, K., 103, 129
Woodley, M., 200

## X

Xiao, M., 278

## Y

Yamamoto, K., 158

## Z

Zhang, X., 256
Zimmermann, F., 20, 200, 205